Insect Molecular Virology

Advances and Emerging Trends

https://doi.org/10.21775/9781912530083

Edited by

Bryony C. Bonning

Department of Entomology and Nematology
University of Florida
Gainesville, FL
USA

Caister Academic Press

Caister Academic Press
Norfolk, UK

www.caister.com

British Library Cataloguing-in-Publication Data
A catalogue record for this book is available from the British Library

ISBN: 978-1-912530-08-3 (paperback)
ISBN: 978-1-912530-09-0 (ebook)

Cover design by Susan E. Rodriguez, adapted in part from Figures 5.3 and 5.4. 3D structure of CrPV-1A monomer obtained by X-ray crystallography at a resolution of 2.6Å (PDB ID: 6C3R). Reprinted from Nayak, A., Kim, D.Y., Trnka, M.J., Kerr, C.H., Lidsky, P.V., Stanley, D.J., Rivera, B.M., Li, K.H., Burlingame, A.L., Jan, E., *et al.* (2018). A viral protein restricts *Drosophila* RNAi immunity by regulating argonaute activity and stability. Cell Host Microbe *24*, 542–557.e9, with permission from Elsevier.

Contents

Current Books of Interest

Bats and Viruses: Current Research and Future Trends	2020
SUMOylation and Ubiquitination: Current and Emerging Concepts	2019
Avian Virology: Current Research and Future Trends	2019
Pathogenic Streptococci: From Genomics to Systems Biology and Control	2019
Insect Molecular Virology: Advances and Emerging Trends	2019
Methylotrophs and Methylotroph Communities	2019
Prions: Current Progress in Advanced Research (Second Edition)	2019
Microbiota: Current Research and Emerging Trends	2019
Microbial Ecology: Current Advances from Genomics, Metagenomics and Other Omics	2019
Porcine Viruses: From Pathogenesis to Strategies for Control	2019
Lactobacillus Genomics and Metabolic Engineering	2019
Cyanobacteria: Signaling and Regulation Systems	2018
Viruses of Microorganisms	2018
Protozoan Parasitism: From Omics to Prevention and Control	2018
Genes, Genetics and Transgenics for Virus Resistance in Plants	2018
Plant-Microbe Interactions in the Rhizosphere	2018
DNA Tumour Viruses: Virology, Pathogenesis and Vaccines	2018
Pathogenic *Escherichia coli*: Evolution, Omics, Detection and Control	2018
Postgraduate Handbook: A Comprehensive Guide for PhD and Master's Students and their Supervisors	2018
Enteroviruses: Omics, Molecular Biology, and Control	2018
Molecular Biology of Kinetoplastid Parasites	2018
Bacterial Evasion of the Host Immune System	2017
Illustrated Dictionary of Parasitology in the Post-Genomic Era	2017
Next-generation Sequencing and Bioinformatics for Plant Science	2017
Brewing Microbiology: Current Research, Omics and Microbial Ecology	2017
Metagenomics: Current Advances and Emerging Concepts	2017
The CRISPR/Cas System: Emerging Technology and Application	2017
Bacillus: Cellular and Molecular Biology (Third edition)	2017
Cyanobacteria: Omics and Manipulation	2017
Foot-and-Mouth Disease Virus: Current Research and Emerging Trends	2017
Staphylococcus: Genetics and Physiology	2016
Microbial Biodegradation: From Omics to Function and Application	2016
Influenza: Current Research	2016
Chloroplasts: Current Research and Future Trends	2016

Preface

The importance of insect-associated viruses is highlighted by their significant impacts on our environment, both positive and negative. On the positive side, insect viruses have proved to be useful tools for the suppression of damaging insect pest populations as well as for biotechnological applications, including protein expression, human gene therapy and vaccine production. Improvements in the baculovirus expression vector system have allowed for multiple applications. This strategy for efficient expression of a single or multiple proteins has utility for vaccine production, the production of virus-like particles for use in gene therapy and the surface display of proteins for use in diagnostic platforms.

In contrast to the positive impacts described above, insect viruses include those that harm beneficial insects, including butterflies, honey bees and other pollinators. Some insects serve as vectors for viruses that negatively impact crops and livestock, resulting in millions of dollars of damage annually. Insects also serve as vectors for viruses that affect humans (as in mosquito transmission of the dengue and Zika viruses), resulting in human morbidity and mortality. Research into the molecular mechanisms of virus–insect associations has been driven in large part by the need to reduce the costs associated with honey bee decline, crop losses, livestock illness and adverse effects on human health. Significant investment has been made into research on the underlying causes of honey bee declines and to understand mosquito transmission of human viruses in particular. Increased understanding of the molecular interactions between insects and associated viruses has led, in some cases, to transmission-blocking strategies to prevent or mitigate viral disease.

Each chapter of this book, written by experts in their respective areas, provides a rigorous review and outlines current trends and future needs, to expedite progress in the field. The book opens with a description of the insect virome, with details of the virus-derived sequences from *Drosophila* spp. and mosquitoes providing an indication of the sheer number of viruses associated with a given insect group. This review also addresses the challenges presented by insect viruses, including the potential impacts of viruses in cell lines commonly used for research. Subsequent chapters address endogenous viral elements (non-retrovirus-derived viral sequences) incorporated into the insect genome and their potential role in antiviral immunity, and mechanisms of insect antiviral defence, with particular emphasis on the role of RNA interference and small RNA in the modulation of virus infection. The use of 'omics to provide insight into molecular interactions between viruses and insects is described for insect vectors of plant viruses, along with strategies to block plant virus transmission. In addition, advances in tetravirus, polydnavirus and baculovirus molecular research are reviewed, followed by reviews on the diverse utility of baculoviruses, including for protein expression, protein display and gene therapy.

This volume on recent advances at the cutting edge of insect molecular virology was inspired in part by a symposium on *Virus–Insect Interactions* held at the International Congress of Entomology in Orlando, Florida, in 2016. I would like to thank the organizers and participants of that symposium, in addition to all others who have contributed to this book. The significant advances in insect molecular virology seen in recent years, resulting in part from enhanced sequencing technologies that paved the way for the discovery of sequences derived from myriad new viruses, and the discovery of RNA interference as a primary mechanism of antiviral defence, are sure to continue.

Dedication

This book is dedicated to the Society for Invertebrate Pathology.

The Insect Virome: Opportunities and Challenges

1

Bryony C. Bonning*

Department of Entomology and Nematology, University of Florida, Gainesville, FL, USA.

*Correspondence: bbonning@ufl.edu

https://doi.org/10.21775/9781912530083.01

Abstract

The insect virome is composed of a myriad of viruses. Both field populations and laboratory colonies of insects harbour diverse viruses, including viruses that infect the insect itself, viruses of microbes associated with the insect, and viruses associated with ingested materials. Metagenomics analysis for identification of virus-derived sequences has allowed for new appreciation of the extent and diversity of the insect virome. The complex interactions between insect viruses and host antiviral immune pathways (RNA interference and apoptosis), and between viruses and other members of the microbiome (e.g. *Wolbachia*) are becoming apparent. In this chapter, an overview of the diversity of viruses in insects and recent virus discovery research for specific insects and insect-derived cell lines is provided. The opportunities and challenges associated with the insect virome, including the potential impacts of viruses on both research and insect management programmes are also addressed.

Introduction

The ubiquity and abundance of viruses in the environment have long been recognized. In recent years, the analyses of virus sequences identified from metagenomics data are changing our perception of virus evolution and the roles played by viruses in organismal biology (Zhang *et al.*, 2018). Our rapidly expanding knowledge of the insect virome is no exception. While the foundational knowledge of the insect virome was based on characterization of viruses that caused distinct phenotypes using conventional approaches (Liu *et al.*, 2011), high throughput sequencing has allowed for the identification of sequences from viruses that are asymptomatic, covert or latent, and of virus sequences incorporated into the host genome (Varghese and van Rij, 2018). In this new era of virus discovery, genomic technologies are used for identification of virus-derived sequences, with examination of virus phenotype only conducted for viruses of particular interest.

While virus sequences can provide valuable insight into evolutionary processes, additional experimentation is essential to confirm the presence of an actual virus, rather than just virus-derived sequence. Such validation would include filling of sequence gaps and sequence confirmation for a putative virus genome, testing for virus replication (dsRNA detection for RNA viruses, small RNA sequencing), visualization of virus particles, and virus isolation for infectivity tests in host insects or cell lines. Many of the viruses from metagenomics analyses described below are based on sequence data alone and therefore represent putative viruses. While a case has been made for the International Committee on Taxonomy of Viruses to include the sequences of putative viruses in their classification for the purposes of sequence-based taxonomy (Simmonds *et al.*, 2017), maintenance of a clear distinction between validated viruses, virus-derived sequence and putative viruses is important for the integrity of the field.

Transcriptome sequencing can be used for initial characterization of an insect virome for detection of virus-derived genome- (for RNA viruses) or transcript-sequences (for RNA and DNA viruses). The use of a single round of polyA purification (rather than the more typical two rounds) increases the likelihood of detection of RNA viruses that lack polyA tails (Liu *et al.*, 2016). Virus-derived sequences incorporated into the genome of the host insect (which may or may not represent an active virus) can be identified from insect genome sequences. Sequencing of small RNA (sRNA) libraries provides greater sensitivity for identification of virus-derived sRNA against the host RNA background (Wu *et al.*, 2010), as does virus-particle purification followed by deep sequencing. No single method will provide an exhaustive overview of the insect virome.

The invertebrate virome

The greatest increase in our knowledge of virus diversity as a result of metagenomics has been in the invertebrates (Webster *et al.*, 2015; Shi *et al.*, 2016a), with arthropods, representing 80% of all known animal species, being a primary component (Shi *et al.*, 2016b). The abundance of RNA viruses is consistently higher in invertebrates than in vertebrates, with apparent tolerance by invertebrates of significant virus loads. Analysis of viral sequences identified from 220 invertebrate species highlighted the dynamic nature of the virus genome with frequent gene loss or gain, genomic rearrangement, recombination and lateral gene transfer between viruses and hosts (Shi *et al.*, 2016a). Some RNA viruses of invertebrates were found to be ancestral to those of vertebrates (Marklewitz *et al.*, 2015). Similarly, in a study focused on viruses with negative-sense RNA genomes, sequencing of 70 arthropod species resulted in identification of sequences derived from 112 novel viruses including those ancestral to viruses that cause disease in plants and vertebrates (Li *et al.*, 2015). This study highlighted the key role of arthropods as reservoirs for virus recombination and genetic exchange resulting in virus genome evolution.

Abundance of RNA viruses

In addition to allowing for identification of virus-derived sequence, metagenomics provides a measure of relative virus abundance based on the proportion of virus-derived transcripts (excluding rRNA) that map to a given virus. This measure can provide an indication of whether the host organism from which the virus sequence was isolated is likely to be a host for the virus, rather than being associated with food material or host associated microorganisms. The validation of a given virus, and assignment to a particular host requires careful additional analysis as described previously however (Carrillo-Tripp *et al.*, 2015).

The diversity of viruses in insects

For any insect population, whether field or laboratory colony-derived, a transcriptomic analysis will reveal a plethora of virus-derived sequences. These sequences will include those derived from *bona fide* insect viruses (Table 1.1), viruses of microorganisms associated with the insect of interest, and viruses of ingested materials including plants, microorganisms, and those found in animal blood for haematophagous insects. The greatest diversity of virus types in the insects shown in Table 1.1 is for drosophilids, which may simply reflect extensive study of this group. Even this list is likely to be incomplete however with some viruses underrepresented due to limitations in methodology (see below). Overall, insect virus discovery results support the observation of a preponderance of positive-sense RNA viruses in eukaryotes (Koonin *et al.*, 2015).

In addition to viruses present as distinct entities, virus-derived sequences may originate from the host genome itself. Virus-derived sequences are abundant in arthropod genomes originating either from endogenized DNA virus genomes (see Chapter 8), or from partial genome sequences (Suzuki *et al.*, 2017; see also Chapter 2). Some endogenized virus-derived sequences function in antiviral immunity (Goic *et al.*, 2013, 2016; Poirier *et al.*, 2018) (Fig. 1.1). The presence of virus-derived sequences in the host genome (endogenous viral elements, EVEs) may contribute to the variation in susceptibility to infection between populations [e.g. Israeli acute paralysis virus of honey bees (Maori *et al.*, 2007)], and has important implications for arbovirus transmission (Schultz *et al.*, 2018).

The discovery of virus-derived sequences from insects provides opportunities for (1) insight into both virus diversity and virus evolution, (2) knowledge of the pervasive presence and dynamic nature

Table 1.1 Diversity of viruses associated with insects

Genome	Virus	*Drosophila* spp.	Mosquito	Honey bee
dsDNA	*Baculoviridae*		×	
	Entomopoxvirinae			
	Iridoviridae		×	
	Nudiviridae	×		
	Polydnaviridae			
	Ascoviridae			
	Hytrosaviridae		×	×
ssDNA	*Parvoviridae*		×	
	(*Densovirinae*)			
	Bidnaviridae	×	×	
	Circoviridae		×	
dsRNA	*Retroviridae*	×		
	Cypovirus			
	Reoviridae	×		
	Partitiviridae	×	×	
	Birnaviridae	×	×	
(-)ssRNA	*Plasmaviridae*	×		
	Bunyavirales		×	×
	Rhabdoviridae	×	×	×
(+)ssRNA	*Dicistroviridae*	×	×	×
	Iflaviridae	×	×	×
	Caliciviridae			
	Permutotetraviridae	×		
	Flaviviridae	×	×	×
	Negevirus	×		
	Nodaviridae	×		
	Nora virus	×		
	Unclassified RNA viruses	20		
Various	Unclassified DNA viruses	1		

Viruses associated with insects include viruses with single-stranded (ss) or double-stranded (ds) genomes comprised of either DNA or RNA. ssRNA viruses have positive (+) or negative (–) sense genomes. Shown is an indication of the breadth of virus diversity found in *Drosophila* spp., mosquito species and in the honey bee, *Apis mellifera*, at the time of writing. Not represented in the table are viruses associated with these insects that are known to infect plants, vertebrates or microorganisms.

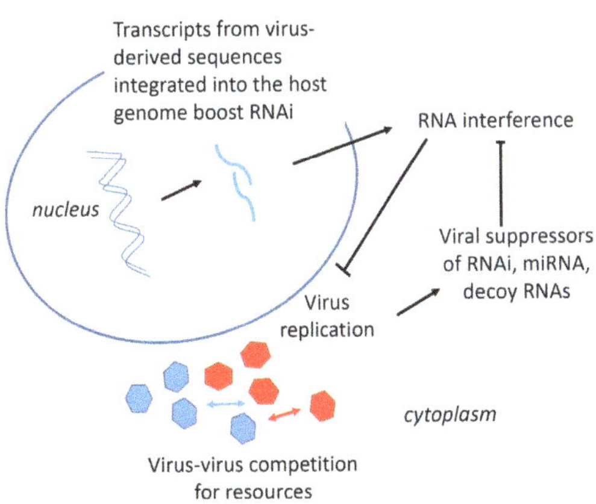

Figure 1.1 Multiple forces drive the insect virome. The outcome of infection by a given virus depends on multiple factors including competition with other viruses present in the host, and interaction with the RNAi pathway. Antiviral immunity may be boosted as shown by endogenous viral elements (EVEs) derived from the virus. An infecting virus may benefit from a strong suppressor of RNAi produced by other resident viruses. Not depicted are the potential impacts of the presence of other microbes, and other components of the immune system.

of virus incidence, (3) information on virus–virus and virus–microbe interactions. Distinguishing sequences derived from viruses that infect the insect under study, from those derived from associated organisms or ingested materials, and from virus-derived EVEs, presents a challenge. The abundance of a given virus-derived sequence is not sufficient to confirm the presence of a replicating insect virus, and additional analyses and experimentation are commonly required for this. Consequently, appropriate naming of viruses identified from metagenomics analysis can be confounded by the lack of knowledge of the host of a given virus.

In the sections below, an overview of the virus diversity for specific insects is provided, for (1) *Drosophila* spp., representing an important model organism in addition to economically important pest species, (2) mosquitoes including species of significant public health importance, and (3) the honey bee, *Apis mellifera*, a beneficial insect of particular importance for pollination services.

The *Drosophila* virome

As a model organism for numerous fundamental physiological systems including antiviral immunity (Huszar and Imler, 2008), *D. melanogaster* is likely the most closely studied of all insects. The power of metagenomics was underscored by the discovery of more than 20 new putative RNA viruses and a new DNA virus (nudivirus) of *Drosophila melanogaster* (Webster *et al.*, 2015, 2016). Analysis of more than 2000 individual wild-caught adult flies showed that more than one third carried detectable virus, with 6% of individuals testing positive for multiple viruses (Webster *et al.*, 2015). This study demonstrated the abundance of viruses in field populations and highlighted the disparity between the virome of field versus laboratory populations of insects: Only a sub-set of the newly discovered viruses was widespread in laboratory colonies, and notably also ubiquitous in cell culture. Characterization of the *D. melanogaster* virome provides an important basis providing an ecological and evolutionary context for use of this species as a model for virus research.

For this work, small RNA was sequenced as a complementary approach for virus sequence discovery. During active virus replication, dsRNA produced as a replication intermediate for RNA viruses, or resulting from RNA secondary structure, is processed by the enzyme Dicer-2 (Dcr2) into small interfering RNA (siRNA). These siRNA are bound by Argonaute (Ago) proteins for use in the RNA interference (RNAi)-mediated, antiviral immune pathway (Ding, 2010; Vijayendran *et al.*, 2013; Mongelli and Saleh, 2016). Research in *Drosophila* demonstrated that transcripts from virus-derived DNA can feed into the RNAi pathway and be processed into siRNA as a means to boost antiviral immunity (Goic *et al.*, 2013; Poirier *et al.*, 2018).

The presence of siRNA that spans the virus genome provides evidence for replication of viruses identified by metagenomics analysis, on the basis that only actively replicating viruses would be processed into siRNA derived from across the viral genome (Wu *et al.*, 2010; Webster *et al.*, 2015). Assembly of siRNAs resulted in identification of novel 'siRNA candidate' viruses although their classification remains to be determined.

Interestingly, the siRNA derived from *Twyford virus*, an iflavirus with a positive-sense ssRNA genome, differed from those of other positive-sense ssRNA virus-derived siRNA, in being negative strand biased, and 22 to 23 nt rather than 21 nt in length. Most of these siRNA also had uridine at the 5′ end, a feature more typical of piRNA. These factors suggest that *Twyford virus* RNA is not processed by *Drosophila* Ago2-Dcr2. The authors suggest that the virus may infect a eukaryotic commensal of *Drosophila* rather than the fly itself, or that a novel pathway is involved in siRNA production (Webster *et al.*, 2015). This case highlights the potential use of siRNA profiles to discriminate between viruses of the target insect versus associated organisms or ingested material.

Combined with a metagenomic analysis of various CO_2-sensitive *Drosophila* spp. for identification of rhabdoviruses, some 85 viruses have now been identified from Drosophilidae (Longdon *et al.*, 2015; Webster *et al.*, 2016). By far the majority of these are RNA viruses. Viruses with positive-sense (+)ssRNA, genomes predominate (50%), followed by those with dsRNA (28%) and negative-sense (−)ssRNA (21%) genomes (Webster *et al.*, 2016). Despite this abundance of drosophilid viruses, some viruses may not be represented due to low abundance (below detection levels), the use of polyA purification of mRNA with associated bias against viral RNAs (such as those from *Nodaviridae*) that lack a polyA tail, and the potential impact of virulent viruses on flight resulting in underrepresentation in collection traps.

It is interesting that only three DNA viruses have been identified from *Drosophila* spp. Given that transcripts derived from DNA viruses can readily be identified in the transcriptomes and small RNA libraries from other insects, it is possible that DNA viruses are underrepresented in the *Drosophila* virome, rather than underrepresentation due to technological limitations.

From these studies, several observations were made including (1) more closely related host insects share more viruses, (2) many viruses infect more than one host and are widely distributed, (3) few viruses are common, many are rare. Only three of 16 viruses surveyed by PCR exceeded 50% prevalence, with most only exceeding 10% prevalence in a few of the populations surveyed.

The mosquito virome

The viruses of mosquitoes have also received considerable attention on account of the role of mosquitoes in disease transmission. In addition to the arthropod-borne viruses (arboviruses) transmitted by mosquitoes to humans and animals, many other viruses are associated with the mosquito vector, including insect-specific viruses that are also members of the arbovirus families (Halbach *et al.*, 2017; Roundy *et al.*, 2017). A primary question has been whether the presence of other viruses in the mosquito vector impacts the ability of the vector to transmit disease (Parry and Asgari, 2018). Given the wide variation in virus infection levels, dependent in part on the efficacy of the viral suppressor of RNAi (VSR) (Nayak *et al.*, 2010; Guo *et al.*, 2018), and direct competition for resources between closely related viruses (Carrillo-Tripp *et al.*, 2016) (Fig. 1.1), the biological impact of resident mosquito viruses on arbovirus transmission has been difficult to pin down. Practical applications of mosquito-specific viruses include their potential use as biological control agents for suppression of mosquito populations, or as novel vaccine agents (Bolling *et al.*, 2015).

As mosquitoes are both haematophagous and nectar feeding, they are potentially exposed to a greater diversity of viruses than is typical (i.e. from both animals and plants), and surveillance is important to monitor for emerging zoonotic disease. Of viral sequences identified from four mosquito species in Hubei, China, the majority (88%) were attributed on the basis of sequence homology as insect viruses, with vertebrate (3.6%), plant (0.8%), bacteriophage (1.87%) and mycovirus (0.03%) also represented (Shi *et al.*, 2015). Sequences derived from insect viruses represented the vast majority of those identified in *Armigeres subalbatus*, about half of those identified from *Culex tritaeniorhynchus*, while the majority from *Anopheles sinensis* were derived from vertebrate viruses (*Anelloviridae* and Parvovirinae).

Mosquitoes in the genus *Culex* include some of the most important vectors of human pathogens. Analysis of culicine species from California, USA, resulted in the identification of 32 novel virus genomes (Sadeghi *et al.*, 2018). The total estimated number of virus families represented in *Culex* in California is 21, with several additional unclassified DNA and RNA viruses. These included six viruses with (–)ssRNA genomes (*Bunyavirales* and *Rhabdoviridae*), and 12 with (+)ssRNA genomes (*Dicistroviridae*, *Iflaviridae*, *Flaviviridae*). In addition, sequences derived from four virus families only known to infect plants were also identified, again highlighting the potential for misinterpretation of host species based on metagenomics data, without careful analysis.

The honey bee virome

Given the economic importance of the honey bee, *Apis mellifera*, which provides both pollination services for diverse crops, in addition to hive-derived products, honey bee pathogens have been extensively studied. One of the earliest studies to reveal the extent of microorganisms associated with a given insect was a metagenomic analysis conducted to identify pathogens associated with the so-called colony collapse disorder of the honey bee, *Apis mellifera* (Cox-Foster *et al.*, 2007). Seven viruses with (+)ssRNA genomes were identified in this study with *Israeli acute paralysis virus* (IAPV) identified as a significant marker of CCD. Subsequent research highlighted that multiple factors are involved with colony collapse, and also the dynamic nature of the honey bee virome with weekly and monthly variation in viruses detected (Runckel *et al.*, 2011). Although the situation among social insects such as the honey bee is rather unique, with frequent contact and opportunity for virus exchange both within and between hives, the number of viruses identified that are associated with the well-studied honey bee is likely to be representative. Around 30 viruses infect or are associated with honey bees (Chen and Siede, 2007; Remnant *et al.*, 2017), most of which are positive-sense RNA viruses in the families *Dicistroviridae* and *Iflaviridae* (de Miranda *et al.*, 2010; de Miranda and Genersch, 2010).

The spread of the *Varroa* mite, which serves as a vector for several viruses of the honey bee has exacerbated colony losses (Wilfert *et al.*, 2016; Ryabov *et al.*, 2017). In particular, the *Varroa* mite alters the honey bee virome increasing the prevalence of a specific strain of *Deformed wing virus* (DWV) (Martin *et al.*, 2012), and decreasing the abundance of other viruses (Roberts *et al.*, 2018). Recent analysis of honey bee colonies from Europe, Africa and the Pacific resulted in genomic evidence for seven viruses associated with honey bees (Remnant *et al.*, 2017). Notably, sequences derived from four viruses with negative-sense ssRNA genomes were identified, although based on sequence similarity, one of these (ABV-1) is likely to infect protozoan parasites of the honey bee rather than the honey bee itself. This analysis included *Apis mellifera capensis* from South Africa and the Pacific islands of Tonga that are resistant to the *Varroa* mite on the basis that high abundance of *Varroa*-associated viruses in colonies elsewhere may have outcompeted other viruses resulting in reduced virus diversity. This study reported the first sequence-based evidence for a flavivirus of honey bees (Remnant *et al.*, 2017). Along these same lines, sequences derived from 42 putative new viruses were identified from honey bees in Australia, which is *Varroa*-free (Roberts *et al.*, 2017, 2018). Virus-derived sequences included those from 11 putative new dicistroviruses, 11 putative new iflaviruses and 20 other putative new small RNA viruses (Roberts *et al.*, 2018). Sequences from viruses associated with drosophilids and hemipterans were also identified, although no validation was presented.

Sequencing of siRNA derived from the honey bee provided support for replication of two newly identified rhabdoviruses in the honey bee, with the typical 21 or 22 nt siRNA derived from both positive and negative-sense RNA strands (Remnant *et al.*, 2017). In contrast, siRNA derived from these same viruses in the mite were predominantly 24 nt from the negative-sense RNA, with a range of siRNA sizes derived from the less abundant positive strand (Remnant *et al.*, 2017). This study provides another example of the potential use of siRNA to address the host organism for a given virus, although whether these rhabdoviruses replicate in the mite remains to be confirmed.

Implications of virus abundance in field populations

The use of viruses for biological control

The examples above provide an indication of both the diversity and abundance of viruses in insect populations in the field. Some viruses naturally regulate insect populations, in some cases resulting in dramatic epizootics under either field or colony-based conditions (Szelei *et al.*, 2011; Myers and Cory, 2016). Indeed, some of these viruses have been employed for pest management purposes with lepidopteran-specific nucleopolyhedroviruses (*Alphabaculovirus* spp.) and granuloviruses (*Betabaculovirus* spp.) being produced commercially, along with a few additional viruses registered for small scale production in China (Lacey *et al.*, 2015). In these cases, the level of exposure to the viral control agents was sufficient to overcome any potential for competition by viruses resident in the targeted pest insect. In broad terms, viruses use different ecological strategies: Those that are relatively virulent (e.g. the paralytic dicistroviruses – *Cricket paralysis virus, Israeli acute paralysis virus*) infect multiple hosts and rely primarily on horizontal transmission (i.e. among individuals of the same generation). In contrast, viruses with low virulence without obvious negative fitness consequences, tend to be restricted to a single or few closely related host species, with a greater propensity for vertical transmission (i.e. from mother to offspring; e.g. *Drosophila C virus*) (Bonning and Miller, 2010). This latter strategy is consistent with long-term adaptation a given host.

The use of dsRNA for pest suppression

The use of dsRNA for pest suppression through RNAi-mediated silencing of essential genes in a target insect has been adopted as an alternative strategy for pest management (Baum *et al.*, 2007; Zhang *et al.*, 2017). A pending question is whether the presence of myriad RNA viruses within a given insect pest will have any impact on the efficacy of dsRNA used for suppression of damaging populations, given that RNA viruses commonly encode suppressors of the host RNAi pathway. As these suppressors target different components of the RNAi pathway (dsRNA, Dcr2, Ago2) (van Mierlo *et al.*, 2014), suppression mediated by one virus will impact the RNAi-mediated antiviral defence against other viruses present in the host. For viruses that result in chronic infection, an equilibrium is maintained between the VSR and the host RNAi pathway. Viruses with highly efficient VSR cause acute infection (Nayak *et al.*, 2010). Based on published reports, the expression levels of dsRNA by transgenic crop plants are sufficient to overcome the impact of VSR from viruses resident in the targeted pest (e.g. Baum *et al.*, 2007), which vary considerably in diversity and abundance.

In parallel with the adoption of RNAi for insect pest management, attention has turned to the use of small RNA viruses as potential delivery systems for silencing RNAs to specific crop pests. As these viruses have evolved to overcome the enzymatic challenges faced by dsRNA on exposure to the saliva or gut milieu, there is potential for use of the viral particle as a protective dsRNA delivery system. To this end, viruses have been identified in soybean aphid (Liu *et al.*, 2016; Feng *et al.*, 2017), western corn rootworm (Liu *et al.*, 2017a–c), stink bugs (Liu *et al.*, 2015) and leafhoppers (Chen *et al.*, 2015) among others. The availability of suitable, virus-free cell lines for development of infectious clones of viruses of such crop pests presents a significant limitation however. Along these same lines, a proof of concept study demonstrated the utility of *Flock House virus* as a virus-induced gene silencing (VIGS) vector for delivery of dsRNA to insect cells (Taning *et al.*, 2018).

Alteration of immune response

An additional implication of abundant viruses in field populations of insects is the potential for alteration of the immune response to other pathogens. Conversely, the presence of other microorganisms may impact antiviral immunity. *Wolbachia* has been shown under laboratory conditions to negatively impact RNA viruses of *Drosophila* (Hedges *et al.*, 2008; Teixeira *et al.*, 2008; Martinez *et al.*, 2017). However, no correlation was detected between the presence of any virus or infection level with *Wolbachia* infection in field caught populations (Webster *et al.*, 2015). This result may reflect the limits of relatively small sample sizes or may reflect more complex interactions when multiple viruses are present in a given host (Webster *et al.*, 2015). A more recent study of *D. melanogaster* field populations with or without *Wolbachia* (*w*Mel) in Australia supported this idea, with no impact of *w*Mel on viral abundance for any of the nine families and floating genera present (Shi *et al.*, 2018). However, the viruses in which *w*Mel-associated reduction is the greatest, were absent (*Drosophila C virus*), or at low abundance (Nora virus) in only one dataset.

The presence of diverse viruses in wild populations may have influenced the antiviral impact of *Wolbachia* in field populations of *Drosophila* that was expected based on laboratory studies (Webster *et al.*, 2015). Indeed, the detailed molecular interactions between resident viruses and bacteria is an area of study ripe for investigation with our increased awareness of the complexity of the microbiome. Knowledge of such interactions in relation to arboviruses will be particularly important for development of potential disease-mitigating strategies.

Similarly, the complexity of the interactions between the many viruses that infect honey bees combined with the propensity of investigators to consider a single pathogen in isolation, has confounded clear elucidation of the potential role of any pathogen in colony collapse. Studies in this area highlight the need for a comprehensive systems approach, i.e. considering individual organisms in the context of their environment and observing the relationships between them, rather than focusing on a single component.

Viruses in insect cell lines

The challenge associated with the presence of cryptic viruses in cell lines used for laboratory research is exemplified by the analysis of 26 commonly used *Drosophila* cell lines and detection of virus-derived reads from 37 different viruses (Fig. S9 in Webster *et al.*, 2015). The numbers of reads detected ranged from < 1% to 50% viral reads in the cell line DmBG1-c1. Table 1.2 lists some of the viruses that have been identified in insect-derived cell lines.

Implications of adventitious viruses in cell lines

For many applications, the presence of cryptic viruses in a cell line is of little consequence, but there are exceptions to this. First, as viruses encode suppressors of antiviral immune systems [both RNA interference (Guo *et al.*, 2018) and apoptosis (Clem, 2015)], resident viruses are likely to alter the immune responsiveness of cultured cells. Second, for virological studies, the presence of an adventitious virus in a cell line may alter the behaviour of viruses used to inoculate the cell line, through competition or complementation. The outcome of infection will depend on the particular virus present: IAPV outcompeted *Sacbrood virus* in the honey bee cell line, AmE-711, unless *Kashmir bee virus* (KBV) was present (Carrillo-Tripp *et al.*, 2016), while the titre of baculovirus is reduced by the presence of rhabdovirus in Sf9 and Sf21 cell lines (Maghodia *et al.*, 2016). Complementation or competition between resident viruses and test viruses may result in false positives or false negatives respectively for the ability of cell lines to support replication of the test virus, which can be particularly problematic for generation of infectious clones of insect RNA viruses (Carrillo-Tripp *et al.*, 2015). Third, virus replication may impact the expression of housekeeping genes commonly used as reference for RNA quantification. For quantification of RNA viruses, virus genome equivalents in relation to total RNA are typically used as a result of virus-mediated disruption of the transcription of housekeeping genes. Fourth, cryptic viruses in cell lines used for baculovirus expression of antigens for use in vaccines, may result in the presence of virus in the antigen sample. In the case of rhabdovirus contamination of *Spodoptera frugiperda* (Sf) cell lines, the virus host range was found to be narrow thereby minimizing risk (Maghodia and Jarvis, 2017), but rhabdovirus-free cell lines were subsequently generated to resolve the problem (Maghodia *et al.*, 2016).

Table 1.2 Adventitious viruses identified in insect-derived cell lines

Insect cell line	Species	Virus	Reference
Lepidoptera			
Sf9, Sf21	*Spodoptera frugiperda*	Rhabdovirus (Sf)	Ma *et al.* (2014)
SL221	*Spodoptera litura*	Rhabdovirus (Sf)	Geisler and Jarvis (2018)
BCIRL-HS-AM1	*Heliothis subflexa*	Rhabdovirus* (Sf)	McIntosh (1989)
Bm-N	*Bombyx mori*	Rhabdovirus* (Sf) *Maculavirus* (plant)	Katsuma *et al.* (2005); Geisler and Jarvis (2018)
High Five™, Tn368, Tn Pro™	*Trichoplusia ni*	*Alphanodavirus* (TnCLV)	Li *et al.* (2007)
Tn368	*Trichoplusia ni*	Nudivirus (HzNV-1)	Lin *et al.* (1999)
Ld652Y	*Lymantria dispar*	*Iflavirus*	Carrillo-Tripp *et al.* (2014)
Diptera			
Drosophila 26 cell lines	*Drosophila melanogaster*	Multiple	Webster *et al.* (2015)
68K S2-GMR	*Drosophila melanogaster*	*Birnavirus*, *Totivirus*, *Tetravirus*	Teninges *et al.* (1979); Wu *et al.* (2010)
S1	*Drosophila melanogaster*	*Alphanodavirus*	Friesen *et al.* (1980)
Aag2	*Aedes aegypti*	*Anphevirus* *Bunyavirus* *Flavivirus*	Zhang *et al.* (2016); Di Giallonardo *et al.* (2018); Schultz *et al.* (2018)
Hymenoptera			
AmE-711	*Apis mellifera*	*Iflavirus**	Carrillo-Tripp *et al.* (2016)

While some viruses are likely to be derived from source material, others are known to result from contamination*. After Geisler and Jarvis (2018).

Viral suppressors of RNAi can be used to indicate the presence of adventitious viruses in cell lines as shown in Fig. 1.2. However, sequencing of the cell line transcriptome or small RNA is recommended for the identification of covert viruses. Establishment of virus-free cell lines is particularly important for production of cell line-derived commercial products, and for immunological and virological studies as described above. Given the high potential for cell line contamination within the laboratory setting (Table 1.2), clean cell lines should be maintained separately from both infected cells and from insect colonies.

Future directions

The exponential increase in novel viral sequence discovery presents a challenge for virus nomenclature. Few of the discovered sequences receive further attention, and some viruses have been named without the necessary validation (Carrillo-Tripp *et al.*, 2015) and without knowledge of host range. Viruses derived from bat faecal material have been named bat viruses for example, even though they were known to include insect and plant viruses (Li *et al.*, 2010). A recent appeal for the International Committee on Taxonomy of Viruses (ICTV) to include metagenomics sequence data in sequence-based virus taxonomy to benefit analysis of virus evolution in particular, has been made (Simmonds *et al.*, 2017).

Several avenues present themselves for future research, building on our expanding knowledge of the insect virome. While the increase in discovery of novel insect viruses provides opportunity for further understanding of the insect virome and for practical applications, the lack of appropriate cell lines for *in vitro* virological studies is limiting. There is a significant need for the establishment of virus-free cell lines for further study of viruses of interest (particularly for pest species) and for production of infectious virus clones for potential use as dsRNA delivery vehicles against pests that are recalcitrant to the impact of dsRNA (Terenius *et al.*, 2011; Nandety *et al.*, 2015). Similarly, the establishment of honey bee cell lines will allow for continued analysis of interactions between honey bee viruses. Knowledge of insect virus interactions with other viruses and other microbes within the host will be of paramount importance for more comprehensive understanding of disease and has potentially important implications for arbovirus transmission.

Acknowledgements

The author acknowledges the extensive work of Dr Sijun Liu, Iowa State University, on insect virus discovery, Dr Jimena Carrillo-Tripp for supporting research including the use of VSR for detection of latent viruses in cell lines, and Dr John VanDyk for IT support. This material is based upon work supported by the National Science Foundation I/UCRC, the Center for Arthropod Management Technologies under Grant No. IIP-1821914 and by industry partners. Additional support was provided by USDA NIFA Hatch Act funds.

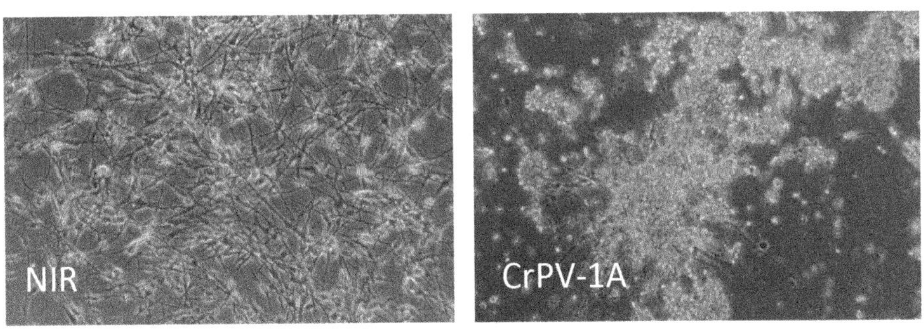

Figure 1.2 Infection of a honey bee-derived cell line was detected following transfection of the cells with the *Cricket paralysis virus* (CrPV)-derived 1A suppressor of RNAi. Cells transfected with non-infective RNA (NIR) retained the typical fibroblast-like structure, while cells transfected with CrPV-1A (Nayak *et al.*, 2010) rounded and clumped in a manner typical of virus infection. Analysis of relative titres of *Deformed wing virus* (DWV) by RT-qPCR confirmed partial release of DWV from suppression by the cell RNAi pathway. For further details, see Carrillo-Tripp *et al.* (2016). Infection with DWV ultimately contributed to the loss of this honey bee-derived cell line.

References

Baum, J.A., Bogaert, T., Clinton, W., Heck, G.R., Feldmann, P., Ilagan, O., Johnson, S., Plaetinck, G., Munyikwa, T., Pleau, M., *et al.* (2007). Control of coleopteran insect pests through RNA interference. Nat. Biotechnol. *25*, 1322–1326.

Bolling, B.G., Weaver, S.C., Tesh, R.B., and Vasilakis, N. (2015). Insect-specific virus discovery: significance for the arbovirus community. Viruses 7, 4911–4928. https://doi.org/10.3390/v7092851

Bonning, B.C., and Miller, W.A. (2010). Dicistroviruses. Annu. Rev. Entomol. *55*, 129–150. https://doi.org/10.1146/annurev-ento-112408-085457

Carrillo-Tripp, J., Krueger, E.N., Harrison, R.L., Toth, A.L., Miller, W.A., and Bonning, B.C. (2014). Lymantria dispar iflavirus 1 (LdIV1), a new model to study iflaviral persistence in lepidopterans. J. Gen. Virol. *95*, 2285–2296. https://doi.org/10.1099/vir.0.067710-0.

Carrillo-Tripp, J., Bonning, B.C., and Miller, W.A. (2015). Challenges associated with research on RNA viruses of insects. Curr. Opin. Insect. Sci. *8*, 62–68.

Carrillo-Tripp, J., Dolezal, A.G., Goblirsch, M.J., Miller, W.A., Toth, A.L., and Bonning, B.C. (2016). In vivo and in vitro infection dynamics of honey bee viruses. Sci. Rep. *6*, 22265. https://doi.org/10.1038/srep22265

Chen, Y., Liu, S., and Bonning, B.C. (2015). Genome sequence of a novel iflavirus from the leafhopper *Graminella nigrifrons*. Genome Announc. 3, e00323-15. https://doi.org/10.1128/genomeA.00323-15

Chen, Y.P., and Siede, R. (2007). Honey bee viruses. Adv. Virus Res. *70*, 33–80. https://doi.org/10.1016/S0065-3527(07)70002-7.

Clem, R.J. (2015). Viral IAPs, then and now. Semin. Cell Dev. Biol. *39*, 72–79. https://doi.org/10.1016/j.semcdb.2015.01.011

Cox-Foster, D.L., Conlan, S., Holmes, E.C., Palacios, G., Evans, J.D., Moran, N.A., Quan, P.L., Briese, T., Hornig, M., Geiser, D.M., *et al.* (2007). A metagenomic survey of microbes in honey bee colony collapse disorder. Science *318*, 283–287.

de Miranda, J.R., and Genersch, E. (2010). Deformed wing virus. J. Invertebr. Pathol. 103 (Suppl. 1), S48–61. https://doi.org/10.1016/j.jip.2009.06.012

de Miranda, J.R., Cordoni, G., and Budge, G. (2010). The Acute bee paralysis virus-Kashmir bee virus-Israeli acute paralysis virus complex. J. Invertebr. Pathol. 103 (Suppl. 1), S30–47. https://doi.org/10.1016/j.jip.2009.06.014

Di Giallonardo, F., Audsley, M.D., Shi, M., Young, P.R., McGraw, E.A., and Holmes, E.C. (2018). Complete genome of *Aedes aegypti* anphevirus in the Aag2 mosquito cell line. J. Gen. Virol. *99*, 832–836. https://doi.org/10.1099/jgv.0.001079

Ding, S.W. (2010). RNA-based antiviral immunity. Nat. Rev. Immunol. *10*, 632–644. https://doi.org/10.1038/nri2824

Feng, Y., Krueger, E.N., Liu, S., Dorman, K., Bonning, B.C., and Miller, W.A. (2017). Discovery of known and novel viral genomes in soybean aphid by deep sequencing. Phytobiomes *1*, 36–45.

Friesen, P., Scotti, P., Longworth, J., and Rueckert, R. (1980). Black beetle virus: propagation in *Drosophila* line 1 cells and an infection-resistant subline carrying endogenous black beetle virus-related particles. J. Virol. *35*, 741–747.

Geisler, C., and Jarvis, D.L. (2018). Adventitious viruses in insect cell lines used for recombinant protein expression. Protein Expr. Purif. *144*, 25–32. https://doi.org/10.1016/j.pep.2017.11.002.

Goic, B., Vodovar, N., Mondotte, J.A., Monot, C., Frangeul, L., Blanc, H., Gausson, V., Vera-Otarola, J., Cristofari, G., and Saleh, M.C. (2013). RNA-mediated interference and reverse transcription control the persistence of RNA viruses in the insect model *Drosophila*. Nat. Immunol. *14*, 396–403. https://doi.org/10.1038/ni.2542

Goic, B., Stapleford, K.A., Frangeul, L., Doucet, A.J., Gausson, V., Blanc, H., Schemmel-Jofre, N., Cristofari, G., Lambrechts, L., Vignuzzi, M., *et al.* (2016). Virus-derived DNA drives mosquito vector tolerance to arboviral infection. Nat. Commun. 7, 12410. https://doi.org/10.1038/ncomms12410

Guo, Z., Li, Y., and Ding, S.W. (2018). Small RNA-based antimicrobial immunity. Nat. Rev. Immunol. https://doi.org/10.1038/s41577-018-0071-x.

Halbach, R., Junglen, S., and van Rij, R.P. (2017). Mosquito-specific and mosquito-borne viruses: evolution, infection, and host defense. Curr. Opin. Insect Sci. *22*, 16–27. https://doi.org/10.1016/j.cois.2017.05.004

Hedges, L.M., Brownlie, J.C., O'Neill, S.L., and Johnson, K.N. (2008). Wolbachia and virus protection in insects. Science *322*, 702. https://doi.org/10.1126/science.1162418

Huszar, T., and Imler, J.L. (2008). *Drosophila* viruses and the study of antiviral host-defense. Adv. Virus Res. *72*, 227–265. https://doi.org/10.1016/S0065-3527(08)00406-5

Katsuma, S., Tanaka, S., Omuro, N., Takabuchi, L., Daimon, T., Imanishi, S., Yamashita, S., Iwanaga, M., Mita, K., Maeda, S., *et al.* (2005). Novel macula-like virus identified in Bombyx mori cultured cells. J. Virol. *79*, 5577–5584. https://doi.org/10.1128/JVI.79.9.5577-5584.2005

Koonin, E.V., Dolja, V.V., and Krupovic, M. (2015). Origins and evolution of viruses of eukaryotes: The ultimate modularity. Virology 479-480, 2–25. https://doi.org/10.1016/j.virol.2015.02.039

Lacey, L.A., Grzywacz, D., Shapiro-Ilan, D.I., Frutos, R., Brownbridge, M., and Goettel, M.S. (2015). Insect pathogens as biological control agents: Back to the future. J. Invertebr. Pathol. *132*, 1–41. https://doi.org/10.1016/j.jip.2015.07.009

Li, C.X., Shi, M., Tian, J.H., Lin, X.D., Kang, Y.J., Chen, L.J., Qin, X.C., Xu, J., Holmes, E.C., and Zhang, Y.Z. (2015). Unprecedented genomic diversity of RNA viruses in arthropods reveals the ancestry of negative-sense RNA viruses. Elife *4*. https://doi.org/10.7554/eLife.05378

Li, L., Victoria, J.G., Wang, C., Jones, M., Fellers, G.M., Kunz, T.H., and Delwart, E. (2010). Bat guano virome: predominance of dietary viruses from insects and plants plus novel mammalian viruses. J. Virol. *84*, 6955–6965. https://doi.org/10.1128/JVI.00501-10

Li, T.C., Scotti, P.D., Miyamura, T., and Takeda, N. (2007). Latent infection of a new alphanodavirus in an insect cell line. J. Virol. *81*, 10890-10896. https://doi.org/10.1128/JVI.00807-07.

Lin, C.L., Lee, J.C., Chen, S.S., Wood, H.A., Li, M.L., Li, C.F., and Chao, Y.C. (1999). Persistent Hz-1 virus infection in insect cells: evidence for insertion of viral DNA into host chromosomes and viral infection in a latent status. J. Virol. *73*, 128–139.

Liu, S., Vijayendran, D., and Bonning, B.C. (2011). Next generation sequencing technologies for insect virus discovery. Viruses *3*, 1849–1869. https://doi.org/10.3390/v3101849

Liu, S.J., Chen, Y.T., and Bonning, B.C. (2015). RNA virus discovery in insects. Curr. Opin. Insect Sci. *8*, 54–61.

Liu, S., Vijayendran, D., Chen, Y., and Bonning, B.C. (2016). Aphis Glycines virus 2, a novel insect virus with a unique genome structure. Viruses *8*, E315. https://doi.org/10.3390/v8110315

Liu, S., Chen, Y., Sappington, T.W., and Bonning, B.C. (2017a). Genome sequence of a novel positive-sense, single-stranded RNA virus isolated from western corn rootworm, *Diabrotica virgifera virgifera* LeConte. Genome Announc. *5*, e00366–17.

Liu, S., Chen, Y., Sappington, T.W., and Bonning, B.C. (2017b). Genome sequence of Diabrotica virgifera virgifera virus 2, a novel small RNA virus of the Western corn rootworm, *Diabrotica virgifera virgifera* LeConte. Genome Announc. *5*, e00365-17. https://doi.org/10.1128/genomeA.00365-17.

Liu, S., Chen, Y., Sappington, T.W., and Bonning, B.C. (2017c). Genome sequence of the first coleopteran iflavirus isolated from western corn rootworm, *Diabrotica virgifera virgifera* LeConte. Genome Announc. *5*, e01530-16. https://doi.org/10.1128/genomeA.01530-16.

Longdon, B., Murray, G.G., Palmer, W.J., Day, J.P., Parker, D.J., Welch, J.J., Obbard, D.J., and Jiggins, F.M. (2015). The evolution, diversity, and host associations of rhabdoviruses. Virus Evol. *1*, vev014. https://doi.org/10.1093/ve/vev014

Ma, H., Galvin, T.A., Glasner, D.R., Shaheduzzaman, S., and Khan, A.S. (2014). Identification of a novel rhabdovirus in *Spodoptera frugiperda* cell lines. J. Virol. *88*, 6576–6585. https://doi.org/10.1128/JVI.00780-14

Maghodia, A.B., and Jarvis, D.L. (2017). Infectivity of Sf-rhabdovirus variants in insect and mammalian cell lines. Virology *512*, 234–245. https://doi.org/10.1016/j.virol.2017.09.025

Maghodia, A.B., Geisler, C., and Jarvis, D.L. (2016). Characterization of an Sf-rhabdovirus-negative *Spodoptera frugiperda* cell line as an alternative host for recombinant protein production in the baculovirus-insect cell system. Protein Expr. Purif. *122*, 45–55. https://doi.org/10.1016/j.pep.2016.02.014

Maori, E., Tanne, E., and Sela, I. (2007). Reciprocal sequence exchange between non-retro viruses and hosts leading to the appearance of new host phenotypes. Virology *362*, 342–349. https://doi.org/10.1016/j.virol.2006.11.038

Marklewitz, M., Zirkel, F., Kurth, A., Drosten, C., and Junglen, S. (2015). Evolutionary and phenotypic analysis of live virus isolates suggests arthropod origin of a pathogenic RNA virus family. Proc. Natl. Acad. Sci. U.S.A. *112*, 7536–7541. https://doi.org/10.1073/pnas.1502036112

Martin, S.J., Highfield, A.C., Brettell, L., Villalobos, E.M., Budge, G.E., Powell, M., Nikaido, S., and Schroeder, D.C. (2012). Global honey bee viral landscape altered by a parasitic mite. Science *336*, 1304–1306. https://doi.org/10.1126/science.1220941

Martinez, J., Tolosana, I., Ok, S., Smith, S., Snoeck, K., Day, J.P., and Jiggins, F.M. (2017). Symbiont strain is the main determinant of variation in Wolbachia-mediated protection against viruses across *Drosophila* species. Mol. Ecol. *26*, 4072–4084. https://doi.org/10.1111/mec.14164

McIntosh, A.H., and Igonoff, C.M. (1989). Replication of *Autographa californica nuclear polyhedrosis virus* in five lepidopteran cell lines. J. Invertebr. Pathol. *54*, 97–102.

Mongelli, V., and Saleh, M.C. (2016). Bugs are not to be silenced: Small RNA pathways and antiviral responses in insects. Annu. Rev. Virol. *3*, 573–589. https://doi.org/10.1146/annurev-virology-110615-042447

Myers, J.H., and Cory, J.S. (2016). Ecology and evolution of pathogens in natural populations of Lepidoptera. Evol. Appl. *9*, 231–247. https://doi.org/10.1111/eva.12328

Nandety, R.S., Kuo, Y.W., Nouri, S., and Falk, B.W. (2015). Emerging strategies for RNA interference (RNAi) applications in insects. Bioengineered *6*, 8–19. https://doi.org/10.4161/21655979.2014.979701

Nayak, A., Berry, B., Tassetto, M., Kunitomi, M., Acevedo, A., Deng, C., Krutchinsky, A., Gross, J., Antoniewski, C., and Andino, R. (2010). Cricket paralysis virus antagonizes Argonaute 2 to modulate antiviral defense in *Drosophila*. Nat. Struct. Mol. Biol. *17*, 547–554. https://doi.org/10.1038/nsmb.1810

Parry, R., and Asgari, S. (2018). *Aedes anphevirus* (AeAV): an insect-specific virus distributed worldwide in *Aedes aegypti* mosquitoes that has complex interplays with *Wolbachia* and dengue virus infection in cells. J. Virol. *92*, e00224-18. https://doi.org/10.1128/JVI.00224-18.

Poirier, E.Z., Goic, B., Tomé-Poderti, L., Frangeul, L., Boussier, J., Gausson, V., Blanc, H., Vallet, T., Loyd, H., Levi, L.I., *et al.* (2018). Dicer-2-dependent generation of viral DNA from defective genomes of RNA viruses modulates antiviral immunity in insects. Cell Host Microbe *23*, 353–365.e8. https://doi.org/10.1016/j.chom.2018.02.001

Remnant, E.J., Shi, M., Buchmann, G., Blacquière, T., Holmes, E.C., Beekman, M., and Ashe, A. (2017). A diverse range of novel RNA viruses in geographically distinct honey bee populations. J. Virol. *91*, e00158–17.

Roberts, J.M.K., Anderson, D.L., and Durr, P.A. (2017). Absence of *Deformed wing virus* and *Varroa destructor* in Australia provides unique perspectives on honeybee viral landscapes and colony losses. Sci. Rep. *7*, 6925. https://doi.org/10.1038/s41598-017-07290-w

Roberts, J.M.K., Anderson, D.L., and Durr, P.A. (2018). Metagenomic analysis of *Varroa*-free Australian honey bees (*Apis mellifera*) shows a diverse Picornavirales virome. J. Gen. Virol. *99*, 818–826. https://doi.org/10.1099/jgv.0.001073

Roundy, C.M., Azar, S.R., Rossi, S.L., Weaver, S.C., and Vasilakis, N. (2017). Insect-Specific Viruses: A Historical Overview and Recent Developments. Adv. Virus Res. *98*, 119–146. https://doi.org/10.1016/bs.aivir.2016.10.001

Runckel, C., Flenniken, M.L., Engel, J.C., Ruby, J.G., Ganem, D., Andino, R., and DeRisi, J.L. (2011). Temporal analysis of the honey bee microbiome reveals four novel viruses and seasonal prevalence of known viruses, *Nosema*, and *Crithidia*. PLOS ONE *6*, e20656. https://doi.org/10.1371/journal.pone.0020656

Ryabov, E.V., Childers, A.K., Chen, Y., Madella, S., Nessa, A., vanEngelsdorp, D., and Evans, J.D. (2017). Recent spread of *Varroa destructor virus-1*, a honey bee pathogen, in the United States. Sci. Rep. *7*, 17447. https://doi.org/10.1038/s41598-017-17802-3

Sadeghi, M., Altan, E., Deng, X., Barker, C.M., Fang, Y., Coffey, L.L., and Delwart, E. (2018). Virome of > 12 thousand Culex mosquitoes from throughout California. Virology *523*, 74–88.

Schultz, M.J., Frydman, H.M., and Connor, J.H. (2018). Dual Insect specific virus infection limits Arbovirus replication in Aedes mosquito cells. Virology *518*, 406–413. https://doi.org/10.1016/j.virol.2018.03.022

Shi, C., Liu, Y., Hu, X., Xiong, J., Zhang, B., and Yuan, Z. (2015). A metagenomic survey of viral abundance and diversity in mosquitoes from Hubei province. PLOS ONE *10*, e0129845. https://doi.org/10.1371/journal.pone.0129845

Shi, M., Lin, X.D., Tian, J.H., Chen, L.J., Chen, X., Li, C.X., Qin, X.C., Li, J., Cao, J.P., Eden, J.S., *et al.* (2016a). Redefining the invertebrate RNA virosphere. Nature [Epub ahead of print]. https://doi.org/10.1038/nature20167

Shi, M., Lin, X.D., Vasilakis, N., Tian, J.H., Li, C.X., Chen, L.J., Eastwood, G., Diao, X.N., Chen, M.H., Chen, X., *et al.* (2016b). Divergent viruses discovered in arthropods and vertebrates revise the evolutionary history of the Flaviviridae and related viruses. J. Virol. *90*, 659–669. https://doi.org/10.1128/JVI.02036-15

Shi, M., White, V.L., Schlub, T., Eden, J.S., Hoffmann, A.A., and Holmes, E.C. (2018). No detectable effect of Wolbachia wMel on the prevalence and abundance of the RNA virome of *Drosophila* melanogaster. Proc. Biol. Sci. *285*, 20181165. https://doi.org/10.1098/rspb.2018.1165.

Simmonds, P., Adams, M.J., Benkő, M., Breitbart, M., Brister, J.R., Carstens, E.B., Davison, A.J., Delwart, E., Gorbalenya, A.E., Harrach, B., *et al.* (2017). Consensus statement: Virus taxonomy in the age of metagenomics. Nat. Rev. Microbiol. *15*, 161–168. https://doi.org/10.1038/nrmicro.2016.177

Suzuki, Y., Frangeul, L., Dickson, L.B., Blanc, H., Verdier, Y., Vinh, J., Lambrechts, L., and Saleh, M.C. (2017). Uncovering the repertoire of endogenous flaviviral elements in *Aedes* mosquito genomes. J. Virol. *91*, e00571–17.

Szelei, J., Woodring, J., Goettel, M.S., Duke, G., Jousset, F.X., Liu, K.Y., Zadori, Z., Li, Y., Styer, E., Boucias, D.G., *et al.* (2011). Susceptibility of North-American and European crickets to Acheta domesticus densovirus (AdDNV) and associated epizootics. J. Invertebr. Pathol. *106*, 394–399. https://doi.org/10.1016/j.jip.2010.12.009

Taning, C.N.T., Christiaens, O., Li, X., Swevers, L., Casteels, H., Maes, M., and Smagghe, G. (2018). Engineered Flock House Virus for targeted gene suppression through RNAi in fruit flies *Drosophila melanogaster in vitro* and *in vivo* Front. Physiol. *9*, 805. https://doi.org/10.3389/fphys.2018.00805

Teixeira, L., Ferreira, A., and Ashburner, M. (2008). The bacterial symbiont Wolbachia induces resistance to RNA viral infections in *Drosophila* melanogaster. PLOS Biol. *6*, e2. https://doi.org/10.1371/journal.pbio.1000002

Teninges, D., Ohanessian, A., Richardmolard, C., and Contamine, D. (1979). Isolation and biological properties of *Drosophila* X-virus. J. Gen. Virol. *42*, 241–254.

Terenius, O., Papanicolaou, A., Garbutt, J.S., Eleftherianos, I., Huvenne, H., Kanginakudru, S., Albrechtsen, M., An, C., Aymeric, J.L., Barthel, A., *et al.* (2011). RNA interference in Lepidoptera: an overview of successful and unsuccessful studies and implications for experimental design. J. Insect Physiol. *57*, 231–245. https://doi.org/10.1016/j.jinsphys.2010.11.006

van Mierlo, J.T., Overheul, G.J., Obadia, B., van Cleef, K.W., Webster, C.L., Saleh, M.C., Obbard, D.J., and van Rij, R.P. (2014). Novel *Drosophila* viruses encode host-specific suppressors of RNAi. PLOS Pathog. *10*, e1004256. https://doi.org/10.1371/journal.ppat.1004256

Varghese, F.S., and van Rij, R.P. (2018). Insect virus discovery by metagenomic and cell culture-based approaches. Methods Mol. Biol. *1746*, 197–213. https://doi.org/10.1007/978-1-4939-7683-6_16

Vijayendran, D., Airs, P.M., Dolezal, K., and Bonning, B.C. (2013). Arthropod viruses and small RNAs. J. Invertebr. Pathol. *114*, 186–195. https://doi.org/10.1016/j.jip.2013.07.006

Webster, C.L., Waldron, F.M., Robertson, S., Crowson, D., Ferrari, G., Quintana, J.F., Brouqui, J.M., Bayne, E.H., Longdon, B., Buck, A.H., *et al.* (2015). The discovery, distribution, and evolution of viruses associated with *Drosophila melanogaster*. PLOS Biol. *13*, e1002210. https://doi.org/10.1371/journal.pbio.1002210

Webster, C.L., Longdon, B., Lewis, S.H., and Obbard, D.J. (2016). Twenty-five new viruses associated with the Drosophilidae (Diptera). Evol. Bioinform. Online *12*, 13–25. https://doi.org/10.4137/EBO.S39454.

Wilfert, L., Long, G., Leggett, H.C., Schmid-Hempel, P., Butlin, R., Martin, S.J., and Boots, M. (2016). Deformed wing virus is a recent global epidemic in honeybees driven by *Varroa* mites. Science *351*, 594–597. https://doi.org/10.1126/science.aac9976

Wu, Q., Luo, Y., Lu, R., Lau, N., Lai, E.C., Li, W.X., and Ding, S.W. (2010). Virus discovery by deep sequencing and assembly of virus-derived small silencing RNAs. Proc. Natl. Acad. Sci. U.S.A. *107*, 1606–1611. https://doi.org/10.1073/pnas.0911353107

Zhang, G., Etebari, K., and Asgari, S. (2016). Wolbachia suppresses cell fusing agent virus in mosquito cells. J. Gen. Virol. *97*, 3427–3432. https://doi.org/10.1099/jgv.0.000653

Zhang, J., Khan, S.A., Heckel, D.G., and Bock, R. (2017). Next-generation insect-resistant plants: RNAi-Mediated crop protection. trends Biotechnol. 35, 871–882.

Zhang, Y.Z., Shi, M., and Holmes, E.C. (2018). Using metagenomics to characterize an expanding virosphere. Cell *172*, 1168–1172. https://doi.org/10.1016/j.cell.2018.02.043.

The Widespread Occurrence and Potential Biological Roles of Endogenous Viral Elements in Insect Genomes

2

Carol D. Blair[1], Ken E. Olson[1] and Mariangela Bonizzoni[2*]

[1]Arthropod-borne and Infectious Diseases Laboratory, Department of Microbiology, Immunology and Pathology, Colorado State University, Fort Collins, CO, USA.
[2]Department of Biology and Biotechnology, University of Pavia, Pavia, Italy.

*Correspondence: mariangela.bonizzoni@unipv.it

https://doi.org/10.21775/9781912530083.02

Abstract

Modern genomic sequencing and bioinformatics approaches have detected numerous examples of DNA sequences derived from DNA and RNA virus genomes integrated into both vertebrate and insect genomes. Retroviruses encode RNA-dependent DNA polymerases (reverse transcriptases) and integrases that convert their RNA viral genomes into DNA proviruses and facilitate proviral DNA integration into the host genome. Surprisingly, DNA sequences derived from RNA viruses that do not encode these enzymes also occur in host genomes. Non-retroviral integrated RNA virus sequences (NIRVS) occur at relatively high frequency in the genomes of the arboviral vectors *Aedes aegypti* and *Aedes albopictus*, are not distributed randomly and possibly contribute to mosquito antiviral immunity, suggesting these mosquitoes could serve as a model system for unravelling the function of NIRVS. Here we address the following questions: What drives DNA synthesis from the genomes of non-retroviral RNA viruses? How does integration of virus cDNA into host DNA occur, and what is its biological function (if any)? We review current knowledge of viral integrations in insect genomes, hypothesize mechanisms of NIRVS formation and their potential impact on insect biology, particularly antiviral immunity, and suggest directions for future research.

Introduction: definition of endogenous viral elements and the surprising discovery of those from non-retroviral RNA viruses

Interactions between viruses and their hosts occur at many levels. Viruses evolve rapidly through genetic drift and/or natural selection and can influence the genetic structures of host populations, especially if viral infection results in disease and mortality. A classic example of virus and host co-evolution is the natural experiment initiated by the release of highly pathogenic myxoma virus into the naive Australian wild rabbit population in 1950, followed by rapid development of resistance in rabbits coincident with natural attenuation of the virus (Best and Kerr, 2000; Kerr, 2012). Another mechanism through which viruses affect the genetics of the host is viral genome integration. Integrations of viral nucleic acid sequences can occur in both somatic and germline cells. Somatic integrations are frequently associated with host genome instability (e.g. Chen *et al.*, 2014).

Additionally, complete viral sequence integration into host genomes can favour the persistence of the infection (Cohn *et al.*, 2015). If integrations occur in germline cells, they can be inherited by the next generation. The persistence and outcome of these integrations in host populations depend on their effects on host fitness. If deleterious, integrations are likely lost. Alternatively, viral sequences can be functionally adopted by the host and exert beneficial functions (Frank and Feschotte, 2017). For instance, the product of the murine retrovirus restriction gene, *Fv1*, protects mice against infection with murine leukaemia virus (MLV). *Fv1* genes in *Mus musculus* and other *Mus* spp. were derived from the *gag* genes of ancient endogenous retroviruses, some of which are distantly related to MLV, and protect against other retroviruses, including lentiviruses and spumaviruses (Yap *et al.*, 2014). A third outcome is found when integrations are in chromosomal locations that are not transcribed or lack regulatory regions. In this case, viral integrations may persist but accumulate mutations at the host rate, which is typically much slower than that of exogenous viruses. These endogenous viral elements (EVEs) thus represent a magnifying lens into the past, allowing researchers to more closely examine the evolutionary history of viruses and virus–host co-evolution (Aiewsakun and Katzourakis, 2015; Katzourakis, 2017).

Integration of genome sequences from DNA viruses and retroviruses is a relatively common phenomenon, as evidenced by the abundance of virus-derived sequences in the genomes of various organisms. For instance, genetic code from retroviruses constitutes about 8% of the human genome (Griffiths *et al.*, 2001). Genome integrations are essential events in the retrovirus life cycle. Because non-retroviral RNA viruses lack coding for reverse transcriptase, the machinery needed for successful conversion to DNA and integration into host genomes, their potential to endogenize has been considered minimal. However, an increasing number of studies show genome integrations into both somatic and germline host cells from non-retroviral RNA viruses, including both single-stranded (positive and negative) and double-stranded RNA viruses. The acronym non-retroviral integrated RNA virus sequences (NIRVS) has been proposed to emphasize the unusual viral origin of these EVEs (Ballinger *et al.*, 2012; Kondo *et al.*, 2013; Ballinger *et al.*, 2014; Olson and Bonizzoni, 2017; Palatini *et al.*, 2017).

Here we describe current knowledge on the origin, widespread occurrence and genomic context of EVEs in insect genomes. Additionally, we report examples of the influence of EVEs on both vertebrate and insect host physiology and immunity.

Integration of genomes from DNA viruses

Polydnaviruses (PDV) provide a well-known example of DNA virus integration into insect genomes (Table 2.1A; for a complete, authoritative review of polydnaviruses and their interactions with their insect hosts, see Chapter 8). PDV are recognized by the International Committee on Taxonomy of Viruses (ICTV; https://talk.ictvonline.org/ictv-reports/ictv_9th_report/dsdna-viruses-2011/w/dsdna_viruses/127/polydnaviridae) as the family *Polydnaviridae*, divided into the genera *Bracovirus* (BV) and *Ichnovirus* (IV), which are associated with braconid and ichneumonid parasitoid wasps, respectively (Strand and Burke, 2015; Gauthier *et al.*, 2018). Bracoviruses have been domesticated by wasps, and their genomes are integrated as proviral segments dispersed throughout the genome in every cell of both female and male wasps (Desjardins *et al.*, 2008). They replicate by producing defective virus-like particles (VLPs) in the calyx of the wasp ovary. Bracovirus genomes in VLPs consist of multiple, diverse, circular dsDNAs depleted of replication genes, with one DNA molecule per VLP. Parasitoid wasps reproduce by depositing their eggs in or on other insects, which are consumed by juvenile stages of the wasp as they develop (Burke *et al.*, 2014). The defective VLPs are deposited into the wasp's hemipteran or dipteran host along with its eggs, their defective genomes migrate to the host cell nucleus, where they subsequently express virulence factors that alter host physiology and allow the development of wasp progeny (Strand and Burke, 2015; Gauthier *et al.*, 2018). PDVs cannot be transmitted horizontally; rather, the integrated proviruses are transmitted vertically in wasps. Transcript studies suggest that BVs evolved from a single whole genome integration of an ancient ancestor of the insect virus family *Nudiviridae* (Bezier *et al.*, 2009; Burke and Strand, 2012). Lineages of ichneumonid wasps with integrated IVs are polyphyletic and integrated IVs are diverse, suggesting multiple evolutionary integration events (Gauthier *et al.*, 2018).

Analyses of a recently discovered endogenization involving an alphanudivirus and the braconid wasp *Fopius arisanus* are shedding light on the mechanisms of integration. Alphanudiviruses are members of the family *Nudiviridae* with a single circular dsDNA genome that usually cause virulent infections in insects. *Fopius arisanus* belongs to the Opiinae subfamily of Braconidae. Wasp species in this subfamily usually lack integrations from bracoviruses (Burke *et al.*, 2018). Consequently, the identification of integrated alphanudivirus sequences in the genome of *F. arisanus* suggests that this integration is a recent event. This event shows that endogenization is accompanied with gene-loss in the viral genome, and that changes in its architecture occur before expression of the endogenized virus genes falls under the control of the wasp genome (Burke *et al.*, 2018).

Genome endogenization has also been observed for the betanudivirus HzNV-1, which naturally infects the corn earworm *Helicoverpa zea*. HzNV-1 can exhibit a latent phase of infection during which its DNA is integrated into the host genome (Gauthier *et al.*, 2018). Moreover, HzNV-1 genome integrations were observed in persistently infected cells of other lepidopteran species (*Trichoplusia ni* and *Spodoptera frugiperda*). Interestingly, cells of the fall armyworm carrying endogenous HzNV-1 appeared resistant to superinfections (Lin *et al.*, 1999).

Integration of non-retroviral RNA virus genomes: discovery and association with retroelements

The first discovery of integrations of non-retroviral RNA virus genome sequences in insect genomes was the unexpected finding in 2004 of four regions similar to RNAs of cell fusing agent (CFAV) and Kamiti River (KRV) viruses in the genomes of *Aedes aegypti* and *Aedes albopictus*. These sequences, which were amplified using degenerate flavivirus primers, were called cell silent agents (CSA) because of their similarity to CFAV (Crochu *et al.*, 2004). Both CFAV and KRV are insect-specific viruses (ISVs) that are genetically related to arboviruses but exclusively infect insect cells. In 2009, additional unexpected integrations from ISVs were detected in wild-caught *Aedes* and *Ochlerotatus* genomes by PCR using flavivirus-specific primers (Roiz *et al.*, 2009; Rizzo *et al.*, 2009; Sanchez-Seco *et al.*, 2010).

Descriptions of RNA virus genome sequence integrations into insect genomes increased with the advent of next-generation sequencing technologies and metagenomic analyses in the early 2000s. Bioinformatics-based identification of viral integrations uses BLAST+ in the form of BLASTx or tBLASTn (Altschul *et al.*, 1990). Researchers parse positive BLAST hits using cut-off values of between 10^{-3} and 10^{-6} (Kondo *et al.*, 2019; Whitfield *et al.*, 2017; ter Horst *et al.*, 2018). Additional criteria (i.e. presence of viral open reading frames or minimum size restrictions) are applied to further reduce false positive results that occur primarily due to low-complexity reads with short tandem repeats or homopolymers. The complexity of the viral database used to search for viral integrations may influence the number of identified BLAST hits. Researchers have used a number of virus genome databases to identify viral integrations in host genomes. These viral sequence databases include sequences from single virus families (Fort *et al.*, 2012; Chen *et al.*, 2015), from plant viruses transmitted by insects (Cui and Holmes, 2012), all known RNA viruses, including those most-recently identified (i.e. Whitfield *et al.*, 2017; Palatini *et al.*, 2017) and from a mixture of DNA and RNA viruses (Katzourakis and Gifford, 2010; ter Horst *et al.*, 2018). These analyses provided the number of viral integrations in sequenced insect genomes and insight into the diversity and richness of their viral origins. The results were influenced by the breadth of viral genomes analysed and the stringency of the criteria used to define an insect sequence as a viral integration (i.e. cut off BLAST e-value; reverse BLAST; analyses of low complexity sequences; length of sequences; inclusion of newly discovered and unclassified viruses).

Currently-known insect genome viral integrations are largely from non-retroviral RNA viruses with sequences also derived from single-stranded DNA viruses, e.g. from densoviruses (*Parvoviridae*) in the genome of *Acyrthosiphon pisum* (ter Horst *et al.*, 2018). NIRVS have been characterized from the genomes of both haematophagous and non-bloodsucking insects (Table 2.1B and C). A review

Table 2.1 Characteristics of EVEs that have been identified in insect genomes including parasitic wasps (A), haematophagous species (B) and non-bloodsucking insects (C)

Host		EVEs					
Family	Genus, species	Viral oigin/R, D	N	Context	Expression	Discovery method	References
A. Parasitic wasps							
Hymenoptera	Ichneumonid and braconid subfamilies	Polydna, D	Full viral genome	Dispersed within wasp genome	Y		Burke and Strand (2012), Gauthier *et al.* (2018)
Lepidoptera	*Helicoverpa*; *Trichoplusia ni* and *Spodoptera frugiperda* cells	*Betanudivirus* HzNV, D	not c	ND			Lin *et al.* (1999)
Hymenoptera	*Fopius arisanus*	*Alphanudivirus*, D	55	Dispersed within host genome		*In silico* screening; PCR	Burke *et al.* (2018)
B. Haematophagous insects							
Culicidae	*Aedes albopictus*, *Aedes aegypti*, C6/36 cell lines	Flavi[1], R	6	Repetitive DNA[3]	Y	PCR	Crochu *et al.* (2004)
Culicidae	*Ae. albopictus*	Flavi[2], R	Not c	ND	ND	PCR	Roiz *et al.* (2009)
Culicidae	*Ochlerotatus* and *Aedes*	Flavi, R	Not c	ND	ND	PCR	Sanchez-Seco *et al.* (2009)
Culicidae	*Ochlerotatus geniculatus*, *Aedes vexans*	Flavi, R	Not c	ND	ND	PCR	Rizzo *et al.* (2009)
Culicidae	*Ae. aegypti*, *Culex quinquefasciatus*	Rhabdo, R	143	ND	ND	*In silico* screening	Katzourakis and Gifford (2010)
		Reo, R	1	ND	ND	*In silico* screening	
		Flavi, R	5	ND	ND	*In silico* screening	
Culicidae	*Ae. vexans*, *Och. caspius*, *Och. detritus*, *Culiseta annulata*	Flavi, R	not c	ND	ND	PCR	Vasquez *et al.* (2012)
Culicidae	*Cx. quinquefasciatus*, *Ae. aegypti*, wild mosquitoes	Rhabdo, R	112 in *Ae. aegypti*, 1 in *Cu. quinquefasciatus*	ND	Y	*In silico* screening; PCR	Fort *et al.* (2012)
Culicidae	*Ae. aegypti*	Plant RNA viruses[6], R	2	*Ace1* gene	ND	*In silico* screening	Cui and Holmes (2012)

Hemiptera	*Rhodnius prolixus*	Benyvirus, R	4	repetitive DNA[7]	ND	*In silico* screening	Kondo *et al*. (2013)
Culicidae	*Aedes aegypti*, *Anopheles spp.*	Chu, Rhabdo[8], Bunya, Phlebovirus-like, *Quaranjavirus*, R	1–12[9]		ND	*In silico* screening, RNA-seq	Li *et al*. (2015)[10]
	Culex quinquefasciatus	Chu, Rhabdo[8], R			ND	*In silico* screening, RNA-seq	
Hemiptera	*Rhodnius prolixus*	Chu, Bunya, MonoN[11]			ND	*In silico* screening, RNA-seq	
Culicidae	*Anopheles minimus*, *Anopheles sinensis*[4]	Flavi, R	2	Repetitive DNA	Y	*In silico* screening	Lequime and Lambrechts (2017)
Culicidae	*Ae. aegypti*, *Ae. albopictus*, wild mosquitoes	Flavi, R	At least 8	Repetitive DNA	Y[12]	PCR; *in silico* screening	Suzuki *et al*. (2017)
Culicidae	*Aedes*, *Culex* and *Anopheles* spp. with genome sequence available	Rhabdo, R	Up to 88 in Aedine; 7 in Anophelinae	Repetitive DNA[13]	Y[14]	*In silico* screening	Palatini *et al*. (2017)
		Flavi, R	Up to 32 in Aedine; 1 in Anophelinae				
		Bunya, R	1				
		Reo, R	1				
Culicidae	*Ae. aegypti* Aag2 cell line	Rhabdo, Flavi, Chu, Virg, Bunya, R	368	repetitive DNA[13]	ND	*In silico* screening	Whitfield *et al*. (2017)
C. Non-blood-sucking insects							
Hymenoptera	*Apis mellifera*	Discistro[15], R	1	Intergenic	Y	PCR	Maori *et al*. (2007)
Drosophilidae	Various species[16]	Sigmavirus, R	<5 per genome	Repetitive DNA[17]	Y[18]	*In silico* screening	Ballinger *et al*. (2012)
Drosophilidae	Various species[16]	Rhabdo, R	1	ND	ND	*In silico* screening	Fort *et al*. (2012)
Homoptera	*Acyrthosiphon pisum*	Rhabdo, R	1	ND	ND	*In silico* screening	Fort *et al*. (2012)
Hymenoptera; Diptera;	Various species[19]	Plant RNA viruses[5], R	<20 per genome	Genes or ND	ND	*In silico* screening	Cui and Holmes (2012)
Drosophilidae	Various species[16]	Bunya, R	2	Intergenic	Y	*In silico* screening	Ballinger *et al*. (2014)

Table 2.1 Continued

Host		EVEs					
Family	Genus, species	Viral oigin/R, D	N	Context	Expression	Discovery method	References
Coleoptera	*Dendroctonus ponderosae*	Chu, Bunya, R				*In silico* screening, RNA-seq	Li *et al.* (2015)
	Tribolium castaneum	Chu, R	3		ND	*In silico* screening, RNA-seq	
Drosophilidae	*Drosophila spp.*	Rhabdo[8,20], Bunya, R			ND	*In silico* screening, RNA-seq	
Isoptera	*Zootermopsis nevadensis*	Chu, R			ND	*In silico* screening, RNA-seq	
Hemiptera	*Acyrthosiphon pisum*	Chu, Rhabdo[8], Phlebovirus-like; *Quaranjavirus*, MonoN[11], R	9		ND	*In silico* screening, RNA-seq	
Hymenoptera	*Atta cephalotes*	MonoN[9], R			ND	*In silico* screening, RNA-seq	
	Acromyrmex echinatior	Chu, MonoN[11]	8		ND	*In silico* screening, RNA-seq	
	Camponotus floridanus	Chu, MonoN[11], Rhabdo[8], R	3		ND	*In silico* screening, RNA-seq	
	Harpegnathos saltator	Chu, R	7		ND	*In silico* screening, RNA-seq	
	Linepithema humile	Chu, R	11		ND	*In silico* screening, RNA-seq	
	Nasonia vitripennis, N. giraulti, N. longicornis	Chu, R	4–20[21]		ND	*In silico* screening, RNA-seq	
	Pogonomyrmex barbarus	Chu, R	23		ND	*In silico* screening, RNA-seq	
	Solenopsis invicta	Chu, MonoN[11], R	12		ND	*In silico* screening, RNA-seq	
Lepidoptera	*Bombyx mori*	Chu, Quaranjavirus, Rhabdo[8], R			ND	*In silico* screening, RNA-seq	
	Melitaea cinxia	Rhabdo[20], Quaranjavirus, R			ND	*In silico* screening, RNA-seq	
	Plutella xylostella	Rhabdo[8], R			ND	*In silico* screening, RNA-seq	
	Spodoptera frugiperda	Phlebovirus-like, R			ND	*In silico* screening, RNA-seq	

Lepidoptera	*Spodoptera frugiperda* cell lines, *Bombyx mori*	Rhabdo[22], R	4[23]	ND	Y	*In silico* screening	Geisler and Jarvis (2016)
Hymenoptera; Thysanoptera, Lepidoptera, Diptera	*Bombus terrestris and B. ignites; Frankliniella occidentalis*, *Heliconius*	Virga/nege-like[24], R	<10 per genome	repetitive DNA	ND	*In silico* screening; PCR	Kondo *et al*. (2017)

Only data from peer-reviewed publications are reported.
[1]Cell fusing agent virus, Kamiti River virus.
[2]Aedes flavivirus.
[3]Repetitive DNA: LTR-retrotransposons of the Pao/Ninja and Copia lineages and an *An. gambiae* putative gene.
[4]*In silico* screening was done across all 24 currently available *Anophelinae* genomes.
[5]Repetitive DNA, LTR-retrotransposons of the Copia class.
[6]Passion fruit mosaic virus (*Virgaviridae*), pelargonium zonate spot virus (*Bromoviridae*), carrot red leaf virus (*Luteoviridae*), beet yellow virus (*Closteroviridae*), cardamine chlorotic fleck virus (*Tombusviridae*), bamboo mosaic virus (*Alphaflexiviridae*), blueberry nectrotic ring blotch virus and citrus leprosis virus c (unclassified).
[7]Repetitive DNA, transposable elements as *mariner* and *mos*.
[8]Dirmarhabdovirus.
[9]One NIRVS in *A. arabiensis*, *An. epiroticus*, *An. atroparvus* and *An. nilii*, two in *An. quandriannulatus*, three in *An. funestus*, *An. farauti*, *An. dirus*, *An. gambiae* and *An. minimus*; four in *An. stephensi*; 12 in *Aedes aegypti*, five in *Cx. Quinquefasciatus*.
[10]NIRVS copy number was shown only for NIRVS from Chuvirus.
[11]Unclassified Mononegavirus.
[12]mRNA expression was detected, but no proteins. Production of small RNA was also observed.
[13]Repetitive DNA, non-LTR retroelements, primarily of the Ty3/Gypsy and Pao Bel lineages.
[14]Limited mRNA expression; piRNA production.
[15]Israeli acute paralysis virus.
[16]Various species within the *Sophophora* and *Drosophila* subgenera.
[17]Repetitive DNA, non-LTR retroelements, primarily of the Ty3/Gypsy lineage.
[18]Expression was verified for one X-linked Sigma virus-P like NIRVS of *Drosophila yakuba*.
[19]NIRVS were identified in 14 out of 53-tested insect genomes, including *Drosophila rhopaloa*, *Dr. ficusphila*, *Dr. anannasse*, *Bombus terrestris*, *B. impatiens*, *Megachile rotundata*, *Pogonomyrmex barbatus*, *Acyrthosiphon pisum*, *Bombyx mori*, *Danaus plexippus*, *Nosonia longicornis*, *Nosonia vitripennis*, *Nosonia giraulti*.
[20]Unclassified rhabdovirus.
[21]Four NIRVS in *Nasonia giraulti*, five in *N. longcornis*, 20 in *N. vitripennis*.
[22]Sf-rhabdovirus.
[23]Four integrations were found in the genome of *S. frugiperda* and three in that of *B. mori*.
[24]Virga/nege like viruses are newly-recognized insect-specific viruses.
Key to headings: context, genomic context of integration; D, DNA-based genome; N, number; R, RNA-based genome; viral or., viral origin.
Key to table: Bunya, *Peribunyaviridae*; Chu, *Chuviridae*; Dicistro, *Dicistroviridae*; Flavi, *Flaviviridae*; N, no; ND, not determined; not c, not certain; Polydna, *Polydnaviridae*; Reo, *Reoviridae*; Rhabdo, *Rhabdoviridae*; Virg, *Virgaviridae*; Y, yes.

of the widespread occurrence of NIRVS in insect genomes suggests similar mechanisms of biogenesis and integration occurred across different insect species; however, striking differences were found in the number of NIRVS and the viral origins of those in *Aedes* mosquitoes relative to other insects.

In contrast to endogenization of polydnavirus sequences originating from entire rearranged viral genomes, NIRVS consist of complete or interrupted viral RNA open-reading frames (ORFs). Most NIRVS have similarity to coding regions of non-retroviral nucleoproteins, glycoproteins, and/or RNA-dependent-RNA polymerases (Crochu *et al.*, 2004; Katzourakis and Gifford, 2010; Fort *et al.*, 2012; Ballinger *et al.*, 2012; Cui and Holmes, 2012; Kondo *et al.*, 2013; Ballinger *et al.*, 2014; Geisler and Jarvis, 2016; Lequime and Lambrechts, 2017; Suzuki *et al.*, 2017; Palatini *et al.*, 2017; Whitfield *et al.*, 2017).

A striking similarity among NIRVS detected across different insect species and from different virus families is that integrations are frequently found in regions of repetitive DNA, in association with long terminal repeat (LTR) retroelements. Retrotransposons of the Pao/Ninja, BEL/Pao, Copia and Ty3/Gypsy lineages and, less frequently, Tc1/mariner, flank NIRVS derived from flaviviruses and rhabdoviruses in both Culicidae and Drosophilidae (Crochu *et al.*, 2004; Ballinger *et al.*, 2012; Goic *et al.*, 2013; Lequime and Lambrechts, 2017; Palatini *et al.*, 2017; Whitfield *et al.*, 2017). Retroelements also flank NIRVS from plant RNA viruses and newly-characterized alphanudivirus-like ISVs in examples from Hemiptera, Hymenoptera, Thysanoptera, Homoptera, Lepidoptera, and from Drosophilidae (Cui and Holmes, 2012; Kondo *et al.*, 2013, 2019).

The origin of NIRVS in insect genomes requires the formation of cDNA from the non-retroviral RNA template with the help of reverse transcriptases. Since NIRVS come from viruses that lack this enzyme, retrotransposons are a likely source within insect cells. Naturally occurring cDNAs of non-retroviral RNA sequences, vDNAs, have been found in cells infected with these viruses (Goic *et al.*, 2013; Nag and Kramer, 2017). A reduction in formation of *Chikungunya virus* (CHIKV; *Alphavirus*) vDNA in *Ae. albopictus* treated with a reverse transcriptase inhibitor has been observed (Goic *et al.*, 2016). This finding and the observed physical contiguity between NIRVS and retroelements in genomes of insects has led to the hypothesis that integration events in insect genomes occur through non-homologous recombination between the genomes of infecting non-retroviral RNA viruses and LTR retroelements during their reverse transcription (Ballinger *et al.*, 2012; Goic *et al.*, 2013). The hypothesis is supported by similar findings in mammalian genomes (Geuking *et al.*, 2009) and by the fact that LTR retroelement reverse transcription occurs in the cytoplasm, where most non-retroviral RNA viruses complete their life-cycle (Havecker *et al.*, 2004; Geuking *et al.*, 2009; Horie *et al.*, 2010; Ballinger *et al.*, 2012). Furthermore, LTR retroelements are active in germline cells, which could favour inheritance of newly acquired viral sequences (Morozov *et al.*, 2017).

The association between NIRVS and retroelements is also supported by the recent finding that some retroelements harbour viral-like domains (Ballinger *et al.*, 2012). Viral-like helicase domains of recently characterized ISVs were found as part of TRAS and Jockey non-LTR retrotransposon proteins in the genomes of Lepidoptera (i.e. *Plutella xylostella*), Hemiptera (i.e. *Nilaparvata lugens*, *Homalodisca vitripennis*) Orthoptera (species of the *Ceutophilius* genus), Hymenoptera (*Leptopilina boulardi*) and Culicidae (*Ae. aegypti*) species (Lazareva *et al.*, 2015; Shi *et al.*, 2016; Morozov *et al.*, 2017). Interestingly, expression of viral-insect fused proteins in some cases increased the evolutionary fitness of the harbouring retroelements by functioning as viral suppressors of RNA silencing (Lazareva *et al.*, 2015). Helicases containing a viral motif could stabilize retrotransposon transcripts and increase their transposition efficiency. An excess of stabilizing helicase activity could also negatively impact non-retroviral RNA virus infection by altering the balance between virus transcription and replication (Morozov *et al.*, 2017).

Integration of non-retroviral RNA virus genomes: viral origin and number of NIRVS in insect genomes

Because most current discoveries of NIRVS come from *in silico* studies of sequenced insect genomes with limited data from natural specimens, knowledge of the widespread occurrence of NIRVS in insect genomes is biased by comparisons across species as compared to among populations within species. With this limitation in mind, ISVs and other viruses known to establish persistent infections are well represented in the list of NIRVS (Table 2.1B).

Currently known insect NIRVS are similar to the sequences of viruses from the *Flavivirus*, *Benyvirus*, and *Quaranjavirus* genera; the *Reoviridae*, *Rhabdoviridae* and *Peribunyaviridae* families and the recently characterized family *Chuviridae* and phlebovirus-like and virga/nege-like viruses. *Chuviridae* is a newly identified monophyletic family of negative-sense RNA viruses, with diverse genome structures (i.e. unsegmented, bi-segmented and circular forms) (Li *et al.*, 2015). *Quaranjavirus* and phlebovirus-like viruses have negative-sense, segmented RNA genomes, belong to the *Orthomyxoviridae* and *Phenuiviridae* families, respectively, and have been recently identified in arthropods (Li *et al.*, 2015). 'Virga/nege-like viruses' refers to a still poorly characterized group of positive-strand RNA viruses that appear to belong to the alphavirus-like superfamily (Kondo *et al.*, 2019). *Virgaviridae* is a family of plant alpha-like viruses, while negevirus refers to newly recognized ISVs from sandflies and mosquitoes, which are distantly phylogenetically-related to plant cileviruses (Vasilakis *et al.*, 2013; Carapeta *et al.*, 2015; Nunes *et al.*, 2017).

NIRVS from flaviviruses (F-NIRVS) have been found almost exclusively in the genomes of culicine mosquitoes (Table 2.1). The Culicidae is a large mosquito family that includes arboviral vectors such as *Aedes* spp., *Ochlerotatus* spp. and *Culex* spp. and protozoan vectors such as those in the subfamily Anophelinae (Chen *et al.*, 2015). F-NIRVS are found in the genomes of medically important arboviral vectors (*Ae. aegypti* and *Ae. albopictus*), in the genomes of the minor arboviral vectors *Ochlerotatus* spp. (i.e. *Oc. caspius* and *Oc. detritus*) and *Ae. vexans*, but not in the genome of *Cx. quinquefasciatus* (Vasquez *et al.*, 2012, Palatini *et al.*, 2017; Whitfield *et al.*, 2017; Blagrove *et al.*, 2017). Among the 19 species of Anophelinae for which a genome sequence is available, F-NIRVS were identified only in *Anopheles minimus* and *Anopheles sinensis* (Palatini *et al.*, 2017; Lequime and Lambrechts, 2017). Interestingly, the number of F-NIRVS in the *Ae. aegypti* and *Ae. albopictus* genomes is thirty times greater than that detected in any other culicine genome. Although the precise number of F-NIRVS will vary with improvements in genome annotations and with sequencing of novel culicine genomes, the large number of F-NIRVS in these two mosquito species was obtained by applying the same conservative bioinformatics pipeline across the genomes of *Aedes*, *Culex* and *Anopheles* spp., and hence the analysis was methodologically unbiased.

Current data, which come from studies using different bioinformatics approaches and analysing different insect species, show that NIRVS from reovirus, bunyavirus, sigmavirus, benyvirus, orthomyxovirus and phlebovirus genomes are sporadic, as they are found in both limited numbers and distribution across insect species (Table 2.1). Even with the paucity of data currently available, it is interesting to note that haematophagous species like *Ae. aegypti* and the kissing bug *Rhodnius prolixus* harbour not only arbovirus-derived NIRVS, but also NIRVS with similarity to the genomes of plant RNA viruses (Cui and Holmes, 2012; Kondo *et al.*, 2013). Non-blood-sucking insect species harbour NIRVS primarily from *Rhabdoviridae* and *Chuviridae*, which are also in the genomes of haematophagous insects (Table 2.1). The *Rhabdoviridae* family includes viruses that infect vertebrates, invertebrates and plants and that are extremely variable in their genomic organization (Dietzgen *et al.*, 2017; Geoghegan *et al.*, 2017). The ecological diversity of rhabdoviruses is also reflected in their frequent cross-species transmission between hosts (Geoghegan *et al.* 2017). While the ecology of *Chuviridae* is not completely understood, viruses in this family were found in arthropod species that share the same ecological niche, indicative of possible horizontal transfer (Li *et al.*, 2015). Integrations from *Rhabdoviridae* and *Chuviridae* may appear to be more frequent than NIRVS from other viruses simply because these viruses are ecologically diverse and have wide geographical distributions. Alternatively, the infection capabilities of these viruses and their capacity to transmit between host species, genera, and even host kingdoms in the case of rhabdoviruses, depending on their ecological and geographic proximity, could select for the emergence of generalist protection mechanisms, which may include viral sequence integrations.

Integration of RNA virus genomes: focus on *Aedes* spp.

Among all the insect species in which NIRVS have been identified to date, *Ae. aegypti* and *Ae. albopictus* are notable for their high number of NIRVS and the diversity of their viral origins. In the genomes of these two mosquito species, NIRVS derived from genomes of flaviviruses, rhabdoviruses, reoviruses, bunyaviruses, phleboviruses, and quaranjavirus have been identified and NIRVS have been characterized from the *Virgaviridae* and *Chuviridae* virus families. (Crochu *et al.*, 2004; Roiz *et al.*, 2009; Katzourakis and Gifford, 2010; Fort *et al.*, 2012; Cui and Holmes, 2012; Li *et al.*, 2015; Suzuki *et al.*, 2017; Palatini *et al.*, 2017; Whitfield *et al.*, 2017; ter Horst *et al.*, 2018). All these studies agree on hundreds of individual NIRVS in *Aedes* spp., although the exact number of NIRVS from each viral type differs among studies depending on the bioinformatics parameters, sequencing and assembly method used, and source of data (i.e. genome sequence from the Aag2 cell line or adult mosquitoes). This richness and diversity of NIRVS in *Ae. aegypti* and *Ae. albopictus* is striking because these species are among the most medically important vectors of arboviruses such as dengue, Zika, yellow fever and chikungunya. This observation has prompted ongoing research efforts to understand the distribution of NIRVS in natural populations and whether NIRVS influence mosquito physiology and arbovirus infection and transmission.

NIRVS are not distributed randomly in the genomes of *Ae. aegypti* and *Ae. albopictus*, but rather are statistically significantly enriched in regions corresponding to PIWI-interacting RNA (piRNA)-generating loci, called piRNA clusters (Palatini *et al.*, 2017; Whitfield *et al.*, 2017). NIRVS are also differentially distributed in geographic populations (Pischedda *et al.*, 2018). A difference between F-NIRVS and NIRVS with similarity to rhabdoviruses (R-NIRVS) was seen in the first whole-genome survey of NIRVS across five geographical populations of *Ae. albopictus*: R-NIRVS were more widespread and included more ancient integrations based on accumulation of sequence changes than F-NIRVS (Pischedda *et al.*, 2018), an interesting result considering that negative-strand RNA viruses like rhabdoviruses are thought to be evolutionarily more recent than positive-strand flaviviruses (Koonin *et al.*, 2015).

Exogenous virus resistance mediated by endogenous viral elements

The most comprehensive current data on the effects of NIRVS on host physiology are from studies of endogenous bornavirus-like nucleoprotein-encoding sequences (EBLNs) in human, mouse and squirrel genomes (Honda and Tomonaga, 2016). A recombinant EBLN from the ground squirrel genome was expressed in human oligodendroglioma cells, where it co-localized with bornaviral sites of replication in the nucleus and inhibited the activity of infecting Bornavirus polymerase (Fujino *et al.*, 2014). This dominant negative effect was assumed to be dependent on the relatively high-degree of amino-acid sequence identity (77%) between the protein expressed from the integrated complete viral ORF from the ground squirrel genome and the nucleocapsid protein of the Bornavirus (Fujino *et al.*, 2014). However, the minimum degree of sequence-identity between the NIRVS-expressed protein and that of cognate viruses essential to exert competitive or dominant interaction was not established.

A transcript of an EBLN of the human genome (i.e. hsEBLN-1) has been proposed to function as a long non-coding RNA and control the expression of the neighbouring COMMD3 gene. COMMD3 encodes a protein that inhibits the NFκB immune pathway (Burstein *et al.*, 2005; Sofuku *et al.*, 2015). As such, hsEBLN-1 is considered an epigenetic immune regulator (Sofuku *et al.*, 2015; Honda and Tomonaga, 2016). This type of function has not been observed for insect NIRVS, likely because, in addition to differences between the insect and vertebrate antiviral immune pathways, insect NIRVS are predominantly found in repetitive regions of the genome and correspond to partial viral ORFs.

The first evidence of a biological role for an insect NIRVS came when a ≈ 420 bp sequence from the *Israeli acute paralysis virus* (IAPV) was found integrated into the genome of bees (*Apis mellifera*), resulting in their resistance to subsequent infection with IAPV (Maori *et al.*, 2007). The molecular mechanisms underlying this phenomenon were not elucidated however (Maori *et al.*, 2007; see Chapter 3 for details of similar observations in *Drosophila*).

Endogenous viral elements and the piRNA pathway in mosquitoes

A possible biological function for NIRVS in mosquitos may be to enhance antiviral defence (Meisen *et al.*, 2016b; Blair and Olson, 2017; Olson and Bonizzoni, 2017). An antiviral function was suggested by two observations. First, detection of RNA virus genome-derived piRNAs was shown in both uninfected and infected mosquitoes and mosquito cells (Scott *et al.*, 2010; Brackney *et al.*, 2010; Arensburger *et al.*, 2011; Morazzoni *et al.*, 2012; Vodovar *et al.*, 2012; Schnettler *et al.*, 2013; Miesen *et al.*, 2016a,b). Second, integration of most NIRVS occurred in or near retroelements in piRNA clusters of *Ae. aegypti* and *Ae. albopictus* genomes. These observations not only suggest a mechanism by which non-retroviral RNA genomes were converted to cDNA and inserted into host genomes, but also that NIRVS function similarly to piRNA clusters, the genomic loci that express transposon-derived piRNAs implicated in defending host genome integrity by restricting transposon activity.

The small interfering RNA (siRNA) pathway of the RNA interference (RNAi) system is the major antiviral defence system in acute arboviral infections of *Aedes* spp. mosquitoes (Sánchez-Vargas *et al.*, 2009; Scott *et al.*, 2010; Blair, 2011). In mosquitoes, the siRNA pathway is a cell-autonomous, cytoplasmic system in which the dsRNA-specific RNase III family member Dicer 2 (Dcr2) recognizes viral dsRNA produced during replication. Dcr2 binds the dsRNA and cleaves it into 21-nt siRNA duplexes with perfect base-pairing and 2-nt overhangs at the 3′ ends. In complex with the dsRNA-binding protein R2D2, Dcr2 loads the siRNA duplex onto Argonaute 2 (Ago2), an endonuclease that is part of the multi-component RNA-induced silencing complex (RISC). Ago2 cleaves and releases one strand of the siRNA duplex, retaining the second strand as a guide to hybridize to a complementary coding sequence in viral genome/mRNA, which it then cleaves in the centre of the complementary region. Transiently knocking down expression of Dcr2, R2D2, or Ago2 results in greatly increased arbovirus replication, indicating their central role in siRNA-mediated antiviral defence (Keene *et al.*, 2004; Sánchez-Vargas *et al.*, 2009). In either *Ae. aegypti* mosquitoes or *Ae aegypti* embryo-derived Aag2 cell cultures, infection with the flavivirus dengue (DENV) results in production of DENV RNA-derived 21 nt siRNAs (Scott *et al.*, 2010). However, in *Ae albopictus*-derived C6/36 cells, which have a single-nucleotide deletion in their Dcr2 gene (Scott *et al.*, 2010), arbovirus infection drives production of virus RNA-derived 27-nt piRNAs (Scott *et al.*, 2010, Brackney *et al.*, 2010). The more robust replication of DENV in exo-siRNA-deficient mosquito cells emphasizes the importance of this pathway in antiviral defence in *Aedes* spp., but production of piRNAs suggests that this pathway might have a secondary role in anti-arbovirus defence.

In *Drosophila*, which has served as a model for the mosquito RNAi system, production of mature piRNAs is carried out by the three members of the Piwi subfamily of the Argonaute protein family, Piwi, Aubergine (Aub) and Ago3, and is Dcr-independent (Brennecke, 2007, Gunawardane, 2007). These proteins are expressed in germline cells of the fly and their major function is to protect the genome by silencing transposable element (TE) activity. The substrates for production of primary piRNAs are nuclear transcripts from piRNA clusters in the genome that contain defective TE sequences. The transcripts are processed in the nucleus to produce mature, antisense Piwi- or Aub-associated 24–27 nt piRNAs, known as piRNA-induced silencing complexes (piRISC), that hybridize to and cleave complementary sense-strand RNA in the cytoplasm. This 'slicer' action of the piRISC results in positive-sense, Ago3-bound piRNA that can initiate a 'ping-pong' amplification loop (Siomi, 2011).

The Piwi protein subfamily has undergone expansion to eight proteins, Ago3 and Piwi1–7, in *Ae. aegypti* (Campbell *et al.*, 2008), suggesting additional functional roles of these proteins. Also, the piRNA pathway is active in somatic as well as germline tissues of *Ae. aegypti* (Morazzani *et al.*, 2012). Further studies are needed to determine which of the eight *Aedes* Piwi-family proteins may be involved in binding, processing and cleavage of the piRNA precursor. Determining whether a particular Piwi-family protein affects virus replication could depend on the mosquito tissue or cultured cells analysed, the time post-infection at which analysis is done, and the infecting arbovirus (flavivirus or alphavirus) (Vodovar *et al.*, 2012; Morazzani *et al.*, 2012; Schnettler *et al.*, 2013; Miesen *et al.*, 2015, 2016a; Goic *et al.*, 2016). In addition to piRNAs derived from NIRVS in *Aedes* spp., newly generated piRNAs derived from the infecting arbovirus genome occur during acute infections of *Aedes* spp. cells and mosquitoes by flaviviruses, alphaviruses, and bunyaviruses (Scott, 2010; Morazzani, 2012; Léger, 2013; Schnettler, 2013; Miesen, 2015, 2016a,b). However, their role in antiviral defence is uncertain and the template for transcription of virus genome-derived piRNAs (vpiRNAs) and protein(s) involved in their biogenesis are not fully understood.

Several studies have shown that *Aedes* mosquitoes and mosquito cells are capable of synthesizing short segments of cDNA (vDNA) from alphavirus, flavivirus, or bunyavirus genome templates, using endogenous reverse transcriptase (RT), likely through recombination between viral RNA and a retrotransposon transcript during reverse transcription (Vodovar *et al.*, 2012; Nag *et al.*, 2016; Nag and Kramer, 2017). By analogy to the proviral DNA form of retrovirus genomes, this hybrid DNA is termed 'proviral'. Proviral DNA is hypothesized to enter the nucleus where it is transcribed into primary piRNA precursors from vDNA episomes or after integration into a piRNA cluster in the genome using the retroelement integrase. Although robust production of virus genome-derived piRNAs has been observed in alphavirus-infected mosquitoes (Morazzani *et al.*, 2012), no alphavirus genome-related NIRVS have been reported in mosquitoes. The evolutionary conservation of NIRVS and their variation in different populations of *Aedes* spp. provide support for their acquisition during acute infection. NIRVS acquisition and any long-term antiviral effects for a particular arbovirus would depend on infection of mosquito germ-line tissue and integration into the vector genome of NIRVS derived from the infecting viral RNA. Recent bioinformatics analyses have detected NIRVS derived from *Dengue virus-1* in the genome of a distinct *Ae. albopictus* population from China; this may signify a recent NIRVS acquisition (Chen *et al.*, 2015). Furthermore, germ-line infections and vertical transmission of dengue viruses in *Ae. aegypti* probably occur more frequently than previously thought (Sánchez-Vargas *et al.*, 2018).

Concluding remarks and future directions

The first experimental evidence that non-retroviral RNA viruses can integrate genome sequences into host DNA genomes surfaced in the 1970s (Zhdanov, 1975) but this evidence was ignored until the development of powerful and inexpensive next-generation sequencing technology that enabled the sequencing of many host genomes. Analyses of genome data are revealing an increasing number of viral sequence integrations in different organisms including insects. Although current data are mostly descriptive, preliminary observations are already transforming our perception of virus–host co-evolution, allowing for the generation of hypotheses about connections between virus evolution, insect ecology, insect genomic structure and insect antiviral immunity.

Understanding the significance of the widespread occurrence of viral integrations within insect genomes will require analysis of the genomes of more species (Fig. 2.1). It would be beneficial for the search for – and characterization of – viral integrations to be standard aims during the annotation of insect genomes. This would help answer whether integration of viral sequences is a common phenomenon within insects, or if it depends on the structure of the host genome (e.g. on the content of LTR transposons). It will also be important to verify which viral species/genera can efficiently integrate, especially as non-retroviral RNA viruses have been proposed as delivery tools for gene therapy (Mogler and Kamrud, 2015). This type of medical treatment is based on the transient presence of non-retroviral RNA viruses in host cells. For safety, it is essential to verify whether integrations into host genomes occur because of the properties of the host genome or of the virus. The use of long-read technologies to refine the genome assembly of species where NIRVS have been characterized will help resolve repetitive regions and unravel the landscape of NIRVS, including their numbers and their positioning within the genome (Whitfield *et al.*, 2017; Matthews *et al.*, 2018). Pertinent to this aim, a new,

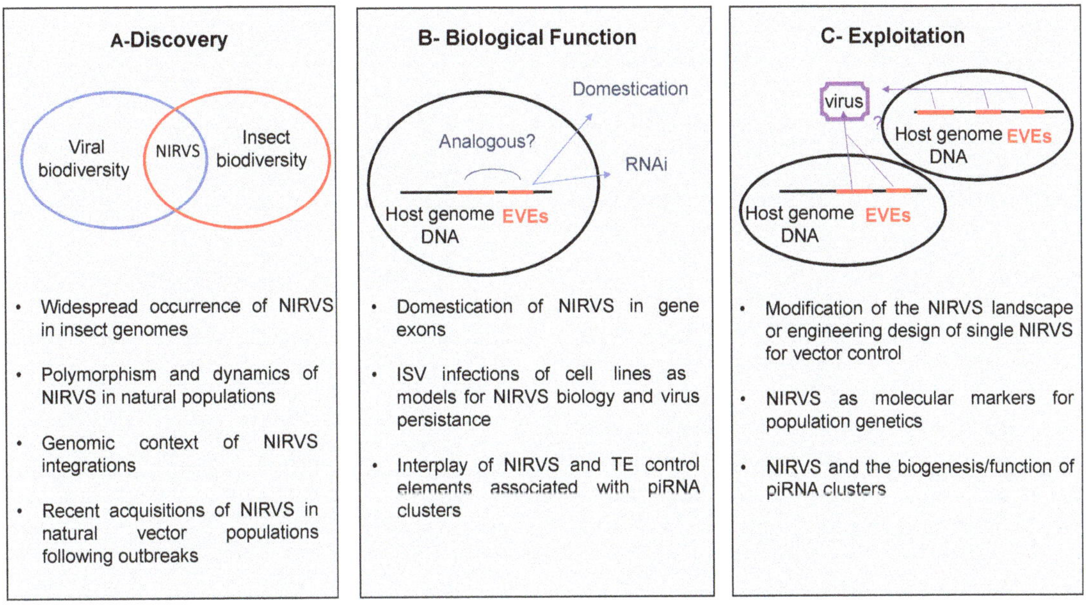

Figure 2.1 Current knowledge gaps and areas for future research on endogenous viral elements from non-retroviral RNA viruses (i.e. non-retroviral integrated RNA virus sequences or NIRVS) in insect genomes. (A) NIRVS Discovery, a better understanding of the widespread occurrence of NIRVS in insect genomes, the timing of integrations and the genomic context in which integrations occur will lead to understanding of the process of endogenization and its impact on insect genomes. (B) Biological Function, indirect evidence on the distribution of NIRVS in the genomes of the mosquitoes *Ae. aegypti* and *Ae. albopictus* and their viral origin point to an interplay with TE control elements associated with piRNA clusters and mainly ISVs, which could be experimentally addressed in cell-line based systems. Cases of NIRVS domestication could also have occurred given the finding of several NIRVS as complete ORFs or part of annotated exons. (C) Exploitation, depending on the differential distribution of NIRVS in insect populations, NIRVS could be co-opted as novel molecular markers to depict population genetic structure or trace recent invasions. Additionally, given that in *Ae. aegypti* and *Ae. albopictus* NIRVS are statistically significantly enriched in piRNA clusters, unravelling the process of endogenization could offer insights on the biogenesis and functions of piRNA clusters in non-model organisms. Finally, NIRVS may be harnessed for vector control by modifying the landscape of NIRVS within the vector genome or using NIRVS sites as anchors to drive expression of effector molecules. In this regard, studies need to determine whether NIRVS have a direct role in anti-viral immunity, how the level of sequence-similarity between NIRVS and incoming viruses can attain antiviral activity, and if it is possible to genetically modify mosquitoes by introducing novel NIRVS into the host genome.

more complete and accurate annotation of the *Ae. aegypti* genome, designated AaegL5, has been released to resolve additional NIRVS and piRNA cluster sites (Matthews *et al.*, 2018). This information could enable testable predictions regarding the mechanisms and entities involved in the integration process (e.g. by confirming NIRVS arrangements with respect to specific TE lineages). At the same time, elucidating the sites of integrations could help design experiments to test for NIRVS functions. If integrations are numerous, if NIRVS are duplicated and located primarily within piRNA clusters, which are composed of TE fragments, using genetic engineering strategies that require unique anchor sequences to modify NIRVS (i.e. knock-out NIRVS or introduce novel viral sequences) may be difficult. Examining the genomic landscape and polymorphism of NIRVS in natural specimens will enable exploration of the pervasiveness and micro-evolution of integrations. Finally, the similarity of most NIRVS to the genomes of ISVs, most of which are easily cultivated in insect cells, point to the promise of using ISVs and insect hosts as 'animal model' systems to elucidate the endogenization process and study the impact of NIRVS on exogenous virus infections.

The major controller of acute arbovirus infection in the vector is the siRNA pathway of RNAi. We (and others) have observed that when the RNAi inhibitor B2 (from *Flock House virus*) is expressed by the arbovirus Sindbis (genus *Alphavirus*), Sindbis virus infection in the vector is greatly increased and the fitness of the vector is significantly lowered (Myles *et al.*, 2008; Cirimotich *et al.*, 2009). We hypothesize that NIRVS act by expressing virus genome-derived piRNAs in the vector to target and control infections. Thus, NIRVS may represent a fortuitous co-adaptation between arboviruses and vectors (Blair and Olson, 2015; Olson and Blair, 2016; Palatini *et al.*, 2017). NIRVS probably do not prevent or ablate infections but help maintain virus persistence in the vectors.

We have developed RNAi- or ribozyme-based antiviral effector genes. Integration and expression of either of these effector genes in *Ae. aegypti* at the time of infection profoundly lowers or eliminates vector competence for several arboviruses (Franz *et al.*, 2006; Mishra *et al.*, 2016). However, these antiviral transgenes are integrated randomly in the genome using mariner or piggyBAC transposable elements, and their expression varies according to their insertion site. As we improve our understanding of NIRVS integration, expression, and biological functions, NIRVS could provide valuable insights into how we might optimally engineer antiviral transgenes into the vector genome to maintain antiviral gene cassettes for multiple generations, as observed for piRNA clusters. As NIRVS reveal naturally selected genetic engineering of the vector to control infection outcomes, NIRVS may provide the perfect road map for how to successfully engineer antiviral transgenes.

References

Aiewsakun, P., and Katzourakis, A. (2015). Endogenous viruses: connecting recent and ancient viral evolution. Virology *479–480*, 26–37. https://doi.org/10.1016/j.virol.2015.02.011

Altschul, S.F., Gish. W., Miller, W., Myers, E.W., and Lipman, D.J. (1990). Basic local 645 alignment search tool. J. Mol. Biol. *215*, 403–410.

Arensburger, P., Hice, R.H., Wright, J.A., Craig, N.L., and Atkinson, P.W. (2011). The mosquito *Aedes aegypti* has a large genome size and high transposable element load but contains a low proportion of transposon-specific piRNAs. BMC Genomics *12*, 606. https://doi.org/10.1186/1471-2164-12-606

Ballinger, M.J., Bruenn, J.A., and Taylor, D.J. (2012). Phylogeny, integration and expression of sigma-virus-like genes in *Drosophila*. Mol. Phylogenet. Evol. *65*, 251–258. https://doi.org/10.1016/j.ympev.2012.06.008

Ballinger, M.J., Bruenn, J.A., Hay, J., Czechowski, D., and Taylor, D.J. (2014). Discovery and evolution of bunyavidids in artic phantom midges and ancient bunyavirid-like sequences in insect genomes. J. Virol. *88*, 8783–8794. https://doi.org/10.1128/JVI.00531-14

Best, S.M., and Kerr, P. (2000). Coevolution of host and virus: the pathogenesis of virulent and attenuated strains of myxoma virus in resistant and susceptible european rabbits. Virology *267*, 36–48.

Bézier, A., Annaheim, M., Herbiniere, J., Wetterwald, C., Gyapay, G., Bernard-Samain, S., Wincker, P., Roditi, I., Heller, M., Belghazi, M., *et al.* (2009). Polydnaviruses of braconid wasps derive from an ancestral nudivirus. Science *323*, 926–930. https://doi.org/10.1126/science.1166788.

Blagrove, M.S., Sherlock, K., Chapman, G.E., Impoivil, D.E., McCall, P.J., Medlock, J.M., Lycett, G., Solomon, T., and Baylis, M. (2017). Evaluation of vector competence of a native UK mosquito *Ochlerotatus detritus* (*Aedes detritus*) for dengue, Chikungunya and West Nile viruses. Parasit. Vectors *9*, 452. https://doi.org/10.1186/s13071-016-1739-3

Blair, C.D. (2011). Mosquito RNAi is the major innate immune pathway controlling arbovirus infection and transmission. Future Microbiol. *6*, 265–277. https://doi.org/10.2217/fmb.11.11

Blair, C.D., and Olson, K.E. (2015). The role of RNA interference (RNAi) in arbovirus–vector interactions. Viruses *7*, 820–843. https://doi.org/10.3390/v7020820

Blair, C.D., and Olson, K.E. (2017). Chapter 12 – Targeting dengue virus replication in mosquitoes. In Arthropod Vector: Controller of Disease Transmission, Volume 1, Wikel, S.K., Aksoy, S., Dimopoulos, G., eds. (Academic Press, San Diego), pp. 201–217.

Brackney, D.E., Scott, J.C., Sagawa, F., Woodward, J.E., Miller, N.A., Schilkey, F.D., Mudge, J., Wilusz, J., Olson, K.E., Blair, C.D., *et al.* (2010). C6/36 *Aedes albopictus* cells have a dysfunctional antiviral RNA interference response. PLOS Neglect. Trop. Dis. *4*, e856. https://doi.org/1 0.1371/journal.pntd.0000856

Brennecke, J., Aravin, A.A., Stark, A., Dus, M., Kellis, M., Sachidanandam, R., and Hannon, G.J. (2007). Discrete small RNA-generating loci as master regulators of transposon activity in *Drosophila*. Cell *128*, 1089–1103.

Burke, G.R., and Strand, M.R. (2012). Polydnaviruses of parasitic wasps: domestication of viruses to act as gene delivery vectors. Insects 3, 91–119. https://doi.org/10.3390/insects3010091

Burke, G.R., Walden, K.K.O., Whitfield, J.B., Robertson, H.M., and Strand, M.R. (2014). Widespread genome reorganization of an obligate virus mutualist. PLOS Genetics *10*, e1004660. https://doi.org/10.1371/journal.pgen.1004660

Burke, G.R., Simmonds, T.J., Sharanowski, B.J., and Beib, S.M. (2018). Rapid viral symbiogenesis via changes in parasitoid wasp genome architecture. Mol. Biol. Evol. *35*, 2463–2474. https://doi.org/10.1093/molbev/msy148

Burstein, E., Hoberg, J.E., Wilkinson, A.S., Rumble, J.M., Csomos, R.A., Komarck, C.M., Maine, G.N., Wilkinson, J.C., Mayo, M.W., and Duckett, C.S. (2005). COMMD proteins, a novel family of structural and functional homologs of MURR1. J Biol Chem *280*, 22222–22232.

Campbell, C.L., Black, W.C. IV, Hess, A.M., and Foy, B.D. (2008). Comparative genomics of small RNA regulatory pathway components in vector mosquitoes. BMC

Genomics 9, 425. https://doi.org/10.1186/1471-2164-9-425

Carapeta, S., do Bem, B., McGuinness, J., Esteves, A., Abecasis, A., Lopes, Â., de Matos, A.P., Piedade, J., António, P.G., de Almeida, A.P.G., and Parreira, R. (2015). Negeviruses found in multiple species of mosquitoes from southern Portugal: isolation, genetic diversity, and replication in insect cell culture. Virology *483*, 318–328.

Chen, X.G., Jiang, X., Gu, J., Xu, M., Wu, Y., Deng, Y., Zhang, C., Bonizzoni, M., Dermauw, W., Vontas, J. *et al.* (2015). Genome sequence of the Asian Tiger mosquito, *Aedes albopictus*, reveals insights into its biology, genetics, and evolution. Proc. Natl. Acad. Sci. U.S.A. *112*, E5907–5915. https://doi.org/10.1073/pnas.1516410112

Chen, Y., Willimans, V., Filippova, M., Filippov, V., and Duerksen-Hughes, P. (2014). Viral carcinogenesis: factors inducing DNA damage and virus integration. Cancers 6, 2155–2186. https://doi.org/10.3390/cancers6042155

Cirimotich, C.M., Scott, J.C., Phillips, A.T., Geiss, B.J., and Olson, K.E. (2009). Suppression of RNA interference increases Alphavirus replication and virus-associated mortality in *Aedes aegypti* mosquitoes. BMC Microbiol. *9*, 49. https://doi.org/10.1186/1471-2180-9-49

Cohn, L.B., Silva, I.T., Oliveira, T.Y., Rosales, R.A., Parrish, E.H., Learn, G.H., Hahn, B.H., Czartoski, J.L., McElrath, M.J., Lehmann, C., *et al.* (2015). HIV-1 integration landscape during latent and active infection. Cell *160*, 420–432. https://doi.org/10.1016/j.cell.2015.01.020

Crochu, S., Cook, S., Attoui, H., Charrel, R.N., De Chesse, R., Belhouchet, M., Lemasson, J.J., de Micco, P., and de Lamballerie, X. (2004). Sequences of flavivirus-related RNA viruses persist in DNA form integrated in the genome of *Aedes* spp. mosquitoes. J. Gen. Virol. *85*, 1971–1980.

Cui, J., and Holmes, E.C., (2012). Endogenous RNA viruses of plants in insect genomes. Virology *427*, 77–79. https://doi.org/10.1016/j.virol.2012.02.014

Desjardin, C.A., Gundersen-Rindal, D.E., Hostetler, J.B., Tallon, L.J., Fadrosh, D.W., Fuester, R.W., Pedroni, M.J., Haas, B.J., Schatz, M.C., Jones, K.M., *et al.* (2008). Comparative genomics of mutualistic viruses of *Glyptapanteles* parasitic wasps. Genome Biol 9, R183. https://doi.org/10.1186/gb-2008-9-12-r183

Dietzgen, R.G., Kondo, H., Goodin, M.M., Kurath, G., and Vasilakis, N. (2017). The family Rhabdoviridae: mono- and bipartite negative-sense RNA viruses with diverse genome organization and common evolutionary origins. Virus Res. *227*, 158–170. https://doi.org/10.1016/j.virusres.2016.10.010

Fort, P., Albertini, A., Van-Hua, A., Berthomieu, A., Roche, S., Delsuc, F., Pasteur, N., Capy, P., Gaudin, Y., and Weill, M. (2012). Fossil rhabdoviral sequences integrated into arthropod genomes: ontogeny, evolution, and potential functionality. Mol. Biol. Evol. 29, 381–390. https://doi.org/10.1093/molbev/msr226

Frank, J.A., and Feschotte, C. (2017). Co-option of endogenous viral sequences for host cell function. Curr. Opin. Virol. 25, 81–89. https://doi.org/10.1016/j.coviro.2017.07.021

Franz, A.W.E., Sanchez-Vargas, I., Adelman, Z.N., Blair, C.D., Beaty, B.J., James, A.A., and Olson, K.E. (2006). Engineering RNA interference-based resistance to dengue virus type 2 in genetically modified *Aedes aegypti*. Proc. Natl. Acad. Sci. U.S.A. *103*, 4198–4203.

Fujino, K., Horie, M., Honda, T., Merriman, D.K., and Tomonaga, K. (2014). Inhibition of Borna disease virus replication by an endogenous bornavirus-like element in the ground squirrel genome. Proc. Natl. Acad. Sci. U.S.A. *111*, 13175–13180.

Gauthier, J., Drezen, J.M., and Herniou, E.A. (2018). The recurrent domestication of viruses: major evolutionary transitions in parasitic wasps. Parasitology *145*, 713–723. https://doi.org/10.1017/S0031182017000725

Geisler, C., and Jarvis, D.L. (2016). Rhabdovirus-like endogenous viral elements in the genome of Spodoptera fruigiperta insect cells are actively transcribed: implications for adventitious virus detection. Biologicals *44*, 219–225. https://doi.org/10.1016/j.biologicals.2016.04.004

Geoghegan, J.L., Duchene, S., and Holmes, E.C. (2017). Comparative analyses estimates the relative frequencies of co-divergence and cross-species transmission within viral families. PLOS Pathog. *13*, e1006215. https://doi.org/10.1371/journal.ppat.1006215

Geuking, M.B., Weber, J., Dewannieux, M., Gorelik, E., Heidmann, T., Hengartner, H., Zinkernagel, R.M., and Hangartne, L. (2009). Recombination of retrotransposon and exogenous RNA virus results in non-retroviral cDNA integration. Science 323, 393–396. https://doi.org/10.1126/science.1167375

Goic, B., Vodovar, N., Mondotte, J.A., Monot, C., Frangeul, L., Blanc, H., Gausson, V., Vera-Otarola, J., Cristofari, G., and Saleh, M.-C. (2013). RNA-mediated interference and reverse transcription control the persistence of RNA viruses in the insect model *Drosophila*. Nat. Immunol. *14*, 396–403. https://doi.org/10.1038/ni.2542

Goic, B., Stapleford, K.A., Frangeul, L., Doucet, A.J., Gausson, V., Blanc, H., Schemmel-Jofre, N., Cristofari, G., Lambrechts, L., Vignuzzi, M., *et al.* (2016). Virus-derived DNA drives mosquito vector tolerance to arboviral infection. Nat. Commun. 7, 12410. https:/https://doi.org/10.1038/ncomms12410

Griffiths, D.J. (2001). Endogenous retroviruses in the human genome sequence. Genome Biol Rev. *2*, 1017.1–1017.5.

Gunawardane, L.S., Saito, K., Nishida, K.M., Miyoshi, K., Kawamura, Y., Nagami, T., Siomi, H., and Siomi, M.C. (2007). A slicer-mediated mechanism for repeat-associated siRNA 5′ end formation in *Drosophila*. Science *315*, 1587–1590.

Havecker, E.R., Gao, X., and Voytas, D.F. (2004). The diversity of LTR retrotransposons. Genome Biol. 5, 225.

Honda, T., and Tomonaga, K. (2016). Endogenous non-retroviral RNA virus elements evidence a novel type of antiviral immunity. Mob. Genet. Elements 6, e1165785. https://doi.org/10.1080/2159256X.2016.1165785

Horie, M., Honda, T., Suzuki, Y., Kobayashi, Y., Daito, T., Oshida, T., Ikuta, K., Jern, P., Gojobori, T., Coffin, J.M., *et al.* (2010). Endogenous non-retroviral RNA virus elements in mammalian genomes. Nature *463*, 84–87.

Katzourakis, A. (2017). Editorial overview: Paleovirology: the genomic fossil record, and consequences of ancient viral infections. Curr. Opin. Virol. *25*, ix–xi. https://doi.org/10.1016/j.coviro.2017.08.002

Katzourakis, A., and Gifford, R.J. (2010). Endogenous viral elements in animal genomes. PLOS Genet. 6, e1001191. https://doi.org/10.1371/journal.pgen.1001191

Keene, K.M., Foy, B.D., Sanchez-Vargas, I., Beaty, B.J., Blair, C.D., and Olson, K.E. (2004). RNA interference acts as a natural antiviral response to O'nyong-nyong virus (*Alphavirus*; *Togaviridae*) infection of *Anopheles gambiae*. Proc. Natl. Acad. Sci. U.S.A. *101*, 17240–17245.

Kerr, P.J. (2012). Myxomatosis in Australia and Europe: A model for emerging infectious diseases. Antiviral Res. *93*, 387–415. https://doi.org/10.1016/j.antiviral.2012.01.009

Kondo, H., Hirano, S., Chiba, S., Andika, I.B., Hirai, M., Maeda, T., and Tamada, T. (2013). Characterization of burdock mottle virus, a novel member of the genus *Benyvirus*, and the identification of benyvirus-related sequences in the plant and insect genomes. Virus Res. *177*, 75–86. https://doi.org/10.1016/j.virusres.2013.07.015

Kondo, H., Chiba, S., Maruyama, K., Andika, I.B., and Suzuki, N. (2019). A novel insect-infecting virga/nege-like virus group and its pervasive endogenization into insect genomes. Virus Res. *262*, 37–47. https://doi.org/10.1016/j.virusres.2017.11.020

Koonin, E.V., Dolja, V.V., and Krupovic, M. (2015). Origins and evolution of viruses of eukaryotes: the ultimate modularity. Virology *479–480*, 2–25. https://doi.org/10.1016/j.virol.2015.02.039

Lazareva, E., Lezzhov, A., Vassetzky, N., Solovyev, A., and Morozov, S. (2015). Acquisition of full-length viral helicase domains by insect retrotransposon-encoded polypeptides. Front. Microbiol. *6*, 1447. https://doi.org/10.3389/fmicb.2015.01447

Léger, P., Lara, E., Jagla, B., Sismeiro., O., Mansuroglu, Z., Coppée, J.Y., Bonnefoy, E., and Bouloy, M. (2013). Dicer-2- and Piwi-mediated RNA interference in Rift Valley Fever Virus-infected mosquito cells. J. Virol. *87*, 1631–1648. https://doi.org/10.1128/JVI.02795-12

Lequime, S., and Lambretchs, L. (2017). Discovery of flavivirus-derived endogenous viral elements in *Anopheles* mosquito genomes supports the existence of *Anopheles*-associated insect-specific flaviviruses. Virus Evol. 3, vew035. https://doi.org/10.1093/ve/vew035

Li, C.X., Shi, M., Tian, J.H., Lin, X.D., Kang, K.Y., Chen, L.J., Quin, X.C., Xu, J., Holmes, E.C., and Zhang, Y.Z. (2015). Unprecedented genomic diversity of RNA viruses in arthropods reveals the ancestry of negative-sense RNA viruses. Elife *4*, e05378. https://doi.org/10.7554/eLife.05378

Lin, C.L., Lee, J.C., Chen, S.S., Wood, H.A., Li, M.L., Li, C.F., and Chao, Y.C. (1999). Persistent Hz-1 virus infection in insect cells: evidence for insertion of viral DNA into host chromosomes and viral infection in a latent status. J. Virol. 73, 128–139.

Maori, E., Tanne. E., and Sela. I. (2017). Reciprocal sequence exchange between non-retro virus and hosts leading to the appearance of new host phenotypes. Virology 362, 342–349.

Matthews, B.J., Dudchenko, O., Kingan, S.B., Koren, S., Antoshechkin, I., Crawford, J.E., Glassford, W.J., Herre, M., Redmond, S.N., Rose, N.H. *et al.* (2018). Improved reference genome of *Aedes aegypti* informs arbovirus vector control. Nature *563*, 501–507. https://doi.org/10.1038/s41586-018-0692-z

Miesen, P., Girardi, E., and van Rij, R.P. (2015). Distinct sets of PIWI proteins produce arbovirus and transposon-derived piRNAs in *Aedes aegypti* mosquito cells. Nucleic Acids Res. *43*, 6545–6556. https://doi.org/10.1093/nar/gkv590

Miesen, P., Ivens, A., Buck, A.H., and van Rij, R.P. (2016a). Small RNA profiling in Dengue Virus 2-infected *Aedes* mosquito cells reveals viral piRNAs and novel host miRNAs. PLOS Negl. Trop. Dis. *10*, e0004452. https://doi.org/10.1371/journal.pntd.0004452

Miesen, P., Joosten. J., and van Rij, R.P. (2016b). PIWIs go viral: arbovirus-derived piRNAs in vector mosquitoes. PLOS Pathogens *12*, e1006017. https://doi.org/10.1371/journal.ppat.1006017

Mishra, P., Furey, C., Balaraman, V., and Fraser, M. (2016). Antiviral hammerhead ribozymes are effective for developing transgenic suppression of Chikungunya virus in *Aedes aegypti* mosquitoes. Viruses *8*, E163. https://doi.org/10.3390/v8060163

Mogler, M.A., and Kamrud, K.I. (2015). RNA-based viral vectors. Expert Rev. Vaccines *14*, 283–312. https://doi.org/10.1586/14760584.2015.979798

Morazzani, E.M., Wiley, M.R., Murreddu, M.G., Adelman, Z.N., and Myles, K.M. (2012). Production of virus-derived ping-pong-dependent piRNA-like small RNAs in the mosquito soma. PLOS Pathog. *8*, e1002470. https://doi.org/10.1371/journal.ppat.1002470

Morozov, S.Y., Lazareva, E.A., and Solovyev, A.G. (2017). RNA helicase domains of viral origin in proteins of insect retrotransposons: possible source for evolutionary advantages. PeerJ *5*, e3673. https://doi.org/10.7717/peerj.3673

Myles, K.M., Wiley, M.R., Morazzani, E.M., and Adelman, Z.N. (2008). Alphavirus-derived small RNAs modulate pathogenesis in disease vector mosquitoes. Proc. Natl. Acad. Sci. U.S.A. *105*, 19938–19943. https://doi.org/10.1073/pnas.0803408105

Nag, D.K., and Kramer, L.D. (2017). Patchy DNA forms of the Zika virus RNA genome are generated following infection in mosquito cell cultures and in mosquitoes. J. Gen. Virol. *98*, 2731–2737. https://doi.org/10.1099/jgv.0.000945

Nag, D.K., Brecher, M., and Kramer, L.D. (2016). DNA forms of arboviral RNA genomes are generated following infection in mosquito cell cultures. Virology *498*, 164–171.

Nunes, M.R.T., Contreras-Gutierrez, M.A., Guzman, H., Martins, L.C., Barbirato, M.F., Savit, C., Balta, V., Uribe, S., Vivero, R., Suaza, J.D., *et al.* (2017). Genetic characterization, molecular epidemiology, and phylogenetic relationships of insect-specific viruses in the taxon Negevirus. Virology *504*, 152–167. https://doi.org/10.1016/j.virol.2017.01.022

Olson, K.E., and Blair, C.D. (2015). Arbovirus–mosquito interactions: RNAi pathway. Curr. Opin. Virol. *15*, 119–126. https://doi.org/10.1016/j.coviro.2015.10.001

Olson, K.E., and Bonizzoni, M. (2017). Nonretrovial Integrated RNA viruses in arthropod vectors: an occasional event or something more? Curr. Opin. Insect Sci. *22*, 45–53. https://doi.org/10.1016/j.cois.2017.05.010

Palatini, U., Miesen, P., Carballar-Lejarazu, R., Ometto, L., Rizzo, E., Tu, Z., van Rij, R.P., and Bonizzoni, M. (2017). Comparative genomics shows that viral integrations are abundant and express piRNAs in the arboviral vectors *Aedes aegypti* and *Aedes albopictus*. BMC Genomics *18*, 512. https://doi.org/10.1186/s12864-017-3903-3

Pischedda, E., Scolari, F., Valerio, F., Carballar-Lejarazu, R., Catapano, P.l., Waterhouse, R.M., and Bonizzoni, M. (2018). Evolutionary landscape of mosquito viral integrations. bioRxiv. https://doi.org/10.1101/385666

Rizzo, F., Cerutti, F., Ballardini, M., Mosca, A., Vitale, N., Radaelli, M.C., Desiato, R., Prearo, M., Pautasso, A., Casalone, C., *et al.* (2014). Molecular characterization of flavivirus from field-collected mosquitoes in northwestern Italy, 2011–2012. Parasit. Vectors *7*, 395. https://doi.org/10.1186/1756-3305-7-395

Roiz, D., Vazquez, A., Sanchez Seco, M.P., Tenorio, A., and Rizzoli, A. (2009). Detection of novel insect flavivirus sequences integrated in *Aedes albopictus* (Diptera: Culicidae) in northern Italy. Virol. J. *6*, 93. https://doi.org/10.1186/1743-422X-6-93

Sanchez-Seco, M.P., Vazquez, A., Collao, X., Hernandez, L., Aranda, C., Ruiz, S., Escosa, H., Marques, E., Bustillo, M.A., Molero, F., *et al.* (2010). Surveilance of arboviruses in Spanish wetlands: detection of new Flavi- and Phleboviruses. Vector Borne Zoonotic Dis. *10*, 203–206. https://doi.org/10.1089/vbz.2008.0188

Sánchez-Vargas, I., Scott, J.C., Poole-Smith, B.K., Franz, A.W., Barbosa-Solomieu, V., Wilusz, J., Olson, K.E., and Blair, C.D. (2009). Dengue virus type 2 infections of *Aedes aegypti* are modulated by the mosquito's RNA interference pathway. PLOS Pathog. *5*, e1000299. https://doi.org/10.1371/journal.ppat.1000299

Sánchez-Vargas, I., Harrington, L., Doty, J.B., Black, W.C., and Olson, K.E. (2018). Demonstration of efficient vertical and venereal transmission of dengue virus type-2 in a genetically diverse laboratory strain of *Aedes aegypti*. PLOS Negl. Trop. Dis. *12*, e0006754. https://doi.org/10.1371/journal.pntd.0006754

Schnettler, E., Donald, C.L., Human, S., Watson, M., Siu, R.W.C., McFarlane, M., Fazakerley, J.K., Kohl, A., and Fragkoudis, R. (2013). Knockdown of piRNA pathway proteins results in enhanced Semliki Forest virus production in mosquito cells. J. Gen. Virol. *94*, 1680–1689. https://doi.org/10.1099/vir.0.053850-0

Scott, J.C., Brackney, D.E., Campbell, C.L., Bondu-Hawkins, V., Hjelle, B., Ebel, G.D., Olson, K.E., and Blair, C.D. (2010). Comparison of Dengue virus type 2-specific small RNAs from RNA interference-competent and -incompetent mosquito cells. PLOS Negl. Trop. Dis. *4*, e848.

Shi, M., Lin, X.D., Tian, J.H., Chen, L.J., Chen, X., Li, C.X., Qin, X.C., Li, J., Cao, J.P., Eden, J.S., *et al.* (2016). Redefining the invertebrate RNA virosphere. Nature *540*, 539–543. https://doi.org/10.1038/nature20167

Siomi, M.C., Sato, K., Pezic, D., and Aravin, A.A., 2011. PIWI-interacting small RNAs: the vanguard of genome defence. Nat. Rev. Mol. Cell Biol. *12*, 246–258. https://doi.org/10.1038/nrm3089

Sofuku, K., Parrish, N.F., Honda, T., and Tomonaga, K. (2015). Transcription profiling demonstrates epigenetic control of non-retroviral RNA virus-derived elements in the human genome. Cell Rep. *12*, 1548–1554. https://doi.org/10.1016/j.celrep.2015.08.007

Strand, M.R., and Burke, G.R. (2015). Polydnaviruses: from discovery to current insights. Virology *479–480*, 393–402. https://doi.org/10.1016/j.virol.2015.01.018

Suzuki, Y., Frangeul, L., Dickson, L.B., Blanc, H., Verdier, Y., Vinh, J., Lambrechts, L., and Saleh, M.C. (2017). Uncovering the repertoire of endogenous flaviviral elements in *Aedes* mosquito genomes. Virol. J. *91*, e00571-17. https://doi.org/10.1128/JVI.00571-17

ter Horst, A.M., Nigg, J.C., and Flak, B.W. (2018). Endogenous viral elements are widespread in arthropod genomes and commonly give rise to piRNAs. bioRxiv. https://doi.org/10.1101/396382

Vasilakis, N., Forrester, N.L., Palacios, G., Nasar, F., Savji, N., Rossi, S.L., Guzman, H., Wood, T.G., Popov, V., Gorchakov, R., Gonzalez, A.V., *et al.* (2013). Negevirus: a proposed new taxon of insect-specific viruses with wide geographic distribution. Virol. J. *87*, 2475–2488. https://doi.org/10.1128/JVI.00776-12

Vasquez, A., Sanchez-Seco, M.P., Palacio, G., Molero, F., Reyes, N., Ruiz, S., Aranda, C., Marques, E., Escosa, R., Moreno, J., *et al.* (2012). Novel flaviviruses detected in different species of mosquitoes in Spain. Vector Borne Zoonotic Dis. *12*, 223–229. https://doi.org/10.1089/vbz.2011.0687

Vodovar, N., Bronkhorst, A.W., van Cleef, K.W.R., Miesen, P., Blanc, H., van Rij, R.P., and Saleh, M.C., 2012. Arbovirus-derived piRNAs exhibit a ping-pong signature in mosquito cells. PLOS ONE *7*, e30861. https://doi.org/10.1371/journal.pone.0030861

Whitfield, Z.J., Dolan, P.T., Kunitomi, M., Tassetto, M., Seetin, M.G., Oh, S., Heiner, C., Paxinos, E., and Andino, R. (2017). The diversity, structure, and function of heritable adaptive immunity sequences in the *Aedes aegypti* genome. Curr. Biol. *27*, 3511–3519.e7. https://doi.org/10.1016/j.cub.2017.09.067

Yap, M.W., Colbeck, E., Ellis, S.A., and Stoye, J.P. (2014). Evolution of the retroviral restriction gene *FV1*: inhibition of non-MLV retroviruses. PLOS Pathog. *10*, e1003968. https://doi.org/10.1371/journal.ppat.1003968

Zhdanov, V.M. (1975). Integration of viral genomes. Nature *256*, 471–473.

Sensing Viral Infections in Insects: A Dearth of Pathway Receptors

3

Loïc Talide[1], Jean-Luc Imler[1,2] and Carine Meignin[1,2*]

[1]Université de Strasbourg, Centre National de la Recherche Scientifique, Modèles Insectes de l'Immunité Innée (M3i; UPR9022), Strasbourg, France.
[2]Université de Strasbourg, Faculté des Sciences de la Vie, Strasbourg, France.

*Correspondence: c.meignin@ibmc-cnrs.unistra.fr

https://doi.org/10.21775/9781912530083.03

Abstract

Insects, the most diverse group of animals, can be infected by an extraordinary diversity of viruses. Among them, arthropod-borne viruses can be transmitted to humans, while bee and silkworm viruses cause important economic losses. Like all invertebrates, insects rely solely on innate immunity to counter viral infections. Protein-based mechanisms, involving restriction factors and evolutionarily conserved signalling pathways regulating transcription factors of the NF-κB and STAT families, participate in the control of viral infections in insects. In addition, RNA-based responses play a major role in the silencing of viral RNAs. We review here our current state of knowledge on insect antiviral defence mechanisms, which include conserved as well as adaptive, insect-specific strategies. Identification of the innate immunity receptors that sense viral infection in insects remains a major challenge for the field.

Introduction

With more than 1 million known species, insects are the largest group of multicellular organisms, representing over 70% of animal species. Dating back to the Early Ordovician (about 480 million years ago, Mya), they were among the first animals to colonize terrestrial and freshwater ecosystems and have undergone major expansions, culminating in the spectacular diversification of holometabolous insects (Hymenoptera, Diptera, Lepidoptera) during the Early Cretaceous (about 120 Mya) (Misof *et al.*, 2014). Insects can be credited with major adaptations, such as flight and establishment of social groups. Importantly, they have undoubtedly contributed to shape the planet's biota and actively interact with other multicellular eukaryotes such as plants and vertebrates. Like them, insects are exposed to a large panel of infectious microorganisms, which they control through their innate immune system (Hoffmann *et al.*, 1999).

Among infectious microbes, viruses represent a particular threat because they offer few intrinsic targets for inhibition by antiviral molecules. This is because they consist in their simplest form of a nucleic acid encapsidated in a protein shell, and hijack molecular machineries from host cells to complete their replication cycle. Therefore, viruses exert great selective pressure on their host to evolve resistance pathways. These, in turn, favour the adaptation of viruses to escape antiviral mechanisms. This arms race results in the diversification of both host-defence and virus escape mechanisms. As a result, it can be highly instructive to broaden the study of antiviral immunity to non-mammalian models. In light of their diversity, insects represent an interesting group of animals for this type of comparative study (Marques and Imler, 2016; Martins *et al.*, 2016).

Recent advances in high-throughput sequencing technologies have opened the way to the characterization of the virome of insects (i.e. the genetic diversity of viruses in a biological sample; see also Chapter 1). In a landmark article, Yong-Zhen Zhang and colleagues analysed the transcriptome of more than 220 invertebrate species covering nine animal phyla and reported the identification of close to 1500 new viruses (Shi *et al.*, 2016). Thus, infection by one or several viruses is common in invertebrates. In addition, the genetic diversity of these viruses surpassed that described previously. Many newly identified viruses fell between families and genera from the current virus classification, filling major phylogenetic gaps and revealing that viruses form a continual spectrum of phylogenetic diversity (Shi *et al.*, 2016). A more detailed analysis focusing on 70 arthropod species and negative-sense RNA viruses, which include important pathogens causing a variety of diseases in humans (flu, rabies, encephalitis, haemorrhagic fever), led to the discovery of 112 new viruses (Li *et al.*, 2015). This study revealed that much of the diversity of negative-sense RNA viruses found in plants and vertebrate animals falls within the genetic diversity of viruses associated with arthropods (Dudas and Obbard, 2015; Li *et al.*, 2015). Of note, arthropods (and insects in particular) can live in large and dense populations, facilitating propagation and transmission of viruses. The close interaction between many insects and plants or vertebrate animals further support the hypothesis that negative-sense RNA viruses, including vertebrate-specific ones, are derived from arthropod dependent viruses (Li *et al.*, 2015).

There are several specific reasons to study virus–host interactions in insects. First, infection of insects can cause important economic losses (e.g. viral diseases of silkworms; contribution to colony-collapse in honey bees) (Bradshaw *et al.*, 2016; Carrillo-Tripp *et al.*, 2016). Second, haematophagous insects such as *Aedes* or *Anopheles* mosquitoes can transmit viral diseases to mammalian hosts. These viruses, the arthropod-borne viruses or arboviruses, include *Dengue* (DENV), *Yellow fever* (YFV), and *West Nile virus* (WNV) (Molina-Cruz *et al.*, 2016; Powers and Waterman, 2017). Third, microbial pathogens (e.g. baculoviruses) can be used as biological control agents against insect pests, which necessitates some knowledge of the host response to these microorganisms (Popham *et al.*, 2016). Fourth, insects such as the genetically tractable model organism *Drosophila melanogaster* can be used to decipher evolutionarily conserved innate immune mechanisms.

Antiviral immunity in insects

NF-κB and STAT dependent inducible responses

Innate immunity is the first line of defence that multicellular organisms deploy to limit pathogen infections. In vertebrates, the innate immune response also regulates the production of cytokines and co-stimulatory molecules, which shape the subsequent adaptive immune response (Hoffmann *et al.*, 1999). Studies on innate immunity in *Drosophila* initially focused on bacterial and fungal infections and revealed that the systemic production of antimicrobial peptides (AMPs) (humoral response) plays an important role in host defence (Steiner *et al.*, 1981; Imler and Bulet, 2005). Also, in the haemolymph, proteolytic cascades involving sequential activation of serine proteases participate in the clotting and melanisation responses to wounding (Binggeli *et al.*, 2014; Theopold *et al.*, 2014). In addition, cellular responses involving both circulating and sessile haemocytes participate in antimicrobial host defence in flies, in particular *via* phagocytosis of bacteria or virus infected cells by macrophage-like plasmatocytes and, in larvae, encapsulation of parasitic wasp eggs by lamellocytes (Gold and Brückner, 2015; Letourneau *et al.*, 2016; Weavers *et al.*, 2016). In the case of viruses, it is now well established that the cell intrinsic mechanism of RNA interference (RNAi) plays a central role in the control of viral infections in insects, as it does in plants and other invertebrates (see below) (Ding, 2010). In addition, inducible responses and restriction factors also contribute to resistance to viral infections (reviewed in Mussabekova *et al.*, 2017).

Expression of AMPs is controlled by the evolutionarily conserved signalling pathways Toll and IMD (immune deficiency), which regulate the activity of transcription factors of the NF-κB family (reviewed in Hoffmann, 2003). These pathways are activated by pattern recognition receptors (PRRs) that sense components of the bacterial or fungal cell wall such as peptidoglycan in the case of bacteria or β-glucans in the case of fungi (Steiner, 2004; Royet *et al.*, 2011; Rao *et al.*, 2018). Based on the transcriptomic signature of virus-infected insects and the phenotype of flies or mosquitoes with genes encoding important components of the pathway mutated or silenced, both of these pathways are proposed to also participate in antiviral immunity (Avadhanula *et al.*, 2009; Costa *et al.*, 2009; Ferreira *et al.*, 2014; McFarlane *et al.*, 2014; Paradkar *et al.*, 2014; Carissimo *et al.*, 2015; Lamiable *et al.*, 2016a; Fig. 3.1). A third evolutionarily conserved pathway connected to inflammation in mammals, the Jak/STAT pathway, has also been proposed to play a role in insect antiviral immunity (Dostert *et al.*, 2005; Fragkoudis *et al.*, 2008; Souza-Neto *et al.*, 2009; Paradkar *et al.*, 2012; Kemp *et al.*, 2013; Barribeau *et al.*, 2015; Carissimo *et al.*, 2015; Merkling *et al.*, 2015a; Zhang *et al.*, 2016; Jupatanakul *et al.*, 2017; West and Silverman, 2018). This pathway is activated by cytokines of the Unpaired (Upd) family, which are upregulated by viral infection or stress in *Drosophila* (Jiang *et al.*, 2009; Kemp *et al.*, 2013; Gordon *et al.*, 2018; Fig. 3.1). In *Culex* mosquitoes and bumble bees, a single von Willebrand factor type C domain secreted factor, related to the *Drosophila* antiviral factor Vago (Deddouche *et al.*, 2008), appears to activate the Jak/STAT pathway and antiviral immunity (Paradkar *et al.*, 2012; Wang *et al.*, 2017). Two major questions in the field arise at this stage: (1) How are these pathways activated by viruses? And (2) what are the antiviral effectors they regulate and how do they counter viruses?

Activation of the Toll and IMD pathways by viruses

The Toll and the IMD pathways are activated by PRRs of the peptidoglycan recognition protein (PGRP) and β-glucan recognition protein (GNBP) families, which sense components of the bacterial and fungal cell walls (reviewed in Ferrandon *et al.*, 2007; Royet *et al.*, 2011). Interestingly, both pathways can also be activated by virulence factors. In the case of the Toll pathway, the circulating zymogen Persephone senses protease activity independently of microbial patterns and triggers a proteolytic cascade that culminates in the processing of the cytokine Spaetzle to generate an active Toll ligand (Gottar *et al.*, 2006; El Chamy *et al.*, 2008; Issa *et al.*, 2018; Fig. 3.1). In a conceptually similar manner, deamidation by bacterial toxins of a critical glutamine residue in the Rho GTPase Rac2 can be sensed by IMD, resulting in activation of the pathway independently from PRRs (Boyer *et al.*, 2011).

For the moment, it remains unclear how viruses activate the Toll and IMD pathways, and whether this involves PRRs or other sensors. In mammals,

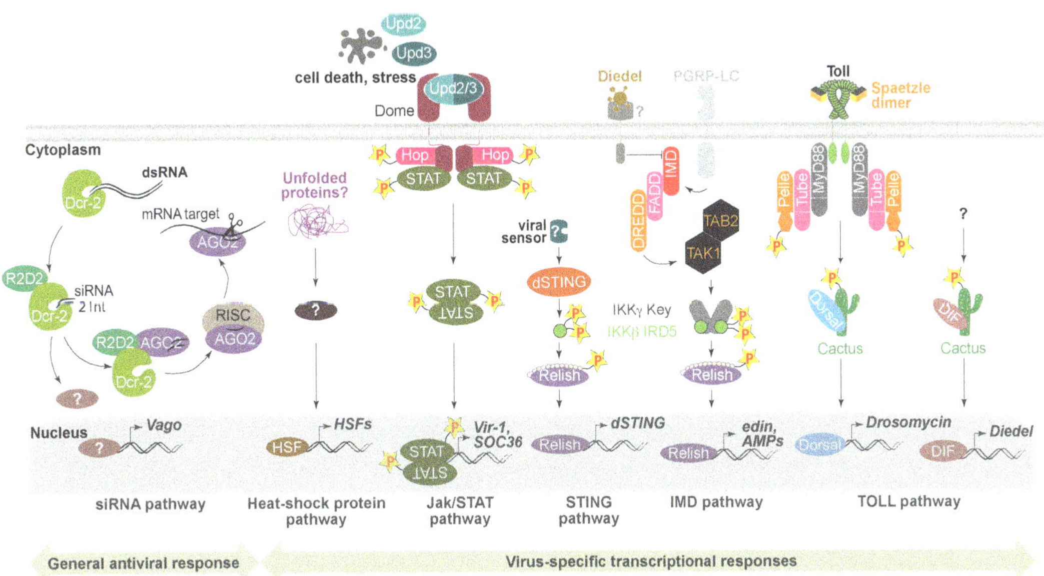

Figure 3.1 Antiviral transcriptional responses in *Drosophila melanogaster*. The signalling pathways activated during viral infection in *Drosophila* are illustrated.

PRRs belonging to different structural families [e.g. Toll like receptors (TLRs), RIG-I-like receptors (RLRs), nucleotide binding domain and leucine-rich repeat containing receptors (NLRs)] sense viral nucleic acids and trigger expression of the antiviral cytokines of the interferon family (Goubau *et al.*, 2013; Roers *et al.*, 2016). At this stage, the only identified receptor for viruses in insects is Dicer-2, which interacts with viral double-stranded (ds) RNA. Interestingly, Dicer-2 has been proposed to serve a dual function in antiviral immunity, activating both antiviral RNAi and signalling leading to expression of antiviral molecules (e.g. Vago) (Deddouche *et al.*, 2008; Paradkar *et al.*, 2012; Wang *et al.*, 2017; Asad *et al.*, 2018; Fig. 3.1). Viruses also affect the physiology of the cells, which can trigger a response. For example, infection by *Drosophila C virus* (DCV) induces a heat shock response, possibly in reaction to the accumulation of unfolded viral proteins in the cytosol of infected cells (Merkling *et al.*, 2015b). Viruses are also notorious for inhibiting cellular translation or altering cellular membranes, which may trigger a cellular response, such as apoptosis or autophagy (Shelly *et al.*, 2009; Nainu *et al.*, 2015; Lamiable *et al.*, 2016b; Khong *et al.*, 2017).

Of note, it is becoming apparent that components of the canonical Toll or IMD pathways may participate in novel pathways, responding to different cues and activating distinct transcriptional programs. For example, two closely related NF-κB proteins, Dorsal and DIF, are regulated by the Toll pathway in fruit flies (Tanji *et al.*, 2010). In spite of strong sequence conservation, DIF lacks important features of Dorsal-mediated pattern formation in the *Drosophila* embryo (Stein *et al.*, 1998). Conversely, DIF, but not Dorsal, is required to regulate expression of the antifungal peptide Drosomycin to resist fungal infections in adult flies, although both proteins are expressed (Lemaitre *et al.*, 1995; Rutschmann *et al.*, 2000a). Interestingly, in a context of oral infection of adult flies, it is Dorsal, rather than DIF, that is required for resistance to DCV (Ferreira *et al.*, 2014; Fig. 3.1). The other components of the Toll pathway, up to the cytokine Spaetzle, are also involved in resistance to DCV. An intriguing unresolved question pertains to the proteases acting upstream of Spaetzle in the context of DCV oral infection (Ferreira *et al.*, 2014). Interestingly, although not essential in the context of DCV infection, DIF is required for the strong up-regulation of the cytokine Diedel (see below) triggered by *Sindbis virus* (SINV) and *Vesicular stomatitis virus* (VSV). However, MyD88, a key signalling adapter protein of the Toll pathway, is dispensable for inductio n of Diedel (Lamiable *et al.*, 2016a). This suggests that, in the context of some viral infections, DIF can be activated in a MyD88-independent manner and, presumably, by a mechanism distinct from the canonical Toll pathway.

The IMD pathway can be activated by SINV in *Drosophila*, by a mechanism that remains unclear (Avadhanula *et al.*, 2009; Lamiable *et al.*, 2016a). Some antimicrobial peptides regulated by this pathway have been proposed to participate in the control of this virus (Huang *et al.*, 2013). The relevance of this pathway in the context of viral infections is highlighted by the observation that members of different families of DNA viruses have hijacked a gene called *Diedel*, which encodes an immunomodulatory cytokine down-regulating the IMD pathway and antagonizing apoptosis (Coste *et al.*, 2012; Lamiable *et al.*, 2016a; Mlih *et al.*, 2018). This observation prompted a reanalysis of the contribution of the IMD pathway to antiviral immunity. Unexpectedly, this study revealed that two components of the pathway, the kinase IKKβ and the NF-κB transcription factor Relish, restrict replication of the dicistroviruses DCV and Cricket paralysis virus (CrPV) in *Drosophila* (Goto *et al.*, 2018). Strikingly, the other components of the canonical IMD pathway, including IKKγ, the regulatory subunit of the IKK complex, are not required for virus suppression. Further analysis revealed that in the context of viral infection, the kinase IKKβ is activated by a different pathway, involving the *Drosophila* homologue of the mammalian gene STING (Stimulator of Interferon genes), a critical component of the cytosolic DNA sensing pathway in mammals (Fig. 3.1). The genes regulated by this alternative new pathway are different from the antibacterial peptides regulated by IMD, suggesting that Relish interacts with an additional transcription factor (Goto *et al.*, 2018). This would be conceptually similar to the cooperation of NF-κB with IRF3 to regulate antiviral genes in mammals (Ikushima *et al.*, 2013). Also pointing to an involvement of components of the IMD pathway in antiviral immunity, in Culex mosquitoes, a Dicer-2-dependent pathway regulates expression of the gene *Vago* upon activation of a TRAF factor and the homologue of Relish, REL2 (Paradkar *et al.*, 2014). In summary, it appears that NF-κB pathways are more diverse than initially thought, and that these evolutionarily conserved transcription factors can be activated by alternative branches of the canonical Toll and IMD pathways initially characterized in the context of bacterial and fungal infections. Consistent with these findings, vankyrins form a family of Iκβ-like molecules encoded by polydnaviruses that can antagonize NF-κB-dependent responses (Kroemer and Webb, 2005, 2006; Gueguen *et al.*, 2013).

Control of viruses by restriction factors

Constitutively expressed restriction factors also participate in the control of viruses in insects. Some of them are evolutionarily conserved and the functions of their mammalian homologues point to mechanisms for virus inhibition. For example, *Drosophila* ref(2)P, a restriction factor for *Sigma virus* (i.e. a Rhabdovirus that is a natural pathogen of *Drosophila*), is a homologue of human p62/sequestosome-1, pointing to possible involvement of autophagy in the control of this virus (Carré-Mlouka *et al.*, 2007; Ktistakis and Tooze, 2016). Interestingly, other restriction factors are not conserved and represent insect-specific adaptations. One example is the gene *CHKov1*, which encodes another restriction factor for Sigma virus (Magwire *et al.*, 2011). The function of CHKov1 is unknown, but the protein contains a choline kinase domain, which is intriguing in light of the function of choline kinase as a host factor for *Hepatitis C virus* (HCV) in human hepatocytes (Wong and Chen, 2016, 2017). The gene *pastrel* is another example of a non-conserved gene having a potent restricting activity on DCV and CrPV in *Drosophila* (Magwire *et al.*, 2012; Cao *et al.*, 2017). How this protein functions to block viral replication, and if its action involves interaction with viral RNA remains unknown.

In summary, it is clear that protein-based mechanisms, involving evolutionarily conserved genes and pathways, are involved in insect antiviral immunity. However, the receptors that sense viruses and trigger these responses are unknown. As a result, Dicer-2 remains the only well characterized sensor for viral infection in insect cells (Fig. 3.1).

The siRNA pathway of RNA interference: mechanism and regulation

RNAi pathways in insects

Fire, Mello and coworkers coined the term RNA interference, or 'RNAi', to describe the observation that dsRNA can block gene expression when introduced into *Caenorhabditis* nematodes (Fire *et al.*, 1998). This discovery was rapidly followed by biochemical characterization of RNAi in fruit flies using embryos and the S2 cell line (Hammond *et al.*, 2000; Zamore *et al.*, 2000; Bernstein *et al.*, 2001; Liu *et al.*, 2003). We currently know that different RNAi mechanisms play important regulatory roles in development, maintenance of genome stability, gene expression and antiviral defence. These RNA-based mechanisms involve proteins of the Argonaute family combined with small regulatory RNAs ranging from 20–30 nt length (Treiber *et al.*, 2019). Argonaute proteins associate with small RNAs that guide them towards target mRNAs leading to inhibition of their translation or direct cleavage catalysed by their RNAse-H like domain (Song *et al.*, 2004; Ma *et al.*, 2005; Yuan *et al.*, 2005). In insects, three RNAi pathways have been well documented, involving small RNAs of 21–22 nt (small interfering RNAs or siRNAs), 22–23 nt (micro RNAs or miRNAs) or 24–30 nt (Piwi-interacting RNAs or piRNAs). Whereas siRNAs and miRNAs are produced from dsRNA precursors by RNAseIII proteins such as Dicers and Drosha, piRNAs are processed independently of these enzymes (see section on piwi pathway below for more detail). *Drosophila* genetics defined two well-separated pathways involving Dicer-1/AGO1 for miRNAs and Dicer-2/AGO2 for siRNAs (Lee *et al.*, 2004; Okamura *et al.*, 2004). A different clade of AGO proteins involving PIWI, AGO3 and Aubergine (Aub) in *Drosophila* regulates the production and activity of piRNAs (reviewed in Huang *et al.*, 2017). miRNAs are essentially produced from nuclear precursors and participate in tight regulation of gene expression during development or cellular homeostasis (e.g. Posadas and Carthew, 2014). Of note, some DNA viruses use virus-encoded or cellular miRNAs to regulate their own gene expression or to modulate host cell transcriptome (Müller and Imler, 2007; Hussain and Asgari, 2014; see also Chapter 4). However, control of viral infections in insects mostly relies on siRNAs and, in some cases might also involve piRNAs.

Production of small non-coding RNAs by Dicer enzymes

The antiviral RNAi pathway, the siRNA pathway, is triggered by long dsRNAs, which are processed in insects by Dicer-2 (Dcr-2). This cytoplasmic enzyme is composed of a N-terminal Duplex RNA activated ATPase (DRA) domain, a central double-stranded RNA-binding domain (dsRBD), a platform-PAZ domain, two RNAseIII domains and a C-terminal dsRBD (Fig. 3.2, left). Interestingly, Dicer enzymes share a common phylogenetic origin with the vertebrate RLRs, which sense viral nucleic acids in the cytosol and trigger synthesis of interferons. Although Dicer-2 and RLRs have different sizes and domain composition, they all contain a conserved DRA domain (Paro *et al.*, 2015). Our understanding of the contribution of the DRA domain of Dicer-2 to virus sensing remains limited and is largely based on *in vitro* studies.

In vitro, Dicer-2 is efficient at processing long dsRNAs in addition to shorter structured RNAs such as pre-miRNAs which are *bona fide* targets of Dicer-1 (Cenik *et al.*, 2011). *In vivo*, it appears that the activity of Dicer enzymes is regulated by cofactors containing two or three dsRBDs, e.g. R2D2 and Loquacious (Loqs) (Fig. 3.2, right). Dicer-2 forms a stable heterodimer with R2D2, which restricts its cellular localization to cytoplasmic D2 bodies (Nishida *et al.*, 2013) and is mandatory for the efficient loading of siRNAs on AGO2 (Liu *et al.*, 2003; Lee *et al.*, 2004; Tomari *et al.*, 2004a). Based on *in vitro* experiments, a role for R2D2 in preventing processing of pre-miRNAs by Dicer-2 has also been proposed (Cenik *et al.*, 2011). However, this was not confirmed by *in vivo* experiments (Marques *et al.*, 2013). Furthermore, inorganic phosphate has been proposed as a factor restricting the panel of targets for Dicer-2 *in vitro*, precluding it from processing aberrant pre-miRNA targets and short dsRNAs (Fukunaga *et al.*, 2014). The other cofactor, Loqs has two main isoforms, Loqs-PB and Loqs-PD. Whereas Loqs-PB functions as a cofactor of Dicer-1 in the miRNA pathway (Förstemann *et al.*, 2005; Jiang *et al.*, 2005; Saito *et al.*, 2005), Loqs-PD enhances production of siRNAs by Dicer-2, especially from synthetic dsRNA and endogenously encoded dsRNAs derived from structured loci,

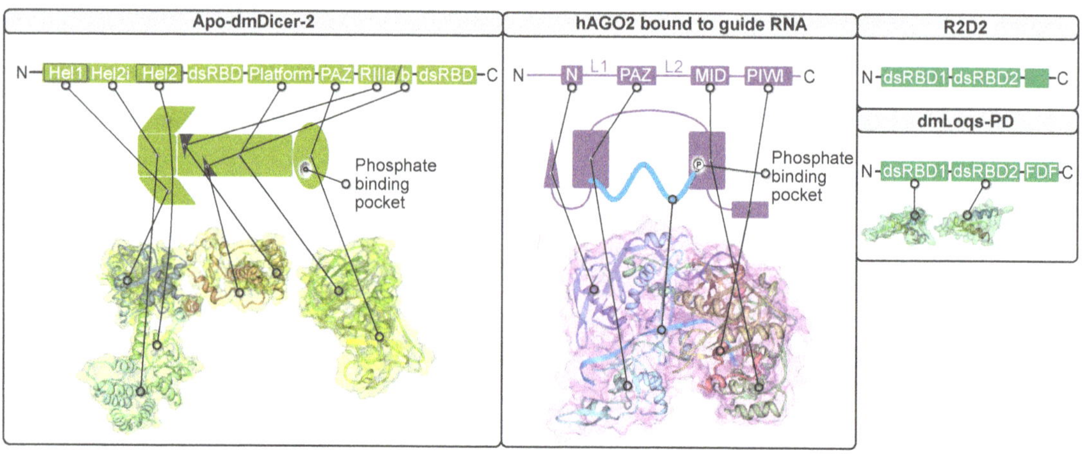

Figure 3.2 Structural organization of the canonical components of the siRNA pathway. The domain organization is shown at the top, above a schematic representation of the proteins and experimentally determined 3D structures. (Left) Apo-dmDicer-2 3D structure was obtained by Cryo-electron microscopy at a resolution of 7.1Å (Sinha *et al.*, 2018; PDB ID: 6BUA). (Centre) Guide RNA-loaded human AGO2 structure was resolved at 2.9Å using X-Ray crystallography (Schirle *et al.*, 2014, PDB ID: 4W5N). (Right) No 3D structures were obtained so far for Loqs-PD and R2D2 full proteins. However, the 3D structures of the two dsRBDs of dmLoqs, identical between all isoforms, were individually predicted using solution Nuclear Magnetic Resonance spectroscopy (Tants *et al.*, 2017; PDB IDs: 5NPG, 5NPA).

sense–antisense pairs and transposable elements (Zhou *et al.*, 2009; Marques *et al.*, 2010; Miyoshi *et al.*, 2010a).

Once produced, the siRNA duplex is loaded on the effector protein AGO2 in a highly coordinated but energetically unfavourable series of events (Fig. 3.3). In a first step, the Dcr-2/R2D2 complex, which displays intrinsically low affinity for duplex siRNAs, associates with TAF11, a TATA-box binding protein Associated Factor. Colocalized with Dicer-2 and R2D2 in D2 bodies, TAF11 acts as a chaperone facilitating the tetramerization of the Dicer-2–R2D2 heterodimer and increasing affinity for siRNAs by tenfold (Liang *et al.*, 2015). R2D2 tends to preferentially bind the extremity of the siRNA duplex showing the strongest stability, creating an asymmetry in the complex (Tomari *et al.*, 2004b). This asymmetry determines the preferential loading of the strand featuring the least stable 5′ extremity in AGO2 to serve as guide siRNA (Fig. 3.3B and C).

Structure–function of AGO proteins

AGO proteins are composed of four globular domains named N, PAZ, MID and PIWI (Fig. 3.2, centre). They adopt a closed, flexible and unstable conformation not suitable to accept the siRNA duplex in their 'Apo' form. A chaperone machinery composed of the Hsp70 system (Hsp40 + Hsp70) and Hsp90 system (Hop, Hsp90 and p23) is required for the efficient loading of siRNA duplexes. Briefly, Hsp70 opens the structure of AGO2 while Hsp90 is required to extend the duration of this opened state to allow sufficient time for the recognition of the 5′ phosphate at the extremity of the guide strand and subsequent loading of the entire duplex (Miyoshi *et al.*, 2010b; Iwasaki *et al.*, 2010, 2015; Tsuboyama *et al.*, 2018). Both processes require ATP hydrolysis. The coordinated action of the heat shock proteins was proposed as the trigger for Dicer-2–R2D2 tetramer destabilization and transfer of the siRNA duplex (Fig. 3.3C). Intriguingly, Hsp proteins can be induced by viral infections in *Drosophila*. However, the siRNA pathway remains functional in flies mutant for the Heat Shock factor (Merkling *et al.*, 2015b). In spite of this progress, the exact mechanism of AGO2 loading remains unclear especially because a 3D structure of the protein in its 'Apo' form is lacking. The semi-closed AGO2 protein loaded with the siRNA duplex is rigid and stable, and constitutes the pre-RNA Induced Silencing Complex (RISC). Studies with AGO proteins from eukaryotic (human) or prokaryotic (*Pyrococcus furiosus*) systems indicate that the strand of the duplex showing the less stable 5′ phosphate extremity is anchored to the phosphate binding pocket of the AGO2 MID domain while its 3′ extremity is bound to the hydrophobic cavity of the PAZ domain (Ma *et al.*, 2004; Song *et al.*, 2004; Fig. 3.3C). In the pre-RISC, the passenger strand occupies the same position as the future target RNAs (Kim *et al.*, 2007). The full-length crystal structure of human AGO2 complexed with RNA reveals large structural differences between Argonautes from different kingdoms of life, even if individual domains superimpose reasonably well (Schirle and MacRae, 2012).

Maturation and slicing

After loading, the pre-RISC is matured through two essential steps required for downstream RNA silencing activity. First, the passenger strand is discarded. This is achieved by the coordinated action of the N-terminal domain of AGO2, acting as a wedge to unwind the siRNA duplex (Kwak and Tomari, 2012) and the PIWI domain RNAseH-like catalytic core, which cleaves the passenger strand in two small RNAs of 9 and 12 nt (Kim *et al.*, 2007; Matranga *et al.*, 2005; Miyoshi *et al.*, 2005; Rand *et al.*, 2005). The Component 3 Promoter Of RISC (C3PO) complex then helps AGO2 to get rid of the unstable cleavage products (Liu *et al.*, 2009; Ye *et al.*, 2011; Mo *et al.*, 2018) (Fig. 3.3C).

A slicer-independent ejection model, relying on the thermal dynamics of the PAZ domain may also participate in discarding of the passenger strand (Gu *et al.*, 2012; Park and Shin, 2015; Nakanishi, 2016). The final maturation step resides in the 2′O methylation of the 3′ extremity guide of the siRNA by the Hen1 enzyme (Horwich *et al.*, 2007). This methylation step is crucial for the protection of the small guide RNA from 3′ uridylation and further 3′–5′ degradation (Li *et al.*, 2005). The mature RISC, programmed with a guide RNA, functions as a Mg^{2+}-dependent, multiple-turnover enzyme that will recognize its mRNA target by perfect base-paired complementarity. The RISC will then slice the mRNA target and release the degradation product in an ATP assisted manner (Hutvágner and Zamore, 2002; Tang *et al.*, 2003; Haley and Zamore,

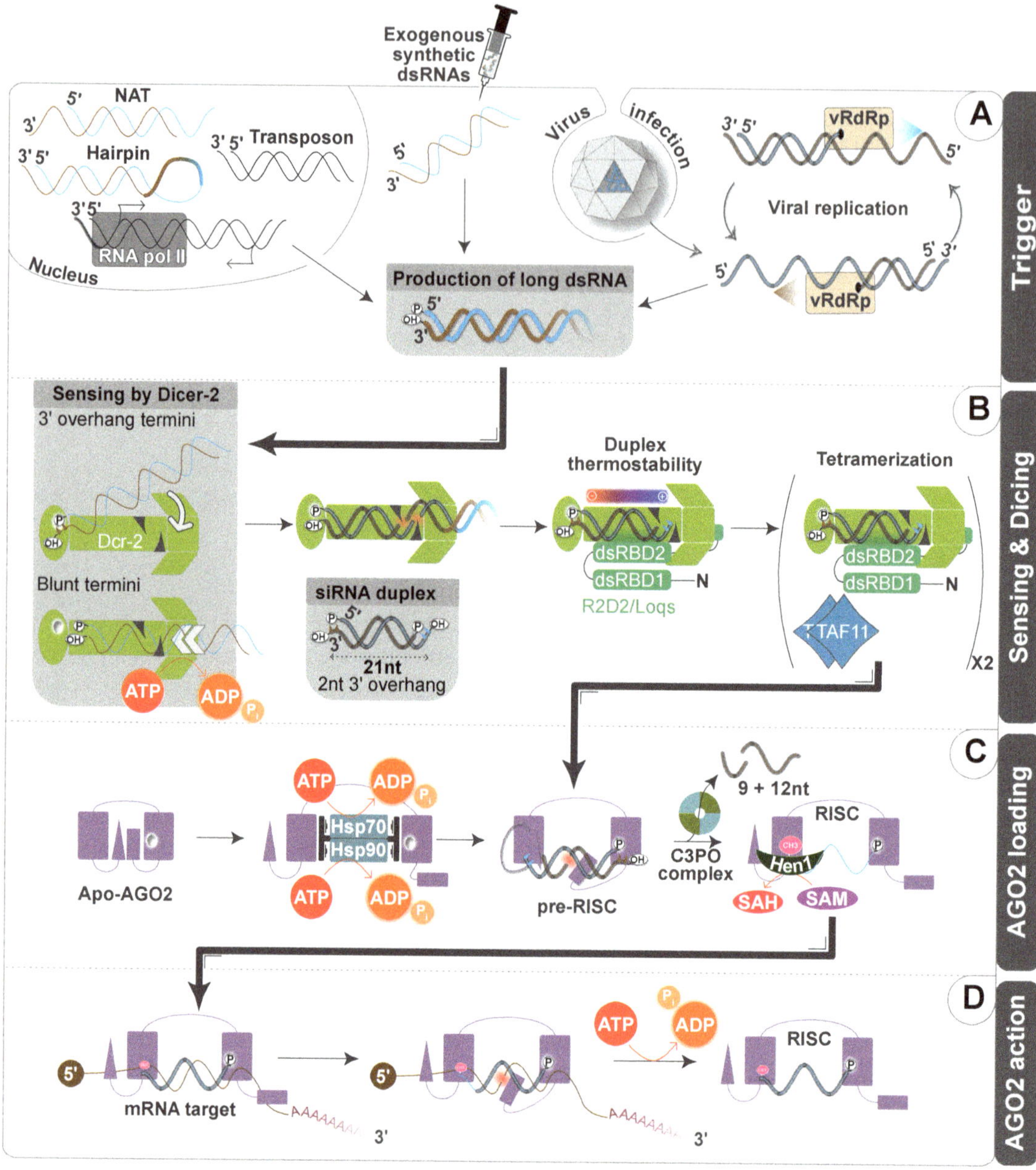

Figure 3.3 The siRNA pathway in *Drosophila melanogaster*. (A) The siRNA pathway can be activated by diverse dsRNA substrates encompassing endogenous transcription of Natural Antisense Transcripts (NATs), transposons, convergent transcripts or structured RNA (hairpins). Experimentally delivered synthetic dsRNAs and virus intermediates of replication represent an exogenous source of dsRNA molecules and triggers RNAi. Blue and brown strands represent RNAs while black strands represent DNA. (B) Endogenous and exogenous long dsRNAs are sensed and diced by Dicer-2. Because Dicer-2 preferentially accesses its substrates at extremities, dsRNA termini are crucial determinants of the mode of action of Dicer-2. Indeed, while blunt end dsRNA triggers an efficient, DRA domain and ATP-dependent processive activity of Dicer-2, a dsRNA molecule with 3′ overhang termini promotes a slow, ATP-independent distributive activity. Generated siRNA duplexes are dsRNA molecules with a characteristic 2nt 3′ overhang signature and are 21nt long. The relative thermostability of siRNA extremities determines the orientation of the RNA duplex in Dicer-2. (C) Loading of AGO2 relies on a highly coordinated series of events. The closed structure of Apo-AGO2 is opened by two ATP-dependent chaperone complexes (Hsp70 and Hsp90). The siRNA duplex is transferred from the Dicer-2–R2D2–TAF11 complex to the opened form of AGO2, forming the pre-RISC. Maturation of the RISC includes cleavage and removal of the passenger strand. The 3′ extremity of the remaining guide strand is 2′O-methylated. (D) Mature RISC sequence specifically targets and cleaves complementary mRNA targets.

2004; Schwarz *et al.*, 2004) (Fig. 3.3D). In the case of viruses, viral mRNAs seem to be preferentially targeted compared to the viral genome (Marques *et al.*, 2013).

Control of viruses by the siRNA pathway

Genetic evidence in *Drosophila*

The discovery that small RNAs are produced in plants after viral infection predates the characterization of RNAi pathways in flies (Hamilton and Baulcombe, 1999). Evidence that RNAi operates against viruses in *Drosophila* was initially provided in S2 cells infected with *Flock House virus* (FHV) (Li *et al.*, 2002). Of note, null mutants for the three main components of the siRNA pathway in flies, namely Dicer-2, AGO2 or R2D2 are homozygous viable (Liu *et al.*, 2003; Lee *et al.*, 2004; Okamura *et al.*, 2004). As a result, adult mutant flies can be infected with viruses and both the survival rate and the viral load can easily be monitored. These experiments established that flies mutant for the siRNA pathway are susceptible to a variety of viruses with RNA or DNA genomes (e.g. Galiana-Arnoux *et al.*, 2006; van Rij *et al.*, 2006; Wang *et al.*, 2006; Mueller *et al.*, 2010; Bronkhorst *et al.*, 2012; Kemp *et al.*, 2013). Interestingly, Dicer-2, R2D2 and AGO2 are among the 3% fastest evolving of all *Drosophila* genes highlighting the high selection pressure on the RNAi pathway (Obbard *et al.*, 2006). Population genomic analysis in multiple invertebrates shows that RNAi genes display a greater rate of adaptive protein substitution than other genes, most likely reflecting their function at the forefront of defence against viruses and transposable elements (TEs) (Palmer *et al.*, 2018).

Control of viruses by the siRNA pathway in other insects

Antiviral RNAi has been investigated in other insects, including vector insects. Many mosquito-borne viruses are associated with human and animal diseases, raising interest in mosquito antiviral immunity (Aguiar *et al.*, 2016; Powers and Waterman, 2017). Over the past 50 years, the *Aedes albopictus* cell line C6/36, isolated from larvae, has been commonly used for amplification of arboviruses but also to study virus–vector interactions. Recently, the genome of C6/36 cells has been sequenced and null mutations in the *dicer-2* gene were identified, which makes them incompetent for production of siRNAs (Morazzani *et al.*, 2012; Miller *et al.*, 2018). As a result, C6/36 cells support viral replication to high titres, confirming the important antiviral function of Dicer-2 in mosquitoes.

Injection of dsRNA in the body cavity efficiently silences gene expression in a sequence-specific manner and has been used in pioneer experiments to knock-down expression of components of the siRNA pathway in *Aedes aegypti*. This led to significantly increased SINV and DENV titres, without compromising insect survival (Campbell *et al.*, 2008; Sánchez-Vargas *et al.*, 2009; Khoo *et al.*, 2010). Interestingly, in genotype–phenotype association studies, Lambrechts *et al.* (2013) found that the *dicer-2* genotype is associated with resistance to DENV in a virus isolate-specific manner. By contrast, no such association is found for flanking loci, suggesting that the *dicer-2* gene and the siRNA pathway are important determinants of virus suppression in *Aedes*. Mutants of the core RNAi components have subsequently been established in *Ae. aegypti* mosquitoes using genome editing approaches such as TALEN (Transcription Activator-Like Effector Nuclease) or CRISPR (Clustered Regularly Interspaced Short Palindromic Repeats) mutagenesis (Basu *et al.*, 2015). In the context of viral infections, *dicer-2* null mutant mosquitoes exhibit a decrease in survival rate compared to wild-type mosquito infection, that correlates with high *Yellow fever virus* titres (Samuel *et al.*, 2016). The midgut epithelium is a critical barrier for viruses in insect vectors, as it becomes infected after acquiring an infectious blood meal from the host. After successfully replicating in the midgut, viruses reach the haemolymph and disseminate systemically. Viruses then reach the salivary glands from which they are transmitted to a naïve host upon blood feeding. An early study revealed that silencing *dicer-2* in the midgut of female mosquitoes increased infection and dissemination of SINV (Khoo *et al.*, 2010). Insects can also be vectors for plant viruses (Chapter 6). For example, the small brown planthopper (SBPH) is an incompetent vector for *Southern rice black streaked dwarf virus* (SRBSDV), a plant virus. SRBSDV is restricted in the midgut epithelium of SBPH. Knock-down of either *dicer-2* or *AGO2* in SBPH results in the dissemination of the virus to the salivary glands and facilitates transmission to rice plants, revealing the importance of the siRNA pathway in the control of vector competence (Lan *et al.*, 2016). However, the siRNA pathway was found to be largely dispensable for antiviral immunity in the midgut of *Anopheles* mosquitoes exposed to the *O'nyong-nyong virus*, and to become operative only during the systemic stage of infection (Carissimo *et al.*, 2015). Interestingly, a recent study indicates that the siRNA pathway also fails to efficiently silence DENV in the midgut of *Aedes aegypti*, even though the canonical components of the pathway are expressed in this tissue, and that the pathway is functional when triggered by endogenous (e.g. control of TE by endo siRNAs) and exogenous (e.g. intrathoracic injection of long dsRNA) dsRNAs (Olmo *et al.*, 2018). This discrepancy between antiviral and conventional siRNA pathways in the midgut results from the lack of expression in this tissue of Loqs2, an *Aedes*-specific paralogue of Loqs and R2D2. Importantly, ectopic expression of Loqs2 in the midgut results in restriction of DENV replication and dissemination (Olmo *et al.*, 2018). Altogether, these results point to an additional level of complexity in the insect siRNA pathway when it comes to the control of viruses (see below).

siRNA as a footprint of antiviral immunity

As mentioned above, virus-derived 21–22nt-siRNAs (vsiRNAs) produced in the course of viral infection in insects, which can be revealed by High Throughput Sequencing (HTS), provide an excellent read-out of Dicer-2 activity (Box 3.1). These vsiRNAs are strongly reduced or abolished in *dicer-2* mutant flies (Aliyari *et al.*, 2008; Mueller *et al.*, 2010; Bronkhorst *et al.*, 2012; Kemp *et al.*, 2013; Marques *et al.*, 2013). For many RNA viruses, vsiRNAs cover the whole viral genome and the ratio between the number of siRNAs matching the (+) strand and the (–) strand of the genome is close to one (Aliyari *et al.*, 2008; Myles *et al.*, 2008;

Box 3.1 High-throughput sequencing

The emergence of next-generation sequencing technologies set a milestone in the development of a tremendous number of high-throughput 'omics' approaches. Until 2010, RNA sequencing methods were all relying on sequencing by synthesis or semi-conductor technologies (second generation). The advent of the third generation of sequencing with PacBio and Oxford Nanopore opened the way to direct DNA and RNA sequencing without any amplification step. These ever-evolving technologies allow the detection of events that can only be bioinformatically predicted with second generation sequencing methods such as splicing or defective genomes in the case of viruses. However, to detect the specific signature of Dicer-2 (siRNA duplexes), small RNA sequencing HTS (second generation) remains the method of choice because of the high-throughput number of reads generated.

	Advantages	Limitations	Main technologies
2nd generation	• High-throughput number of reads (4M – 20B) • Low error rate (~0,1 - 1%) • Low cost	• Library preparation bias including RT, PCR and optional size exclusion steps • Short reads (max ~400bp) • Sample bleeding due to multiplexing (2-5%: check numbers)	• iSeq – NovaSeq 6000 (Illumina) • Ion PGM (Life Technologies) • SOLiD DNA sequencer (Thermo Fisher Scientific) • 454 GS FLX (Roche)
3rd generation	• Direct RNA/DNA sequencing • Long reads (up to 2Mb so far) allowing precise detection of splicing events for instance • Portability	• High error rate (5 – 15%) • Low number of reads (50k – 1M)	• PacBio • Oxford Nanopore

Mueller *et al.*, 2010; Marques *et al.*, 2013; Ferreira *et al.*, 2018), as would be expected from the processing of long dsRNA formed during viral replication. By contrast, in the case of DNA viruses, hotspots of vsiRNAs are observed on specific regions of the viral genome. These vsiRNAs match both strands of the genome, suggesting the siRNA pathway targets regions transcribed on both strands and producing dsRNA (Bronkhorst *et al.*, 2012, 2013; Jayachandran *et al.*, 2012; Kemp *et al.*, 2013). Secondary structures in the genome or antigenome may also be targeted by Dicer-2 in some viruses (Sabin *et al.*, 2013).

Based on *in vitro* experiments, AGO2 has been suggested to impair viral replication by cleaving the viral RNA through its slicer activity (van Mierlo *et al.*, 2012a). *In vivo* experiments further suggest that viral mRNAs are the primary target of vsiRNA-loaded AGO2 (Marques *et al.*, 2013). As genetics are still tricky in most non-model organisms, HTS provides a convenient readout to monitor activity of RNAi pathways in insects infected with viruses (e.g. Chejanovsky *et al.*, 2014; Zografidis *et al.*, 2015; Ferreira *et al.*, 2018). For example, HTS of small RNAs isolated from SINV-infected *Ae. aegypti* mosquitoes revealed the presence of 21 nt-long vsiRNAs and brought the first evidence that the siRNA pathway is activated during viral infection in vector mosquitoes (Myles *et al.*, 2008). This technique has also been successfully used for virus identification in both plants and insects (Kreuze *et al.*, 2009; Wu *et al.*, 2010). Indeed, large fragments of viral genomes can be reconstituted upon assembly of contigs from sequenced vsiRNAs. Furthermore, such contigs are enriched for viral sequences compared to long RNA sequencing reads, because they are by-products of the detection of viral replication by the insect immune system. Thus, HTS of small RNAs can be used to determine the virome of laboratory and wild populations of insects and possibly also other multicellular eukaryotes (Aguiar *et al.*, 2015; Waldron *et al.*, 2018). As such, HTS of small RNAs represents a powerful tool for virus surveillance in populations of vector insects.

Viral suppressors of RNAi (VSRs)

The study of viruses themselves can provide interesting insight into antiviral defence in insects. Indeed, in the course of their interaction with host cells, viruses have evolved to counter antiviral defence. Thus, many insect viruses, including *Drosophila* viruses, encode viral suppressors of RNAi (VSRs) (Bronkhorst and van Rij, 2014). Some VSRs (e.g. DCV-1A, FHV-B2, IIV6-340R, *Drosophila X virus* VP3 and *Culex Y virus* VP3) directly bind long dsRNA through canonical dsRNA binding domains, dsRBDs and prevent processing by Dicer-2 (Li *et al.*, 2002; van Rij *et al.*, 2006; Bronkhorst *et al.*, 2014; van Cleef *et al.*, 2014; see also Fig. 3.4A and B). Interestingly, with the exception of DCV-1A, these VSRs bind siRNA duplexes as well, suggesting that they also inhibit the pathway after long dsRNAs have been processed into siRNAs (Morazzani *et al.*, 2012; Valli *et al.*, 2012; Bronkhorst and van Rij, 2014). Other VSRs (e.g. CrPV-1A and Nora-VP1) bind directly to AGO2 and inhibit its endonuclease activity (Nayak *et al.*, 2010; van Mierlo *et al.*, 2012b). In addition, CrPV-1A also targets AGO2 to the proteasome through the K48 polyubiquitination pathway (Nayak *et al.*, 2018). This is reminiscent of the mode of action of P0, a VSR from poleroviruses, which triggers degradation of AGO1 in plant cells (Baumberger *et al.*, 2007; Bortolamiol *et al.*, 2007).

The importance of VSRs for viruses has been particularly well illustrated in the case of FHV-B2. *Nodaviridae* have small bipartite RNA genomes that are easy to manipulate genetically. One segment of the genome, RNA1, encodes the replicase, whereas the second, RNA2, encodes the capsid proteins. A third RNA transcript, RNA3, is also produced from RNA1 once it has replicated and encodes the VSR B2 (Chao *et al.*, 2005). Whereas wild-type FHV is highly pathogenic upon injection into the body cavity of flies, viral mutants unable to express B2 are completely attenuated (Galiana-Arnoux *et al.*, 2006; Wang *et al.*, 2006; Fig. 3.4C). As expected, the virus regains virulence when injected into *dicer-2* or *AGO2* mutant flies (Han *et al.*, 2011; Petrillo *et al.*, 2013). Similarly, mutation of residue Phe 114 into Ala in the flexible loop of CrPV-1A, which is involved in the interaction with AGO2 results in an attenuated virus in wild type flies but not in AGO2 mutant flies (Nayak *et al.*, 2018). Of note however, even in RNAi mutants, the B2 deficient virus exhibits reduced virulence compared to wild-type FHV. Indeed, an additional function of B2 is to bind double stranded regions of RNA2, thus preventing its recruitment into poorly characterized cytoplasmic RNA granules where its

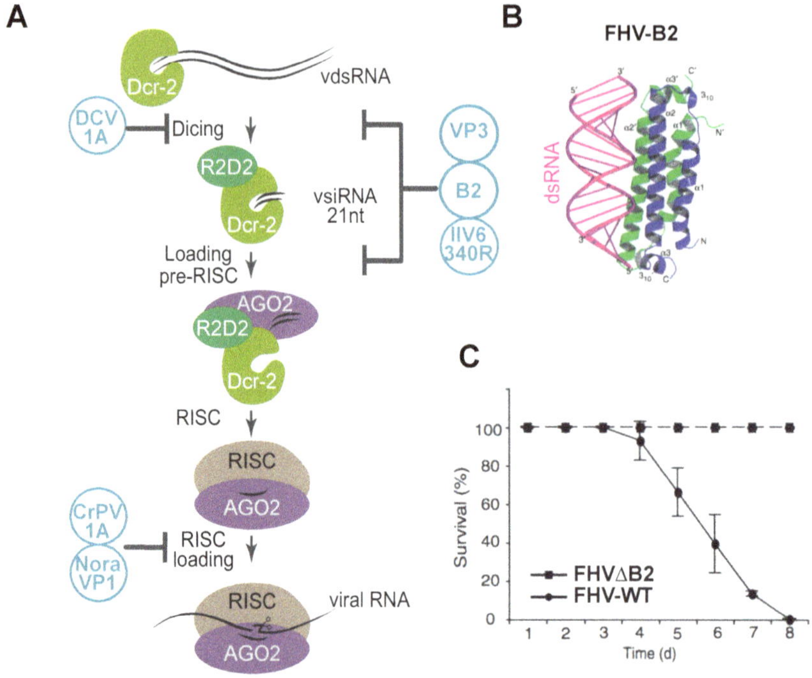

Figure 3.4 Viral suppressors of RNAi neutralize the siRNA pathway in insects. (A) Schematic representation of the action of a set of insects VSRs. (B) Crystal structure of FHV-B2 dimer associated with dsRNA. Reprinted by permission from Springer Nature, *Nature Structural & Molecular Biology*, 12, 952–957. Dual modes of RNA-silencing suppression by *Flock House virus* protein B2, Chao, J.A., Lee, J.H., Chapados, B.R., Debler, E.W., Schneemann, A., and Williamson, J.R. Copyright 2005. (C) Survival curve of flies injected with WT FHV or a mutant version deleted for B2 (adapted from Galiana-Arnoux *et al.*, 2006).

translation would be repressed. As a result, translation of the capsid protein is impaired in the absence of B2, even in RNAi deficient cells (Petrillo *et al.*, 2013). Interestingly, the FHV-B2 VSR can be used to neutralize the siRNA pathway upon ectopic expression. For example, when *Ae. aegypti* mosquitoes were challenged with SINV recombined with FHV-B2 (SINV-B2) either by injection or infected blood meal, the virus titre was increased compared to the control virus. The recombinant SINV-B2 caused high mortality among the mosquitoes at 4–6 days post-infection, highlighting that the RNAi pathway is essential to control arboviruses replication (Myles *et al.*, 2008; Cirimotich *et al.*, 2009). In *Ae. aegypti* mosquitoes, the constitutive and ubiquitous expression of FHV-B2 impaired the siRNA pathway and enhanced replication of both SINV and DENV (Khoo *et al.*, 2013).

One consequence of the direct interaction of some VSRs with protein factors of the siRNA pathway to alter their ability to neutralize viral RNAs or trigger their degradation (Singh *et al.*, 2009; Nayak *et al.*, 2010, 2018; van Mierlo *et al.*, 2012b, 2014) is that both VSRs and components of the siRNA pathway evolve rapidly (Obbard *et al.*, 2006). As a result, the activity of viral suppressors can be host-specific. For example, the VP1 protein from a divergent Nora virus isolated from *D. immigrans* interacts with and suppresses *D. immigrans* AGO2, but not *D. melanogaster* AGO2 (van Mierlo *et al.*, 2014). This provides an excellent example for the co-evolution of the host RNAi machinery and viral suppression mechanisms. Because of their intimate association with key components of the siRNA pathway, VSRs provide promising tools to decipher the regulation and molecular mechanisms of antiviral RNAi. For example, development of single-molecule approaches can shed light on VSR mode of action and how they discriminate viral from cellular RNA (Fareh *et al.*, 2018). VSRs may also be used to visualize and track dsRNA in live plant and animal cells, as recently shown for an FHV-B2–GFP fusion protein (Monsion *et al.*, 2018).

Systemic RNAi in insects?

In both plants and *C. elegans*, systemic RNAi contributes to the control of viral infections. The mechanism at play involves spreading of siRNAs generated in infected cells to neighbour healthy cells. There, these siRNAs prime the synthesis of dsRNAs by host-encoded RNA-dependent RNA polymerases (RdRPs) (Ding, 2010). Insect genomes do not encode such RdRPs and clonal analyses in *Drosophila* revealed that the siRNA pathway is a cell autonomous pathway (Roignant *et al.*, 2003). Nevertheless, the spread of antiviral RNAi has been proposed to contribute to the control of viral infections in insects. Indeed, injection of exogenous dsRNA in the body cavity of most insects, or even feeding dsRNA, leads to gene silencing through the siRNA pathway (reviewed in Zotti *et al.*, 2018). Thus, viral infection, which is known to trigger transcriptional responses, may induce a mechanism of systemic RNAi. In *Drosophila*, uptake of exogenous dsRNA is mediated by the endocytic pathway (Saleh *et al.*, 2006). Nanotube like structures have also been reported to transfer dsRNA and components of the RNAi machinery between cells (Karlikow *et al.*, 2016). Thus, dsRNA released from infected dying cells may trigger RNAi in distant cells, upon internalization by the dsRNA uptake pathway (Saleh *et al.*, 2009). Surprisingly, however, infection of flies with a sublethal dose of DCV, which should prime the antiviral siRNA pathway, did not induce protection against a challenge with a lethal dose of virus (Longdon *et al.*, 2013). Of note, viral RNA can be reverse transcribed into DNA (vDNA) in *Drosophila* and in mosquitoes, through the action of the reverse transcriptase from transposable elements (TEs) (Goic *et al.*, 2013, 2016), acting together with Dicer-2 (Poirier *et al.*, 2018) or AGO2 (Tassetto *et al.*, 2017). Transcription of vDNA has been proposed to result in the production of secondary siRNAs bearing a 5′-triphosphate mark, associated with systemic antiviral effect (Tassetto *et al.*, 2017). However, the existence of such secondary siRNAs was not confirmed by an independent study (Mondotte *et al.*, 2018). Overall, the mechanisms involved in systemic antiviral RNAi remain to be characterized and genetic evidence for the importance of the contribution of vDNA and secondary siRNAs in antiviral immunity is still lacking.

In summary, it now appears that the antiviral siRNA pathway in insects is more complex than previously thought. A number of host and viral factors, which affect the stability, the binding affinity to viral nucleic acids, and even the cytoplasmic localization of ribonucleoprotein complexes mediating RNA silencing can influence antiviral defence. Biochemistry, genetics and live imaging will clarify the mode of action of the components of the siRNA pathway in the context of infected cells, in different types of tissues and in various insects.

The piRNA pathway in antiviral immunity

Production of primary and secondary piRNAs

Another small RNA pathway, the piRNA pathway is able to sense foreign nucleic acids and has been proposed to participate in antiviral immunity (Miesen *et al.*, 2016a). This pathway involves 24–30 nt small RNAs and Argonaute proteins from the PIWI clade. The founding member of this subfamily, PIWI, was initially characterized as a *Drosophila* gene essential for male fertility (Lin and Spradling, 1997). This phenotype results from derepression and mobilization of transposable elements in the germline. It was subsequently found that PIWI works with two other Argonaute proteins from the same clade, namely Aub and AGO3 (Saito *et al.*, 2006; Vagin *et al.*, 2006; Brennecke *et al.*, 2007; Gunawardane *et al.*, 2007). The PIWI-mediated mechanism of genome maintenance in the germline is conserved in all animals. The piRNAs were initially identified in mouse testis and in *Drosophila*, where they were first called repeat-associated small interfering (rasi)RNAs. They interact with PIWI proteins (Aravin *et al.*, 2006; Girard *et al.*, 2006; Grivna *et al.*, 2006; Vagin *et al.*, 2006). *Drosophila* genetics coupled with HTS, have shed light on the molecular mechanism generating piRNAs. This involves a primary processing pathway that primes the production of phased piRNAs during the so-called ping-pong amplification mechanism (Czech and Hannon, 2016; Fig. 3.5). Of note, HTS of cell lines or mosquitoes revealed the existence of virus-derived piRNAs (Wu *et al.*, 2010; Hess *et al.*, 2011; Morazzani *et al.*, 2012; Vodovar *et al.*, 2012; Léger *et al.*, 2013; Aguiar *et al.*, 2015).

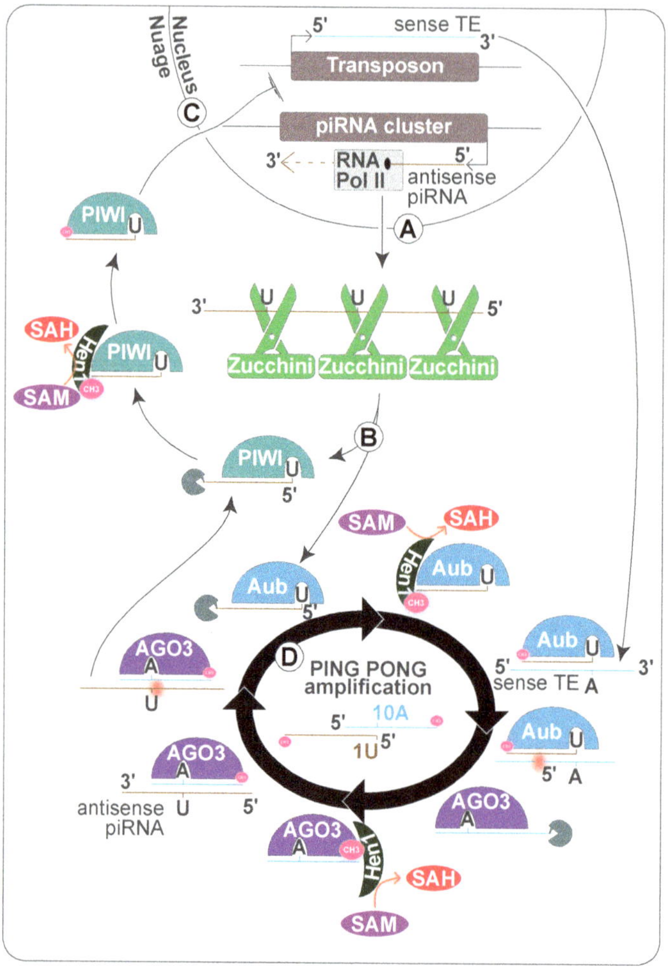

Figure 3.5 The piRNA pathway in *Drosophila* ovaries. piRNA clusters are mainly composed of defective transposon copies integrated in the genome and serve as a genetic memory of transposon exposition in a population. Transcription of these genomic clusters by the host RNA polymerase II generates long RNAs that are exported in the cytoplasm (A). There, they are processed by Zucchini, an endonuclease with a strong preference for cleavage 5′ of a uridine. This leads to the production of phased piRNAs, which can either be loaded on PIWI or on Aub (B). PIWI- and Aub-loaded piRNAs undergo trimming and 2′O methylation of their 3′ extremity. piRNA-loaded PIWI translocates to the nucleus where it participates in transcriptional silencing of transposons through deposition of repressive histone modifications (C). On the other hand, piRNA-loaded Aub will initiate the Ping-Pong amplification cycle (D). Briefly, loaded Aub will recognize a complementary transposon transcript and induce endonucleolytic slicing of the target between nucleotide 10 and 11 of the piRNA. This slicing generates the 5′ end of a new sense piRNA with a 10 nt 5′ overlap with the initial antisense piRNA and an adenosine residue at position 10. This newly formed piRNA is loaded on AGO3, trimmed and 2′O methylated at its 3′ extremity. Finally, piRNA-loaded AGO3, using a similar mechanism, will generate Ago-bound piRNAs from piRNA clusters.

The primary processing pathway targets Pol II-dependent transcripts generated from genomic loci rich in transposon remnants known as piRNA cluster (Bucheton, 1995; Brennecke *et al.*, 2007; Pélisson *et al.*, 2007). These precursors are processed by the endonuclease Zucchini, which preferentially cleaves 5′ of a uridine residue. As a result, the piRNA intermediates produced are enriched for 5′ uridine residues (1U) (Pane *et al.*, 2007; Han *et al.*, 2015; Mohn *et al.*, 2015; Fig. 3.5A). In fruit flies, these piRNA intermediates are loaded onto PIWI and Aub in an electron dense perinuclear region of the germline called 'nuage'. Indeed, the binding pocket of the MID domain of these two proteins preferentially accommodates 5′ uridine residues (Cora *et al.*, 2014; Matsumoto *et al.*, 2016; Fig. 3.5B). piRNA-loaded PIWI translocates to the nucleus where it participates in transcriptional silencing of transposons through deposition of repressive histone modifications (Sienski *et al.*, 2012; Dönertas *et al.*, 2013; Le Thomas *et al.*, 2013; Ohtani *et al.*, 2013; Sienski *et al.*, 2015; Yu *et al.*, 2015; Fig. 3.5C). By contrast, piRNA-loaded Aub remains in the nuage and initiates the ping-pong amplification mechanism (Fig. 3.5D). The Aub pi-RISC recognizes and cleaves complementary transposon mRNAs. The resulting cleavage product corresponds to the precursor of a secondary piRNA and associates with AGO3 (Brennecke *et al.*, 2007; Gunawardane *et al.*, 2007; Lim and Kai, 2007). Of note, because cleavage mediated by enzymes of the PIWI clade occurs specifically between nucleotide 10 and 11, AGO3 bound piRNAs are enriched for adenosine residues in position 10 (10A). piRNAs loaded into AGO3 target and cleave antisense piRNA precursors, thus generating the 5′ end of new sense piRNAs and resulting in the ping-pong amplification cycle (Fig. 3.5D). In summary, the combination of slicer activity of PIWI proteins together with the activity of endo- and exonucleases explains the Dicer-independent production of 24–30 nt long piRNAs. Of note, the 1U/10A signature characteristic of the ping-pong amplification provides a convenient way to monitor the activity of this pathway by HTS and was instrumental in revealing the existence of virus-derived piRNAs.

Virus-derived piRNA

Transposable elements share with viruses the property of being selfish genetic units encoding proteins that enable their proliferation and spread. Hence, one can wonder whether the piRNA pathway also participates in antiviral immunity. Indeed, transposable elements are targeted by the siRNA pathway in *Drosophila* somatic tissues as revealed by accumulation of siRNAs matching transposable elements in HTS analysis (Chung *et al.*, 2008). Conversely, virus-derived piRNAs could be observed in the OSS cell line derived from *Drosophila* ovarian tissue (Wu *et al.*, 2010). However, in *Drosophila*, activity of the piRNA pathway is restricted to the germline and neither genetics nor HTS support an antiviral function of the piRNA pathway in *Drosophila* (Petit *et al.*, 2016; van den Beek *et al.*, 2018). Nonetheless, there are significant differences between piRNA pathways in *Drosophila* and other insects. Indeed, a recent study investigating 20 species across the arthropod phylum revealed that TEs are commonly targeted by somatic piRNAs unlike in *Drosophila* (Lewis *et al.*, 2018). Consistent with this observation, the piRNA pathway components differ between insect species. Notably, it is apparent that the PIWI clade has significantly expanded in *Ae. aegypti* mosquitoes, where it contains seven PIWIs (instead of two in *Drosophila*, Piwi and Aub) and one AGO3 (reviewed in Miesen *et al.*, 2016a). This, together with identification of virus derived piRNAs in mosquito Aag2 and C6/36 cell lines, led to the suggestion that the piRNA pathway could form a second layer of antiviral defence in mosquitoes (Wu *et al.*, 2010; Hess *et al.*, 2011; Morazzani *et al.*, 2012; Vodovar *et al.*, 2012; Léger *et al.*, 2013; Aguiar *et al.*, 2015). Depletion in AGO3 and PIWI5 resulting in the decreased production of viral piRNAs revealed that these enzymes mediate recognition and processing of SINV RNAs, although the functional consequence of this processing on viral replication was not reported (Miesen *et al.*, 2015). A subsequent report focusing on DENV indicated that the knockdown of PIWI proteins did not significantly affect viral RNA levels (Miesen *et al.*, 2016b). An independent study proposed a role for PIWI4 in antiviral immunity in Aag2 cells, an *Aedes aegypti* derived cell line. Interestingly, PIWI4 behaves as an atypical member of the PIWI family and is not involved in piRNA production but associates with the siRNA pathway core components AGO2 and Dicer-2 (Varjak *et al.*, 2017a,b, 2018). This protein may reveal a cross-talk between piRNA and siRNA pathways in the context of viral infections. The presence of virus derived somatic piRNAs in other arthropods was tested by Jiggins and colleagues in their landmark paper (Lewis *et al.*, 2018) and virus derived siRNAs could be identified in 9 of the 20 species investigated. Among them, 5 species also produced 24–30 nt 5′U biased small RNAs derived from viruses. However, only in *Aedes aegypti* did these piRNAs bare the 1U/10A signature of ping-pong amplification. In the four other species, piRNAs mapped to one strand only, similar to primary piRNAs (Lewis *et al.*, 2018). Therefore, the predominant role of siRNAs compared to piRNAs in antiviral defence observed in *Drosophila* and *Lutzomyia* is probably relevant across arthropods, with *Aedes* mosquitoes representing a notable exception (Ferreira *et al.*, 2018).

Sensing viral RNA in insects

Viral nucleic acid sensors

In mammals a major molecular pattern associated with viral infection is long dsRNA, generated as a by-product of viral replication. The Toll receptor TLR3 is localized in the endosome compartment and is activated upon binding of long dsRNA (Kawai and Akira, 2011). This receptor probes the content of the endosomes for the presence of endocytosed signs of viral infection in the extracellular milieu. Cross-linking of two TLR3 subunits by dsRNA triggers TRIF-dependent interferon activation. Of note, two other endosomal TLRs, TLR7 and TLR8 detect UG rich short ssRNA fragments (Maeda and Akira, 2016). In the cytosol another family of pattern recognition receptors, the RIG-like receptors, sense viral RNA (Goubau *et al.*, 2013). As mentioned above, a hallmark of these receptors is the presence of a DRA domain phylogenetically related to the Dicer proteins (Paro *et al.*, 2015). This domain is followed by a C-terminal domain (CTD), which participates in RNA binding. In addition, two of the three RLRs, RIG-I and MDA5 contain amino-terminal caspase recruitment domains (CARDs) (Kawai and Akira, 2011). This enables them to recruit the signalling adaptor MAVS and to activate expression of interferon genes. Although the three RLRs can bind dsRNA *in vitro*, other molecular features found on viral RNA are necessary to activate RIG-I. Indeed, the CTD of RIG-I, which forms a tighter pocket than the one from MDA5, detects the presence of 5′ di- or triphosphate at the extremity of viral RNAs (Kowalinski *et al.*, 2011; Goubau *et al.*, 2014). This provides an efficient means to discriminate between capped cellular mRNAs and uncapped viral RNAs. This biochemical distinction is supported by genetic data that point to non-redundant functions of RIG-I and MDA5. Indeed, RIG-I mutant mice have impaired interferon responses following infection by viruses such as influenza, VSV, or *Japanese encephalitis virus* (Kato *et al.*, 2005, 2006). Of note, most of these viruses have ssRNA genomes of negative polarity and do not produce detectable amounts of long dsRNA in infected cells (Weber *et al.*, 2006). By contrast, RIG-I mutant mice respond normally to picornaviruses. The 5′ end of the genome and antigenome strand of these viruses is covalently linked to a VpG protein, which prevents recognition of the termini by RIG-I. On the other hand, MDA5 mutant mice are highly susceptible to picornaviruses, which generate large quantities of dsRNA (Kato *et al.*, 2006). The third RLR, LGP2, functions together with MDA5 (Deddouche *et al.*, 2014).

In insects, the only viral nucleic acid sensor identified so far is Dicer-2, suggesting that long dsRNA is the major molecular pattern used to detect viral infection (Fig. 3.3A). Importantly, even in the case of the negative strand RNA virus VSV, the profile of virus-derived siRNAs reveals a typical long dsRNA signature, with siRNAs covering in equal amounts the whole length of both genome and antigenome strands (Ferreira *et al.*, 2018; Marques *et al.*, 2013; Mueller *et al.*, 2010). However, sensing of viral RNA by Dicer-2 appears not to be limited only to the recognition of dsRNA.

Sensing viral RNA by Dicer-2

In vitro dicing assays using recombinant versions of Dicer-2 with or without its cofactors incubated with diverse dsRNA showed that Dicer-2 preferentially accesses its substrate at the extremities (Sinha *et al.*, 2015). Furthermore, dsRNA termini are crucial determinants of Dicer-2 mode of action *in vitro* (Fig. 3.3B). Indeed, while blunt end dsRNA triggers an efficient, DRA domain and ATP-dependent processive activity of Dicer-2, a dsRNA molecule with 3′ overhang termini promotes a slow, ATP-independent distributive activity (Cenik *et al.*, 2011; Welker *et al.*, 2011). Cryo-EM studies revealed that dsRNAs presenting 3′ overhangs are repeatedly bound by Dicer-2 platform-PAZ domain to be sequentially cleaved while blunt dsRNAs are threaded through the DRA domain in an ATP-dependent manner and successively diced to generate many phased siRNAs (Sinha *et al.*, 2018a). This distinct mode of dicing is driven by an extensive conformational change of the protein upon binding of a blunt dsRNA extremity, which cannot occur if the extremity harbours a 3′ overhang (Fig. 3.3B). Loqs-PD interacts with the DRA domain of Dicer-2 in an RNA independent manner through its C-terminal FDF motif and allows the enzyme to process RNA substrates normally refractory to cleavage, such as dsRNA with blocked, structured or frayed ends (Sinha *et al.*, 2015; Trettin *et al.*, 2017). In summary, *in vitro* experiments with purified recombinant proteins point to a model where subtle changes in the substrate result in tremendous differences in the dicing mechanism. How this model can be reconciled with the complexity of Dicer-2 natural substrates in the cellular context is an important challenge for future studies. This is particularly relevant in the context of viral infection as the RNA extremities of viruses and viral replication complexes are highly variable. A striking example is the case of picornaviruses and dicistroviruses, which, as mentioned above, display a covalently linked VpG protein at the 5′ extremities of their genome and antigenome (Virgen-Slane *et al.*, 2012). This covalent modification at the extremities of the genome from *Dicistroviridae* such as DCV or CrPV is expected to impact sensing by Dicer-2. One asset to solve this important question is the characteristic siRNA signature of Dicer-2, which is amenable to bioinformatic analysis following small RNA HTS. Bioinformatic analysis of the pattern of vsiRNAs produced in wild-type or mutant flies (e.g. inactivated ATP binding site in the DRA domain of Dicer-2) may provide insights on an alternative access point of Dicer-2 on DCV or CrPV RNA (see Box 3.1) (e.g. Aliyari *et al.*, 2008; Mueller *et al.*, 2010; Bronkhorst *et al.*, 2012; Sabin *et al.*, 2013; Aguiar *et al.*, 2015).

Another indication that features other than double strandedness are sensed lies in the importance of the cofactors acting together with Dicer-2. For example, Loqs-PD is not required for silencing viral RNA, in spite of its essential role in RNA interference triggered by endogenous or *in vitro* synthesized dsRNA (Marques *et al.*, 2010, 2013). This points to the existence of differences between viral RNAs and other dsRNAs, produced from endogenous sources or synthesized *in vitro*. Of note, similar observations were made in the nematode *Caenorhabditis elegans*, where a RIG-I-like factor known as DRH1 is essential for antiviral RNAi yet dispensable for the other silencing pathways (Ashe *et al.*, 2013; Guo *et al.*, 2013), but also in *Ae. aegypti* mosquitoes. As mentioned before, the siRNA pathway silences targets of endogenous and exogenous dsRNAs in the midgut of these mosquitoes but fails to suppress viruses (Olmo *et al.*, 2018). These results confirm the existence of differences between silencing triggered by exogenous or endogenous dsRNA and viral RNA, with the latter specifically requiring the dsRBP Loqs2. We note that an isoform of Staufen that evolved in coleopterans is another example of a species-specific dsRBP that regulates the siRNA pathway (Yoon *et al.*, 2018). The fact that *Drosophila* (Loqs-PD), *Aedes* (Loqs2) and coleopteran insects (StaufenC) all require specific cofactors to define the activity of siRNA pathway is intriguing and highlights that an important facet of this pathway remains ill characterized.

The discrimination between the dsRNA precursors of endo- or exo-siRNAs and the dsRNA generated during viral infection may reflect differences either in the receptor complex sensing dsRNAs or in the viral RNA itself. On the protein side, it will be important to characterize biochemically the role of Loqs and R2D2 proteins in *Drosophila* and *Aedes*. On the RNA side, the pervasiveness of RNA modifications begs the question of how much they contribute to the discrimination of viral RNA by the siRNA pathway (Gokhale and Horner, 2017; Helm and Motorin, 2017; Fig. 3.6). For example, ADAR is an RNA editing enzyme catalysing the conversion of A to I within dsRNA regions of cellular RNAs, which prevents unwanted activation of immunity in mice and *C. elegans* (Liddicoat *et al.*, 2015; Reich *et al.*, 2018). In mammals, the presence of epitranscriptomic marks such as 2′O-methylation on viral RNA affects sensing by TLR7 or MDA5 whereas N6-methyladenosine (m6A) in dsRNA reduces activation of TLR3 (Karikó *et al.*, 2005; Gonzales-van Horn and Sarnow, 2017). Therefore, it will be interesting to investigate the presence of post transcriptional modifications in viral RNAs in insect cells and their impact on antiviral RNAi. This will be particularly interesting in the case of flaviviruses such as dengue and Zika viruses, which have been shown to contain m6A modified nucleosides when grown in mammalian cells (Gokhale *et al.*, 2016; Fig. 3.6).

Future directions

A great deal of progress was made over the past decade on the genetic characterization of antiviral innate immunity in insects. Yet, lots of questions remain, paramount among them the identification of the receptors that sense viral infections. Among insects, the fruit fly *D. melanogaster* offers a fantastic

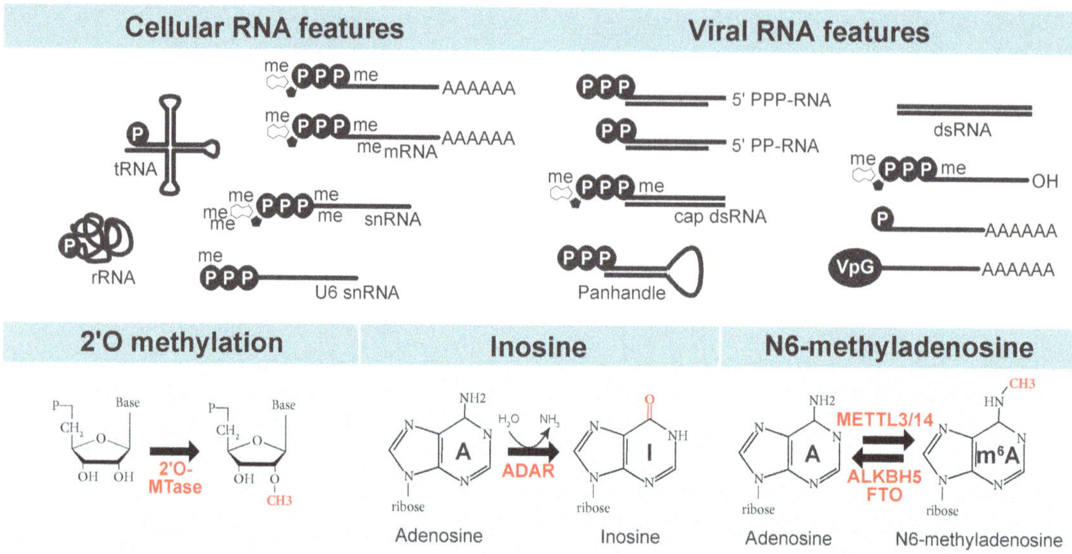

Figure 3.6 Overview of cellular and viral RNAs. Schematic representation of self and non-self RNAs adapted from Gebhardt *et al.* (2017). 2′O methylation, Inosine and N6-methyladenosine are the three main RNA modifications described on viral RNAs affecting their recognition by the innate immune system.

model for unbiased, large scale, mutagenesis screens (Wieschaus and Nüsslein-Volhard, 2016). Indeed, such screens were instrumental in defining the components of the IMD and Toll pathways of innate immunity, and in identifying the PRRs activating them in response to bacterial or fungal infections (e.g. Leulier *et al.*, 2000; Rutschmann *et al.*, 2000a,b; Lu *et al.*, 2001; Vidal *et al.*, 2001; Choe *et al.*, 2002; Gottar *et al.*, 2002, 2006). Genetic screens are however time consuming, especially when viruses must be injected into the body cavity of the flies (Merkling and van Rij, 2015). This limitation could, to some extent, be bypassed by using natural infections with fly pathogens, although this would require production of large amounts of virus and the control of viral uptake would be challenging. Alternatively, transgenic viral replicons expressing fluorescent proteins could be used as a proxy for infection, with the caveat that critical steps of the viral cycle (binding, entry, uncoating, assembly and budding of viral particles) would be bypassed (Avadhanula *et al.*, 2009; Wernet *et al.*, 2014). The recent advances in mass spectrometry (MS) technologies provide other opportunities to decipher antiviral innate immunity in insects. These methods can be used to further characterize the antiviral siRNA pathway by defining the interactome of the canonical components Dicer-2, R2D2 and AGO2, but also to identify PRRs sensing nucleic acids.

The genetic characterization of antiviral RNAi in both *Drosophila* and *Aedes* mosquitoes points to differences between the siRNA-dependent response triggered by dsRNA of endogenous (endo-siRNA pathway) or exogenous (exo-siRNA pathway) origin on one hand, and the antiviral siRNA pathway on the other (Marques *et al.*, 2013; Olmo *et al.*, 2018). Cell imaging of the canonical components Dicer-2, AGO2, R2D2 and Loqs in the context of cells treated with exogenous dsRNA or infected with viruses is likely to reveal differences in the dynamics of these proteins. This information can subsequently be used to define the interactome of the siRNA pathway at critical steps of the infection cycle, to identify novel regulatory co-factors of the pathway participating in the sensing or neutralization of viral RNAs. Of note, this approach requires synchronized infections, and will have to be carried out in tissue culture cells, a caveat considering the possible existence of tissue specific regulators (e.g. Olmo *et al* 2018).

MS can also be used to identify host proteins that sense nucleic acids of viral origin. Some foreign nucleic acids bear specific marks absent in cellular nucleic acids, such as, in the case of RNA, long double strandedness, the presence of

5′ triphosphates or missing methylation marks (reviewed in Habjan and Pichlmair, 2015; Gebhardt *et al.*, 2017). Protein binding to these marked, viral-like nucleic acids can be identified by nucleic acid affinity purification coupled to MS-based identification of proteins. For example, such an approach in mammals resulted in the identification of the host proteins IFIT1 and AIM2 as viral RNA binding protein with antiviral function and a DNA sensor of the innate immune system, respectively (Bürckstümmer *et al.*, 2009; Pichlmair *et al.*, 2011). Applied to insect cell lines or whole animals, this strategy, coupled to RNAi or CRISPR/Cas9 functional screens may reveal novel nucleic acid binding proteins involved in antiviral immunity in insects.

Finally, a fascinating question for future work pertains to the sensing of DNA in insect cells. Indeed, activation of the IMD pathway in flies mutant for the enzyme DNaseII (significant but moderate compared to a bacterial infection) suggests that cytosolic DNA can activate an NF-κB-dependent pathway in insect cells (Mukae *et al.*, 2002; Liu *et al.*, 2012). In mammals, a dedicated pathway, the cGAS-STING pathway, activates synthesis of interferons upon detecting cytosolic DNA (Hornung *et al.*, 2014). So far, the data available indicate that viral DNA produces dsRNAs and activation of the siRNA pathway, akin to the sensing of herpes virus derived dsRNA by TLR3 in mammals (Tabeta *et al.*, 2004; Zhang *et al.*, 2007; Bronkhorst *et al.*, 2012; Kemp *et al.*, 2013). Yet, important questions remain. For example, is viral DNA, which is expected to be transcribed in the nucleus possibly by a host RNA polymerase, activating the endo-siRNA pathway (Loqs-PD dependent) or the antiviral siRNA pathway (Loqs-PD independent)? Is there another pathway sensing DNA in the cytosol, as in mammals? The cGAS-like enzymes identified in insect genomes so far do not contain the zinc ribbon motif mediating interaction with DNA in mammalian cGAS (Wu *et al.*, 2014; Margolis *et al.*, 2017). However, the recent discovery that STING carries immune functions in some insects, in particular against DNA viruses in *Bombyx mori*, raises the possibility that dedicated receptors operate in the cytosol of insect cells to sense DNA (Goto *et al.*, 2018; Hua *et al.*, 2018; Liu *et al.*, 2018; Martin *et al.*, 2018). Characterization of the ligands activating the insect STINGs now at hand, which do not appear to bind cyclic dinucleotides (Kranzusch *et al.*, 2015), should help to identify these receptors.

Acknowledgements

We thank Ludmila dos Santos Silva and Joao Marques for critical reading of the manuscript and insightful suggestions. Work in our laboratory is supported by the CNRS, the ANR, the Fondation ARC and the Programme Investissements d'Avenir (Labex NetRNA and Equipex I2MC). LT acknowledges financial support from the Hoffmann Infinitus Program.

References

Aguiar, E.R., Olmo, R.P., Paro, S., Ferreira, F.V., de Faria, I.J., Todjro, Y.M., Lobo, F.P., Kroon, E.G., Meignin, C., Gatherer, D., *et al.* (2015). Sequence-independent characterization of viruses based on the pattern of viral small RNAs produced by the host. Nucleic Acids Res. *43*, 6191–6206. https://doi.org/10.1093/nar/gkv587

Aguiar, E.R., Olmo, R.P., and Marques, J.T. (2016). Virus-derived small RNAs: molecular footprints of host-pathogen interactions. Wiley Interdiscip. Rev. RNA *7*, 824–837. https://doi.org/10.1002/wrna.1361

Aliyari, R., Wu, Q., Li, H.W., Wang, X.H., Li, F., Green, L.D., Han, C.S., Li, W.X., and Ding, S.W. (2008). Mechanism of induction and suppression of antiviral immunity directed by virus-derived small RNAs in *Drosophila*. Cell Host Microbe *4*, 387–397. https://doi.org/10.1016/j.chom.2008.09.001

Aravin, A., Gaidatzis, D., Pfeffer, S., Lagos-Quintana, M., Landgraf, P., Iovino, N., Morris, P., Brownstein, M.J., Kuramochi-Miyagawa, S., Nakano, T., *et al.* (2006). A novel class of small RNAs bind to MILI protein in mouse testes. Nature *442*, 203–207.

Asad, S., Parry, R., and Asgari, S. (2018). Upregulation of Aedes aegypti Vago1 by Wolbachia and its effect on dengue virus replication. Insect Biochem. Mol. Biol. *92*, 45–52.

Ashe, A., Bélicard, T., Le Pen, J., Sarkies, P., Frézal, L., Lehrbach, N.J., Félix, M.A., and Miska, E.A. (2013). A deletion polymorphism in the Caenorhabditis elegans RIG-I homolog disables viral RNA dicing and antiviral immunity. Elife *2*, e00994. https://doi.org/10.7554/eLife.00994

Avadhanula, V., Weasner, B.P., Hardy, G.G., Kumar, J.P., and Hardy, R.W. (2009). A novel system for the launch of alphavirus RNA synthesis reveals a role for the Imd pathway in arthropod antiviral response. PLOS Pathog. *5*, e1000582. https://doi.org/10.1371/journal.ppat.1000582

Barribeau, S.M., Sadd, B.M., du Plessis, L., Brown, M.J., Buechel, S.D., Cappelle, K., Carolan, J.C., Christiaens, O., Colgan, T.J., Erler, S., *et al.* (2015). A depauperate immune repertoire precedes evolution of sociality in bees. Genome Biol. *16*, 83. https://doi.org/10.1186/s13059-015-0628-y

Basu, S., Aryan, A., Overcash, J.M., Samuel, G.H., Anderson, M.A., Dahlem, T.J., Myles, K.M., and Adelman, Z.N.

(2015). Silencing of end-joining repair for efficient site-specific gene insertion after TALEN/CRISPR mutagenesis in Aedes aegypti. Proc. Natl. Acad. Sci. U.S.A. *112*, 4038–4043. https://doi.org/10.1073/pnas.1502370112

Baumberger, N., Tsai, C.H., Lie, M., Havecker, E., and Baulcombe, D.C. (2007). The Polerovirus silencing suppressor P0 targets ARGONAUTE proteins for degradation. Curr. Biol. *17*, 1609–1614.

Bernstein, E., Caudy, A.A., Hammond, S.M., and Hannon, G.J. (2001). Role for a bidentate ribonuclease in the initiation step of RNA interference. Nature *409*, 363–366. https://doi.org/10.1038/35053110

Binggeli, O., Neyen, C., Poidevin, M., and Lemaitre, B. (2014). Prophenoloxidase activation is required for survival to microbial infections in *Drosophila*. PLOS Pathog. *10*, e1004067. https://doi.org/10.1371/journal.ppat.1004067

Bortolamiol, D., Pazhouhandeh, M., Marrocco, K., Genschik, P., and Ziegler-Graff, V. (2007). The Polerovirus F box protein P0 targets ARGONAUTE1 to suppress RNA silencing. Curr. Biol. *17*, 1615–1621.

Boyer, L., Magoc, L., Dejardin, S., Cappillino, M., Paquette, N., Hinault, C., Charriere, G.M., Ip, W.K., Fracchia, S., Hennessy, E., *et al.* (2011). Pathogen-derived effectors trigger protective immunity via activation of the Rac2 enzyme and the IMD or Rip kinase signaling pathway. Immunity 35, 536–549. https://doi.org/10.1016/j.immuni.2011.08.015

Bradshaw, C.J., Leroy, B., Bellard, C., Roiz, D., Albert, C., Fournier, A., Barbet-Massin, M., Salles, J.M., Simard, F., and Courchamp, F. (2016). Massive yet grossly underestimated global costs of invasive insects. Nat. Commun. 7, 12986. https://doi.org/10.1038/ncomms12986

Brennecke, J., Aravin, A.A., Stark, A., Dus, M., Kellis, M., Sachidanandam, R., and Hannon, G.J. (2007). Discrete small RNA-generating loci as master regulators of transposon activity in *Drosophila*. Cell *128*, 1089–1103.

Bronkhorst, A.W., and van Rij, R.P. (2014). The long and short of antiviral defense: small RNA-based immunity in insects. Curr. Opin. Virol. 7, 19–28. https://doi.org/10.1016/j.coviro.2014.03.010

Bronkhorst, A.W., van Cleef, K.W., Vodovar, N., Ince, I.A., Blanc, H., Vlak, J.M., Saleh, M.C., and van Rij, R.P. (2012). The DNA virus Invertebrate iridescent virus 6 is a target of the *Drosophila* RNAi machinery. Proc. Natl. Acad. Sci. U.S.A. *109*, E3604-13. https://doi.org/10.1073/pnas.1207213109

Bronkhorst, A.W., Miesen, P., and van Rij, R.P. (2013). Small RNAs tackle large viruses: RNA interference-based antiviral defense against DNA viruses in insects. Fly *7*, 216–223. https://doi.org/10.4161/fly.25708

Bronkhorst, A.W., van Cleef, K.W., Venselaar, H., and van Rij, R.P. (2014). A dsRNA-binding protein of a complex invertebrate DNA virus suppresses the *Drosophila* RNAi response. Nucleic Acids Res. *42*, 12237–12248. https://doi.org/10.1093/nar/gku910

Bucheton, A. (1995). The relationship between the flamenco gene and gypsy in *Drosophila*: how to tame a retrovirus. Trends Genet. *11*, 349–353.

Bürckstümmer, T., Baumann, C., Blüml, S., Dixit, E., Dürnberger, G., Jahn, H., Planyavsky, M., Bilban, M., Colinge, J., Bennett, K.L., *et al.* (2009). An orthogonal proteomic-genomic screen identifies AIM2 as a cytoplasmic DNA sensor for the inflammasome. Nat. Immunol. *10*, 266–272. https://doi.org/10.1038/ni.1702

Campbell, C.L., Keene, K.M., Brackney, D.E., Olson, K.E., Blair, C.D., Wilusz, J., and Foy, B.D. (2008). Aedes aegypti uses RNA interference in defense against Sindbis virus infection. BMC Microbiol. *8*, 47. https://doi.org/10.1186/1471-2180-8-47

Cao, C., Cogni, R., Barbier, V., and Jiggins, F.M. (2017). Complex Coding and Regulatory Polymorphisms in a Restriction Factor Determine the Susceptibility of *Drosophila* to Viral Infection. Genetics *206*, 2159–2173. https://doi.org/10.1534/genetics.117.201970

Carissimo, G., Pondeville, E., McFarlane, M., Dietrich, I., Mitri, C., Bischoff, E., Antoniewski, C., Bourgouin, C., Failloux, A.B., Kohl, A., *et al.* (2015). Antiviral immunity of Anopheles gambiae is highly compartmentalized, with distinct roles for RNA interference and gut microbiota. Proc. Natl. Acad. Sci. U.S.A. *112*, E176–185. https://doi.org/10.1073/pnas.1412984112

Carré-Mlouka, A., Gaumer, S., Gay, P., Petitjean, A.M., Coulondre, C., Dru, P., Bras, F., Dezélée, S., and Contamine, D. (2007). Control of sigma virus multiplication by the ref(2)P gene of *Drosophila* melanogaster: an in vivo study of the PB1 domain of Ref(2)P. Genetics *176*, 409–419.

Carrillo-Tripp, J., Dolezal, A.G., Goblirsch, M.J., Miller, W.A., Toth, A.L., and Bonning, B.C. (2016). In vivo and in vitro infection dynamics of honey bee viruses. Sci. Rep. *6*, 22265. https://doi.org/10.1038/srep22265

Cenik, E.S., Fukunaga, R., Lu, G., Dutcher, R., Wang, Y., Tanaka Hall, T.M., and Zamore, P.D. (2011). Phosphate and R2D2 restrict the substrate specificity of Dicer-2, an ATP-driven ribonuclease. Mol. Cell *42*, 172–184. https://doi.org/10.1016/j.molcel.2011.03.002

Chao, J.A., Lee, J.H., Chapados, B.R., Debler, E.W., Schneemann, A., and Williamson, J.R. (2005). Dual modes of RNA-silencing suppression by Flock House virus protein B2. Nat. Struct. Mol. Biol. *12*, 952–957.

Chejanovsky, N., Ophir, R., Schwager, M.S., Slabezki, Y., Grossman, S., and Cox-Foster, D. (2014). Characterization of viral siRNA populations in honey bee colony collapse disorder. Virology 454-455, 176–183. https://doi.org/10.1016/j.virol.2014.02.012

Choe, K.M., Werner, T., Stöven, S., Hultmark, D., and Anderson, K.V. (2002). Requirement for a peptidoglycan recognition protein (PGRP) in Relish activation and antibacterial immune responses in *Drosophila*. Science *296*, 359–362. https://doi.org/10.1126/science.1070216

Chung, W.J., Okamura, K., Martin, R., and Lai, E.C. (2008). Endogenous RNA interference provides a somatic defense against *Drosophila* transposons. Curr. Biol. *18*, 795–802. https://doi.org/10.1016/j.cub.2008.05.006

Cirimotich, C.M., Scott, J.C., Phillips, A.T., Geiss, B.J., and Olson, K.E. (2009). Suppression of RNA interference increases alphavirus replication and virus-associated mortality in Aedes aegypti mosquitoes. BMC Microbiol. 9, 49. https://doi.org/10.1186/1471-2180-9-49

Cora, E., Pandey, R.R., Xiol, J., Taylor, J., Sachidanandam, R., McCarthy, A.A., and Pillai, R.S. (2014). The MID-PIWI module of Piwi proteins specifies nucleotide- and strand-biases of piRNAs. RNA *20*, 773–781. https://doi.org/10.1261/rna.044701.114

Costa, A., Jan, E., Sarnow, P., and Schneider, D. (2009). The Imd pathway is involved in antiviral immune responses in *Drosophila*. PLOS ONE *4*, e7436. https://doi.org/10.1371/journal.pone.0007436

Coste, F., Kemp, C., Bobezeau, V., Hetru, C., Kellenberger, C., Imler, J.L., and Roussel, A. (2012). Crystal structure of Diedel, a marker of the immune response of *Drosophila* melanogaster. PLOS ONE 7, e33416. https://doi.org/10.1371/journal.pone.0033416

Czech, B., and Hannon, G.J. (2016). One Loop to Rule Them All: The Ping-Pong Cycle and piRNA-Guided Silencing. Trends Biochem. Sci. *41*, 324–337.

Deddouche, S., Matt, N., Budd, A., Mueller, S., Kemp, C., Galiana-Arnoux, D., Dostert, C., Antoniewski, C., Hoffmann, J.A., and Imler, J.L. (2008). The DExD/H-box helicase Dicer-2 mediates the induction of antiviral activity in *drosophila*. Nat. Immunol. *9*, 1425–1432. https://doi.org/10.1038/ni.1664

Deddouche, S., Goubau, D., Rehwinkel, J., Chakravarty, P., Begum, S., Maillard, P.V., Borg, A., Matthews, N., Feng, Q., van Kuppeveld, F.J., *et al.* (2014). Identification of an LGP2-associated MDA5 agonist in picornavirus-infected cells. Elife 3, e01535. https://doi.org/10.7554/eLife.01535

Ding, S.W. (2010). RNA-based antiviral immunity. Nat. Rev. Immunol. *10*, 632–644. https://doi.org/10.1038/nri2824

Dönertas, D., Sienski, G., and Brennecke, J. (2013). *Drosophila* Gtsf1 is an essential component of the Piwi-mediated transcriptional silencing complex. Genes Dev. *27*, 1693–1705. https://doi.org/10.1101/gad.221150.113

Dostert, C., Jouanguy, E., Irving, P., Troxler, L., Galiana-Arnoux, D., Hetru, C., Hoffmann, J.A., and Imler, J.-L. (2005). The Jak-STAT signaling pathway is required but not sufficient for the antiviral response of *Drosophila*. Nat. Immunol. *6*, 946–953. https://doi.org/10.1038/ni1237

Dudas, G., and Obbard, D.J. (2015). Are arthropods at the heart of virus evolution? ELife 4 https://doi.org/10.7554/eLife.06837

El Chamy, L., Leclerc, V., Caldelari, I., and Reichhart, J.-M. (2008). Sensing of 'danger signals' and pathogen-associated molecular patterns defines binary signaling pathways 'upstream' of Toll. Nat. Immunol. 9, 1165–1170 https://doi.org/10.1038/ni.1643

Fareh, M., van Lopik, J., Katechis, I., Bronkhorst, A.W., Haagsma, A.C., van Rij, R.P., and Joo, C. (2018). Viral suppressors of RNAi employ a rapid screening mode to discriminate viral RNA from cellular small RNA. Nucleic Acids Res. *46*, 3187–3197. https://doi.org/10.1093/nar/gkx1316

Ferrandon, D., Imler, J.L., Hetru, C., and Hoffmann, J.A. (2007). The *Drosophila* systemic immune response: sensing and signalling during bacterial and fungal infections. Nat. Rev. Immunol. 7, 862–874.

Ferreira, Á.G., Naylor, H., Esteves, S.S., Pais, I.S., Martins, N.E., and Teixeira, L. (2014). The Toll-dorsal pathway is required for resistance to viral oral infection in *Drosophila*. PLOS Pathog. *10*, e1004507. https://doi.org/10.1371/journal.ppat.1004507

Ferreira, F.V., Aguiar, E.R.G.R., Olmo, R.P., de Oliveira, K.P.V., Silva, E.G., Sant'Anna, M.R.V., Gontijo, N.F., Kroon, E.G., Imler, J.L., and Marques, J.T. (2018). The small non-coding RNA response to virus infection in the Leishmania vector Lutzomyia longipalpis. PLOS Negl. Trop. Dis. *12*, e0006569. https://doi.org/10.1371/journal.pntd.0006569

Fire, A., Xu, S., Montgomery, M.K., Kostas, S.A., Driver, S.E., and Mello, C.C. (1998). Potent and specific genetic interference by double-stranded RNA in Caenorhabditis elegans. Nature *391*, 806–811. https://doi.org/10.1038/35888

Förstemann, K., Tomari, Y., Du, T., Vagin, V.V., Denli, A.M., Bratu, D.P., Klattenhoff, C., Theurkauf, W.E., and Zamore, P.D. (2005). Normal microRNA maturation and germ-line stem cell maintenance requires Loquacious, a double-stranded RNA-binding domain protein. PLOS Biol. 3, e236.

Fragkoudis, R., Chi, Y., Siu, R.W., Barry, G., Attarzadeh-Yazdi, G., Merits, A., Nash, A.A., Fazakerley, J.K., and Kohl, A. (2008). Semliki Forest virus strongly reduces mosquito host defence signaling. Insect Mol. Biol. *17*, 647–656. https://doi.org/10.1111/j.1365-2583.2008.00834.x

Fukunaga, R., Colpan, C., Han, B.W., and Zamore, P.D. (2014). Inorganic phosphate blocks binding of pre-miRNA to Dicer-2 via its PAZ domain. EMBO J. 33, 371–384. https://doi.org/10.1002/embj.201387176

Galiana-Arnoux, D., Dostert, C., Schneemann, A., Hoffmann, J.A., and Imler, J.L. (2006). Essential function in vivo for Dicer-2 in host defense against RNA viruses in *drosophila*. Nat. Immunol. 7, 590–597.

Gebhardt, A., Laudenbach, B.T., and Pichlmair, A. (2017). Discrimination of self and non-self ribonucleic acids. J. Interferon Cytokine Res. *37*, 184–197. https://doi.org/10.1089/jir.2016.0092

Girard, A., Sachidanandam, R., Hannon, G.J., and Carmell, M.A. (2006). A germline-specific class of small RNAs binds mammalian Piwi proteins. Nature *442*, 199–202.

Goic, B., Vodovar, N., Mondotte, J.A., Monot, C., Frangeul, L., Blanc, H., Gausson, V., Vera-Otarola, J., Cristofari, G., and Saleh, M.C. (2013). RNA-mediated interference and reverse transcription control the persistence of RNA viruses in the insect model *Drosophila*. Nat. Immunol. *14*, 396–403. https://doi.org/10.1038/ni.2542

Goic, B., Stapleford, K.A., Frangeul, L., Doucet, A.J., Gausson, V., Blanc, H., Schemmel-Jofre, N., Cristofari, G., Lambrechts, L., Vignuzzi, M., *et al.* (2016). Virus-derived DNA drives mosquito vector tolerance to arboviral infection. Nat. Commun. 7, 12410. https://doi.org/10.1038/ncomms12410

Gokhale, N.S., and Horner, S.M. (2017). RNA modifications go viral. PLOS Pathog. *13*, e1006188. https://doi.org/10.1371/journal.ppat.1006188

Gokhale, N.S., McIntyre, A.B.R., McFadden, M.J., Roder, A.E., Kennedy, E.M., Gandara, J.A., Hopcraft, S.E., Quicke, K.M., Vazquez, C., Willer, J., *et al.* (2016). N6-Methyladenosine in Flaviviridae Viral RNA Genomes Regulates Infection. Cell Host Microbe *20*, 654–665.

Gold, K.S., and Brückner, K. (2015). Macrophages and cellular immunity in *Drosophila* melanogaster. Semin.

Immunol. 27, 357–368. https://doi.org/10.1016/j.smim.2016.03.010

Gonzales-van Horn, S.R., and Sarnow, P. (2017). Making the Mark: The Role of Adenosine Modifications in the Life Cycle of RNA Viruses. Cell Host Microbe *21*, 661–669.

Gordon, O., Henry, C.M., Srinivasan, N., Ahrens, S., Franz, A., Deddouche, S., Chakravarty, P., Phillips, D., George, R., Kjaer, S., et al. (2018). α-actinin accounts for the bioactivity of actin preparations in inducing STAT target genes in *Drosophila* melanogaster. ELife 7 https://doi.org/10.7554/eLife.38636.

Goto, A., Okado, K., Martins, N., Cai, H., Barbier, V., Lamiable, O., Troxler, L., Santiago, E., Kuhn, L., Paik, D., *et al.* (2018). The Kinase IKKβ Regulates a STING- and NF-κB-Dependent Antiviral Response Pathway in *Drosophila*. Immunity *49*, 225–234.e4.

Gottar, M., Gobert, V., Michel, T., Belvin, M., Duyk, G., Hoffmann, J.A., Ferrandon, D., and Royet, J. (2002). The *Drosophila* immune response against Gram-negative bacteria is mediated by a peptidoglycan recognition protein. Nature *416*, 640–644. https://doi.org/10.1038/nature734

Gottar, M., Gobert, V., Matskevich, A.A., Reichhart, J.M., Wang, C., Butt, T.M., Belvin, M., Hoffmann, J.A., and Ferrandon, D. (2006). Dual detection of fungal infections in *Drosophila* via recognition of glucans and sensing of virulence factors. Cell *127*, 1425–1437.

Goubau, D., Deddouche, S., and Reis e Sousa, C. (2013). Cytosolic sensing of viruses. Immunity *38*, 855–869. https://doi.org/10.1016/j.immuni.2013.05.007

Goubau, D., Schlee, M., Deddouche, S., Pruijssers, A.J., Zillinger, T., Goldeck, M., Schuberth, C., Van der Veen, A.G., Fujimura, T., Rehwinkel, J., *et al.* (2014). Antiviral immunity via RIG-I-mediated recognition of RNA bearing 5'-diphosphates. Nature *514*, 372–375. https://doi.org/10.1038/nature13590

Grivna, S.T., Pyhtila, B., and Lin, H. (2006). MIWI associates with translational machinery and PIWI-interacting RNAs (piRNAs) in regulating spermatogenesis. Proc. Natl. Acad. Sci. U.S.A. *103*, 13415–13420.

Gu, S., Jin, L., Huang, Y., Zhang, F., and Kay, M.A. (2012). Slicing-independent RISC activation requires the argonaute PAZ domain. Curr. Biol. *22*, 1536–1542. https://doi.org/10.1016/j.cub.2012.06.040

Gueguen, G., Kalamarz, M.E., Ramroop, J., Uribe, J., and Govind, S. (2013). Polydnaviral ankyrin proteins aid parasitic wasp survival by coordinate and selective inhibition of hematopoietic and immune NF-kappa B signaling in insect hosts. PLOS Pathog. *9*, e1003580. https://doi.org/10.1371/journal.ppat.1003580

Gunawardane, L.S., Saito, K., Nishida, K.M., Miyoshi, K., Kawamura, Y., Nagami, T., Siomi, H., and Siomi, M.C. (2007). A slicer-mediated mechanism for repeat-associated siRNA 5' end formation in *Drosophila*. Science *315*, 1587–1590.

Guo, X., Zhang, R., Wang, J., Ding, S.W., and Lu, R. (2013). Homologous RIG-I-like helicase proteins direct RNAi-mediated antiviral immunity in C. elegans by distinct mechanisms. Proc. Natl. Acad. Sci. U.S.A. *110*, 16085–16090. https://doi.org/10.1073/pnas.1307453110

Habjan, M., and Pichlmair, A. (2015). Cytoplasmic sensing of viral nucleic acids. Curr. Opin. Virol. *11*, 31–37. https://doi.org/10.1016/j.coviro.2015.01.012

Haley, B., and Zamore, P.D. (2004). Kinetic analysis of the RNAi enzyme complex. Nat. Struct. Mol. Biol. *11*, 599–606. https://doi.org/10.1038/nsmb780

Hamilton, A.J., and Baulcombe, D.C. (1999). A species of small antisense RNA in posttranscriptional gene silencing in plants. Science *286*, 950–952.

Hammond, S.M., Bernstein, E., Beach, D., and Hannon, G.J. (2000). An RNA-directed nuclease mediates post-transcriptional gene silencing in *Drosophila* cells. Nature *404*, 293–296. https://doi.org/10.1038/35005107

Han, B.W., Wang, W., Li, C., Weng, Z., and Zamore, P.D. (2015). Noncoding RNA. piRNA-guided transposon cleavage initiates Zucchini-dependent, phased piRNA production. Science *348*, 817–821. https://doi.org/10.1126/science.aaa1264

Han, Y.H., Luo, Y.J., Wu, Q., Jovel, J., Wang, X.H., Aliyari, R., Han, C., Li, W.X., and Ding, S.W. (2011). RNA-based immunity terminates viral infection in adult *Drosophila* in the absence of viral suppression of RNA interference: characterization of viral small interfering RNA populations in wild-type and mutant flies. J. Virol. *85*, 13153–13163. https://doi.org/10.1128/JVI.05518-11

Helm, M., and Motorin, Y. (2017). Detecting RNA modifications in the epitranscriptome: predict and validate. Nat. Rev. Genet. *18*, 275–291. https://doi.org/10.1038/nrg.2016.169

Hess, A.M., Prasad, A.N., Ptitsyn, A., Ebel, G.D., Olson, K.E., Barbacioru, C., Monighetti, C., and Campbell, C.L. (2011). Small RNA profiling of Dengue virus-mosquito interactions implicates the PIWI RNA pathway in anti-viral defense. BMC Microbiol. *11*, 45. https://doi.org/10.1186/1471-2180-11-45

Hoffmann, J.A. (2003). The immune response of *Drosophila*. Nature *426*, 33–38. https://doi.org/10.1038/nature02021

Hoffmann, J.A., Kafatos, F.C., Janeway, C.A., and Ezekowitz, R.A. (1999). Phylogenetic perspectives in innate immunity. Science *284*, 1313–1318.

Hornung, V., Hartmann, R., Ablasser, A., and Hopfner, K.P. (2014). OAS proteins and cGAS: unifying concepts in sensing and responding to cytosolic nucleic acids. Nat. Rev. Immunol. *14*, 521–528. https://doi.org/10.1038/nri3719

Horwich, M.D., Li, C., Matranga, C., Vagin, V., Farley, G., Wang, P., and Zamore, P.D. (2007). The *Drosophila* RNA methyltransferase, DmHen1, modifies germline piRNAs and single-stranded siRNAs in RISC. Curr. Biol. *17*, 1265–1272.

Hua, X., Li, B., Song, L., Hu, C., Li, X., Wang, D., Xiong, Y., Zhao, P., He, H., Xia, Q., *et al.* (2018). Stimulator of interferon genes (STING) provides insect antiviral immunity by promoting Dredd caspase-mediated NF-κB activation. J. Biol. Chem. *293*, 11878–11890. https://doi.org/10.1074/jbc.RA117.000194

Huang, X., Fejes Tóth, K., and Aravin, A.A. (2017). piRNA Biogenesis in *Drosophila* melanogaster. Trends Genet. *33*, 882–894 https://doi.org/10.1016/j.tig.2017.09.002.

Huang, Z., Kingsolver, M.B., Avadhanula, V., and Hardy, R.W. (2013). An antiviral role for antimicrobial peptides during the arthropod response to alphavirus replication. J. Virol. *87*, 4272–4280. https://doi.org/10.1128/JVI.03360-12

Hussain, M., and Asgari, S. (2014). MicroRNAs as mediators of insect host-pathogen interactions and immunity. J. Insect Physiol. *70*, 151–158. https://doi.org/10.1016/j.jinsphys.2014.08.003

Hutvágner, G., and Zamore, P.D. (2002). A microRNA in a multiple-turnover RNAi enzyme complex. Science *297*, 2056–2060. https://doi.org/10.1126/science.1073827

Ikushima, H., Negishi, H., and Taniguchi, T. (2013). The IRF family transcription factors at the interface of innate and adaptive immune responses. Cold Spring Harb. Symp. Quant. Biol. *78*, 105–116. https://doi.org/10.1101/sqb.2013.78.020321

Imler, J.L., and Bulet, P. (2005). Antimicrobial peptides in *Drosophila*: structures, activities and gene regulation. Chem. Immunol. Allergy *86*, 1–21 https://doi.org/10.1159/000086648.

Issa, N., Guillaumot, N., Lauret, E., Matt, N., Schaeffer-Reiss, C., Van Dorsselaer, A., Reichhart, J.M., and Veillard, F. (2018). The Circulating Protease Persephone Is an Immune Sensor for Microbial Proteolytic Activities Upstream of the *Drosophila* Toll Pathway. Mol. Cell *69*, 539–550.e6.

Iwasaki, S., Kobayashi, M., Yoda, M., Sakaguchi, Y., Katsuma, S., Suzuki, T., and Tomari, Y. (2010). Hsc70/Hsp90 chaperone machinery mediates ATP-dependent RISC loading of small RNA duplexes. Mol. Cell *39*, 292–299. https://doi.org/10.1016/j.molcel.2010.05.015

Iwasaki, S., Sasaki, H.M., Sakaguchi, Y., Suzuki, T., Tadakuma, H., and Tomari, Y. (2015). Defining fundamental steps in the assembly of the *Drosophila* RNAi enzyme complex. Nature *521*, 533–536. https://doi.org/10.1038/nature14254

Jayachandran, B., Hussain, M., and Asgari, S. (2012). RNA interference as a cellular defense mechanism against the DNA virus baculovirus. J. Virol. *86*, 13729–13734. https://doi.org/10.1128/JVI.02041-12

Jiang, F., Ye, X., Liu, X., Fincher, L., McKearin, D., and Liu, Q. (2005). Dicer-1 and R3D1-L catalyze microRNA maturation in *Drosophila*. Genes Dev. *19*, 1674–1679 https://doi.org/10.1101/gad.1334005.

Jiang, H., Patel, P.H., Kohlmaier, A., Grenley, M.O., McEwen, D.G., and Edgar, B.A. (2009). Cytokine/Jak/Stat signaling mediates regeneration and homeostasis in the *Drosophila* midgut. Cell *137*, 1343–1355. https://doi.org/10.1016/j.cell.2009.05.014

Jupatanakul, N., Sim, S., Angleró-Rodríguez, Y.I., Souza-Neto, J., Das, S., Poti, K.E., Rossi, S.L., Bergren, N., Vasilakis, N., and Dimopoulos, G. (2017). Engineered Aedes aegypti JAK/STAT Pathway-Mediated Immunity to Dengue Virus. PLOS Negl. Trop. Dis. *11*, e0005187. https://doi.org/10.1371/journal.pntd.0005187

Karikó, K., Buckstein, M., Ni, H., and Weissman, D. (2005). Suppression of RNA recognition by Toll-like receptors: the impact of nucleoside modification and the evolutionary origin of RNA. Immunity 23, 165–175 https://doi.org/10.1016/j.immuni.2005.06.008.

Karlikow, M., Goic, B., Mongelli, V., Salles, A., Schmitt, C., Bonne, I., Zurzolo, C., and Saleh, M.C. (2016). *Drosophila* cells use nanotube-like structures to transfer dsRNA and RNAi machinery between cells. Sci. Rep. *6*, 27085. https://doi.org/10.1038/srep27085

Kato, H., Sato, S., Yoneyama, M., Yamamoto, M., Uematsu, S., Matsui, K., Tsujimura, T., Takeda, K., Fujita, T., Takeuchi, O., *et al.* (2005). Cell type-specific involvement of RIG-I in antiviral response. Immunity 23, 19–28 https://doi.org/10.1016/j.immuni.2005.04.010.

Kato, H., Takeuchi, O., Sato, S., Yoneyama, M., Yamamoto, M., Matsui, K., Uematsu, S., Jung, A., Kawai, T., Ishii, K.J., *et al.* (2006). Differential roles of MDA5 and RIG-I helicases in the recognition of RNA viruses. Nature *441*, 101–105.

Kawai, T., and Akira, S. (2011). Toll-like receptors and their crosstalk with other innate receptors in infection and immunity. Immunity *34*, 637–650. https://doi.org/10.1016/j.immuni.2011.05.006

Kemp, C., Mueller, S., Goto, A., Barbier, V., Paro, S., Bonnay, F., Dostert, C., Troxler, L., Hetru, C., Meignin, C., *et al.* (2013). Broad RNA interference-mediated antiviral immunity and virus-specific inducible responses in *Drosophila*. J. Immunol. *190*, 650–658. https://doi.org/10.4049/jimmunol.1102486

Khong, A., Kerr, C.H., Yeung, C.H., Keatings, K., Nayak, A., Allan, D.W., and Jan, E. (2017). Disruption of stress granule formation by the multifunctional cricket paralysis virus 1A protein. J. Virol. *91*, e01779-16. https://doi.org/10.1128/JVI.01779-16.

Khoo, C.C., Piper, J., Sanchez-Vargas, I., Olson, K.E., and Franz, A.W. (2010). The RNA interference pathway affects midgut infection- and escape barriers for Sindbis virus in Aedes aegypti. BMC Microbiol. *10*, 130. https://doi.org/10.1186/1471-2180-10-130

Khoo, C.C., Doty, J.B., Heersink, M.S., Olson, K.E., and Franz, A.W. (2013). Transgene-mediated suppression of the RNA interference pathway in Aedes aegypti interferes with gene silencing and enhances Sindbis virus and dengue virus type 2 replication. Insect Mol. Biol. *22*, 104–114. https://doi.org/10.1111/imb.12008

Kim, K., Lee, Y.S., and Carthew, R.W. (2007). Conversion of pre-RISC to holo-RISC by Ago2 during assembly of RNAi complexes. RNA *13*, 22–29. https://doi.org/10.1261/rna.283207

Kowalinski, E., Lunardi, T., McCarthy, A.A., Louber, J., Brunel, J., Grigorov, B., Gerlier, D., and Cusack, S. (2011). Structural basis for the activation of innate immune pattern-recognition receptor RIG-I by viral RNA. Cell *147*, 423–435. https://doi.org/10.1016/j.cell.2011.09.039

Kranzusch, P.J., Wilson, S.C., Lee, A.S., Berger, J.M., Doudna, J.A., and Vance, R.E. (2015). Ancient Origin of cGAS-STING Reveals Mechanism of Universal 2′,3′ cGAMP Signaling. Mol. Cell 59, 891–903. https://doi.org/10.1016/j.molcel.2015.07.022

Kreuze, J.F., Perez, A., Untiveros, M., Quispe, D., Fuentes, S., Barker, I., and Simon, R. (2009). Complete viral genome sequence and discovery of novel viruses by deep sequencing of small RNAs: a generic method for diagnosis, discovery and sequencing of viruses. Virology *388*, 1–7. https://doi.org/10.1016/j.virol.2009.03.024

Kroemer, J.A., and Webb, B.A. (2005). Ikappabeta-related vankyrin genes in the Campoletis sonorensis ichnovirus: temporal and tissue-specific patterns of expression in parasitized Heliothis virescens lepidopteran hosts. J. Virol. *79*, 7617–7628.

Kroemer, J.A., and Webb, B.A. (2006). Divergences in protein activity and cellular localization within the Campoletis sonorensis Ichnovirus Vankyrin family. J. Virol. *80*, 12219–12228. https://doi.org/10.1128/JVI.01187-06

Ktistakis, N.T., and Tooze, S.A. (2016). Digesting the expanding mechanisms of autophagy. Trends Cell Biol. *26*, 624–635.

Kwak, P.B., and Tomari, Y. (2012). The N domain of Argonaute drives duplex unwinding during RISC assembly. Nat. Struct. Mol. Biol. *19*, 145–151. https://doi.org/10.1038/nsmb.2232

Lambrechts, L., Quillery, E., Noël, V., Richardson, J.H., Jarman, R.G., Scott, T.W., and Chevillon, C. (2013). Specificity of resistance to dengue virus isolates is associated with genotypes of the mosquito antiviral gene Dicer-2. Proc. Biol. Sci. *280*, 20122437. https://doi.org/10.1098/rspb.2012.2437

Lamiable, O., Kellenberger, C., Kemp, C., Troxler, L., Pelte, N., Boutros, M., Marques, J.T., Daeffler, L., Hoffmann, J.A., Roussel, A., *et al.* (2016a). Cytokine Diedel and a viral homologue suppress the IMD pathway in *Drosophila*. Proc. Natl. Acad. Sci. U.S.A. *113*, 698–703. https://doi.org/10.1073/pnas.1516122113

Lamiable, O., Arnold, J., de Faria, I.J.D.S., Olmo, R.P., Bergami, F., Meignin, C., Hoffmann, J.A., Marques, J.T., and Imler, J.L. (2016b). Analysis of the Contribution of Hemocytes and Autophagy to *Drosophila* Antiviral Immunity. J. Virol. *90*, 5415–5426. https://doi.org/10.1128/JVI.00238-16

Lan, H., Chen, H., Liu, Y., Jiang, C., Mao, Q., Jia, D., Chen, Q., and Wei, T. (2016). Small interfering RNA pathway modulates initial viral infection in midgut epithelium of insect after ingestion of virus. J. Virol. *90*, 917–929. https://doi.org/10.1128/JVI.01835-15

Le Thomas, A., Rogers, A.K., Webster, A., Marinov, G.K., Liao, S.E., Perkins, E.M., Hur, J.K., Aravin, A.A., and Tóth, K.F. (2013). Piwi induces piRNA-guided transcriptional silencing and establishment of a repressive chromatin state. Genes Dev. *27*, 390–399. https://doi.org/10.1101/gad.209841.112

Lee, Y.S., Nakahara, K., Pham, J.W., Kim, K., He, Z., Sontheimer, E.J., and Carthew, R.W. (2004). Distinct roles for *Drosophila* Dicer-1 and Dicer-2 in the siRNA/miRNA silencing pathways. Cell *117*, 69–81.

Léger, P., Lara, E., Jagla, B., Sismeiro, O., Mansuroglu, Z., Coppée, J.Y., Bonnefoy, E., and Bouloy, M. (2013). Dicer-2- and Piwi-mediated RNA interference in Rift Valley fever virus-infected mosquito cells. J. Virol. *87*, 1631–1648. https://doi.org/10.1128/JVI.02795-12

Lemaitre, B., Meister, M., Govind, S., Georgel, P., Steward, R., Reichhart, J.M., and Hoffmann, J.A. (1995). Functional analysis and regulation of nuclear import of dorsal during the immune response in *Drosophila*. EMBO J. *14*, 536–545.

Letourneau, M., Lapraz, F., Sharma, A., Vanzo, N., Waltzer, L., and Crozatier, M. (2016). *Drosophila* hematopoiesis under normal conditions and in response to immune stress. FEBS Lett. *590*, 4034–4051. https://doi.org/10.1002/1873-3468.12327

Leulier, F., Rodriguez, A., Khush, R.S., Abrams, J.M., and Lemaitre, B. (2000). The *Drosophila* caspase Dredd is required to resist gram-negative bacterial infection. EMBO Rep. *1*, 353–358. https://doi.org/10.1093/embo-reports/kvd073

Lewis, S.H., Quarles, K.A., Yang, Y., Tanguy, M., Frézal, L., Smith, S.A., Sharma, P.P., Cordaux, R., Gilbert, C., Giraud, I., *et al.* (2018). Pan-arthropod analysis reveals somatic piRNAs as an ancestral defence against transposable elements. Nat. Ecol. Evol. *2*, 174–181. https://doi.org/10.1038/s41559-017-0403-4

Li, C.X., Shi, M., Tian, J.H., Lin, X.D., Kang, Y.J., Chen, L.J., Qin, X.C., Xu, J., Holmes, E.C., and Zhang, Y.Z. (2015). Unprecedented genomic diversity of RNA viruses in arthropods reveals the ancestry of negative-sense RNA viruses. Elife *4*. https://doi.org/10.7554/eLife.05378

Li, H., Li, W.X., and Ding, S.W. (2002). Induction and suppression of RNA silencing by an animal virus. Science *296*, 1319–1321. https://doi.org/10.1126/science.1070948

Li, J., Yang, Z., Yu, B., Liu, J., and Chen, X. (2005). Methylation protects miRNAs and siRNAs from a 3'-end uridylation activity in Arabidopsis. Curr. Biol. *15*, 1501–1507.

Liang, C., Wang, Y., Murota, Y., Liu, X., Smith, D., Siomi, M.C., and Liu, Q. (2015). TAF11 Assembles the RISC Loading Complex to Enhance RNAi Efficiency. Mol. Cell *59*, 807–818. https://doi.org/10.1016/j.molcel.2015.07.006

Liddicoat, B.J., Piskol, R., Chalk, A.M., Ramaswami, G., Higuchi, M., Hartner, J.C., Li, J.B., Seeburg, P.H., and Walkley, C.R. (2015). RNA editing by ADAR1 prevents MDA5 sensing of endogenous dsRNA as nonself. Science *349*, 1115–1120. https://doi.org/10.1126/science.aac7049

Lim, A.K., and Kai, T. (2007). Unique germ-line organelle, nuage, functions to repress selfish genetic elements in *Drosophila* melanogaster. Proc. Natl. Acad. Sci. U.S.A. *104*, 6714–6719.

Lin, H., and Spradling, A.C. (1997). A novel group of pumilio mutations affects the asymmetric division of germline stem cells in the *Drosophila* ovary. Development *124*, 2463–2476.

Liu, Q., Rand, T.A., Kalidas, S., Du, F., Kim, H.E., Smith, D.P., and Wang, X. (2003). R2D2, a bridge between the initiation and effector steps of the *Drosophila* RNAi pathway. Science *301*, 1921–1925. https://doi.org/10.1126/science.1088710

Liu, X., Sano, T., Guan, Y., Nagata, S., Hoffmann, J.A., and Fukuyama, H. (2012). *Drosophila* EYA regulates the immune response against DNA through an evolutionarily conserved threonine phosphatase motif. PLOS ONE *7*, e42725. https://doi.org/10.1371/journal.pone.0042725

Liu, Y., Ye, X., Jiang, F., Liang, C., Chen, D., Peng, J., Kinch, L.N., Grishin, N.V., and Liu, Q. (2009). C3PO, an endoribonuclease that promotes RNAi by facilitating RISC activation. Science *325*, 750–753. https://doi.org/10.1126/science.1176325

Liu, Y., Gordesky-Gold, B., Leney-Greene, M., Weinbren, N.L., Tudor, M., and Cherry, S. (2018). Inflammation-induced, STING-dependent autophagy restricts Zika virus infection in the *Drosophila* brain. Cell

Host Microbe *24*, 57–68.e3 https://doi.org/10.1016/j.chom.2018.05.022
Longdon, B., Cao, C., Martinez, J., and Jiggins, F.M. (2013). Previous exposure to an RNA virus does not protect against subsequent infection in *Drosophila* melanogaster. PloS One *8*, e73833 https://doi.org/10.1371/journal.pone.0073833.
Lu, Y., Wu, L.P., and Anderson, K.V. (2001). The antibacterial arm of the *drosophila* innate immune response requires an IkappaB kinase. Genes Dev. *15*, 104–110.
Ma, J.B., Ye, K., and Patel, D.J. (2004). Structural basis for overhang-specific small interfering RNA recognition by the PAZ domain. Nature *429*, 318–322. https://doi.org/10.1038/nature02519
Ma, J.B., Yuan, Y.R., Meister, G., Pei, Y., Tuschl, T., and Patel, D.J. (2005). Structural basis for 5'-end-specific recognition of guide RNA by the A. fulgidus Piwi protein. Nature *434*, 666–670.
Maeda, K., and Akira, S. (2016). TLR7 Structure: Cut in Z-Loop. Immunity *45*, 705–707 https://doi.org/10.1016/j.immuni.2016.10.003
Magwire, M.M., Bayer, F., Webster, C.L., Cao, C., and Jiggins, F.M. (2011). Successive increases in the resistance of *Drosophila* to viral infection through a transposon insertion followed by a Duplication. PLOS Genet. *7*, e1002337. https://doi.org/10.1371/journal.pgen.1002337
Magwire, M.M., Fabian, D.K., Schweyen, H., Cao, C., Longdon, B., Bayer, F., and Jiggins, F.M. (2012). Genome-wide association studies reveal a simple genetic basis of resistance to naturally coevolving viruses in *Drosophila* melanogaster. PLOS Genet. *8*, e1003057. https://doi.org/10.1371/journal.pgen.1003057
Margolis, S.R., Wilson, S.C., and Vance, R.E. (2017). Evolutionary Origins of cGAS-STING Signaling. Trends Immunol. *38*, 733–743 https://doi.org/10.1016/j.it.2017.03.004
Marques, J.T., and Imler, J.-L. (2016). The diversity of insect antiviral immunity: insights from viruses. Curr. Opin. Microbiol. *32*, 71–76 https://doi.org/10.1016/j.mib.2016.05.002
Marques, J.T., Kim, K., Wu, P.H., Alleyne, T.M., Jafari, N., and Carthew, R.W. (2010). Loqs and R2D2 act sequentially in the siRNA pathway in *Drosophila*. Nat. Struct. Mol. Biol. *17*, 24–30. https://doi.org/10.1038/nsmb.1735
Marques, J.T., Wang, J.P., Wang, X., de Oliveira, K.P., Gao, C., Aguiar, E.R., Jafari, N., and Carthew, R.W. (2013). Functional specialization of the small interfering RNA pathway in response to virus infection. PLOS Pathog. *9*, e1003579. https://doi.org/10.1371/journal.ppat.1003579
Martin, M., Hiroyasu, A., Guzman, R.M., Roberts, S.A., and Goodman, A.G. (2018). Analysis of *Drosophila* STING Reveals an Evolutionarily Conserved Antimicrobial Function. Cell Rep. *23*, 3537–3550.e6 https://doi.org/10.1016/j.celrep.2018.05.029
Martins, N., Imler, J.L., and Meignin, C. (2016). Discovery of novel targets for antivirals: learning from flies. Curr. Opin. Virol. *20*, 64–70 https://doi.org/10.1016/j.coviro.2016.09.005
Matranga, C., Tomari, Y., Shin, C., Bartel, D.P., and Zamore, P.D. (2005). Passenger-strand cleavage facilitates assembly of siRNA into Ago2-containing RNAi enzyme complexes. Cell *123*, 607–620 https://doi.org/10.1016/j.cell.2005.08.044
Matsumoto, N., Nishimasu, H., Sakakibara, K., Nishida, K.M., Hirano, T., Ishitani, R., Siomi, H., Siomi, M.C., and Nureki, O. (2016). Crystal Structure of Silkworm PIWI-Clade Argonaute Siwi Bound to piRNA. Cell *167*, 484–497.e9 https://doi.org/10.1016/j.cell.2016.09.002
McFarlane, M., Arias-Goeta, C., Martin, E., O'Hara, Z., Lulla, A., Mousson, L., Rainey, S.M., Misbah, S., Schnettler, E., Donald, C.L., *et al.* (2014). Characterization of Aedes aegypti innate-immune pathways that limit Chikungunya virus replication. PLOS Negl. Trop. Dis. *8*, e2994. https://doi.org/10.1371/journal.pntd.0002994
Merkling, S.H., and van Rij, R.P. (2015). Analysis of resistance and tolerance to virus infection in *Drosophila*. Nat. Protoc. *10*, 1084–1097. https://doi.org/10.1038/nprot.2015.071
Merkling, S.H., Bronkhorst, A.W., Kramer, J.M., Overheul, G.J., Schenck, A., and Van Rij, R.P. (2015a). The epigenetic regulator G9a mediates tolerance to RNA virus infection in *Drosophila*. PLOS Pathog. *11*, e1004692. https://doi.org/10.1371/journal.ppat.1004692
Merkling, S.H., Overheul, G.J., van Mierlo, J.T., Arends, D., Gilissen, C., and van Rij, R.P. (2015b). The heat shock response restricts virus infection in *Drosophila*. Sci. Rep. *5*, 12758. https://doi.org/10.1038/srep12758
Miesen, P., Girardi, E., and van Rij, R.P. (2015). Distinct sets of PIWI proteins produce arbovirus and transposon-derived piRNAs in Aedes aegypti mosquito cells. Nucleic Acids Res. *43*, 6545–6556. https://doi.org/10.1093/nar/gkv590
Miesen, P., Joosten, J., and van Rij, R.P. (2016a). PIWIs Go Viral: Arbovirus-Derived piRNAs in Vector Mosquitoes. PLOS Pathog. *12*, e1006017. https://doi.org/10.1371/journal.ppat.1006017
Miesen, P., Ivens, A., Buck, A.H., and van Rij, R.P. (2016b). Small RNA Profiling in Dengue Virus 2-Infected Aedes Mosquito Cells Reveals Viral piRNAs and Novel Host miRNAs. PLOS Negl. Trop. Dis. *10*, e0004452. https://doi.org/10.1371/journal.pntd.0004452
Miller, J.R., Koren, S., Dilley, K.A., Puri, V., Brown, D.M., Harkins, D.M., Thibaud-Nissen, F., Rosen, B., Chen, X.G., Tu, Z., *et al.* (2018). Analysis of the Aedes albopictus C6/36 genome provides insight into cell line utility for viral propagation. Gigascience *7*, 1–13. https://doi.org/10.1093/gigascience/gix135
Misof, B., Liu, S., Meusemann, K., Peters, R.S., Donath, A., Mayer, C., Frandsen, P.B., Ware, J., Flouri, T., Beutel, R.G., *et al.* (2014). Phylogenomics resolves the timing and pattern of insect evolution. Science *346*, 763–767. https://doi.org/10.1126/science.1257570
Miyoshi, K., Tsukumo, H., Nagami, T., Siomi, H., and Siomi, M.C. (2005). Slicer function of *Drosophila* Argonautes and its involvement in RISC formation. Genes Dev. *19*, 2837–2848. https://doi.org/10.1101/gad.1370605
Miyoshi, K., Miyoshi, T., Hartig, J.V., Siomi, H., and Siomi, M.C. (2010a). Molecular mechanisms that funnel RNA precursors into endogenous small-interfering RNA and microRNA biogenesis pathways in *Drosophila*. RNA *16*, 506–515. https://doi.org/10.1261/rna.1952110

Miyoshi, T., Takeuchi, A., Siomi, H., and Siomi, M.C. (2010b). A direct role for Hsp90 in pre-RISC formation in *Drosophila*. Nat. Struct. Mol. Biol. *17*, 1024–1026. https://doi.org/10.1038/nsmb.1875

Mlih, M., Khericha, M., Birdwell, C., West, A.P., and Karpac, J. (2018). A virus-acquired host cytokine controls systemic aging by antagonizing apoptosis. PLOS Biol. *16*, e2005796. https://doi.org/10.1371/journal.pbio.2005796

Mo, X., Yang, X., and Yuan, Y.A. (2018). Structural insights into *Drosophila*-C3PO complex assembly and 'Dynamic Side Port' model in substrate entry and release. Nucleic Acids Res. *46*, 8590–8604. https://doi.org/10.1093/nar/gky465

Mohn, F., Handler, D., and Brennecke, J. (2015). Noncoding RNA. piRNA-guided slicing specifies transcripts for Zucchini-dependent, phased piRNA biogenesis. Science *348*, 812–817. https://doi.org/10.1126/science.aaa1039

Molina-Cruz, A., Zilversmit, M.M., Neafsey, D.E., Hartl, D.L., and Barillas-Mury, C. (2016). Mosquito Vectors and the Globalization of Plasmodium falciparum Malaria. Annu. Rev. Genet. *50*, 447–465. https://doi.org/10.1146/annurev-genet-120215-035211

Mondotte, J.A., Gausson, V., Frangeul, L., Blanc, H., Lambrechts, L., and Saleh, M.C. (2018). Immune priming and clearance of orally acquired RNA viruses in *Drosophila*. Nat. Microbiol. *3*, 1394–1403. https://doi.org/10.1038/s41564-018-0265-9

Monsion, B., Incarbone, M., Hleibieh, K., Poignavent, V., Ghannam, A., Dunoyer, P., Daeffler, L., Tilsner, J., and Ritzenthaler, C. (2018). Efficient Detection of Long dsRNA *in Vitro* and *in Vivo* Using the dsRNA Binding Domain from FHV B2 Protein. Front. Plant Sci. *9*, 70. https://doi.org/10.3389/fpls.2018.00070

Morazzani, E.M., Wiley, M.R., Murreddu, M.G., Adelman, Z.N., and Myles, K.M. (2012). Production of virus-derived ping-pong-dependent piRNA-like small RNAs in the mosquito soma. PLOS Pathog. *8*, e1002470. https://doi.org/10.1371/journal.ppat.1002470

Mueller, S., Gausson, V., Vodovar, N., Deddouche, S., Troxler, L., Perot, J., Pfeffer, S., Hoffmann, J.A., Saleh, M.C., and Imler, J.L. (2010). RNAi-mediated immunity provides strong protection against the negative-strand RNA vesicular stomatitis virus in *Drosophila*. Proc. Natl. Acad. Sci. U.S.A. *107*, 19390–19395. https://doi.org/10.1073/pnas.1014378107

Mukae, N., Yokoyama, H., Yokokura, T., Sakoyama, Y., and Nagata, S. (2002). Activation of the innate immunity in *Drosophila* by endogenous chromosomal DNA that escaped apoptotic degradation. Genes Dev. *16*, 2662–2671. https://doi.org/10.1101/gad.1022802

Müller, S., and Imler, J.L. (2007). Dicing with viruses: microRNAs as antiviral factors. Immunity *27*, 1–3.

Mussabekova, A., Daeffler, L., and Imler, J.L. (2017). Innate and intrinsic antiviral immunity in *Drosophila*. Cell. Mol. Life Sci. *74*, 2039–2054. https://doi.org/10.1007/s00018-017-2453-9

Myles, K.M., Wiley, M.R., Morazzani, E.M., and Adelman, Z.N. (2008). Alphavirus-derived small RNAs modulate pathogenesis in disease vector mosquitoes. Proc. Natl. Acad. Sci. U.S.A. *105*, 19938–19943. https://doi.org/10.1073/pnas.0803408105

Nainu, F., Tanaka, Y., Shiratsuchi, A., and Nakanishi, Y. (2015). Protection of insects against viral infection by apoptosis-dependent phagocytosis. J. Immunol. *195*, 5696–5706. https://doi.org/10.4049/jimmunol.1500613

Nakanishi, K. (2016). Anatomy of RISC: how do small RNAs and chaperones activate Argonaute proteins? Wiley Interdiscip. Rev. RNA *7*, 637–660 https://doi.org/10.1002/wrna.1356.

Nayak, A., Berry, B., Tassetto, M., Kunitomi, M., Acevedo, A., Deng, C., Krutchinsky, A., Gross, J., Antoniewski, C., and Andino, R. (2010). Cricket paralysis virus antagonizes Argonaute 2 to modulate antiviral defense in *Drosophila*. Nat. Struct. Mol. Biol. *17*, 547–554. https://doi.org/10.1038/nsmb.1810

Nayak, A., Kim, D.Y., Trnka, M.J., Kerr, C.H., Lidsky, P.V., Stanley, D.J., Rivera, B.M., Li, K.H., Burlingame, A.L., Jan, E., *et al.* (2018). A Viral Protein Restricts *Drosophila* RNAi Immunity by Regulating Argonaute Activity and Stability. Cell Host Microbe *24*, 542–557.e9 https://doi.org/10.1016/j.chom.2018.09.006

Nishida, K.M., Miyoshi, K., Ogino, A., Miyoshi, T., Siomi, H., and Siomi, M.C. (2013). Roles of R2D2, a cytoplasmic D2 body component, in the endogenous siRNA pathway in *Drosophila*. Mol. Cell *49*, 680–691. https://doi.org/10.1016/j.molcel.2012.12.024

Obbard, D.J., Jiggins, F.M., Halligan, D.L., and Little, T.J. (2006). Natural selection drives extremely rapid evolution in antiviral RNAi genes. Curr. Biol. *16*, 580–585. https://doi.org/10.1016/j.cub.2006.01.065

Ohtani, H., Iwasaki, Y.W., Shibuya, A., Siomi, H., Siomi, M.C., and Saito, K. (2013). DmGTSF1 is necessary for Piwi-piRISC-mediated transcriptional transposon silencing in the *Drosophila* ovary. Genes Dev. *27*, 1656–1661. https://doi.org/10.1101/gad.221515.113

Okamura, K., Ishizuka, A., Siomi, H., and Siomi, M.C. (2004). Distinct roles for Argonaute proteins in small RNA-directed RNA cleavage pathways. Genes Dev. *18*, 1655–1666. https://doi.org/10.1101/gad.1210204

Olmo, R.P., Ferreira, A.G.A., Izidoro-Toledo, T.C., Aguiar, E.R.G.R., de Faria, I.J.S., de Souza, K.P.R., Osório, K.P., Kuhn, L., Hammann, P., de Andrade, E.G., *et al.* (2018). Control of dengue virus in the midgut of Aedes aegypti by ectopic expression of the dsRNA-binding protein Loqs2. Nat. Microbiol. *3*, 1385–1393. https://doi.org/10.1038/s41564-018-0268-6

Palmer, W.H., Hadfield, J.D., and Obbard, D.J. (2018). RNA-Interference Pathways Display High Rates of Adaptive Protein Evolution in Multiple Invertebrates. Genetics *208*, 1585–1599. https://doi.org/10.1534/genetics.117.300567

Pane, A., Wehr, K., and Schüpbach, T. (2007). zucchini and squash encode two putative nucleases required for rasiRNA production in the *Drosophila* germline. Dev. Cell *12*, 851–862. https://doi.org/10.1016/j.devcel.2007.03.022

Paradkar, P.N., Trinidad, L., Voysey, R., Duchemin, J.B., and Walker, P.J. (2012). Secreted Vago restricts West Nile virus infection in Culex mosquito cells by activating the Jak-STAT pathway. Proc. Natl. Acad. Sci. U.S.A. *109*, 18915–18920. https://doi.org/10.1073/pnas.1205231109

Paradkar, P.N., Duchemin, J.B., Voysey, R., and Walker, P.J. (2014). Dicer-2-dependent activation of Culex Vago occurs via the TRAF-Rel2 signaling pathway. PLOS Negl. Trop. Dis. *8*, e2823. https://doi.org/10.1371/journal.pntd.0002823

Park, J.H., and Shin, C. (2015). Slicer-independent mechanism drives small-RNA strand separation during human RISC assembly. Nucleic Acids Res. *43*, 9418–9433. https://doi.org/10.1093/nar/gkv937

Paro, S., Imler, J.L., and Meignin, C. (2015). Sensing viral RNAs by Dicer/RIG-I like ATPases across species. Curr. Opin. Immunol. *32*, 106–113. https://doi.org/10.1016/j.coi.2015.01.009

Pélisson, A., Sarot, E., Payen-Groschêne, G., and Bucheton, A. (2007). A novel repeat-associated small interfering RNA-mediated silencing pathway downregulates complementary sense gypsy transcripts in somatic cells of the *Drosophila* ovary. J. Virol. *81*, 1951–1960. https://doi.org/10.1128/JVI.01980-06

Petit, M., Mongelli, V., Frangeul, L., Blanc, H., Jiggins, F., and Saleh, M.-C. (2016). piRNA pathway is not required for antiviral defense in *Drosophila melanogaster*. Proc. Natl. Acad. Sci. U.S.A. *113*, E4218–4227 https://doi.org/10.1073/pnas.1607952113

Petrillo, J.E., Venter, P.A., Short, J.R., Gopal, R., Deddouche, S., Lamiable, O., Imler, J.L., and Schneemann, A. (2013). Cytoplasmic granule formation and translational inhibition of nodaviral RNAs in the absence of the double-stranded RNA binding protein B2. J. Virol. *87*, 13409–13421. https://doi.org/10.1128/JVI.02362-13

Pichlmair, A., Lassnig, C., Eberle, C.A., Górna, M.W., Baumann, C.L., Burkard, T.R., Bürckstümmer, T., Stefanovic, A., Krieger, S., Bennett, K.L., *et al.* (2011). IFIT1 is an antiviral protein that recognizes 5'-triphosphate RNA. Nat. Immunol. *12*, 624–630. https://doi.org/10.1038/ni.2048

Poirier, E.Z., Goic, B., Tomé-Poderti, L., Frangeul, L., Boussier, J., Gausson, V., Blanc, H., Vallet, T., Loyd, H., Levi, L.I., *et al.* (2018). Dicer-2-Dependent Generation of Viral DNA from Defective Genomes of RNA Viruses Modulates Antiviral Immunity in Insects. Cell Host Microbe *23*, 353–365.e8. https://doi.org/10.1016/j.chom.2018.02.001

Popham, H.J., Ellersieck, M.R., Li, H., and Bonning, B.C. (2016). Evaluation of the Insecticidal Efficacy of Wild Type and Recombinant Baculoviruses. Methods Mol. Biol. *1350*, 407–444. https://doi.org/10.1007/978-1-4939-3043-2_21

Posadas, D.M., and Carthew, R.W. (2014). MicroRNAs and their roles in developmental canalization. Curr. Opin. Genet. Dev. *27*, 1–6. https://doi.org/10.1016/j.gde.2014.03.005

Powers, A.M., and Waterman, S.H. (2017). A decade of arboviral activity-Lessons learned from the trenches. PLOS Negl. Trop. Dis. *11*, e0005421. https://doi.org/10.1371/journal.pntd.0005421

Rand, T.A., Petersen, S., Du, F., and Wang, X. (2005). Argonaute2 cleaves the anti-guide strand of siRNA during RISC activation. Cell *123*, 621–629. https://doi.org/10.1016/j.cell.2005.10.020

Rao, X.J., Zhan, M.Y., Pan, Y.M., Liu, S., Yang, P.J., Yang, L.L., and Yu, X.Q. (2018). Immune functions of insect βGRPs and their potential application. Dev. Comp. Immunol. *83*, 80–88. https://doi.org/10.1016/j.dci.2017.12.007

Reich, D.P., Tyc, K.M., and Bass, B.L. (2018). *C. elegans* ADARs antagonize silencing of cellular dsRNAs by the antiviral RNAi pathway. Genes Dev. *32*, 271–282. https://doi.org/10.1101/gad.310672.117

Roers, A., Hiller, B., and Hornung, V. (2016). Recognition of endogenous nucleic acids by the innate immune System. Immunity *44*, 739–754. https://doi.org/10.1016/j.immuni.2016.04.002

Roignant, J.Y., Carré, C., Mugat, B., Szymczak, D., Lepesant, J.A., and Antoniewski, C. (2003). Absence of transitive and systemic pathways allows cell-specific and isoform-specific RNAi in *Drosophila*. RNA *9*, 299–308.

Royet, J., Gupta, D., and Dziarski, R. (2011). Peptidoglycan recognition proteins: modulators of the microbiome and inflammation. Nat. Rev. Immunol. *11*, 837–851. https://doi.org/10.1038/nri3089

Rutschmann, S., Jung, A.C., Hetru, C., Reichhart, J.M., Hoffmann, J.A., and Ferrandon, D. (2000a). The Rel protein DIF mediates the antifungal but not the antibacterial host defense in *Drosophila*. Immunity *12*, 569–580.

Rutschmann, S., Jung, A.C., Zhou, R., Silverman, N., Hoffmann, J.A., and Ferrandon, D. (2000b). Role of *Drosophila* IKK gamma in a toll-independent antibacterial immune response. Nat. Immunol. *1*, 342–347. https://doi.org/10.1038/79801

Sabin, L.R., Zheng, Q., Thekkat, P., Yang, J., Hannon, G.J., Gregory, B.D., Tudor, M., and Cherry, S. (2013). Dicer-2 processes diverse viral RNA species. PLOS ONE *8*, e55458. https://doi.org/10.1371/journal.pone.0055458

Saito, K., Ishizuka, A., Siomi, H., and Siomi, M.C. (2005). Processing of pre-microRNAs by the Dicer-1-Loquacious complex in *Drosophila* cells. PLOS Biol. *3*, e235. https://doi.org/10.1371/journal.pbio.0030235

Saito, K., Nishida, K.M., Mori, T., Kawamura, Y., Miyoshi, K., Nagami, T., Siomi, H., and Siomi, M.C. (2006). Specific association of Piwi with rasiRNAs derived from retrotransposon and heterochromatic regions in the *Drosophila* genome. Genes Dev. *20*, 2214–2222. https://doi.org/10.1101/gad.1454806

Saleh, M.C., van Rij, R.P., Hekele, A., Gillis, A., Foley, E., O'Farrell, P.H., and Andino, R. (2006). The endocytic pathway mediates cell entry of dsRNA to induce RNAi silencing. Nat. Cell Biol. *8*, 793–802. https://doi.org/10.1038/ncb1439

Saleh, M.C., Tassetto, M., van Rij, R.P., Goic, B., Gausson, V., Berry, B., Jacquier, C., Antoniewski, C., and Andino, R. (2009). Antiviral immunity in *Drosophila* requires systemic RNA interference spread. Nature *458*, 346–350. https://doi.org/10.1038/nature07712

Samuel, G.H., Wiley, M.R., Badawi, A., Adelman, Z.N., and Myles, K.M. (2016). Yellow fever virus capsid protein is a potent suppressor of RNA silencing that binds double-stranded RNA. Proc. Natl. Acad. Sci. U.S.A. *113*, 13863–13868. https://doi.org/10.1073/pnas.1600544113

Sánchez-Vargas, I., Scott, J.C., Poole-Smith, B.K., Franz, A.W., Barbosa-Solomieu, V., Wilusz, J., Olson, K.E., and Blair, C.D. (2009). Dengue virus type 2 infections of Aedes aegypti are modulated by the mosquito's RNA interference pathway. PLOS Pathog. *5*, e1000299. https://doi.org/10.1371/journal.ppat.1000299

Schirle, N.T., and MacRae, I.J. (2012). The crystal structure of human Argonaute2. Science *336*, 1037–1040. https://doi.org/10.1126/science.1221551

Schirle, N.T., Sheu-Gruttadauria, J., and MacRae, I.J. (2014). Structural basis for microRNA targeting. Science *346*, 608–613. https://doi.org/10.1126/science.1258040

Schwarz, D.S., Tomari, Y., and Zamore, P.D. (2004). The RNA-induced silencing complex is a Mg2+-dependent endonuclease. Curr. Biol. *14*, 787–791. https://doi.org/10.1016/j.cub.2004.03.008

Shelly, S., Lukinova, N., Bambina, S., Berman, A., and Cherry, S. (2009). Autophagy is an essential component of *Drosophila* immunity against vesicular stomatitis virus. Immunity *30*, 588–598. https://doi.org/10.1016/j.immuni.2009.02.009

Shi, M., Lin, X.D., Tian, J.H., Chen, L.J., Chen, X., Li, C.X., Qin, X.C., Li, J., Cao, J.P., Eden, J.S., *et al.* (2016). Redefining the invertebrate RNA virosphere. Nature [Epub ahead of print]. https://doi.org/10.1038/nature20167

Sienski, G., Dönertas, D., and Brennecke, J. (2012). Transcriptional silencing of transposons by Piwi and maelstrom and its impact on chromatin state and gene expression. Cell *151*, 964–980. https://doi.org/10.1016/j.cell.2012.10.040

Sienski, G., Batki, J., Senti, K.A., Dönertas, D., Tirian, L., Meixner, K., and Brennecke, J. (2015). Silencio/CG9754 connects the Piwi-piRNA complex to the cellular heterochromatin machinery. Genes Dev. *29*, 2258–2271. https://doi.org/10.1101/gad.271908.115

Singh, G., Popli, S., Hari, Y., Malhotra, P., Mukherjee, S., and Bhatnagar, R.K. (2009). Suppression of RNA silencing by Flock house virus B2 protein is mediated through its interaction with the PAZ domain of Dicer. FASEB J. *23*, 1845–1857. https://doi.org/10.1096/fj.08-125120

Sinha, N.K., Trettin, K.D., Aruscavage, P.J., and Bass, B.L. (2015). *Drosophila* dicer-2 cleavage is mediated by helicase- and dsRNA termini-dependent states that are modulated by Loquacious-PD. Mol. Cell *58*, 406–417. https://doi.org/10.1016/j.molcel.2015.03.012

Sinha, N.K., Iwasa, J., Shen, P.S., and Bass, B.L. (2018). Dicer uses distinct modules for recognizing dsRNA termini. Science *359*, 329–334. https://doi.org/10.1126/science.aaq0921

Song, J.J., Smith, S.K., Hannon, G.J., and Joshua-Tor, L. (2004). Crystal structure of Argonaute and its implications for RISC slicer activity. Science *305*, 1434–1437. https://doi.org/10.1126/science.1102514

Souza-Neto, J.A., Sim, S., and Dimopoulos, G. (2009). An evolutionary conserved function of the JAK-STAT pathway in anti-dengue defense. Proc. Natl. Acad. Sci. U.S.A. *106*, 17841–17846. https://doi.org/10.1073/pnas.0905006106

Stein, D., Goltz, J.S., Jurcsak, J., and Stevens, L. (1998). The Dorsal-related immunity factor (Dif) can define the dorsal-ventral axis of polarity in the *Drosophila* embryo. Development *125*, 2159–2169.

Steiner, H. (2004). Peptidoglycan recognition proteins: on and off switches for innate immunity. Immunol. Rev. *198*, 83–96.

Steiner, H., Hultmark, D., Engström, A., Bennich, H., and Boman, H.G. (1981). Sequence and specificity of two antibacterial proteins involved in insect immunity. Nature *292*, 246–248.

Tabeta, K., Georgel, P., Janssen, E., Du, X., Hoebe, K., Crozat, K., Mudd, S., Shamel, L., Sovath, S., Goode, J., *et al.* (2004). Toll-like receptors 9 and 3 as essential components of innate immune defense against mouse cytomegalovirus infection. Proc. Natl. Acad. Sci. U.S.A. *101*, 3516–3521. https://doi.org/10.1073/pnas.0400525101

Tang, G., Reinhart, B.J., Bartel, D.P., and Zamore, P.D. (2003). A biochemical framework for RNA silencing in plants. Genes Dev. *17*, 49–63. https://doi.org/10.1101/gad.1048103

Tanji, T., Yun, E.Y., and Ip, Y.T. (2010). Heterodimers of NF-kappaB transcription factors DIF and Relish regulate antimicrobial peptide genes in *Drosophila*. Proc. Natl. Acad. Sci. U.S.A. *107*, 14715–14720. https://doi.org/10.1073/pnas.1009473107

Tants, J.N., Fesser, S., Kern, T., Stehle, R., Geerlof, A., Wunderlich, C., Juen, M., Hartlmüller, C., Böttcher, R., Kunzelmann, S., *et al.* (2017). Molecular basis for asymmetry sensing of siRNAs by the *Drosophila* Loqs-PD/Dcr-2 complex in RNA interference. Nucleic Acids Res. *45*, 12536–12550. https://doi.org/10.1093/nar/gkx886

Tassetto, M., Kunitomi, M., and Andino, R. (2017). Circulating Immune Cells Mediate a Systemic RNAi-Based Adaptive Antiviral Response in *Drosophila*. Cell *169*, 314–325.e13. https://doi.org/10.1016/j.cell.2017.03.033

Theopold, U., Krautz, R., and Dushay, M.S. (2014). The *Drosophila* clotting system and its messages for mammals. Dev. Comp. Immunol. *42*, 42–46. https://doi.org/10.1016/j.dci.2013.03.014

Tomari, Y., Du, T., Haley, B., Schwarz, D.S., Bennett, R., Cook, H.A., Koppetsch, B.S., Theurkauf, W.E., and Zamore, P.D. (2004a). RISC assembly defects in the *Drosophila* RNAi mutant armitage. Cell *116*, 831–841.

Tomari, Y., Matranga, C., Haley, B., Martinez, N., and Zamore, P.D. (2004b). A protein sensor for siRNA asymmetry. Science *306*, 1377–1380. https://doi.org/10.1126/science.1102755

Treiber, T., Treiber, N., and Meister, G. (2019). Regulation of microRNA biogenesis and its crosstalk with other cellular pathways. Nat. Rev. Mol. Cell Biol. *20*, 5–20. https://doi.org/10.1038/s41580-018-0059-1

Trettin, K.D., Sinha, N.K., Eckert, D.M., Apple, S.E., and Bass, B.L. (2017). Loquacious-PD facilitates *Drosophila* Dicer-2 cleavage through interactions with the helicase domain and dsRNA. Proc. Natl. Acad. Sci. U.S.A. *114*, E7939–E7948. https://doi.org/10.1073/pnas.1707063114

Tsuboyama, K., Tadakuma, H., and Tomari, Y. (2018). Conformational activation of Argonaute by distinct yet coordinated actions of the Hsp70 and Hsp90 chaperone systems. Mol. Cell *70*, 722–729.e4. https://doi.org/10.1016/j.molcel.2018.04.010

Vagin, V.V., Sigova, A., Li, C., Seitz, H., Gvozdev, V., and Zamore, P.D. (2006). A distinct small RNA pathway silences selfish genetic elements in the germline. Science *313*, 320–324. https://doi.org/10.1126/science.1129333

Valli, A., Busnadiego, I., Maliogka, V., Ferrero, D., Castón, J.R., Rodríguez, J.F., and García, J.A. (2012). The VP3 factor from viruses of Birnaviridae family suppresses RNA silencing by binding both long and small RNA duplexes. PLOS ONE *7*, e45957. https://doi.org/10.1371/journal.pone.0045957

van Cleef, K.W., van Mierlo, J.T., Miesen, P., Overheul, G.J., Fros, J.J., Schuster, S., Marklewitz, M., Pijlman, G.P., Junglen, S., and van Rij, R.P. (2014). Mosquito and *Drosophila* entomobirnaviruses suppress dsRNA- and siRNA-induced RNAi. Nucleic Acids Res. *42*, 8732–8744. https://doi.org/10.1093/nar/gku528

van den Beek, M., da Silva, B., Pouch, J., Ali Chaouche, M.E.A., Carré, C., and Antoniewski, C. (2018). Dual-layer transposon repression in heads of *Drosophila melanogaster*. RNA *24*, 1749–1760. https://doi.org/10.1261/rna.067173.118

van Mierlo, J.T., Bronkhorst, A.W., Overheul, G.J., Sadanandan, S.A., Ekström, J.O., Heestermans, M., Hultmark, D., Antoniewski, C., and van Rij, R.P. (2012a). Convergent evolution of argonaute-2 slicer antagonism in two distinct insect RNA viruses. PLOS Pathog. *8*, e1002872. https://doi.org/10.1371/journal.ppat.1002872

van Mierlo, J.T., Bronkhorst, A.W., Overheul, G.J., Sadanandan, S.A., Ekström, J.O., Heestermans, M., Hultmark, D., Antoniewski, C., and van Rij, R.P. (2012b). Convergent evolution of argonaute-2 slicer antagonism in two distinct insect RNA viruses. PLOS Pathog. *8*, e1002872. https://doi.org/10.1371/journal.ppat.1002872

van Mierlo, J.T., Overheul, G.J., Obadia, B., van Cleef, K.W., Webster, C.L., Saleh, M.C., Obbard, D.J., and van Rij, R.P. (2014). Novel *Drosophila* viruses encode host-specific suppressors of RNAi. PLOS Pathog. *10*, e1004256. https://doi.org/10.1371/journal.ppat.1004256

van Rij, R.P., Saleh, M.C., Berry, B., Foo, C., Houk, A., Antoniewski, C., and Andino, R. (2006). The RNA silencing endonuclease Argonaute 2 mediates specific antiviral immunity in *Drosophila* melanogaster. Genes Dev. *20*, 2985–2995. https://doi.org/10.1101/gad.1482006

Varjak, M., Maringer, K., Watson, M., Sreenu, V.B., Fredericks, A.C., Pondeville, E., Donald, C.L., Sterk, J., Kean, J., Vazeille, M., *et al.* (2017a). *Aedes aegypti* Piwi4 is a noncanonical PIWI protein involved in antiviral responses. MSphere *2*. https://doi.org/10.1128/mSphere.00144-17

Varjak, M., Donald, C.L., Mottram, T.J., Sreenu, V.B., Merits, A., Maringer, K., Schnettler, E., and Kohl, A. (2017b). Characterization of the Zika virus induced small RNA response in Aedes aegypti cells. PLOS Negl. Trop. Dis. *11*, e0006010. https://doi.org/10.1371/journal.pntd.0006010

Varjak, M., Dietrich, I., Sreenu, V.B., Till, B.E., Merits, A., Kohl, A., and Schnettler, E. (2018). Spindle-E Acts Antivirally Against Alphaviruses in Mosquito Cells. Viruses *10*, E88. https://doi.org/10.3390/v10020088

Vidal, S., Khush, R.S., Leulier, F., Tzou, P., Nakamura, M., and Lemaitre, B. (2001). Mutations in the *Drosophila* dTAK1 gene reveal a conserved function for MAPKKKs in the control of rel/NF-kappaB-dependent innate immune responses. Genes Dev. *15*, 1900–1912. https://doi.org/10.1101/gad.203301

Virgen-Slane, R., Rozovics, J.M., Fitzgerald, K.D., Ngo, T., Chou, W., van der Heden van Noort, G.J., Filippov, D.V., Gershon, P.D., and Semler, B.L. (2012). An RNA virus hijacks an incognito function of a DNA repair enzyme. Proc. Natl. Acad. Sci. U.S.A. *109*, 14634–14639. https://doi.org/10.1073/pnas.1208096109

Vodovar, N., Bronkhorst, A.W., van Cleef, K.W., Miesen, P., Blanc, H., van Rij, R.P., and Saleh, M.C. (2012). Arbovirus-derived piRNAs exhibit a ping-pong signature in mosquito cells. PLOS ONE *7*, e30861. https://doi.org/10.1371/journal.pone.0030861

Waldron, F.M., Stone, G.N., and Obbard, D.J. (2018). Metagenomic sequencing suggests a diversity of RNA interference-like responses to viruses across multicellular eukaryotes. PLOS Genet. *14*, e1007533. https://doi.org/10.1371/journal.pgen.1007533

Wang, H., Smagghe, G., and Meeus, I. (2017). The role of a single gene encoding the Single von Willebrand factor C-domain protein (SVC) in bumblebee immunity extends beyond antiviral defense. Insect Biochem. Mol. Biol. *91*, 10–20. https://doi.org/10.1016/j.ibmb.2017.10.002

Wang, X.H., Aliyari, R., Li, W.X., Li, H.W., Kim, K., Carthew, R., Atkinson, P., and Ding, S.W. (2006). RNA interference directs innate immunity against viruses in adult *Drosophila*. Science *312*, 452–454. https://doi.org/10.1126/science.1125694

Weavers, H., Evans, I.R., Martin, P., and Wood, W. (2016). Corpse engulfment generates a molecular memory that primes the macrophage Inflammatory Response. Cell *165*, 1658–1671. https://doi.org/10.1016/j.cell.2016.04.049

Weber, F., Wagner, V., Rasmussen, S.B., Hartmann, R., and Paludan, S.R. (2006). Double-stranded RNA is produced by positive-strand RNA viruses and DNA viruses but not in detectable amounts by negative-strand RNA viruses. J. Virol. *80*, 5059–5064. https://doi.org/10.1128/JVI.80.10.5059-5064.2006

Welker, N.C., Maity, T.S., Ye, X., Aruscavage, P.J., Krauchuk, A.A., Liu, Q., and Bass, B.L. (2011). Dicer's helicase domain discriminates dsRNA termini to promote an altered reaction mode. Mol. Cell *41*, 589–599. https://doi.org/10.1016/j.molcel.2011.02.005

Wernet, M.F., Klovstad, M., and Clandinin, T.R. (2014). Generation of infectious virus particles from inducible transgenic genomes. Curr. Biol. *24*, R107–8. https://doi.org/10.1016/j.cub.2013.12.009

West, C., and Silverman, N. (2018). p38b and JAK-STAT signaling protect against Invertebrate iridescent virus 6 infection in *Drosophila*. PLOS Pathog. *14*, e1007020. https://doi.org/10.1371/journal.ppat.1007020

Wieschaus, E., and Nüsslein-Volhard, C. (2016). The Heidelberg SCREEN FOR PATTERN MUTANTS OF *Drosophila*: A personal account. Annu. Rev. Cell Dev. Biol. *32*, 1–46. https://doi.org/10.1146/annurev-cellbio-113015-023138

Wong, M.T., and Chen, S.S. (2016). Human Choline Kinase-α Promotes Hepatitis C Virus RNA Replication through Modulation of Membranous Viral Replication Complex Formation. J. Virol. *90*, 9075–9095. https://doi.org/10.1128/JVI.00960-16

Wong, M.T., and Chen, S.S. (2017). Hepatitis C Virus Subverts Human Choline Kinase-α To Bridge Phosphatidylinositol-4-Kinase IIIα (PI4KIIIα) and

NS5A and Upregulates PI4KIIIα Activation, Thereby Promoting the Translocation of the Ternary Complex to the Endoplasmic Reticulum for Viral Replication. J. Virol. *91*, e00355-17. https://doi.org/10.1128/JVI.00355-17

Wu, Q., Luo, Y., Lu, R., Lau, N., Lai, E.C., Li, W.X., and Ding, S.W. (2010). Virus discovery by deep sequencing and assembly of virus-derived small silencing RNAs. Proc. Natl. Acad. Sci. U.S.A. *107*, 1606–1611. https://doi.org/10.1073/pnas.0911353107

Wu, X., Wu, F.H., Wang, X., Wang, L., Siedow, J.N., Zhang, W., and Pei, Z.M. (2014). Molecular evolutionary and structural analysis of the cytosolic DNA sensor cGAS and STING. Nucleic Acids Res. *42*, 8243–8257. https://doi.org/10.1093/nar/gku569

Ye, X., Huang, N., Liu, Y., Paroo, Z., Huerta, C., Li, P., Chen, S., Liu, Q., and Zhang, H. (2011). Structure of C3PO and mechanism of human RISC activation. Nat. Struct. Mol. Biol. *18*, 650–657. https://doi.org/10.1038/nsmb.2032

Yoon, J.S., Mogilicherla, K., Gurusamy, D., Chen, X., Chereddy, S.C.R.R., and Palli, S.R. (2018). Double-stranded RNA binding protein, Staufen, is required for the initiation of RNAi in coleopteran insects. Proc. Natl. Acad. Sci. U.S.A. *115*, 8334–8339. https://doi.org/10.1073/pnas.1809381115

Yu, Y., Gu, J., Jin, Y., Luo, Y., Preall, J.B., Ma, J., Czech, B., and Hannon, G.J. (2015). Panoramix enforces piRNA-dependent cotranscriptional silencing. Science *350*, 339–342. https://doi.org/10.1126/science.aab0700

Yuan, Y.R., Pei, Y., Ma, J.B., Kuryavyi, V., Zhadina, M., Meister, G., Chen, H.Y., Dauter, Z., Tuschl, T., and Patel, D.J. (2005). Crystal structure of A. aeolicus argonaute, a site-specific DNA-guided endoribonuclease, provides insights into RISC-mediated mRNA cleavage. Mol. Cell *19*, 405–419. https://doi.org/10.1016/j.molcel.2005.07.011

Zamore, P.D., Tuschl, T., Sharp, P.A., and Bartel, D.P. (2000). RNAi: double-stranded RNA directs the ATP-dependent cleavage of mRNA at 21 to 23 nucleotide intervals. Cell *101*, 25–33. https://doi.org/10.1016/S0092-8674(00)80620-0

Zhang, S.Y., Jouanguy, E., Ugolini, S., Smahi, A., Elain, G., Romero, P., Segal, D., Sancho-Shimizu, V., Lorenzo, L., Puel, A., *et al.* (2007). TLR3 deficiency in patients with herpes simplex encephalitis. Science *317*, 1522–1527. https://doi.org/10.1126/science.1139522

Zhang, X., Guo, R., Kumar, D., Ma, H., Liu, J., Hu, X., Cao, G., Xue, R., and Gong, C. (2016). Identification, gene expression and immune function of the novel Bm-STAT gene in virus-infected Bombyx mori. Gene *577*, 82–88. https://doi.org/10.1016/j.gene.2015.11.027

Zhou, R., Czech, B., Brennecke, J., Sachidanandam, R., Wohlschlegel, J.A., Perrimon, N., and Hannon, G.J. (2009). Processing of *Drosophila* endo-siRNAs depends on a specific Loquacious isoform. RNA *15*, 1886–1895. https://doi.org/10.1261/rna.1611309

Zografidis, A., Van Nieuwerburgh, F., Kolliopoulou, A., Apostolou-Karampelis, K., Head, S.R., Deforce, D., Smagghe, G., and Swevers, L. (2015). Viral Small-RNA Analysis of Bombyx mori larval midgut during persistent and pathogenic cytoplasmic polyhedrosis virus infection. J. Virol. *89*, 11473–11486. https://doi.org/10.1128/JVI.01695-15

Zotti, M., Dos Santos, E.A., Cagliari, D., Christiaens, O., Taning, C.N.T., and Smagghe, G. (2018). RNA interference technology in crop protection against arthropod pests, pathogens and nematodes. Pest Manag. Sci. *74*, 1239–1250. https://doi.org/10.1002/ps.4813

Micro-RNA Modulation of Insect Virus Replication

4

Verna Monsanto-Hearne and Karyn N. Johnson*

School of Biological Sciences, University of Queensland, Brisbane, Australia.

*Correspondence: karynj@uq.edu.au

https://doi.org/10.21775/9781912530083.04

Abstract

The outcome of virus infection in insects is impacted by regulation of both host and virus gene expression. A class of small RNAs called micro-RNAs (miRNA) have emerged as important regulators of gene expression that can influence the outcome of virus infection. miRNA regulation occurs at a comparatively late stage of gene expression, allowing for rapid control and fine-tuning of gene expression levels. Here we discuss the biogenesis of miRNAs from both host and virus genomes, the interactions that lead to regulation of gene expression, and the miRNA–mRNA interactions that lead to either antivirus or provirus consequences in the course of virus infection in insects.

Introduction: insect–virus interaction and microRNAs

The course and outcome of virus infection is reliant on intricate and complex host–virus interactions (Rapicetta *et al.*, 2002; Whitton *et al.*, 2005; Clyde *et al.*, 2006). With small genomes and limited coding capacity, viruses hijack host factors for almost every step of virus infection while simultaneously evading the host antiviral system (Ploegh, 1998; Ahlquist *et al.*, 2003; Cherry and Silverman, 2006). The virus exploits the host's cell factors and the host counters the viral assault by restricting viral access to cell factors and/or mounting its antiviral immune response (Ahlquist *et al.*, 2003, 2005; Cherry and Silverman, 2006; Huszar and Imler, 2008; Kemp and Imler, 2009; Delpeut *et al.*, 2012; Vodovar and Saleh, 2012; Xu and Cherry, 2014; Mussabekova *et al.*, 2017). Many molecular components mediate and are mediated by this host–virus cross-talk, including microRNAs.

microRNAs (miRNAs) are a large class of highly conserved, ≈22 nt non-coding RNAs that regulate gene expression. Compared to the more upstream regulatory mechanisms such as transcriptional regulation and chromatin remodelling, regulation by miRNA occurs at a later stage of gene expression, allowing for rapid control and fine-tuning of gene expression levels (Chen *et al.*, 2013). Complementary binding of at least the seed region (second to eighth nucleotide from the 5′-end) of the ≈22 nt long miRNA with target mRNAs influences mRNA stability and/or translational efficiency and consequently modulates genes involved in a spectrum of important cellular processes (Bartel, 2009; Fabian *et al.*, 2010; Pasquinelli, 2012), including immune response and host–virus interactions (Lindsay, 2008; Lodish *et al.*, 2008; Tsitsiou and Lindsay, 2009; Xiao and Rajewsky, 2009; O'Connell *et al.*, 2010; Li and Rana, 2014).

Changes in miRNA abundance during the host–virus interaction can direct the course of virus pathogenesis (Monsanto-Hearne and Johnson, 2018). There are two distinct motifs in terms of miRNA effect on virus infection: while regulation of some miRNAs have antivirus consequences, regulation of others have provirus consequences. Antivirus consequences can arise by either one of the two mechanisms: binding of host miRNA-host mRNA or host miRNA-virus mRNA. On the other hand, provirus consequences can occur as a result of any of four binding events: host miRNA-host mRNA, host miRNA-virus mRNA, virus miRNA-virus mRNA, and finally virus miRNA-host mRNA. The growing body of literature on insect-virus infection in the context of miRNA regulation shows that there are many differentially regulated miRNAs that can have either antiviral or proviral outcomes. The net effect of cooperative and opposing impacts of the suite of miRNAs on the host–virus interaction then influences whether or not virus infection ensues. This chapter starts with a discussion on miRNA genomics, biogenesis, and regulation of target mRNAs, many of which have been established through the study of insect model *Drosophila melanogaster*. The chapter then details miRNA–mRNA interactions leading to either antivirus or provirus consequences that determine the course of virus infection in insects.

miRNA: genomics, biogenesis to regulation of target

miRNAs are ubiquitous in nature. Regarded as a mere curiosity upon the discovery of the first miRNA in *Caenorhabditis elegans* in 1993 (Bartel, 2018), miRNAs are now known to be encoded in the genome of a full spectrum of organisms, from unicellular flagellates (Zhao *et al.*, 2007) to multicellular plants and animals (Bartel and Chen, 2004; Bartel, 2009). At present, the more than 270 species recorded in the miRNA registry, miRBase, produce over 28,000 mature miRNAs. From these, around 5000 miRNAs have been documented for insects (Table 4.1) (Kozomara and Griffiths-Jones, 2014). The coverage of the insect miRNA groups is highly biased and incomplete (Ylla *et al.*, 2016), with more than a third of the insects and more than 40% of the currently known insect miRNAs in miRBase belonging to the genus *Drosophila* (Griffiths-Jones *et al.*, 2008; Kozomara and Griffiths-Jones, 2014). Viruses also encode miRNAs (Grundhoff and Sullivan, 2011; Kincaid and Sullivan, 2012), with over 500 mature miRNAs encoded in 34 virus genomes recorded (Griffiths-Jones *et al.*, 2008; Kozomara and Griffiths-Jones, 2014).

Table 4.1 Number of mature miRNAs produced by insect species in miRbase

Insect species	Number of mature miRNAs
Aedes aegypti	164
Anopheles gambiae	163
Apis mellifera	262
Acyrthosiphon pisum	97
Bactrocera dorsalis	78
Biston betularia	2
Bombyx mori	563
Culex quinquefasciatus	91
Dinoponera quadriceps	197
Drosophila ananassae	75
Drosophila erecta	120
Drosophila grimshawi	72
Drosophila melanogaster	469
Drosophila mojavensis	71
Drosophila persimilis	69
Drosophila pseudoobscura	271
Drosophila sechellia	120
Drosophila simulans	213
Drosophila virilis	328
Drosophila willistoni	72
Drosophila yakuba	103
Heliconius melpomene	97
Locusta migratoria	14
Manduca sexta	93
Nasonia giraulti	32
Nasonia longicornis	28
Nasonia vitripennis	53
Polistes canadensis	140
Plutella xylostella	127
Spodoptera frugiperda	221
Tribolium castaneum	590

Host miRNA genomics

miRNAs are a widespread component of the insect genome. miRNAs account for an estimated 1–2% of genes of the model insect *D. melanogaster* (Lai *et al.*, 2003; Landgraf *et al.*, 2007; Lim *et al.*, 2003a,b; Ruby *et al.*, 2007), while the number of miRNAs in *Tribolium* is at least 15% larger than that of *Drosophila* (Marco *et al.*, 2010). miRNA loci are found throughout the genome: intergenic miRNAs are embedded between clusters of genes, while intronic miRNAs are found within introns of protein-coding regions. The former is transcribed independently by their own transcription units, whereas the latter are processed either by their host transcription units or by their own transcription units. Some miRNAs are encoded individually, while others are found in clusters with other miRNAs (Lee *et al.*, 2002; Kim and Kim, 2007; Carthew and Sontheimer, 2009).

miRNAs are transcribed from the genome using similar mechanisms as mRNAs. miRNA promoters are typically the same as mRNA-encoding promoters in terms of distance from the miRNA-encoding sequence and presence of recognition elements. The miRNA transcription initiation site can be as close as a few hundreds of base pairs (bp) to the miRNA-encoding sequence or can be as far as 20 kb. miRNA recognition promoter elements can include TATA box, transcription factor II B recognition element (BRE), initiator (Inr) motif 10 element, and downstream promoter element (Ozsolak *et al.*, 2008).

miRNA biogenesis to recognition of targets

Canonical biogenesis of host miRNA

miRNA biogenesis canonically begins with the transcription of miRNA-encoding genomic regions (Rodriguez *et al.*, 2004) by RNA polymerase II (Lee *et al.*, 2004a) (Fig. 4.1). The Pol II-generation of the miRNA primary transcripts (pri-miRNAs) is coupled with pri-miRNA processing. Similar to most other Pol II-generated products, pri-miRNAs are methylguanosine-capped at the 5′-end and polyadenylated at the 3′-end (Bracht *et al.*, 2004; Cai *et al.*, 2004; Lee *et al.*, 2004a). The pri-miRNA folds back on itself to form a hairpin structure and then undergoes sequential nuclear and cytoplasmic processing to yield mature duplex miRNA.

The first step in the sequential processing of pri-miRNAs into mature miRNAs is promoted by the microprocessor comprising RNase III Drosha and its protein partner Pasha (Denli *et al.*, 2004; Gregory *et al.*, 2004; Han *et al.*, 2004; Landthaler *et al.*, 2004; Tomari and Zamore, 2005). Recognition of pri-miRNA by the microprocessor is facilitated by the presence of three RNA elements of the primary miRNA transcript: ≈80-nucleotide (nt) hairpin, unstructured ≥10-nt terminal loop, and single-single stranded RNA extensions beyond the pri-miRNA hairpin (Fig. 4.2) (Lee *et al.*, 2003; Zeng and Cullen, 2003; Zeng and Cullen, 2005; Zeng *et al.*, 2005). Upon recognition of pri-miRNA, Pasha determines the cleavage site at 11 base pairs away from the basal junction between the single-stranded RNA and double-stranded RNA of the pri-miRNA and ≈22 bp away from the apical junction of the terminal loop. Drosha cleaves the pri-miRNA at the Pasha-determined sites to release the 50–70 nt hairpin-structured pre-miRNA with two nucleotide 3′ overhang (Lee *et al.*, 2002; Gregory *et al.*, 2004; Han *et al.*, 2004). Following nuclear processing by the microprocessor, the pre-miRNA is then shuttled out of the nucleus by the nuclear transport factor Exportin-5 (Exp-5) in complex with Ran-GTP (Yi *et al.*, 2003; Bohnsack *et al.*, 2004; Lund *et al.*, 2004).

In the cytoplasm, the pre-miRNA is further processed by the RNase III enzyme Dicer-1 (Dcr1) and its protein partner Loquacious (in flies). Dcr1 is a highly conserved protein that typically bears PAZ and two RNase III domains. The PAZ domain of *D. melanogaster* Dcr1 (Lee *et al.*, 2004b) recognizes the 2-nt 3′ overhang at the base of the pre-miRNA stem, and each of the 2 RNase domain then cleaves one of the strands off the loop to liberate the ≈22-nt-long miRNA duplex with each strand bearing 2-nt overhangs at 3′-end (Lee *et al.*, 2002; MacRae *et al.*, 2007).

Non-canonical biogenesis of host miRNA

Some miRNAs, termed non-canonical miRNAs, bypass one or more of the catalytic steps in the canonical miRNA biogenesis (Xie and Steitz, 2014). Non-canonical miRNAs are produced from introns and various RNAs such as small nucleolar RNAs (snoRNAs), endogenous short hairpin RNAs (shRNAs), and tRNAs (Abdelfattah *et al.*, 2014). The most prominent non-canonical maturation pathway, the mirtron pathway, substitutes Drosha cleavage with splicing. Splicing, carried out by spliceosomes and debranching enzymes, produces miRNA hairpins directly suitable for Dicer cleavage. After splicing, the mirtron pathway then merges with the canonical miRNA pathway at the Exp-5-bound transport stage (Okamura *et al.*, 2007).

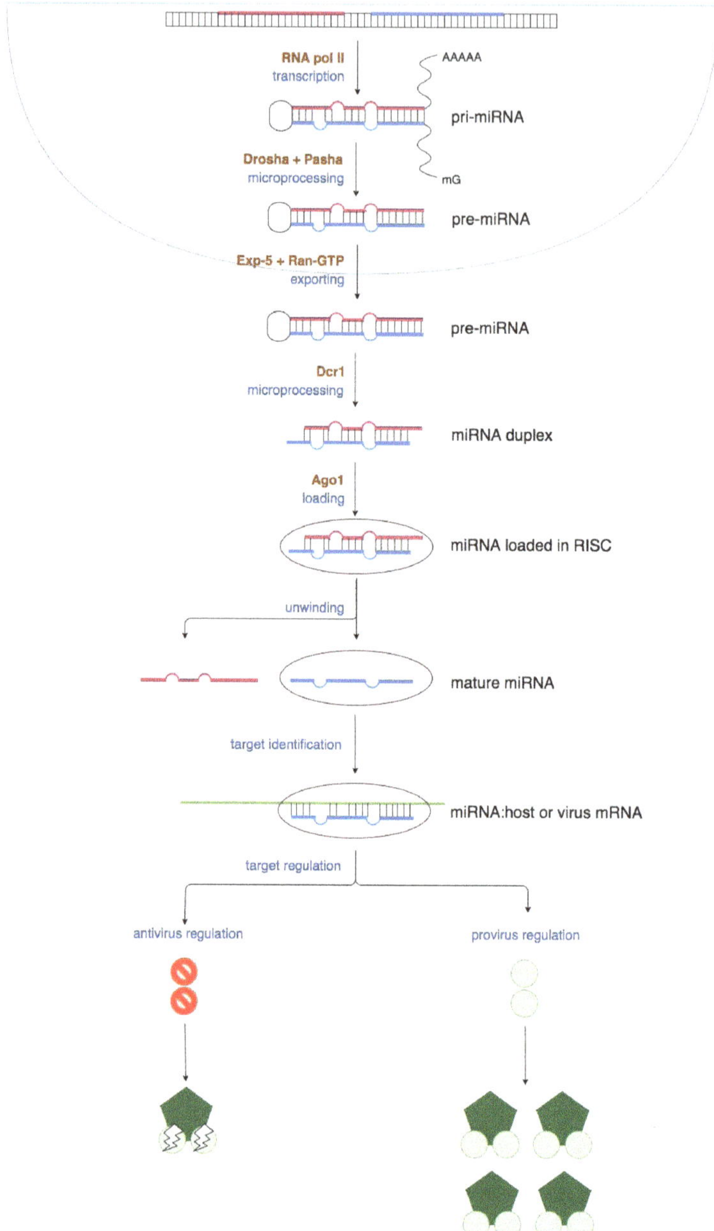

Figure 4.1 From canonical miRNA biogenesis to regulation of virus replication. The canonical biogenesis of miRNA begins with RNA polymerase II (RNA pol II)-mediated transcription of a miRNA gene. The resulting hairpin-containing primary transcript, pri-miRNA, spontaneously folds into a hairpin structure and is methylguanosine-capped at the 5′-end and may be polyadenylated at the 3′-tail. The pri-miRNA is then cleaved by Drosha, in association with Pasha, to generate the pre-miRNA. The pre-miRNA is shuttled out of the nucleus by the nuclear transport factor Exportin-5 (Exp-5), in association with Ran-GTP, for further processing by the cytoplasm-located Dicer-1 (Dcr1) and its partner Loquacious. Dcr1 cleaves the hairpin loop off the pre-miRNA to generate a ≈22 nt long miRNA duplex. The miRNA duplex is then loaded onto the effector molecule Argonaute (Ago) to form the RNA induced silencing complex (RISC). In the RISC, one strand of the miRNA duplex is unwound and discarded, while the other is retained to guide the effector complex to target mRNA. miRNA target recognition occurs through perfect or near perfect complementary binding between miRNA seed region (second to eighth nucleotides from the 5′ end of the miRNA) and the target mRNA. In the host–virus interaction, the target can be either host mRNA or virus mRNA. Upon binding, target mRNA stability and/or translational efficiency is modified leading to regulation of gene expression of virus or host factors which consequently impacts virus replication and/or pathogenesis. Where target regulation is provirus, virus replication is enhanced, conversely, where target regulation is antivirus, virus replication is restricted.

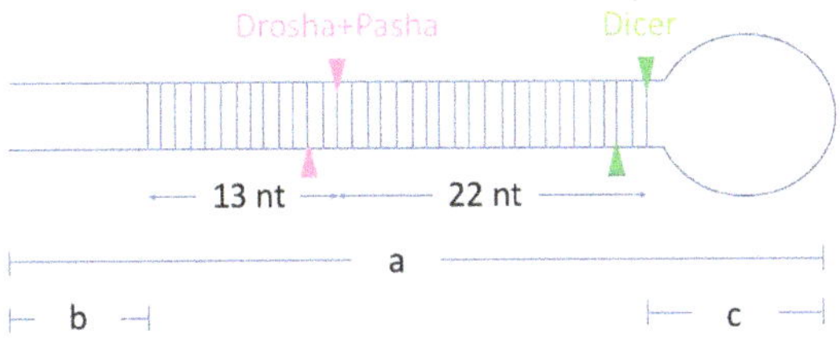

Figure 4.2 RNA elements in pri-miRNA required for Drosha recognition. (a) ≈80-nt long hairpin, (b) single-stranded extension beyond pri-miRNA hairpin, and (c) ≥10 nt terminal loop. (Modified from Bartel, 2018.)

The classical definition of miRNAs anchors heavily on the size of the RNA molecule and the biogenesis. Specifically, (1) the mature miRNA must have the distinct size of ≈ 22 nt, (2) the mature miRNA should occupy the stem part of a hairpin-shaped precursor that does not have large internal bulges, and (3) the mature miRNA should be processed by Dicer (Berezikov *et al.*, 2011). However, there exists a non-canonical miRNA pathway that bypasses Dicer cleavage. Mature miR-451 in humans, mouse, and zebrafish are produced by Drosha-cleavage of pri-miR-451 to generate a pre-miRNA that is too short (≈ 18 bp) to act as substrate for Dicer (Yang *et al.*, 2010). Whether there are other miRNAs produced through this unique miRNA biogenesis pathway is still unknown, and why seemingly only miR-451 is produced this way remains one of the larger unresolved miRNA questions (Bartel, 2018).

Biogenesis of virus miRNA

DNA viruses can encode miRNAs in their genome, the biogenesis of which parallels the biogenesis of host miRNAs. DNA virus miRNAs can be derived from non-coding RNAs and open reading frames of protein-coding mRNAs. In several cases, viral miRNAs are expressed from polycistronic transcripts, thereby ensuring their co-regulation. Similar to host miRNAs, most known DNA virus miRNAs are derived from Pol II transcripts. However, some miRNAs produced by viruses such as murine gammaherpesvirus 68 (MHV68), *Bovine leukaemia virus* (BLV), and *Simian foamy virus* (SFV), appear to be derived from transcripts produced by RNA polymerase III (Pol III). Most virus pri-miRNA transcripts then undergo Drosha and then Dicer cleavage. However, as with host miRNAs, some virus miRNAs are produced via the non-canonical pathway which bypasses the Drosha cleavage (Tycowski *et al.*, 2015).

Most known virus miRNAs are encoded by DNA viruses, although miRNAs/miRNA-like sequences can also be derived from RNA virus genome. While the biogenesis of RNA virus miRNAs has yet to be defined, the production of functional miRNAs from a pre-miRNA sequence engineered into the Influenza virus (nuclear RNA virus) genome suggests that RNA viruses are capable of producing miRNAs. Furthermore, the production of miRNAs capable of conferring RNAi-like activity from pre-miRNAs cloned into *Sindbis virus* (cytoplasmic RNA virus), called virtrons, suggests the existence of an uncharacterised Exp-5-independent, Dcr-dependent pathway capable of processing cytoplasmic hairpins (Shapiro *et al.*, 2010; Varble *et al.*, 2010; Langlois *et al.*, 2012; Asgari, 2015).

RISC assembly and miRNA strand selection

Whether miRNAs mature through the canonical or non-canonical pathway, RNAi-induced silencing complex (RISC) assembly follows the biogenesis of the miRNA duplex (Fig. 4.1). During RISC formation the miRNA duplex is loaded onto an Argonaute (Ago) protein. Ago proteins are highly conserved and specialized ≈ 100 kDa small-RNA-binding modules containing PAZ and PIWI domains. While some organisms such as the fission yeast, *Schizosaccharomyces pombe*, express only one type of Ago, others express multiple Ago family members. The model insect *D. melanogaster* expresses five Ago proteins (Hock and Meister, 2008). In *D. melanogaster*, the miRNA duplex, which typically contains central mismatches, associates with Ago1. Provided the miRNA duplex has no central mismatches, the miRNA is loaded onto the *D. melanogaster* Ago typically used for siRNAs, Ago2 (Okamura *et al.*, 2004; Förstemann *et al.*, 2007; Tomari *et al.*, 2007; Czech *et al.*, 2009; Ghildiyal *et al.*, 2010). Following loading onto the Ago protein, the strand with the more stable base pairs at the 5′-end, excess pyrimidines, and C at nucleotide position 1 is typically removed. The strand that normally gets removed has traditionally been termed miRNA star strand, miRNA*, or miR* or passenger strand (see Box 4.1 for naming

Box 4.1 What's in a mature miRNA name?

Except for the miRNAs bantam, let-7, and lin-4, a typical mature miRNA name follows the xxx-miR-yy-zp convention, where the first x represents the first letter of organism's genus name, and the last two xx represent the first 2 letters of the organism's species name. y indicates the sequential discovery of the specific miRNA. zp may be 5p or 3p, where 5p means that the mature miRNA was derived from the 5′ arm of the pre-miRNA, and 3p means that the mature miRNA came from the 3′ arm of the pre-miRNA (Fig. 4.1).

Mature miRNA names were previously formatted as xxx-miR-yy or xxx-miR-yy*, where the former represents the more abundant strand, and the latter the less abundant strand (miR*). There is generally a bias towards one strand of the miRNA duplex in the mature miRNA pool. This bias stems from the increased miRNA half-life and stability imparted by guide miRNA association with Ago (Winter and Diederichs, 2011), whereby Ago1 structure may shield the 5′ and 3′-ends of miRNA from ribonuclease action, thus protecting it from degradation (Wang *et al.*, 2008). The contrast in the abundance of the two strands of miRNAs led to the misconception that while one arm is used for RNA-induced silencing, the other arm is simply degraded. However, improvements in miRNA profiling technologies allowed for a fraction of the traditionally termed miR* to be easily detected (Okamura *et al.*, 2008). The ratio of each of the strands relative to the other is now known to vary depending on the tissue, developmental stage, and physiological state of the organism. The recognition that either or both strands of the miRNA duplex can serve as guide miRNA depending on the biological state (Okamura *et al.*, 2008; Yang *et al.*, 2011) prompted the shift in miRNA nomenclature from the traditional miR/miR* to miR-3p and miR-5p (Desvignes *et al.*, 2015).

conventions). The strand with the lesser interstrand hydrogen bonding resistance, excess purines, and U at nucleotide position 1 is normally retained. This strand becomes the guide miRNA strand, also known as the miRNA, or miR (Khvorova *et al.*, 2003; Czech *et al.*, 2009; Hu *et al.*, 2009; Okamura *et al.*, 2009; Ghildiyal *et al.*, 2010).

miRNA recognition of targets

With the miRNA strand specified and selected, the miRNA is now ready to guide the RISC complex to recognize mRNA transcripts. Ago interacts with the miRNA phosphate backbone to splay out the seed sequences in a helical conformation, exposing the bases for target binding (Ma *et al.*, 2004; Nakanishi *et al.*, 2012; Schirle and MacRae, 2012; Schirle *et al.*, 2014). miRNA targets are recognized via Watson–Crick base pairing between the miRNA and the mRNA. While there are many types of miRNA–mRNA pairs, the most common type involves miRNA nucleotides 2–8 from the 5′-end (termed miRNA seed) binding with mRNA (Lewis *et al.*, 2003; Brennecke *et al.*, 2005; Krek *et al.*, 2005; Lewis *et al.*, 2005; Bartel, 2009) (Fig. 4.3). The importance of miRNA seed region is associated

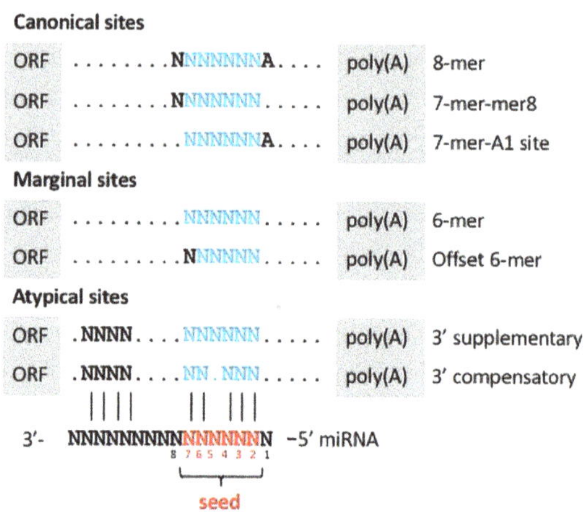

Figure 4.3 miRNA seed types. (Modified from Bartel, 2009.)

with the mechanistic effect of miR seed–target pairing on RISC association. Specifically, during Ago movement along the mRNA target, base pairing of miRNA nucleotides 2–5 contribute to primary RISC binding, while nucleotides 6–8 slow down RISC dissociation after encountering the target. This reduction in the rate of RISC dissociation strengthens the interaction between the miRNA, target, and RISC, resulting in increased efficacy of RISC (Brennecke *et al.*, 2005; Grimson *et al.*, 2007).

miRNA regulation of target

Following mRNA target recognition, miRNA facilitated target regulation occurs. Repression of gene expression is the most common mode of regulation. In the rare case of full complementarity between the miRNA and target, the mRNA is endonucleolitically cleaved by Ago, as is common in plants. In animals, however, the complementarity is most often imperfect, and target repression occurs through mRNA decay and/or mRNA translational repression. mRNA decay has been shown to account for 66–90% of the miRNA-mediated repression, thus contributing substantially to overall silencing (Jonas and Izaurralde, 2015). mRNA decay involves mRNA deadenylation, followed by decapping, and final (in the current model) degradation by cytoplasmic nuclease 5′-to-3′ exoribonuclease (Guo *et al.*, 2010; Huntzinger and Izaurralde, 2011; Djuranovic *et al.*, 2012; Wilczynska and Bushell, 2015; Jonas and Izaurralde, 2015). Mounting evidence suggests that translational repression in association with a 182 kDa phosphoprotein bearing glycine-tryptophan repeats (GW182) precedes mRNA decay (Baek *et al.*, 2008; Selbach *et al.*, 2008; Ingolia *et al.*, 2009; Huntzinger and Izaurralde, 2011; Bazzini *et al.*, 2012; Djuranovic *et al.*, 2012; Wilczynska and Bushell, 2015), however the mechanism(s) through which miRNAs repress translation of mRNAs is still unclear (Iwakawa and Tomari, 2015). Current models propose several concurrent and overlapping mechanisms that could account for translational repression, including: GW182-mediated recruitment of translational repressors and GW182-mediated displacement of poly(A)-binding proteins (PABP), which subsequently interferes with the mRNA circularization and therefore translation (Eulalio *et al.*, 2009; Lian *et al.*, 2009).

While most examples of miRNA-mediated regulation show target repression, miRNAs can also stimulate gene expression (Vasudevan *et al.*, 2007; Ørom *et al.*, 2008; Bruno *et al.*, 2011; Hussain *et al.*, 2011; Vasudevan, 2012). Positive regulation of genes by miRNAs can occur as a result of miRNA targeting RNA decay machineries (Bruno *et al.*, 2011) or as a result of miRNA associations that decrease levels of GW182 (Vasudevan, 2012).

miRNA roles in host–virus interaction

miRNA-mediated differential gene expression is a cornerstone event of biological processes. Each miRNA can potentially target hundreds of mRNAs (Enright *et al.*, 2004; Farh *et al.*, 2005; Krek *et al.*, 2005; Stark *et al.*, 2005; Sood *et al.*, 2006; Baek *et al.*, 2008; Betel *et al.*, 2010; Hashimoto *et al.*, 2013), and miRNA-mediated changes in gene expression have been implicated in a wide variety of functions. In insects, miRNAs have been implicated in growth, muscle and wing development, neurogenesis, apoptosis, sex determination, phenotypic plasticity, oogenesis and embryogenesis, and host–pathogen interactions and immunity (Asgari, 2013; Lucas *et al.*, 2015)

Virus infection changes the host miRNA profile such that the miRNA abundance in uninfected and infected cells, tissues and organisms differ (Scaria *et al.*, 2006; Dykxhoorn, 2007; Pedersen *et al.*, 2007; Gottwein and Cullen, 2008; Lindsay, 2008; Lodish *et al.*, 2008; Tsitsiou and Lindsay, 2009; Xiao and Rajewsky, 2009; O'Connell *et al.*, 2010; Cullen, 2011; Libri *et al.*, 2013; Asgari, 2015; Trobaugh and Klimstra, 2017; Monsanto-Hearne and Johnson, 2018). Indeed, differential abundance of miRNAs during viral infection is often used to identify which miRNAs may function in host–virus interactions (Asgari, 2015; Monsanto-Hearne and Johnson, 2018). The role of the differentially abundant miRNAs is verified based on the impact of miRNA regulation on the host and the virus (Fig. 4.4). The impact of miRNA abundance changes during virus infection is assessed by manipulation of miRNA levels via loss-of-function (LOF) or gain-of-function (GOF) methods (Min and Yoon, 2010; Martinez-Sanchez and Murphy, 2013) often employing miRNA inhibitors and miRNA mimics, respectively. miRNA inhibitors are chemically synthesized single-stranded RNA molecules designed to specifically bind to, inhibit, and artificially down-regulate endogenous miRNAs, while miRNA mimics are chemically synthesized double-stranded RNAs that artificially simulate up-regulation of miRNAs by augmenting endogenous miRNAs (Kuhn *et al.*,

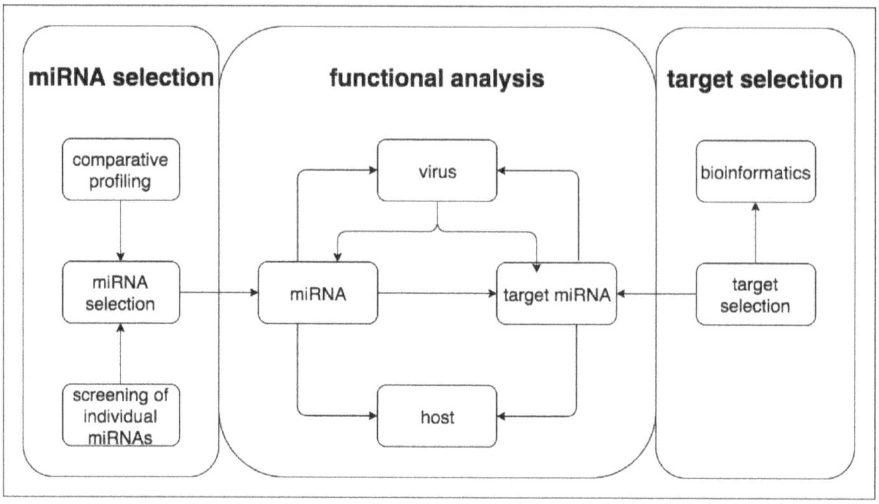

Figure 4.4 Pipeline and criteria for analysis of miRNA function in the host–virus interaction. Comparison of miRNA profiles of uninfected and virus-infected host identifies miRNAs which are differentially regulated during virus infection. Once a miRNA of interest has been identified (left panel), putative targets are identified using various bioinformatics approaches (right panel). Once miRNA–target interaction is confirmed, a closed-loop inter-relationship amongst the (1) miRNA, (2) target gene, (3) virus, and (4) host has to be confirmed (centre panel) so that the miRNA–target pair can be categorically identified as being relevant to the host–virus interaction. Where LOF/GOF mutants exist, identification of relevant miRNAs can start with genetic screens (left panel). (Modified from Monsanto-Hearne and Johnson, 2018.)

2008; Thomson *et al.*, 2011; Chugh and Dittmer, 2012; Ekimler and Sahin, 2014). Other common methods for LOF use miRNA sponges, tough decoys, target protectors, miRNA activity sensors, and target sensors (Chugh and Dittmer, 2012; Steinkraus *et al.*, 2016). Other methods for GOF use expression vectors containing mature miRNA, precursor miRNA, or pri-miRNA sequences (Min and Yoon, 2010; Thomson *et al.*, 2011). To test the effect of changes in miRNA abundance on host–virus interactions, miRNA inhibitors and mimics are introduced into cells by transfection, or to whole organisms *per os* or by injection. Involvement of a miRNA in host–virus interaction is evidenced by LOF and/or GOF resulting in changes in parameters such as cytopathic effects, host survival, and virus titres (Ekimler and Sahin, 2014) (see Box 4.2).

miRNAs exert their physiological effects through target mRNAs. Once a specific miRNA's change in abundance has been confirmed to affect host-virus replication, the next step in functional analysis typically involves identification of putative targets using bioinformatics. Several computational filtering prediction tools that generate a list of putative targets based on miRNA-target seed match and other experimentally defined features are now available. Different miRNA-target prediction algorithms generate different output lists with their associated false-positive and false-negative rates (Ritchie *et al.*, 2009; Thomson *et al.*, 2011). Identifying putative targets that are commonly identified by two or more algorithms is commonly employed to reduce the false-positive rates (Min and Yoon, 2010; Witkos *et al.*, 2011; Martinez-Sanchez and Murphy, 2013; Ekimler and Sahin, 2014; Riffo-Campos *et al.*, 2016). Putative targets are then experimentally validated using small-scale or genome-wide experimental approaches. In small-scale studies, RT-qPCR, *in situ* hybridization, northern blot, western blot, and protein arrays are used to measure changes in the target gene or protein after perturbation in miRNA levels. Because miRNAs potentially regulate hundreds of mRNA targets, large-scale transcriptional and proteomic profiling allows for a simultaneous and global assessment of targets following modulation of miRNA activity. Differences in target levels can then be quantitatively compared between control samples and LOF/GOF samples, and between control samples and virus-infected samples (Steinkraus *et al.*, 2016).

Box 4.2 Functional analysis *in vivo*

The use of cell cultures has substantially contributed to the functional analysis of miRNA roles in host–virus interaction. In particular, *ex vivo* approaches have been useful for documenting the physical interactions between miRNAs and their targets (Steinkraus *et al.*, 2016). However, disrupting miRNA activity *in vivo* has allowed for spatially and temporally synchronized data, providing more evidence on the definitive biological functions of miRNAs (Monsanto-Hearne and Johnson, 2018).

The development of miRNA LOF and GOF in whole organisms already has some success stories. Delivery of inhibitors and mimics *per os* (Jayachandran *et al.*, 2012) or by injection (Joo *et al.*, 2007) have both successfully resulted in change in the magnitude of miRNA effects on targets. There is also now an extensive collection of *D. melanogaster* miRNA stocks with miRNA LOF and GOF. For LOF, knockout mutant flies and lines with transgenes that 'sponge' or sequester miRNAs have been produced. For GOF, lines that use the Gal4-UAS system for miRNA expression have been developed (Bejarano *et al.*, 2012; Schertel *et al.*, 2012).

The final step of miRNA functional analysis is the evaluation of the functional impact of the target mRNA in the host–virus interaction. This requires manipulation of the target mRNA levels by GOF or LOF. The use of RNAi to knock down the target gene is the most commonly used technique in insect miRNA functional studies. Confirmation that the identified target mRNA functions in host–virus interaction occurs when the knockdown of the target mRNA results in changes in determinants of pathogenesis.

Host miRNA roles in insect–virus interactions

miRNA profiling and functional analysis of insect host–virus interactions have shown that miRNA abundance changes in the host can influence the outcome of virus pathogenesis (Asgari, 2015). The most thoroughly investigated insect hosts in the context of virus infection are mosquitoes of the *Aedes* and *Culex* genera. The involvement of miRNAs during virus infection has also been explored in the silkworm *Bombyx mori*, the moths *Helicoverpa armigera*, *Heliothis virescens*, *Spodoptera frugiperda*, and *Spodoptera litura*, a bumble bee *Bombus terrestis*, a plant hopper *Laodelphax striatellus*, the cabbage looper *Trichoplusia ni*, and in vinegar flies *D. melanogaster*. Different insect hosts, tissue sample sources, viruses, methods of infection, and sampling times post-infection have been used in miRNA profiling studies. Despite this, common patterns from the literature on the insect–virus interaction have emerged: (1) some miRNAs are commonly altered by virus infection, (2) the number of differentially regulated miRNAs changes over the course of virus infection, (3) the direction of regulation of a miRNA can change depending on the time post-infection, and (4) the general direction of change in miRNA abundance can vary depending on tissue sample source. Further functional analyses of the differentially abundant miRNAs show that insect host miRNAs can target either host mRNA or virus mRNA and can either have antivirus or provirus consequences (Fig. 4.5) (Monsanto-Hearne and Johnson, 2018). Details on the host miRNAs that have been functionally characterized in insect–virus interactions are described below.

Host miRNA-host mRNA interaction with antivirus consequence

Host miRNAs are first and foremost for the host to use. Various insect miRNAs inhibit viral replication and protect the host from viral attack. Here we summarize the literature on the involvement of insect host miRNA in antiviral mechanisms.

Infection of the Asian tiger mosquito *Aedes albopictus* with the medically important positive-sense ssRNA flavivirus *West Nile virus* (WNV) is regulated by mosquito host miRNA aae-miR-2940-5p. Mosquito aae-miR-2940-5p is selectively down-regulated during virus infection of the *Ae. albopictus* C6/36 cell line. Artificial miRNA down-regulation using synthetic aae-miR-2940-5p inhibitors represses WNV replication (Fig. 4.6). Conversely, artificial miRNA up-regulation using synthetic aae-miR-2940-5p mimic facilitates increase in

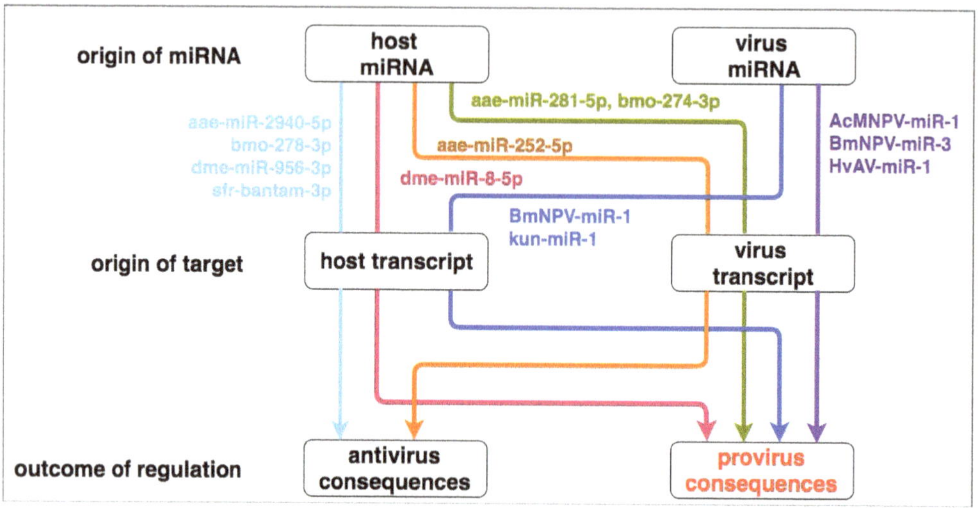

Figure 4.5 Both host and virus-derived miRNAs can target either host or virus mRNA transcripts. Host miRNA can target host or virus miRNAs and can either have pro-host or pro-virus consequences whereas virus-derived miRNAs can target either host or virus miRNAs but lead only to pro-virus consequences. miRNA with described functions are mapped to the pathway of impact.

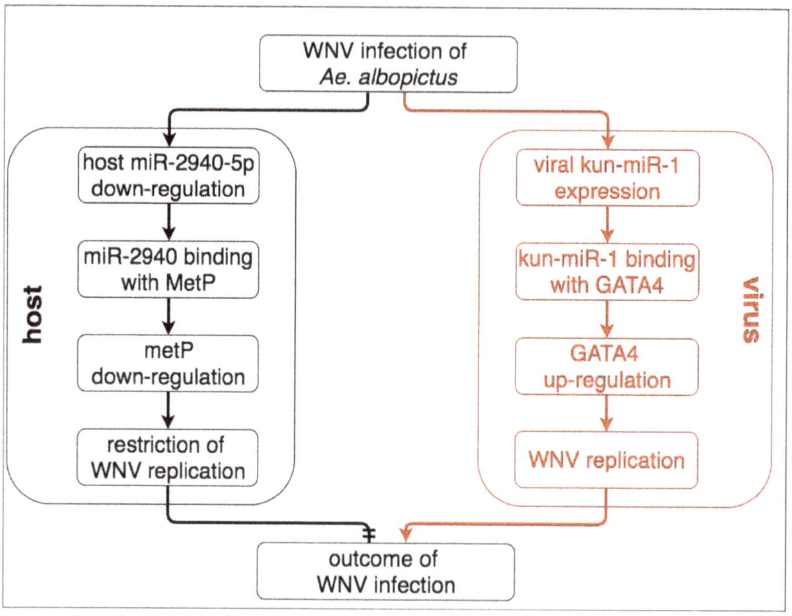

Figure 4.6 The contrasting effects of host miRNA and virus miRNA regulation of host mRNA in *Ae. albopictus*–WNV interaction. Black arrows antivirus pathway, in contrast, red arrows denote provirus pathway.

WNV titres, suggesting that the infection-induced miRNA down-regulation negatively impacts virus infection. Infection-induced decrease in aae-miR-2940-5p levels lowers the expression of *metalloprotease m41 filamentous temperature sensitive H* (*MetP m41 FtsH*). Experimental knockdown of *MetP* also negatively impacts virus infection. The cascade of events beginning with infection-induced down-regulation of miR-2940-5p ends with restriction of WNV infection. Although the molecular mechanism that leads to *MetP*'s negative impact on WNV replication is not yet understood, this cascade of events shows that miR-2940-5p regulation during virus infection is a miRNA-dependent antiviral response (Slonchak *et al.*, 2014).

Bombyx mori cytoplasmic polyhedrosis virus (BmCPV) infection of the mulberry silkworm *B. mori* is regulated by host miRNA that binds to host miRNA. Oral infection of *B. mori* with the segmented dsRNA reovirus BmCPV by feeding newly exuviated fifth instar silkworm larvae down-regulates host bmo-miR-278-3p in the midgut (Fig. 4.7). The infection-induced decrease in miR-278-3p then leads to an increase in transcription of the *insulin-related peptide binding protein 2* (*IBP2*) gene. Conversely, experimental increase in miR-278-3p by injection of miR-278-3p mimic into silkworm larvae decreases *IBP2* transcripts. An increase in miR-278-3p in turn increases the mRNA transcript of virus polyhedrin, which is a proxy for measuring BmCPV replication. Although a direct relationship between IBP2 levels and BmCPV has not yet been established, it is hypothesized that IBP2 plays a role in the insulin-signal pathway, restricting BmCPV replication as part of a silkworm immune response (Wu *et al.*, 2016).

The infection of *S. litura* larvae and Sf9 cells with the baculovirus *Autographa californica multiple nucleopolyhedrovirus* (AcMNPV) increases the abundance of the miRNA bantam. The up-regulation of bantam using an miRNA mimic is associated with a dose-dependent decrease in viral accumulation in cells. In contrast, artificial down-regulation of bantam using a synthetic inhibitor in cells increases viral accumulation in a dose dependent manner. In addition, *per os* delivery of the miRNA bantam antagomir into *S. litura* from the second instar larvae to pre-pupal stages results in accelerated mortality. A number of genes were concomitantly regulated with bantam abundance change during infection, so may also be involved. The detailed nature of the interaction between bantam and the mRNAs requires further experimentation. Additionally, the relationship between the mRNAs and the virus has yet to be tested (Shi *et al.*, 2016).

Injection of the model organism *D. melanogaster* with its natural pathogen *Drosophila C virus* (DCV) down regulates dme-miR-956-3p (Monsanto-Hearne *et al.*, 2017b). The decrease in miR-956-3p

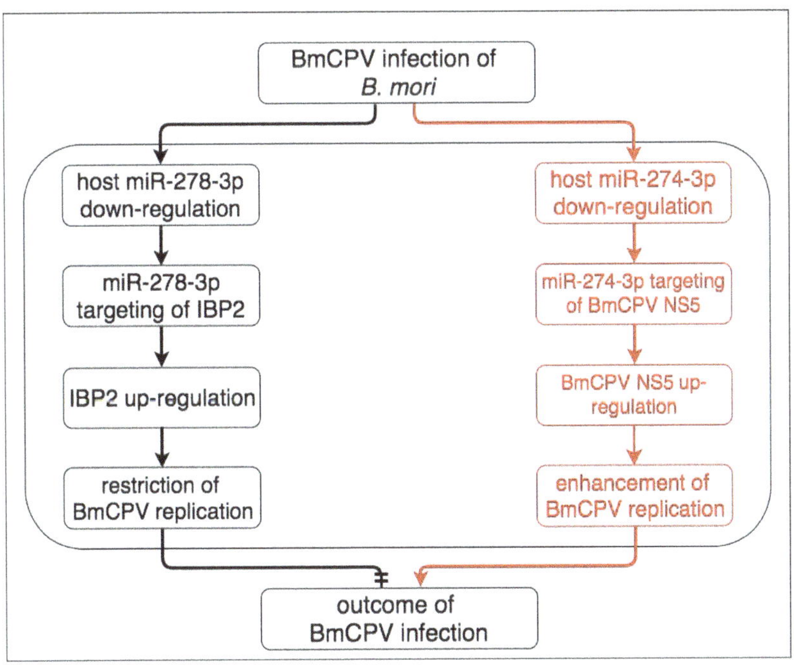

Figure 4.7 The contrasting effects of host miRNA regulation of host mRNA and host miRNA regulation of virus miRNA in *B. mori*–BmCPV interaction. Black arrows antivirus pathway, in contrast, red arrows denote provirus pathway.

abundance during DCV infection up-regulates the expression of the gene *Ectoderm-expressed 4* (*Ect4*), as shown by the higher levels of *Ect4* in mutant flies with a loss-of-function deletion of miR-956 when compared to *Ect4* in wild-type flies with normal levels of miR-956. This DCV-induced miR-956-3p down-regulation, which consequently up-regulates *Ect4*, correlates with slower viral accumulation and slower fly mortality, thus demonstrating that miRNA-mediated host protection is exerted through changes in host gene expression (Fig. 4.8).

Host miRNAs–virus mRNA interaction with antivirus consequences

Host miRNA can also protect hosts against virus infection through host miRNA interaction with virus mRNA. Infection of *Ae. albopictus* C6/36 cells with the flavivirus *Dengue virus* serotype 2 (DENV-2) induces miR-252-5p. Artificial up-regulation of miR-252-5p in cultured cells by means of transfection of synthetic oligonucleotides mimicking miR-252-5p decreases DENV-2 E protein RNA and protein, and therefore DENV-2 replication. Conversely, artificial down-regulation of miR-252 using inhibitors increases DENV-2 E protein RNA and protein, demonstrating an inverse relationship between miR-252 and DENV-2 E protein. This suggests that miR-252-3p up-regulation acts as a repressor of DENV-2 E protein, and consequently of DENV-2 replication (Fig. 4.9) (Yan *et al.*, 2014). It is noteworthy that, although acutely infected C6/36 cells have up-regulated levels of miR-252-5p, persistently infected C6/36 cells have down-regulated miR-252-5p (Avila-Bonilla *et al.*, 2017). It will be interesting to explore the implications of miR-252-5p down-regulation of DENV-2 persistent infection of cells.

Host miRNAs–host mRNA interaction with provirus consequences

Functional analyses of miRNAs during virus infection show that differential abundance of host miRNAs could result from viral manipulation of the host machinery required for virus proliferation. Provirus effects of miRNAs can occur through interaction between host miRNA and cellular mRNA. During DCV infection of *D. melanogaster* adult flies, miR-8-5p is down-regulated (Monsanto-Hearne *et al.*, 2017a). The decrease in miR-8-5p abundance during DCV infection up-regulates the expression of *Jun-related antigen* (*Jra*), also known as *dJun* due to its homology with human *Jun*. The

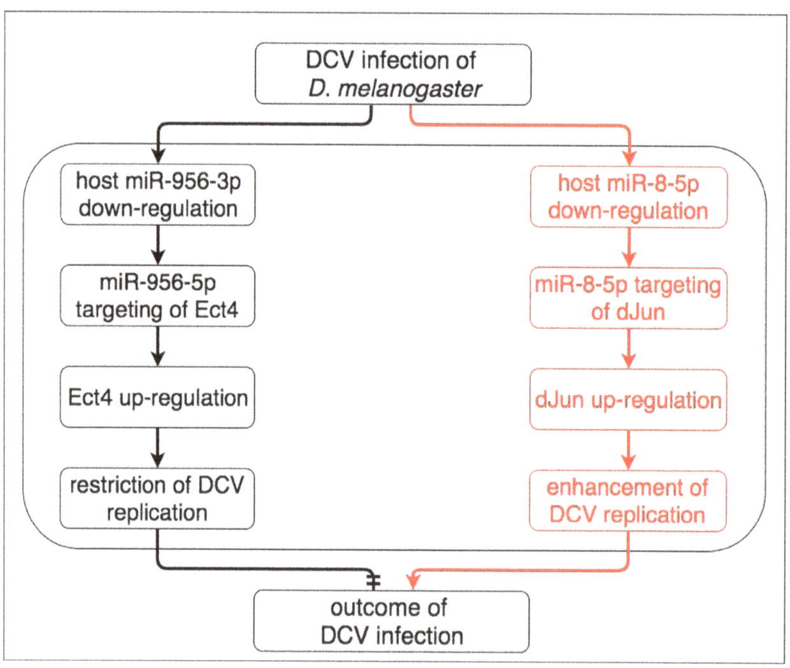

Figure 4.8 The contrasting effects of host miRNA regulation of host mRNA in *D. melanogaster*–DCV interaction. Black arrows antivirus pathway, in contrast, red arrows denote provirus pathway.

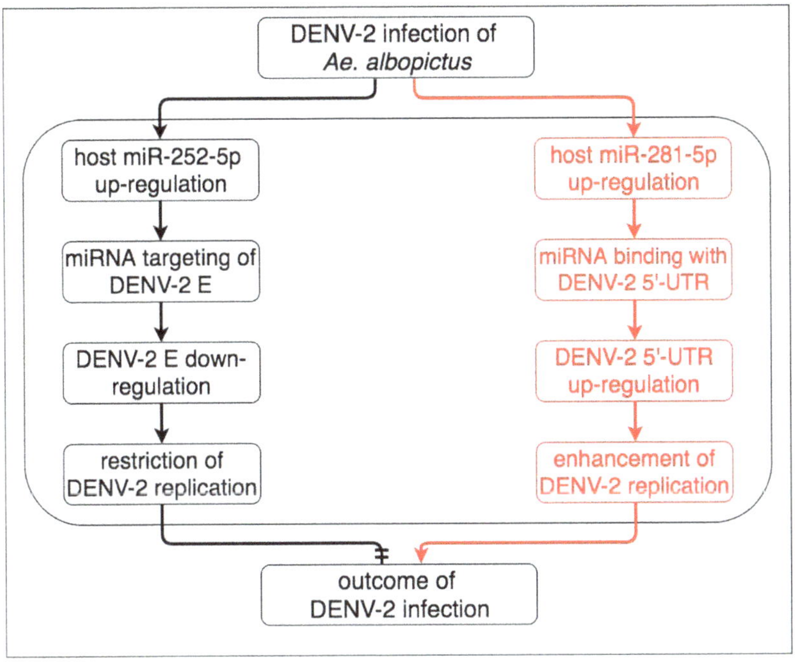

Figure 4.9 The contrasting effects of host miRNA regulation of virus mRNA in *Ae. albopictus*–DENV interaction. Black arrows antivirus pathway, in contrast, red arrows denote provirus pathway.

increase in *dJun* then results in faster viral accumulation. *dJun* expression is widely known to be regulated by protein phosphorylation, which then positively auto regulates its transcription to create more *dJun* (Smeal *et al.*, 1994). *dJun* up-regulation resulting from miR-8-5p down-regulation shows that the miRNA pathway contributes to *dJun* induction during viral infection (Fig. 4.8). Thus, there is complementarity of different regulatory networks determining the gene expression of host factors during viral infection. This demonstrates the ability of viruses to subvert multiple host regulatory pathways to enable viral replication.

Host miRNA–virus mRNA interaction with provirus consequences

Provirus effects of miRNAs can also occur through interaction between host miRNA and virus mRNA. Infection of silkworm with BmCPV is not only regulated by silkworm miRNA targeting host mRNA (see above), but also by host miRNA targeting virus mRNA. Oral infection of silkworm with BmCPV down-regulates miR-274-3p, which up-regulates the non-structural protein RNA of the BmCPV genome segment NS5, as shown by a direct negative interaction between the host miRNA and the virus NS5 segment in cells (Wu *et al.*, 2017). Although NS5 segment RNA was down-regulated by miR-274-3p mimic, the transcript was unaffected by negative mimic control. Additionally, microinjection of miR-274-3p inhibitors into silkworm larvae enhanced NS5 mRNA and protein expression. Importantly, artificial down-regulation of miR-274-3p also increased virus polyhedrin expression, which was used as proxy for virus replication. This suggests that infection-induced miR-274-3p inhibition enhances BmCPV replication. This contrasts with bmo-miR-278-3p, which acts by increasing the abundance of the *IBP2* gene, which is hypothesized to then inhibit BmCPV replication (Fig. 4.7).

The provirus effect of host miRNA regulation of virus mRNA also occurs in the *Ae. albopictus*–DENV-2 interaction. During oral DENV-2 infection of *Ae. albopictus*, miR-281-5p is up-regulated in the mosquito midgut, with miR-281-5p positively interacting with DENV-2 genomic 5′-untranslated region (UTR) to facilitate DENV-2 replication (Fig. 4.9). The artificial up-regulation of miR-281-5p prior to DENV-2 infection increases DENV-2 genomic RNA (gRNA), DENV-2 E protein, and therefore DENV-2 replication in C6/36 cells (Zhou *et al.*, 2014). In contrast, artificial down-regulation of miR-281-5p prior to infection reduces viral gDNA levels, DENV-2E protein, and DENV-2 titres in cells. Knock-down of miR-281-5p in mosquitoes *in vivo* through antagomir-281 injection also reduces viral gDNA, demonstrating that the miR-281-5p impact on DENV-2 occurs in both cells and whole organisms.

Host miRNA–virus mRNA interaction with unknown consequences

During early infection of a *Heliothis virescens* fat body cell line with *Heliothis virescens ascovirus 3a* (HvAV-3a), *H. virescens* miRNA Hz-miR-24 is down-regulated in cells; however, the same miRNA is up-regulated in late infection (Hussain and Asgari, 2010). During late infection, the transcript levels of DNA-dependent RNA polymerase (DdRP, *ORF64*) and DdRPβ (*ORF82*) are decreased, corresponding to higher levels of miR-24-5p. Artificial increase of miR-24-5p by cloning the pre-miRNA sequence of miR-24-5p and ectopically expressing the miRNA decreased the transcript levels of DdRP and DdRPβ compared to the levels in control transfection. Conversely, artificial decrease of miR-24-5p by synthetic miR-24-5p inhibitor increased DdRP expression. Neither artificial miR-24-5p regulation nor silencing of DdRP and DdRPβ impacted virus replication, suggesting that miR-24-5p and its two identified targets are not essential for *in vitro* HvAV-3 replication. It has been hypothesized that the target genes may be important for the expression of late genes and production of virions *in vivo*, but this requires testing (Hussain and Asgari, 2010).

Virus miRNA roles in host–virus interaction

In contrast to host miRNAs, which can either be antivirus or provirus, all virus miRNAs thus far discovered, regardless of whether the target is virus mRNA or host mRNA, are provirus elements. The provirus effects of virus miRNAs occur either through direct regulation of viral products or through directly regulating host transcripts (Aguado and tenOever, 2018).

Virus miRNA–virus mRNA interaction

Virus miRNA can bind to the virus mRNAs to exert provirus functions. This can result in enhanced virus replication. Alternatively, virus miRNA–virus mRNA interaction can regulate virus replication by facilitating transitions from one viral replication stage to another or by preventing early host recognition and immune response. Virus miRNA–virus mRNA binding could also potentially prevent the rapid death of the host, providing the virus with a viable host/virus reservoir for a longer period of time.

HvAV-3 produces HvAV-miR-1 from the 3′ stem-loop (3′SL) in the ORF coding for capsid protein at 96 hours post-infection. At the same time point, the viral DNA pol I (*orf1*) transcript is sharply down-regulated as a result of experimentally validated, direct binding of HvAV-miR-1 and the *orf1* transcript. Ectopic expression of HvAV-miR-1 in cells decreases viral DNA levels, suggesting that HvAV-miR-1 mediates inhibition of viral replication by specifically binding with *orf1* transcript. It is hypothesized that virus HvAV-miR-1 down-regulation of its own gene and its own replication delays host death, therefore conferring an advantage to long-term virus propagation (Hussain *et al.*, 2008).

The baculovirus AcMNPV also produces a miRNA, AcMNPV-miR-1. First identified bioinformatically, AcMNPV-miR-1 production was later experimentally validated. The experimentally validated miRNA perfectly matches and directly negatively targets a segment in the viral gene ODV-E25, reducing the level of infectious budded virions (Zhu *et al.*, 2013).

Finally, *Bombyx mori nucleopolyhedrovirus* (BmNPV) produces BmNPV-miR-3 during the early stage of infection of silkworm larvae. BmNPV-miR-3 negatively affects P6.9 expression and restricts viral replication such that P6.9 levels and viral replication decrease upon synthetic up-regulation of BmNPV-3. In contrast, P6.9 levels and virus replication increase upon synthetic down-regulation of the miRNA (Fig. 4.10). The control of P6.9 (and other late genes) by virus miRNA helps the virus escape early detection by the host, and therefore early immune response attack (Singh *et al.*, 2014).

Virus miRNA–host mRNA interaction

While some virus miRNAs bind to viral mRNA to exert provirus functions, other virus miRNAs achieve their provirus functions by regulating host

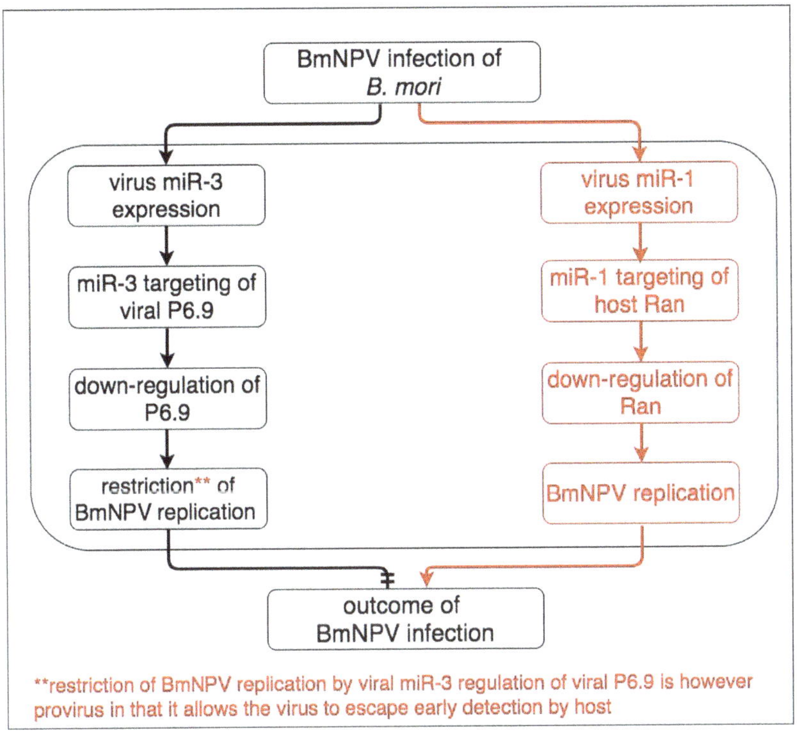

Figure 4.10 The contrasting effects of virus miRNA regulation of virus mRNA and virus miRNA regulation of host mRNA in *B. mori*–BmNPV interaction. Black arrows antivirus pathway, in contrast, red arrows denote provirus pathway.

gene expression. WNV encodes viral miRNA like kun-miR-1 in WNV 3′SL. Using bioinformatics and cloning approaches, host *GATA4* was identified and confirmed as a kun-miR-1 target. Kun-miR-1 and *GATA4* interaction specifically up-regulates *GATA4*. Both the artificial decrease in kun-miR-1 levels by use of inhibitor and knockdown of *GATA4* by RNAi decreased WNV short flavivirus RNA (sfRNA), demonstrating that kun-miR-1 expression up-regulates *GATA4*, which then enhances WNV infection (Fig. 4.6) (Hussain *et al.*, 2012).

Virus miRNA–host miRNA biogenesis pathway interaction

Virus infection can lead to changes in the global miRNA profile through regulation of miRNA biogenesis. The infection of silkworms with BmNPV involves targeting of a host gene by baculovirus miRNA BmNPV-miR-1 (Fig. 4.10). BmNPV-miR-1 specifically targets the host gene *Ran*, which is involved in the export of pre-miRNAs from the nucleus to the cytoplasm during miRNA biogenesis (Singh *et al.*, 2010). The resulting impairment of miRNA biogenesis following BmNPV-miR-1:*Ran* interaction thus increases virus proliferation. This mechanism employed by BmNPV is particularly interesting because it changes the miRNA biogenesis itself. miRNA level changes through alteration of components of the miRNA biogenesis pathway is also seen during infection of *Bombus terrestris* with the avirulent *Slow bee paralysis virus* (SBPV). In this system, SBPV infection increases expression of Dcr1 and Ago1, demonstrating the importance of miRNAs in host–virus interaction (Niu *et al.*, 2017). Furthermore, it raises the possibility that down-regulation of miRNAs during DCV infection of *D. melanogaster* could be due to the impact of DCV-infection on the miRNA biogenesis pathway (Monsanto-Hearne *et al.*, 2017b).

Future directions

Current data on functional analysis of miRNAs during insect–virus interaction show that miRNAs are important determinants of host–virus interactions. Detailed knowledge of miRNAs in this context may lead to important future applications. Examples include: interventions to address virus-infection of economically important insects such as silkworm and bees; as the basis for design of interventions for vector-borne diseases in humans, livestock, and crops; and finally, particularly for the insect model *D. melanogaster*, as the potential basis for understanding host–virus interaction in humans.

References

Abdelfattah, A.M., Park, C., and Choi, M.Y. (2014). Update on non-canonical microRNAs. Biomol. Concepts *5*, 275–287. https://doi.org/10.1515/bmc-2014-0012

Aguado, L.C., and tenOever, B. (2018). RNA virus building blocks-miRNAs not included. PLOS Pathog. *14*, e1006963. https://doi.org/10.1371/journal.ppat.1006963

Ahlquist, P., Noueiry, A.O., Lee, W.M., Kushner, D.B., and Dye, B.T. (2003). Host factors in positive-strand RNA virus genome replication. J. Virol. *77*, 8181–8186. https://doi.org/10.1128/JVI.77.15.8181

Ahlquist, P., Schwartz, M., Chen, J., Kushner, D., Hao, L., and Dye, B.T. (2005). Viral and host determinants of RNA virus vector replication and expression. Vaccine *23*, 1784–1787. https://doi.org/10.1016/j.vaccine.2004.11.005

Asgari, S. (2013). MicroRNA functions in insects. Insect Biochem. Mol. Biol. *43*, 388–397. https://doi.org/10.1016/j.ibmb.2012.10.005

Asgari, S. (2015). Regulatory role of cellular and viral microRNAs in insect-virus interactions. Curr. Opin. Insect Sci. *8*, 104–110. https://doi.org/10.1016/j.cois.2014.12.008

Avila-Bonilla, R.G., Yocupicio-Monroy, M., Marchat, L.A., De Nova-Ocampo, M.A., Del Ángel, R.M., and Salas-Benito, J.S. (2017). Analysis of the miRNA profile in C6/36 cells persistently infected with dengue virus type 2. Virus Res. *232*, 139–151. https://doi.org/10.1016/j.virusres.2017.03.005

Baek, D., Villén, J., Shin, C., Camargo, F.D., Gygi, S.P., and Bartel, D.P. (2008). The impact of microRNAs on protein output. Nature *455*, 64–71. https://doi.org/10.1038/nature07242

Bartel, D.P. (2009). MicroRNAs: target recognition and regulatory functions. Cell *136*, 215–233. https://doi.org/10.1016/j.cell.2009.01.002

Bartel, D.P. (2018). Metazoan MicroRNAs. Cell *173*, 20–51. https://doi.org/10.1016/j.cell.2018.03.006

Bartel, D.P., and Chen, C.Z. (2004). Micromanagers of gene expression: the potentially widespread influence of metazoan microRNAs. Nat. Rev. Genet. *5*, 396–400. https://doi.org/10.1038/nrg1328

Bazzini, A.A., Lee, M.T., and Giraldez, A.J. (2012). Ribosome profiling shows that miR-430 reduces translation before causing mRNA decay in zebrafish. Science *336*, 233–237. https://doi.org/10.1126/science.1215704

Bejarano, F., Bortolamiol-Becet, D., Dai, Q., Sun, K., Saj, A., Chou, Y.T., Raleigh, D.R., Kim, K., Ni, J.Q., Duan, H., *et al.* (2012). A genome-wide transgenic resource for conditional expression of *Drosophila* microRNAs. Development *139*, 2821–2831. https://doi.org/10.1242/dev.079939

Berezikov, E., Robine, N., Samsonova, A., Westholm, J.O., Naqvi, A., Hung, J.H., Okamura, K., Dai, Q., Bortolamiol-Becet, D., Martin, R., *et al.* (2011). Deep annotation of *Drosophila melanogaster* microRNAs yields insights into their processing, modification, and emergence. Genome Res. *21*, 203–215. https://doi.org/10.1101/gr.116657.110

Betel, D., Koppal, A., Agius, P., Sander, C., and Leslie, C. (2010). Comprehensive modeling of microRNA targets predicts functional non-conserved and non-canonical sites. Genome Biol. *11*, R90. https://doi.org/10.1186/gb-2010-11-8-r90

Bohnsack, M.T., Czaplinski, K., and Gorlich, D. (2004). Exportin 5 is a RanGTP-dependent dsRNA-binding protein that mediates nuclear export of pre-miRNAs. RNA *10*, 185–191. https://doi.org/10.1261/rna.5167604

Bracht, J., Hunter, S., Eachus, R., Weeks, P., and Pasquinelli, A.E. (2004). Trans-splicing and polyadenylation of let-7 microRNA primary transcripts. RNA *10*, 1586–1594. https://doi.org/10.1261/rna.7122604

Brennecke, J., Stark, A., Russell, R.B., and Cohen, S.M. (2005). Principles of microRNA-target recognition. PLOS Biol. *3*, e85. https://doi.org/10.1371/journal.pbio.0030085

Bruno, I.G., Karam, R., Huang, L., Bhardwaj, A., Lou, C.H., Shum, E.Y., Song, H.W., Corbett, M.A., Gifford, W.D., Gecz, J., *et al.* (2011). Identification of a microRNA that activates gene expression by repressing nonsense-mediated RNA decay. Mol. Cell *42*, 500–510. https://doi.org/10.1016/j.molcel.2011.04.018

Cai, X., Hagedorn, C.H., and Cullen, B.R. (2004). Human microRNAs are processed from capped, polyadenylated transcripts that can also function as mRNAs. RNA *10*, 1957–1966. https://doi.org/10.1261/rna.7135204

Carthew, R.W., and Sontheimer, E.J. (2009). Origins and mechanisms of miRNAs and siRNAs. Cell *136*, 642–655. https://doi.org/10.1016/j.cell.2009.01.035

Chen, C.Z., Schaffert, S., Fragoso, R., and Loh, C. (2013). Regulation of immune responses and tolerance: the microRNA perspective. Immunol. Rev. *253*, 112–128. https://doi.org/10.1111/imr.12060

Cherry, S., and Silverman, N. (2006). Host-pathogen interactions in *Drosophila*: new tricks from an old friend. Nat. Immunol. *7*, 911–917. https://doi.org/10.1038/ni1388

Chugh, P., and Dittmer, D.P. (2012). Potential pitfalls in microRNA profiling. Wiley Interdiscip. Rev. RNA *3*, 601–616. https://doi.org/10.1002/wrna.1120

Clyde, K., Kyle, J.L., and Harris, E. (2006). Recent advances in deciphering viral and host determinants of dengue virus replication and pathogenesis. J. Virol. *80*, 11418–11431. https://doi.org/10.1128/JVI.01257-06

Cullen, B.R. (2011). Viruses and microRNAs: RISCy interactions with serious consequences. Genes Dev. *25*, 1881–1894. https://doi.org/10.1101/gad.17352611

Czech, B., Zhou, R., Erlich, Y., Brennecke, J., Binari, R., Villalta, C., Gordon, A., Perrimon, N., and Hannon, G.J. (2009). Hierarchical rules for Argonaute loading in *Drosophila*. Mol. Cell *36*, 445–456. https://doi.org/10.1016/j.molcel.2009.09.028

Delpeut, S., Noyce, R.S., Siu, R.W., and Richardson, C.D. (2012). Host factors and measles virus replication. Curr. Opin. Virol. *2*, 773–783. https://doi.org/10.1016/j.coviro.2012.10.008

Denli, A.M., Tops, B.B., Plasterk, R.H., Ketting, R.F., and Hannon, G.J. (2004). Processing of primary microRNAs by the Microprocessor complex. Nature *432*, 231–235. https://doi.org/10.1038/nature03049

Desvignes, T., Batzel, P., Berezikov, E., Eilbeck, K., Eppig, J.T., McAndrews, M.S., Singer, A., and Postlethwait, J.H. (2015). miRNA nomenclature: a view incorporating genetic origins, biosynthetic pathways, and sequence variants. Trends Genet. *31*, 613–626. https://doi.org/10.1016/j.tig.2015.09.002

Djuranovic, S., Nahvi, A., and Green, R. (2012). miRNA-mediated gene silencing by translational repression followed by mRNA deadenylation and decay. Science *336*, 237–240. https://doi.org/10.1126/science.1215691

Dykxhoorn, D.M. (2007). MicroRNAs in viral replication and pathogenesis. DNA Cell Biol. *26*, 239–249. https://doi.org/10.1089/dna.2006.0559

Ekimler, S., and Sahin, K. (2014). Computational methods for microrna target prediction. Genes *5*, 671–683. https://doi.org/10.3390/genes5030671

Enright, A.J., John, B., Gaul, U., Tuschl, T., Sander, C., and Marks, D.S. (2003). MicroRNA targets in *Drosophila*. Genome Biol. *5*, R1. https://doi.org/10.1186/gb-2003-5-1-r1

Eulalio, A., Helms, S., Fritzsch, C., Fauser, M., and Izaurralde, E. (2009). A C-terminal silencing domain in GW182 is essential for miRNA function. RNA *15*, 1067–1077. https://doi.org/10.1261/rna.1605509

Fabian, M.R., Sonenberg, N., and Filipowicz, W. (2010). Regulation of mRNA translation and stability by microRNAs. Annu. Rev. Biochem. *79*, 351–379. https://doi.org/10.1146/annurev-biochem-060308-103103

Farh, K.K., Grimson, A., Jan, C., Lewis, B.P., Johnston, W.K., Lim, L.P., Burge, C.B., and Bartel, D.P. (2005). The widespread impact of mammalian MicroRNAs on mRNA repression and evolution. Science *310*, 1817–1821. https://doi.org/10.1126/science.1121158

Förstemann, K., Horwich, M.D., Wee, L., Tomari, Y., and Zamore, P.D. (2007). *Drosophila* microRNAs are sorted into functionally distinct argonaute complexes after production by dicer-1. Cell *130*, 287–297. https://doi.org/10.1016/j.cell.2007.05.056

Ghildiyal, M., Xu, J., Seitz, H., Weng, Z., and Zamore, P.D. (2010). Sorting of *Drosophila* small silencing RNAs partitions microRNA* strands into the RNA interference pathway. RNA *16*, 43–56. https://doi.org/10.1261/rna.1972910

Gottwein, E., and Cullen, B.R. (2008). Viral and cellular microRNAs as determinants of viral pathogenesis and immunity. Cell Host Microbe 3, 375–387. https://doi.org/10.1016/j.chom.2008.05.002

Gregory, R.I., Yan, K.P., Amuthan, G., Chendrimada, T., Doratotaj, B., Cooch, N., and Shiekhattar, R. (2004). The microprocessor complex mediates the genesis of microRNAs. Nature *432*, 235–240.

Griffiths-Jones, S., Saini, H.K., van Dongen, S., and Enright, A.J. (2008). miRBase: tools for microRNA genomics. Nucleic Acids Res. 36, D154–D158. https://doi.org/10.1093/nar/gkm952

Grimson, A., Farh, K.K., Johnston, W.K., Garrett-Engele, P., Lim, L.P., and Bartel, D.P. (2007). MicroRNA targeting specificity in mammals: determinants beyond seed pairing. Mol. Cell *27*, 91–105.

Grundhoff, A., and Sullivan, C.S. (2011). Virus-encoded microRNAs. Virology *411*, 325–343. https://doi.org/10.1016/j.virol.2011.01.002

Guo, H., Ingolia, N.T., Weissman, J.S., and Bartel, D.P. (2010). Mammalian microRNAs predominantly act to decrease target mRNA levels. Nature *466*, 835–840. https://doi.org/10.1038/nature09267

Han, J., Lee, Y., Yeom, K.H., Kim, Y.K., Jin, H., and Kim, V.N. (2004). The Drosha-DGCR8 complex in primary microRNA processing. Genes Dev. *18*, 3016–3027. https://doi.org/10.1101/gad.1262504

Hashimoto, Y., Akiyama, Y., and Yuasa, Y. (2013). Multiple-to-multiple relationships between microRNAs and target genes in gastric cancer. PLOS ONE *8*, e62589. https://doi.org/10.1371/journal.pone.0062589

Hu, H.Y., Yan, Z., Xu, Y., Hu, H., Menzel, C., Zhou, Y.H., Chen, W., and Khaitovich, P. (2009). Sequence features associated with microRNA strand selection in humans and flies. BMC Genomics *10*, 413. https://doi.org/10.1186/1471-2164-10-413

Huntzinger, E., and Izaurralde, E. (2011). Gene silencing by microRNAs: contributions of translational repression and mRNA decay. Nat. Rev. Genet. *12*, 99–110. https://doi.org/10.1038/nrg2936

Hussain, M., and Asgari, S. (2010). Functional analysis of a cellular microRNA in insect host-ascovirus interaction. J. Virol. *84*, 612–620. https://doi.org/10.1128/JVI.01794-09

Hussain, M., Taft, R.J., and Asgari, S. (2008). An insect virus-encoded microRNA regulates viral replication. J. Virol. *82*, 9164–9170. https://doi.org/10.1128/JVI.01109-08

Hussain, M., Frentiu, F.D., Moreira, L.A., O'Neill, S.L., and Asgari, S. (2011). *Wolbachia* utilizes host microRNAs to manipulate host gene expression and facilitate colonization of the dengue vector *Aedes aegypti*. Proc. Natl. Acad. Sci. U.S.A. *108*, 9250–9255. https://doi.org/10.1073/pnas.1105469108

Hussain, M., Torres, S., Schnettler, E., Funk, A., Grundhoff, A., Pijlman, G.P., Khromykh, A.A., and Asgari, S. (2012). West Nile virus encodes a microRNA-like small RNA in the 3' untranslated region which up-regulates GATA4 mRNA and facilitates virus replication in mosquito cells. Nucleic Acids Res. *40*, 2210–2223. https://doi.org/10.1093/nar/gkr848

Huszar, T., and Imler, J.L. (2008). *Drosophila* viruses and the study of antiviral host-defense. Adv. Virus Res. 72, 227–265. https://doi.org/10.1016/S0065-3527(08)00406-5

Ingolia, N.T., Ghaemmaghami, S., Newman, J.R., and Weissman, J.S. (2009). Genome-wide analysis *in vivo* of translation with nucleotide resolution using ribosome profiling. Science *324*, 218–223. https://doi.org/10.1126/science.1168978

Iwakawa, H.O., and Tomari, Y. (2015). The functions of microRNAs: mRNA decay and translational repression. Trends Cell Biol. *25*, 651–665.

Jayachandran, B., Hussain, M., and Asgari, S. (2013). An insect trypsin-like serine protease as a target of microRNA: utilization of microRNA mimics and inhibitors by oral feeding. Insect Biochem. Mol. Biol. *43*, 398–406. https://doi.org/10.1016/j.ibmb.2012.10.004

Jonas, S., and Izaurralde, E. (2015). Towards a molecular understanding of microRNA-mediated gene silencing. Nat. Rev. Genet. *16*, 421–433. https://doi.org/10.1038/nrg3965

Joo, C.H., Shin, Y.C., Gack, M., Wu, L., Levy, D., and Jung, J.U. (2007). Inhibition of interferon regulatory factor 7 (IRF7)-mediated interferon signal transduction by the Kaposi's Sarcoma-associated herpesvirus viral IRF homolog vIRF3. J. Virol. *81*, 8282–8292. https://doi.org/10.1128/JVI.00235-07

Kemp, C., and Imler, J.L. (2009). Antiviral immunity in *Drosophila*. Curr. Opin. Immunol. *21*, 3–9. https://doi.org/10.1016/j.coi.2009.01.007

Khvorova, A., Reynolds, A., and Jayasena, S.D. (2003). Functional siRNAs and miRNAs exhibit strand bias. Cell *115*, 209–216. https://doi.org/10.1016/S0092-8674(03)00801-8

Kim, Y.K., and Kim, V.N. (2007). Processing of intronic microRNAs. EMBO J. *26*, 775–783. https://doi.org/10.1038/sj.emboj.7601512

Kincaid, R.P., and Sullivan, C.S. (2012). Virus-encoded microRNAs: an overview and a look to the future. PLOS Pathog. *8*, e1003018. https://doi.org/10.1371/journal.ppat.1003018

Kozomara, A., and Griffiths-Jones, S. (2014). miRBase: annotating high confidence microRNAs using deep sequencing data. Nucleic Acids Res. *42*, D68–D73. https://doi.org/10.1093/nar/gkt1181

Krek, A., Grün, D., Poy, M.N., Wolf, R., Rosenberg, L., Epstein, E.J., MacMenamin, P., da Piedade, I., Gunsalus, K.C., Stoffel, M., *et al.* (2005). Combinatorial microRNA target predictions. Nat. Genet. *37*, 495–500. https://doi.org/10.1038/ng1536

Kuhn, D.E., Martin, M.M., Feldman, D.S., Terry, A.V., Nuovo, G.J., and Elton, T.S. (2008). Experimental validation of miRNA targets. Methods *44*, 47–54. https://doi.org/10.1016/j.ymeth.2007.09.005

Lai, E.C., Tomancak, P., Williams, R.W., and Rubin, G.M. (2003). Computational identification of *Drosophila* microRNA genes. Genome Biol. *4*, R42. https://doi.org/10.1186/gb-2003-4-7-r42

Landgraf, P., Rusu, M., Sheridan, R., Sewer, A., Iovino, N., Aravin, A., Pfeffer, S., Rice, A., Kamphorst, A.O., Landthaler, M., *et al.* (2007). A mammalian microRNA expression atlas based on small RNA library sequencing. Cell *129*, 1401–1414. https://doi.org/10.1016/j.cell.2007.04.040

Landthaler, M., Yalcin, A., and Tuschl, T. (2004). The human DiGeorge syndrome critical region gene 8 and Its *D. melanogaster* homolog are required for miRNA biogenesis. Curr. Biol. *14*, 2162–2167. https://doi.org/10.1016/j.cub.2004.11.001

Langlois, R.A., Shapiro, J.S., Pham, A.M., and tenOever, B.R. (2012). *In vivo* delivery of cytoplasmic RNA virus-derived miRNAs. Mol. Ther. *20*, 367–375. https://doi.org/10.1038/mt.2011.244

Lee, Y., Jeon, K., Lee, J.T., Kim, S., and Kim, V.N. (2002). MicroRNA maturation: stepwise processing and subcellular localization. EMBO J. *21*, 4663–4670. https://doi.org/10.1093/emboj/cdf476

Lee, Y., Ahn, C., Han, J., Choi, H., Kim, J., Yim, J., Lee, J., Provost, P., Rådmark, O., Kim, S., *et al.* (2003). The nuclear RNase III Drosha initiates microRNA processing. Nature *425*, 415–419. https://doi.org/10.1038/nature01957

Lee, Y., Kim, M., Han, J., Yeom, K.H., Lee, S., Baek, S.H., and Kim, V.N. (2004a). MicroRNA genes are transcribed by RNA polymerase II. EMBO J. *23*, 4051–4060. https://doi.org/10.1038/sj.emboj.7600385

Lee, Y.S., Nakahara, K., Pham, J.W., Kim, K., He, Z., Sontheimer, E.J., and Carthew, R.W. (2004b). Distinct roles for *Drosophila* Dicer-1 and Dicer-2 in the siRNA/miRNA silencing pathways. Cell *117*, 69–81.

Lewis, B.P., Shih, I.H., Jones-Rhoades, M.W., Bartel, D.P., and Burge, C.B. (2003). Prediction of mammalian microRNA targets. Cell *115*, 787–798 https://doi.org/10.1016/S0092-8674(03)01018-3

Lewis, B.P., Burge, C.B., and Bartel, D.P. (2005). Conserved seed pairing, often flanked by adenosines, indicates that thousands of human genes are microRNA targets. Cell *120*, 15–20. https://doi.org/10.1016/j.cell.2004.12.035

Li, Z., and Rana, T.M. (2014). Therapeutic targeting of microRNAs: current status and future challenges. Nat. Rev. Drug Discov. *13*, 622–638. https://doi.org/10.1038/nrd4359

Lian, S.L., Li, S., Abadal, G.X., Pauley, B.A., Fritzler, M.J., and Chan, E.K. (2009). The C-terminal half of human Ago2 binds to multiple GW-rich regions of GW182 and requires GW182 to mediate silencing. RNA *15*, 804–813. https://doi.org/10.1261/rna.1229409

Libri, V., Miesen, P., van Rij, R.P., and Buck, A.H. (2013). Regulation of microRNA biogenesis and turnover by animals and their viruses. Cell. Mol. Life Sci. *70*, 3525–3544. https://doi.org/10.1007/s00018-012-1257-1

Lim, L.P., Glasner, M.E., Yekta, S., Burge, C.B., and Bartel, D.P. (2003a). Vertebrate microRNA genes. Science *299*, 1540. https://doi.org/10.1126/science.1080372

Lim, L.P., Lau, N.C., Weinstein, E.G., Abdelhakim, A., Yekta, S., Rhoades, M.W., Burge, C.B., and Bartel, D.P. (2003b). The microRNAs of *Caenorhabditis elegans*. Genes Dev. *17*, 991–1008. https://doi.org/10.1101/gad.1074403

Lindsay, M.A. (2008). microRNAs and the immune response. Trends Immunol. *29*, 343–351. https://doi.org/10.1016/j.it.2008.04.004

Lodish, H.F., Zhou, B., Liu, G., and Chen, C.Z. (2008). Micromanagement of the immune system by microRNAs. Nat. Rev. Immunol. *8*, 120–130. https://doi.org/10.1038/nri2252

Lucas, K.J., Zhao, B., Liu, S., and Raikhel, A.S. (2015). Regulation of physiological processes by microRNAs in insects. Curr. Opin. Insect Sci. *11*, 1–7. https://doi.org/10.1016/j.cois.2015.06.004

Lund, E., Güttinger, S., Calado, A., Dahlberg, J.E., and Kutay, U. (2004). Nuclear export of microRNA precursors. Science *303*, 95–98. https://doi.org/10.1126/science.1090599

Ma, J.B., Ye, K., and Patel, D.J. (2004). Structural basis for overhang-specific small interfering RNA recognition by the PAZ domain. Nature *429*, 318–322. https://doi.org/10.1038/nature02519

MacRae, I.J., Zhou, K., and Doudna, J.A. (2007). Structural determinants of RNA recognition and cleavage by Dicer. Nat. Struct. Mol. Biol. *14*, 934–940. https://doi.org/10.1038/nsmb1293

Marco, A., Hui, J.H., Ronshaugen, M., and Griffiths-Jones, S. (2010). Functional shifts in insect microRNA evolution. Genome Biol. Evol. *2*, 686–696. https://doi.org/10.1093/gbe/evq053

Martinez-Sanchez, A., and Murphy, C.L. (2013). MicroRNA Target Identification-Experimental Approaches. Biology *2*, 189–205. https://doi.org/10.3390/biology2010189

Min, H., and Yoon, S. (2010). Got target? Computational methods for microRNA target prediction and their extension. Exp. Mol. Med. *42*, 233–244. https://doi.org/10.3858/emm.2010.42.4.032

Monsanto-Hearne, V., and Johnson, K.N. (2018). miRNAs in insects infected by animal and plant viruses. Viruses *10*, E354. https://doi.org/10.3390/v10070354

Monsanto-Hearne, V., Asad, S., Asgari, S., and Johnson, K.N. (2017a). *Drosophila* microRNA modulates viral replication by targeting a homologue of mammalian cJun. J. Gen. Virol. *98*, 1904–1912. https://doi.org/10.1099/jgv.0.000831

Monsanto-Hearne, V., Tham, A.L.Y., Wong, Z.S., Asgari, S., and Johnson, K.N. (2017b). *Drosophila* miR-956 suppression modulates Ectoderm-expressed 4 and inhibits viral replication. Virology *502*, 20–27. https://doi.org/10.1016/j.virol.2016.12.009

Mussabekova, A., Daeffler, L., and Imler, J.L. (2017). Innate and intrinsic antiviral immunity in *Drosophila*. Cell. Mol. Life Sci. *74*, 2039–2054. https://doi.org/10.1007/s00018-017-2453-9

Nakanishi, K., Weinberg, D.E., Bartel, D.P., and Patel, D.J. (2012). Structure of yeast Argonaute with guide RNA. Nature *486*, 368–374. https://doi.org/10.1038/nature11211

Niu, J., Meeus, I., De Coninck, D.I., Deforce, D., Etebari, K., Asgari, S., and Smagghe, G. (2017). Infections of virulent and avirulent viruses differentially influenced the expression of dicer-1, ago-1, and microRNAs in *Bombus terrestris*. Sci. Rep. *7*, 45620. https://doi.org/10.1038/srep45620

O'Connell, R.M., Rao, D.S., Chaudhuri, A.A., and Baltimore, D. (2010). Physiological and pathological roles for microRNAs in the immune system. Nat. Rev. Immunol. *10*, 111–122. https://doi.org/10.1038/nri2708

Okamura, K., Ishizuka, A., Siomi, H., and Siomi, M.C. (2004). Distinct roles for Argonaute proteins in small RNA-directed RNA cleavage pathways. Genes Dev. *18*, 1655–1666. https://doi.org/10.1101/gad.1210204

Okamura, K., Hagen, J.W., Duan, H., Tyler, D.M., and Lai, E.C. (2007). The mirtron pathway generates microRNA-class regulatory RNAs in *Drosophila*. Cell *130*, 89–100. https:/https://doi.org/10.1016/j.cell.2007.06.028

Okamura, K., Phillips, M.D., Tyler, D.M., Duan, H., Chou, Y.T., and Lai, E.C. (2008). The regulatory activity of microRNA* species has substantial influence on microRNA and 3' UTR evolution. Nat. Struct. Mol. Biol. *15*, 354–363. https://doi.org/10.1038/nsmb.1409

Okamura, K., Liu, N., and Lai, E.C. (2009). Distinct mechanisms for microRNA strand selection by *Drosophila* Argonautes. Mol. Cell *36*, 431–444. https://doi.org/10.1016/j.molcel.2009.09.027

Ørom, U.A., Nielsen, F.C., and Lund, A.H. (2008). MicroRNA-10a binds the 5'UTR of ribosomal protein mRNAs and enhances their translation. Mol. Cell *30*, 460–471. https://doi.org/10.1016/j.molcel.2008.05.001

Ozsolak, F., Poling, L.L., Wang, Z., Liu, H., Liu, X.S., Roeder, R.G., Zhang, X., Song, J.S., and Fisher, D.E. (2008). Chromatin structure analyses identify miRNA promoters. Genes Dev. *22*, 3172–3183. https://doi.org/10.1101/gad.1706508

Pasquinelli, A.E. (2012). MicroRNAs and their targets: recognition, regulation and an emerging reciprocal relationship. Nat. Rev. Genet. *13*, 271–282. https://doi.org/10.1038/nrg3162

Pedersen, I.M., Cheng, G., Wieland, S., Volinia, S., Croce, C.M., Chisari, F.V., and David, M. (2007). Interferon modulation of cellular microRNAs as an antiviral mechanism. Nature *449*, 919–922. https://doi.org/10.1038/nature06205

Ploegh, H.L. (1998). Viral strategies of immune evasion. Science *280*, 248–253. https://doi.org/10.1126/science.280.5361.248

Rapicetta, M., Ferrari, C., and Levrero, M. (2002). Viral determinants and host immune responses in the pathogenesis of HBV infection. J. Med. Virol. *67*, 454–457. https://doi.org/10.1002/jmv.10096

Riffo-Campos, Á.L., Riquelme, I., and Brebi-Mieville, P. (2016). Tools for sequence-based miRNA target prediction: What to choose? Int. J. Mol. Sci. *17*, E1987. https://doi.org/10.3390/ijms17121987

Ritchie, W., Flamant, S., and Rasko, J.E. (2009). Predicting microRNA targets and functions: traps for the unwary. Nat. Methods *6*, 397–398. https://doi.org/10.1038/nmeth0609-397

Rodriguez, A., Griffiths-Jones, S., Ashurst, J.L., and Bradley, A. (2004). Identification of mammalian microRNA host genes and transcription units. Genome Res. *14*, 1902–1910. https://doi.org/10.1101/gr.2722704

Ruby, J.G., Stark, A., Johnston, W.K., Kellis, M., Bartel, D.P., and Lai, E.C. (2007). Evolution, biogenesis, expression, and target predictions of a substantially expanded set of *Drosophila* microRNAs. Genome Res. *17*, 1850–1864.

Scaria, V., Hariharan, M., Maiti, S., Pillai, B., and Brahmachari, S.K. (2006). Host-virus interaction: a new role for microRNAs. Retrovirology 3, 68. https://doi.org/10.1186/1742-4690-3-68

Schertel, C., Rutishauser, T., Förstemann, K., and Basler, K. (2012). Functional characterization of *Drosophila* microRNAs by a novel *in vivo* library. Genetics *192*, 1543–1552. https://doi.org/10.1534/genetics.112.145383

Schirle, N.T., and MacRae, I.J. (2012). The crystal structure of human Argonaute2. Science *336*, 1037–1040. https://doi.org/10.1126/science.1221551

Schirle, N.T., Sheu-Gruttadauria, J., and MacRae, I.J. (2014). Structural basis for microRNA targeting. Science *346*, 608–613. https://doi.org/10.1126/science.1258040

Selbach, M., Schwanhäusser, B., Thierfelder, N., Fang, Z., Khanin, R., and Rajewsky, N. (2008). Widespread changes in protein synthesis induced by microRNAs. Nature 455, 58–63. https://doi.org/10.1038/nature07228

Shapiro, J.S., Varble, A., Pham, A.M., and Tenoever, B.R. (2010). Noncanonical cytoplasmic processing of viral microRNAs. RNA 16, 2068–2074. https://doi.org/10.1261/rna.2303610

Shi, X., Ran, Z., Li, S., Yin, J., and Zhong, J. (2016). The effect of microRNA bantam on Baculovirus AcMNPV Infection *in* vitro and *in* vivo. Viruses 8, E136. https://doi.org/10.3390/v8050136

Singh, C.P., Vaishna, R.L., Kakkar, A., Arunkumar, K.P., and Nagaraju, J. (2014). Characterization of antiviral and antibacterial activity of *Bombyx mori* seroin proteins. Cell. Microbiol. 16, 1354–1365. https://doi.org/10.1111/cmi.12294

Singh, J., Singh, C.P., Bhavani, A., and Nagaraju, J. (2010). Discovering microRNAs from *Bombyx mori* nucleopolyhedrosis virus. Virology 407, 120–128. https://doi.org/10.1016/j.virol.2010.07.033

Slonchak, A., Hussain, M., Torres, S., Asgari, S., and Khromykh, A.A. (2014). Expression of mosquito microRNA aae-miR-2940-5p is down-regulated in response to West Nile virus infection to restrict viral replication. J. Virol. 88, 8457-8467. https://doi.org/10.1128/JVI.00317-14

Smeal, T., Hibi, M., and Karin, M. (1994). Altering the specificity of signal transduction cascades: positive regulation of c-Jun transcriptional activity by protein kinase A. EMBO J. 13, 6006–6010.

Sood, P., Krek, A., Zavolan, M., Macino, G., and Rajewsky, N. (2006). Cell-type-specific signatures of microRNAs on target mRNA expression. Proc. Natl. Acad. Sci. U.S.A. 103, 2746–2751.

Stark, A., Brennecke, J., Bushati, N., Russell, R.B., and Cohen, S.M. (2005). Animal microRNAs confer robustness to gene expression and have a significant impact on 3'UTR evolution. Cell 123, 1133–1146.

Steinkraus, B.R., Toegel, M., and Fulga, T.A. (2016). Tiny giants of gene regulation: experimental strategies for microRNA functional studies. Wiley Interdiscip. Rev. Dev. Biol. 5, 311–362. https://doi.org/10.1002/wdev.223

Thomson, D.W., Bracken, C.P., and Goodall, G.J. (2011). Experimental strategies for microRNA target identification. Nucleic Acids Res. 39, 6845–6853. https://doi.org/10.1093/nar/gkr330

Tomari, Y., and Zamore, P.D. (2005). MicroRNA biogenesis: drosha can't cut it without a partner. Curr. Biol. 15, R61–R64. https://doi.org/10.1016/j.cub.2004.12.057

Tomari, Y., Du, T., and Zamore, P.D. (2007). Sorting of *Drosophila* small silencing RNAs. Cell 130, 299–308. https://doi.org/10.1016/j.cell.2007.05.057

Trobaugh, D.W., and Klimstra, W.B. (2017). MicroRNA regulation of RNA virus replication and pathogenesis. Trends Mol. Med. 23, 80–93. https://doi.org/10.1016/j.molmed.2016.11.003

Tsitsiou, E., and Lindsay, M.A. (2009). MicroRNAs and the immune response. Curr. Opin. Pharmacol. 9, 514–520. https://doi.org/10.1016/j.coph.2009.05.003

Tycowski, K.T., Guo, Y.E., Lee, N., Moss, W.N., Vallery, T.K., Xie, M., and Steitz, J.A. (2015). Viral noncoding RNAs: more surprises. Genes Dev. 29, 567–584. https://doi.org/10.1101/gad.259077.115

Varble, A., Chua, M.A., Perez, J.T., Manicassamy, B., García-Sastre, A., and tenOever, B.R. (2010). Engineered RNA viral synthesis of microRNAs. Proc. Natl. Acad. Sci. U.S.A. 107, 11519–11524. https://doi.org/10.1073/pnas.1003115107

Vasudevan, S. (2012). Posttranscriptional upregulation by microRNAs. Wiley Interdiscip. Rev. RNA 3, 311–330. https://doi.org/10.1002/wrna.121

Vasudevan, S., Tong, Y., and Steitz, J.A. (2007). Switching from repression to activation: microRNAs can up-regulate translation. Science 318, 1931–1934. https://doi.org/10.1126/science.1149460

Vodovar, N., and Saleh, M.C. (2012). Of insects and viruses: the role of small RNAs in insect defence. In Small RNAs: Their Diversity, Roles and Practical Uses, E.L. Jockusch, ed. (Academic Press, London), pp. 1–36.

Wang, Y., Sheng, G., Juranek, S., Tuschl, T., and Patel, D.J. (2008). Structure of the guide-strand-containing argonaute silencing complex. Nature 456, 209–213. https://doi.org/10.1038/nature07315

Whitton, J.L., Cornell, C.T., and Feuer, R. (2005). Host and virus determinants of picornavirus pathogenesis and tropism. Nat. Rev. Microbiol. 3, 765–776. https://doi.org/10.1038/nrmicro1284

Wilczynska, A., and Bushell, M. (2015). The complexity of miRNA-mediated repression. Cell Death Differ. 22, 22–33. https://doi.org/10.1038/cdd.2014.112

Winter, J., and Diederichs, S. (2011). Argonaute proteins regulate microRNA stability: Increased microRNA abundance by Argonaute proteins is due to microRNA stabilization. RNA Biol. 8, 1149–1157. https://doi.org/10.4161/rna.8.6.17665

Witkos, T.M., Koscianska, E., and Krzyzosiak, W.J. (2011). Practical aspects of microRNA target prediction. Curr. Mol. Med. 11, 93–109. https://doi.org/10.2174/156652411794859250

Wu, P., Qin, G., Qian, H., Chen, T., and Guo, X. (2016). Roles of miR-278-3p in IBP2 regulation and *Bombyx mori* cytoplasmic polyhedrosis virus replication. Gene 575, 264–269. https://doi.org/10.1016/j.gene.2015.09.009

Wu, P., Jiang, X., Sang, Q., Annan, E., Cheng, T., and Guo, X. (2017). Inhibition of miR-274-3p increases BmCPV replication by regulating the expression of BmCPV NS5 gene in *Bombyx mori*. Virus Genes 53, 643–649. https://doi.org/10.1007/s11262-017-1466-7

Xiao, C., and Rajewsky, K. (2009). MicroRNA control in the immune system: basic principles. Cell 136, 26–36. https://doi.org/10.1016/j.cell.2008.12.027

Xie, M., and Steitz, J.A. (2014). Versatile microRNA biogenesis in animals and their viruses. RNA Biol. 11, 673–681. https://doi.org/10.4161/rna.28985

Xu, J., and Cherry, S. (2014). Viruses and antiviral immunity in *Drosophila*. Dev. Comp. Immunol. 42, 67–84. https://doi.org/10.1016/j.dci.2013.05.002

Yan, H., Zhou, Y., Liu, Y., Deng, Y., and Chen, X. (2014). miR-252 of the Asian tiger mosquito *Aedes albopictus* regulates dengue virus replication by suppressing the expression of the dengue virus envelope protein. J.

Med. Virol. *86*, 1428–1436. https://doi.org/10.1002/jmv.23815

Yang, J.S., Maurin, T., Robine, N., Rasmussen, K.D., Jeffrey, K.L., Chandwani, R., Papapetrou, E.P., Sadelain, M., O'Carroll, D., and Lai, E.C. (2010). Conserved vertebrate mir-451 provides a platform for Dicer-independent, Ago2-mediated microRNA biogenesis. Proc. Natl. Acad. Sci. U.S.A. *107*, 15163–15168. https://doi.org/10.1073/pnas.1006432107

Yang, J.S., Phillips, M., Betel, D., Mu, P., Ventura, A., Siepel, A., Chen, K., and Lai, E. (2011). Widespread regulatory activity of vertebrate microRNA* species. RNA *17*, 312–326. https://doi.org/10.1261/rna.2537911

Yi, R., Qin, Y., Macara, I.G., and Cullen, B.R. (2003). Exportin-5 mediates the nuclear export of pre-microRNAs and short hairpin RNAs. Genes Dev. *17*, 3011–3016. https://doi.org/10.1101/gad.1158803

Ylla, G., Fromm, B., Piulachs, M.D., and Belles, X. (2016). The microRNA toolkit of insects. Sci. Rep. *6*, 37736. https://doi.org/10.1038/srep37736

Zeng, Y., and Cullen, B.R. (2003). Sequence requirements for micro RNA processing and function in human cells. RNA *9*, 112–123. https://doi.org/10.1261/rna.2780503

Zeng, Y., and Cullen, B.R. (2005). Efficient processing of primary microRNA hairpins by Drosha requires flanking nonstructured RNA sequences. J. Biol. Chem. *280*, 27595–27603. https://doi.org/10.1074/jbc.M504714200

Zeng, Y., Yi, R., and Cullen, B.R. (2005). Recognition and cleavage of primary microRNA precursors by the nuclear processing enzyme Drosha. EMBO J. *24*, 138–148. https://doi.org/10.1038/sj.emboj.7600491

Zhao, T., Li, G., Mi, S., Li, S., Hannon, G.J., Wang, X.J., and Qi, Y. (2007). A complex system of small RNAs in the unicellular green alga *Chlamydomonas reinhardtii*. Genes Dev. *21*, 1190–1203.

Zhou, Y., Liu, Y., Yan, H., Li, Y., Zhang, H., Xu, J., Puthiyakunnon, S., and Chen, X. (2014). miR-281, an abundant midgut-specific miRNA of the vector mosquito *Aedes albopictus* enhances dengue virus replication. Parasit. Vectors *7*, 488. https://doi.org/10.1186/s13071-014-0488-4

Zhu, M., Wang, J., Deng, R., Xiong, P., Liang, H., and Wang, X. (2013). A microRNA encoded by *Autographa californica* nucleopolyhedrovirus regulates expression of viral gene ODV-E25. J. Virol. *87*, 13029–13034. https://doi.org/10.1128/JVI.02112-13

Dicistrovirus–Host Molecular Interactions

5

Reid Warsaba†, Jibin Sadasivan† and Eric Jan*

Department of Biochemistry and Molecular Biology, University of British Columbia, Vancouver, BC, Canada.

*Correspondence: ej@mail.ubc.ca

†Both authors contributed equally to this work

https://doi.org/10.21775/9781912530083.05

Abstract

Members of the family *Dicistroviridae* are small RNA viruses containing a monopartite positive-sense RNA genome. Dicistroviruses mainly infect arthropods, causing diseases that impact agriculture and the economy. In this chapter, we provide an overview of current and past research on dicistroviruses including the viral life cycle, viral translational control mechanisms, virus structure, and the use of dicistrovirus infection in *Drosophila* as a model to identify insect antiviral responses. We then delve into how research on dicistrovirus mechanisms has yielded insights into ribosome dynamics, RNA structure/function and insect innate immunity signalling. Finally, we highlight the diseases caused by dicistroviruses, their impacts on agriculture including the shrimp and honey bee industries, and the potential use of dicistroviruses as biopesticides. Although knowledge of the mechanisms underlying dicistrovirus virus–host interactions is limited, the establishment of the first infectious clone should accelerate the discovery of new mechanistic insights into dicistrovirus infections and pathogenesis.

Introduction

Dicistroviruses are small plus-strand RNA viruses that primarily infect arthropods. They are classified into three genera: *Cripavirus*, *Aparavirus* and *Triatovirus*. Initially discovered in 1963 with the identification of *Acute bee paralysis virus* in honey bees (Bailey *et al.*, 1964), they were initially thought to be picornaviruses due to the similarly small size of the virion and genome, and disease symptoms (i.e. paralysis). However, it is now apparent that dicistroviruses belong to a unique virus family. Dicistroviruses have been identified in many orders of arthropods: Coleoptera (Reinganum, 1975), Lepidoptera (Reinganum, 1975), Orthoptera (Wilson *et al.*, 2000b), Diptera (Jousset *et al.*, 1972), Hemiptera (D'Arcy *et al.*, 1981a; Muscio *et al.*, 1987; Williamson *et al.*, 1988; Toriyama *et al.*, 1992; Nakashima *et al.*, 1998; Hunnicutt *et al.*, 2006), Hymenoptera (Bailey *et al.*, 1963; Bailey and Woods, 1977; Maori *et al.*, 2007; Valles and Hashimoto, 2009), Decapoda (Hasson *et al.*, 1995). Dicistroviruses provide for an excellent model system for studying fundamental virus–host interactions including viral translational control mechanisms and innate insect immune pathways. *Dicistroviridae* are of economic and agricultural importance; infections can lead to disease and outbreaks in arthropods, including in shrimp and honey bees. As the name implies, dicistroviruses contain a bicistronic monopartite RNA genome encoding two main open reading frames (ORFs). However, recent evidence indicates that some members contain a third ORF, termed ORFx. In this chapter, we provide an overview of dicistrovirus life-cycle, viral translational strategies, host interactions that modulate cellular pathways and innate immune responses, pathogenesis, and the potential use of these viruses as biopesticides.

Genome structure

Dicistroviruses have single-stranded, positive-sense RNA genomes of 8–10 kb in length and encoding two long ORFs (Fig. 5.1). The ORFs are translated as polyproteins that are then processed into the individual mature viral proteins by the virally-encoded protease 3C, a 2A 'stop-go' translation mechanism and other yet to be identified

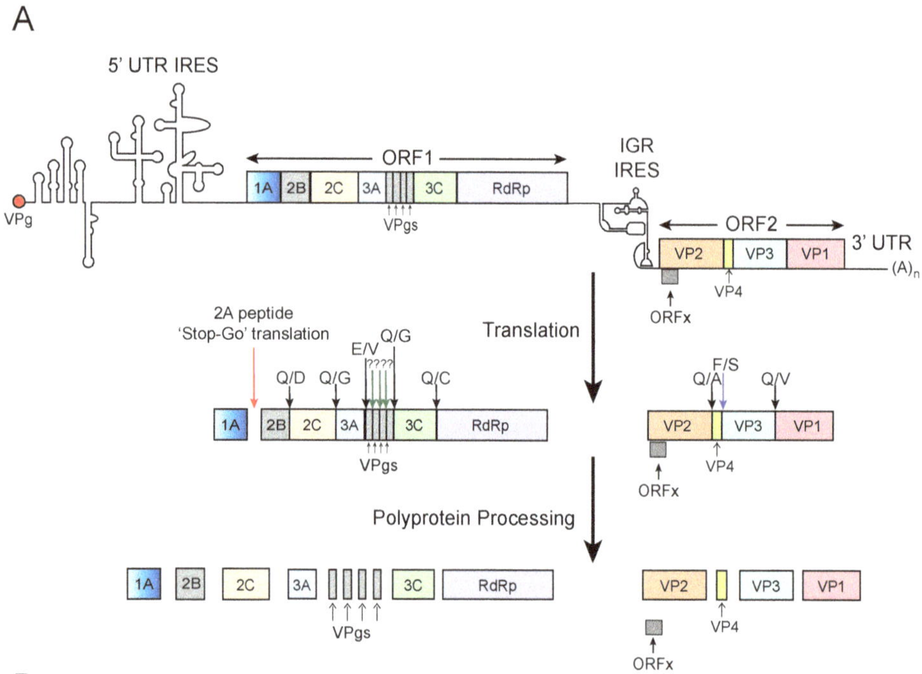

B

ORFx sequences

CrPV: KMKIKDLCQNRKKLYILLAKELPLVLLRSLISLIFQLIIWT (41 amino acids (aa))

IAPV: AIHNKKAILPTYTIRNSLRPLVKTRLRPKKSQPFMMWKLQIGSIPPWLRILHRLGTWMIRTVLFSFY
SAPFSLTTLRSLLEQRPMQTNPLADMC (94 aa)

ABPV: PIKKQILPTYIIRNSRRPLKKTQLKRNKSPPFMMWKLQIGSIPPWLKTLHRLGAWMIRTVLFSFYNAP
YSLTTLRSLLDQQQMITNPSIDMC (92 aa)

KBV: LITKKTILPMYITRNSRRPQRRMPLRRNKSPPFMMWKLQIGSIPPWLRILHRLGAWMIRTVLFSFYN
APFSLTTLRLLQEQLPITTQHSVDMC (93 aa)

SINV-1: LNQTMRILLRRKRQQPLLPTHNRNLARRPQKIPLPDKNSQFSMMLNNLASLFQLLRKRLALLLSLIL
QRQLWIFFLELLSSINSSLFKVNQTITTNPLTQQLLKTPNQPSDSIPCQETFLSWVAS (125 aa)

Figure 5.1 (A) The RNA genome and polyprotein processing of dicistroviruses. The positive sense, single-stranded RNA genome is shown above. The RNA genome is translated by the 5′UTR and IGR IRESs to produce two large polyproteins that are then processed by proteases. Open reading frames are labelled as ORF1 and ORF2, encoding non-structural proteins (1A, 2B, 2C, 3A, 3B, 3C, RdRp (3D), VPgs and structural proteins (VP1-VP4). The CrPV 5′UTR and IGR IRES structures are shown (Gross *et al.*, 2017). Dicistroviruses contain one to six VPgs. Shown is the CrPV genome containing 4 VPgs. A subset of dicistroviruses including IAPV, KBV, ABPV and SINV-1 contains a short +1 frame open reading frame called ORFx that overlaps with the 0 frame ORF2 adjacent to the IGR IRES. The IGR IRES directs both 0 and +1 frame translation. Polyprotein processing occurs co-translationally and post-translationally (red arrow). The 2A peptide, sequence located between 1A and 2B, directs 'stop go' translation separating 1A from polyprotein. Black arrows: Cleavages mediated by the virally encoded 3C protease at the indicated cleavage sites preferring cleavage after a Q residue. Blue arrow: Predicted cleavage site. Green arrows: Processing events that have not been characterized. (B) The ORFx amino acid sequences of CrPV, IAPV, ABPV, KBV and SINV-1 from accession numbers NC_003924.1 (CrPV), EF219380.1 (IAPV), AF150629 (ABPV), AY275710 (KBV), AF277675 (SINV-1).

mechanisms. The upstream ORF, termed ORF1, encodes non-structural proteins that modulate host cellular processes and evade innate immune responses, thereby facilitating virus infection. ORF1 encodes the viral silencing suppressor 1A protein, the 2C RNA helicase, the viral genome-linked protein VPg (King and Morre, 1988), the 3C protease, and the RNA-dependent RNA polymerase (RdRp) 3D (Bonning and Miller, 2010) (Table 5.1). ORF1 also encodes 2B and 3A proteins that, based on sequence homology to other viral proteins, are thought to be involved in the assembly of the viral replication complex membrane and in membrane remodelling. The roles of the 1A protein during infection are beginning to be understood (see below), whereas the functions of many of the other non-structural proteins have yet to be elucidated in detail. The downstream ORF2 encodes the structural proteins VP2, VP1, and VP3 that are involved in encapsidation of the virion, as well as VP4 (Tate *et al.*, 1999; Sanchez-Eugenia *et al.*, 2005). The RNA genome also contains a 5′ linked viral protein VPg and a 3′ poly A tail (King and Moore, 1988).

Translation of the ORFs is mediated by two distinct internal ribosome entry sites (IRES): a 5′ untranslated region (5′UTR) IRES and an intergenic (IGR) IRES, the latter of which has been studied extensively at the biochemical and structural levels. A subset of dicistroviruses, specifically the honey bee viruses, *Israeli acute paralysis virus* (IAPV), *Kashmir bee virus* (KBV) and *Acute bee paralysis virus* (ABPV), the fire ant virus, *Solenopsis invicta virus* (SINV-1) and *Cricket paralysis virus* (CrPV), encode an overlapping +1 frame ORF termed ORFx within ORF2, downstream of the IRES (Firth *et al.*, 2009; Sabath *et al.*, 2009; Ren *et al.*, 2014; Au *et al.*, 2015; Kerr *et al.*, 2018b) (Table 5.2). During dicistrovirus infection, the structural proteins are produced in supramolar excess over the non-structural proteins, suggesting that translation of ORF1 and ORF2 are differentially regulated by the IRESs (Moore *et al.*, 1980; Wilson *et al.*, 2000b; Khong *et al.*, 2016).

Dicistroviruses encode up to six copies of the genome linked protein VPg (King and Moore, 1988; Nakashima and Shibuya, 2006). Based on studies of other RNA viruses, VPgs have direct roles

Table 5.1 Functions of dicistrovirus proteins

Viral proteins	Known or predicted* functions	Reference
1A (CrPV, DCV)	Inhibition of RNA interference	Nayak *et al.* (2010)
1A (CrPV)	Inhibition of SG assembly	Khong *et al.* (2011, 2017)
2B	Membrane remodelling*	Aldabe *et al.* (1996)
2C	RNA helicase, membrane permeabilization*	Gorbalenya *et al.* (1990); Mirzayan *et al.* (1994); Rodriguez *et al.*(1993)
3A	Viral replication complex formation*	Fujita *et al.* (2007)
VPgs	Covalently linked to 5' end of genomic RNA for priming viral replication	King and Morre (1988)
3C (PSIV)	Viral poly protein processing	Nakashima *et al.* (2008)
3D	Negative strand RNA synthesis, viral replication	Bonning and Miller (2010)
ORFx (IAPV, ABPV, KBV, SINV-1)	?	Au *et al.* (2015); Firth *et al.* (2009); Ren *et al.* (2014); Sabath *et al.* (2009)
VP1-3	Capsid proteins, virion structure	Tate *et al.* (1999)
VP4 (TrV)	Membrane permeabilization during viral entry	Sanchez-Eugenia *et al.* (2005)
? (CrPV)	Host translation shut off	Garrey *et al.* (2010)
? (CrPV, DCV)	Induction of JAK-STAT pathway	Dostert *et al.* (2005)
N terminal region of ORF1(TrV)	Induction of apoptosis	Mari *et al.* (2002)
? (DCV)	Induction of heat shock responses	Merkling *et al.* (2015)

*Predicted functions based on sequence alignment.

Table 5.2 Members of *Dicistroviridae*

Genus[1]	Members[2]	Natural host[3]	IGR IRES type[4]	ORF2 start codon[5]	+1 Frame ORFx[6]	ORFx length[7]	Reference[8]
Aparavirus	*Acute bee paralysis virus** (ABPV)	Honey bee	II	GCC (Ala)	Yes (CCG, Pro)	92	Govan *et al.* (2000)
	Israeli acute paralysis virus (IAPV)	Honey bee	II	GGC (Gly)	Yes (GCG, Ala)	94	Maori *et al.* (2007)
	Kashmir bee virus (KBV)	Honey bee	II	GCC (Ala)	Yes (CUG, Leu)	93	de Miranda *et al.* (2004)
	Mud crab dicistrovirus (MCDV-1)	Mud crab	II	GCU (Ala)	? (CUC, Leu)	9	Guo *et al.* (2013)
	Solenopsis invicta virus 1 (SINV-1)	Fire ant	II	GCU (Ala)	Yes (CUC, Leu)	125	Valles *et al.* (2004)
	Taura syndrome virus (TSV)	Penaeid shrimp	II	GCU (Ala)	? (CUA, Leu)	16	Mari *et al.* (2002)
Cripavirus	*Aphid lethal paralysis virus* (ALPV)	Aphid	I	GCU (Ala)	? (CUA, Leu)	7	Van Munster *et al.* (2002)
	*Cricket paralysis virus** (CrPV)	Cricket+	I	GCU (Ala)	Yes (AAA, Lys)	41	Wilson (2000)
	Drosophila C virus (DCV)	Fruit fly	I	GCU (Ala)	? (CUA,Leu)	51	Johnson Christian (1998)
	Rhopalosiphum padi virus (RhPV)	Aphid	I	GCA (Ala)	? (CAA, Glu)	2	Moon *et al.* (1998)
Triatovirus	*Black queen cell virus* (BQCV)	Honey bee	I	GCU (Ala)	? (CUG, Leu)	26	Leat *et al.* (2000)
	Himetobi P virus (HiPV)	Planthopper	I	GCA (Ala)	? (CAA, Glu)	14	Nakashima, Toriyama (1999)
	Homalodisca coagulata virus 1 (HoCV-1)	Leafhopper	I	GCA (Ala)	? (CAA, Glu)	14	Hunnicutt (2006)
	Plautia stali intestine virus (PSIV)	Stink bug	I	CAA (Gln)	? (AAG, Lys)	43	Sasaki *et al.* (1998)
	*Triatoma virus** (TrV)	Triatomine bug	I	GCU (Ala)	? (CUG, Leu)	1	Czibener *et al.* (2000)

[1,2]Genus and members of *Dicistroviridae* based on ICTV classification.
[3]Natural hosts based on experimental and observational evidence.
[4]Type of intergenic internal ribosome entry site (IGR IRES).
[5]Start codon of ORF2 based on experimental evidence and/or phylogenetic analysis.
[6]Yes, experimentally validated; ?, has not been tested.
[7]Predicted ORFx amino acid length.
[8]Reference of first published sequence.
+CrPV has a wide range of hosts, including in the orders Diptera, Hemiptera, Hymenoptera, Lepidoptera, Orthoptera
*Indicates type species.

in replication, acting as a primer, and in viral translation (Goodfellow, 2011). Although it is unclear why there are redundant copies in some dicistroviruses, studies on other multi-VPg-containing RNA viruses, such as the picornavirus, *Foot-and-mouth disease virus*, suggest that the multiple VPgs ensure that each progeny RNA molecule is linked to a VPg for efficient replication (Falk *et al.*, 1992). Because translation of non-structural proteins from ORF1 is much lower than that of ORF2, it may be that multiple copies are needed to compensate for the lower expression of non-structural proteins during dicistrovirus infection (Nakashima and Shibuya, 2006).

Taxonomy

Though originally proposed to be picorna-like viruses, it is now apparent that dicistroviruses belong to a distinct viral family based on their bicistronic genome arrangement as compared to the picornavirus monocistronic genome. Moreover, dicistrovirus non-structural proteins are upstream of the structural proteins, the reverse of picornaviruses. *Dicistroviridae* is one of six families belonging to the order *Picornavirales* of (+) ssRNA viruses. Based on alignments of the RdRp protein sequences, they are most similar to members of the subfamily *Comoviridae*, plant viruses containing two genomic RNAs: one encoding the structural proteins and the other encoding the non-structural proteins (Chen and Siede, 2007; Koonin *et al.*, 2008). At the time of writing, there are three genera within the *Dicistroviridae* based on phylogenetic analysis of the IGR IRES as classified by the International Committee on Taxonomy of Viruses (ICTV): *Cripavirus*, *Aparavirus* and *Triatovirus* (Valles *et al.*, 2017) (Table 5.2 and Fig. 5.2). *Aparavirus* is further divided into distinct, divergent clades: the insect-infecting aparaviruses and the crustacean-infecting aparaviruses. As more dicistrovirus sequences are classified (Shi *et al.*, 2016) *Aparavirus* may be reclassified into two distinct genera based on differences in host, details of the IGR IRES and the presence of an ORFx (Firth *et al.*, 2009; Sabath *et al.*, 2009).

Members of the genus *Aparavirus* include the honey bee infecting IAPV, KBV and ABPV, the red fire ant infecting SINV-1 and the penaeid shrimp-infecting *Taura syndrome virus* (TSV). Several of these aparaviruses have had a negative impact in agriculture and on the economy (see below). Dicistroviruses also have indirectly linked to human health. For example, SINV-1 has been proposed as a possible biopesticide for fire ant infestations (Valles, 2012), and the *Triatoma virus* (TrV), a member of the genus *Triatovirus*, can infect Triatominae, which can transmit *Trypanosoma cruzi*, the parasite that causes Chagas disease (Marti *et al.*, 2017). Members of the genus *Cripavirus*, including *Cricket paralysis virus* (CrPV) and *Drosophila C virus* (DCV), have wide host ranges, including the genetic model organism *Drosophila melanogaster* (Plus *et al.*, 1978; Masoumi *et al.*, 2003). As such, CrPV and DCV infections of *D. melanogaster* are

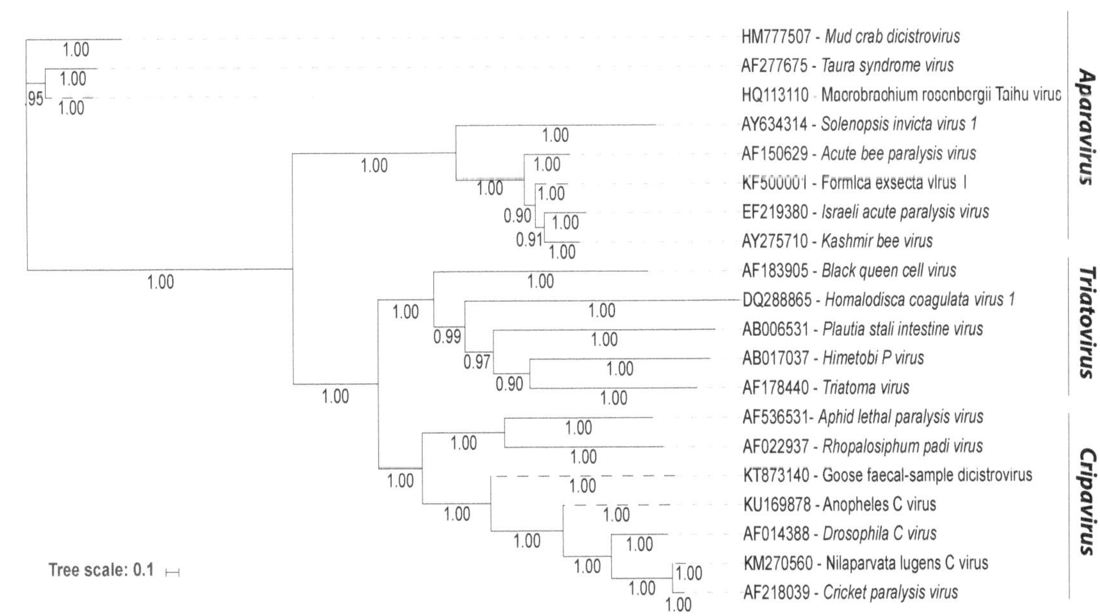

Figure 5.2 Phylogenetic tree of dicistroviruses classified by the ICTV. Midpoint rooted phylogenetic tree constructed from full-length structural polyprotein amino acid sequences of accepted and classified dicistroviruses. In addition, a few unclassified metagenome-derived sequences were included and are rooted using dashed lines. Analysis was performed using Bayesian inference of phylogeny. Bootstrap probability values to two decimal places are given and labelled at the nodes. The current genera, as described by ICTV are shown. Analysis was run by ICTV (Valles *et al.*, 2017) and the tree was manipulated using iTOL (Letunic and Bork, 2016).

used successfully as models for studying fundamental dicistrovirus–host interactions.

In a landmark transcriptomics paper, several hundred invertebrate viruses were discovered from a wide variety of invertebrate species based on sequence alignment using RdRps, including over a hundred novel dicistro-like viruses (Shi *et al.*, 2016). Although the new dicistro-like viruses remain to be classified by ICTV, the identification of these new genomes highlights the diversity of this viral family across a wide spectrum of host species.

Virion structure

The structures of the dicistroviruses CrPV (Tate *et al.*, 1999), TrV (Squires *et al.*, 2013), IAPV (Mullapudi *et al.*, 2016), and BQCV (Spurny *et al.*, 2017) have been solved (Fig. 5.3). Dicistroviruses are non-enveloped viruses that are roughly spherical, approximately 30 nm in diameter and exhibit icosahedral, $T = 3$ symmetry comprising of 60 protomers, each consisting of the three major capsid proteins, VP1, VP2 and VP3. The virions are structurally similar but not identical. Each capsid protein contains a jelly roll structure comprised of eight β sheets in a β barrel arrangement. VP1 subunits form a pentamer around a 5-axes fold, and the VP2 and VP3 form heterohexamers around the icosahedral 3-fold axes. The structures also revealed that the small protein VP4 is attached to the inner surface of the viral asymmetric unit and is located between the protein shell and RNA within the capsid, possibly contacting the viral genome. VP4 is also present in crystallized virions that are infectious.

Dicistroviruses assemble as immature particles containing the precursor protein VP0, which consists of VP3 and VP4. Cleavage of VP0 into VP3 and VP4 is proposed to occur after the particles encapsidate the viral RNA genome, supported by the observation that empty capsids contain unprocessed VP0 (Tate *et al.*, 1999). A conserved Asp-Asp-Phe (DDF) motif within VP1 protein may be involved in this cleavage process (Squires *et al.*, 2013; Mullapudi *et al.*, 2016; Spurny *et al.*, 2017). The DDF motif is in close proximity of the N terminus of VP3 and the C terminus of VP4, suggesting that this motif catalyses this cleavage. Interestingly, TrV VP3 contains a second autoproteolytic DDF motif in addition to the conserved motif within VP1 (Squires *et al.*, 2013). The function of this additional motif has yet to be elucidated.

Dicistrovirus virions are stable in acidic conditions (to pH 3.0) whereas alkaline conditions (pH > 8.0) trigger uncoating by releasing the interactions

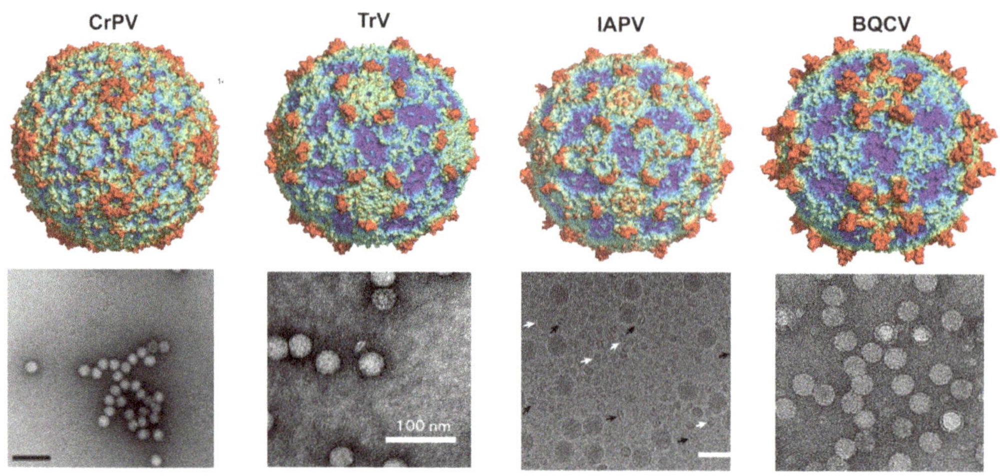

Figure 5.3 Dicistrovirus structures. Shown are crystal structures of capsids from CrPV, TrV, IAPV and BQCV (Spurny *et al.*, 2017). PDB structure ID: 1B35, 3NAP, 5LWG and 5MQC. Negative stained electron micrographs of purified CrPV (Kerr *et al.*, 2015), TrV (Valles *et al.*, 2017), IAPV (Mullapudi *et al.*, 2016) and BQCV (Spurny *et al.*, 2017). Figures are republished with permission from Kerr *et al.* (2015; Copyright © 2015, American Society for Microbiology. All Rights Reserved. doi:10.1128/JVI.00463–15), Mullapudi *et al.* (2016), Spurny *et al.* (2017) and Valles *et al.* (2017).

between capsid proteins, as demonstrated using TrV virions (Snijder *et al.*, 2013). Uncoating occurs through an intermediate step in which the virions disassemble into pentons, with concomitant release of VP4.

Transmission

While the majority of dicistroviruses are transmitted horizontally via a faecal–oral route (Jousset *et al.*, 1972; Gomariz-Zilber *et al.*, 1995; Muscio *et al.*, 2000; Chen and Siede, 2007), some dicistroviruses may be transmitted vertically. For example, CrPV can be transmitted vertically transovum (with the virus on the surface or inside of the egg) (Reinganum *et al.*, 1970). ALPV, *Plautia stali intestine virus* (PSIV), TrV and SINV-1 can all be transmitted vertically by transovarial transmission, i.e. within the ovary with the virus inside the egg (D'Arcy *et al.*, 1981b; Hatfill *et al.*, 1990; Muscio *et al.*, 1997; Nakashima *et al.*, 1998). Viruses of sap-sucking insects (Hemiptera) such as RhPV, can be transmitted via plants. RhPV uses plants as a vehicle for spread, by circulating in the phloem of the host plant from which it can be acquired by other aphids (Gildow and d'Arcy, 1988; Ban *et al.*, 2007). The dicistro-like Big-Sioux River virus (BSRV) has also been found in maize (Ban *et al.*, 2007; Wamonje *et al.*, 2017). These examples highlight the use of plants as vectors for the spread of dicistroviruses to sap-sucking insects.

Viral entry, polyprotein processing and replication

Viral entry

Molecular details of dicistrovirus entry and replication have mostly come from studies of DCV. Using DCV-infected *Drosophila* S2 cells as a model, viral entry occurs through the clathrin-mediated endocytosis pathway and involves nuclear protein Bub1 (Budding uninhibited by benzimidazoles 1), a highly conserved subunit of the kinetochore complex that regulates chromosome congression (Cherry and Perrimon, 2004; Yang *et al.*, 2018). DCV infection results in more Bub1 translocated to the membrane, supporting a model where Bub1 regulates endocytosis to promote viral entry, possibly interacting with a clathrin-dependent endocytic compartment. In support of this, *bub1* deficient flies are resistant to DCV infection (Yang *et al.*, 2018). Cell attachment and entry is likely mediated through an initial attachment with a host cell surface receptor, which has yet to be identified. Following entry, the dicistrovirus uncoats, releasing its genome into the cytoplasm; details of this process are beginning to emerge. Purified VP4 capsid protein from TrV can induce permeabilization by inserting into the membrane and forming discrete pores, which may enable cell entry or genome transfer during virus infection (Sánchez-Eugenia *et al.*, 2015). Cryo-EM analysis of empty and full TrV particles has revealed that RNA release leads to empty capsids of similar size to the native virions that displayed no prominent conformational changes (Agirre *et al.*, 2013). According to the tectonic model for subunit movement, individual capsid proteins rotate within the capsid complex, leading to the disassembly of empty TrV capsids into small symmetrical, lip-shaped particles that are probably dimers of pentamers of capsid protein promoters. Cryo-EM structures of empty IAPV virions and virions that are about to release the viral genome suggest a distinct mechanism of genome release (Mullapudi *et al.*, 2016). The minor capsid proteins also detach from the capsid protein upon genome release. Finally, although structurally similar to the picornavirus capsid proteins, the CrPV capsid lacks the canyon that surrounds the five-fold symmetry identified in the poliovirus structure, which is important for poliovirus entry, suggesting a distinct mechanism of entry for CrPV (Tate *et al.*, 1999).

Polyprotein processing

Upon entry, the positive-strand dicistrovirus RNA genome is released and directly used as a template for viral protein synthesis (see below for detail on viral IRES translation). Indeed, transfection of an *in vitro* transcribed CrPV infectious RNA in *Drosophila* S2 cells can directly lead to viral protein synthesis and productive infection (Kerr *et al.*, 2015). Translation of the two main ORFs results in expression of polyproteins that are then processed into mature viral proteins (Moore *et al.*, 1980). The majority of the processing is mediated by the virally encoded 3C-like protease, a cysteine protease predicted to adopt a similar fold to picornavirus 3C protease (Nakashima and Nakamura, 2008). The 3C cleavage sites in the polyprotein of PSIV have been identified (Nakashima and Ishibashi, 2008,

2010). N-terminal sequencing of processed viral proteins showed cleavage occurs at Q/GWSLW and Q/NGVFD, which correspond to the 2A/3C and 3C/3D cleavage sites in the PSIV ORF1, respectively. The 2B/2C cleavage site has been mapped to amino acids 408–409 (Q/D), which are conserved among dicistroviruses. Q/A has also been observed at the putative cleavage junction of VP2/VP0. Collectively, these results indicate that the dicistrovirus 3C protease preferentially cleaves between Q/A, Q/G, Q/N and Q/D residues. In addition, the conserved Q/C, Q/A and Q/V sites between 3C/RdRp, VP2/VPO and VP0/VP1, respectively, are predicted to be cleaved by the 3C protease. However, this has not been experimentally validated (Fig. 5.1). Importantly, polyprotein processing can be recapitulated by incubating a CrPV infectious clone in an insect translation extract, thus providing a model to study protease-mediated cleavages (Kerr *et al.*, 2015). In a CrPV clone with a mutation that prevents 3C protease expression, ORF2 structural polyprotein processing is blocked, suggesting that ORF2 polyprotein is processed by 3C protease. However, ORF2 polyprotein could also be processed by other means. VP0 may undergo an autocleavage mechanism upon capsid formation involving a DDF motif present in dicistrovirus VP1 and in both VP1 and VP3 in TrV (Tate *et al.*, 1999; Agirre *et al.*, 2013).

Although the ability of 3C protease or a cellular protease to mediate cleavages at other junctions of the viral polyprotein is unclear, a subset of dicistroviruses including CrPV, DCV, IAPV, KBV and ABPV possess a 2A peptide sequence that mediates a 'self-cleavage' process. The 2A peptide sequence, located between 1A and 2B of ORF1, undergoes an unusual 'stop-go' translation mechanism leading to 'cleavage' of the protein (Luke *et al.*, 2008) (Fig. 5.1). The 2A peptide encodes a conserved DxExNPGP sequence, which interacts with the ribosome exit channel to induce stop-codon independent translation termination ending at the second to last glycine residue (Brown and Ryan, 2010). However, instead of the ribosome dissociating, the ribosome continues translation starting with a proline codon. The 2A peptide mechanism, which has been termed as 'stop-go' or 'stop-carry' translational recoding mechanism, is present in several positive-strand RNA viral genomes as a viral strategy to produce two protein products from a single polyprotein ORF. The 2A peptide of CrPV, DCV, IAPV, KBV and ABPV 'stop-go' translation has been experimentally tested *in vitro*, showing 88–95% cleavage activity (Donnelly *et al.*, 2001; Luke *et al.*, 2008). The conserved 2A peptide DxExNPGP sequence is present near the N-terminus of ORF1 in only a subset of dicistroviruses. It is not clear how those dicistroviruses lacking the 2A peptide mediate cleavage between the 1A and 2B non-structural proteins, nor why only a subset of dicistroviruses utilize this mechanism. Another remaining question is why the 2A peptide is utilized to separate 1A and 2B proteins, rather than a protease cleavage site.

Replication

Replication of the dicistrovirus RNA genome is mediated by the RdRp (3D) protein encoded within ORF1. A CrPV infectious clone containing a mutant catalytically-inactive RdRp does not support replication (Khong *et al.*, 2016). Like other RNA viruses, the plus strand RNA genome is used as a template for both translation and replication. RdRp synthesizes the negative strand RNA, which in turn transcribes the positive-strand RNA. DCV has been used as a model for study of dicistrovirus replication. Like other RNA viruses, DCV replication occurs on membrane structures, which are formed in part from remodelling of the Golgi apparatus to form a novel intracellular vesicular compartment where virus replication takes place (Cherry *et al.*, 2006). In support of this model, the coat protein complex 1 and fatty acid biosynthesis pathway are required for DCV replication. The vesicles are non-uniform in size, averaging 115 nm in diameter (Cherry *et al.*, 2006). Co-localization of the DCV helicase protein with the Golgi marker DG13 suggests that this is the site of viral replication (Cherry *et al.*, 2006). The molecular details underlying dicistrovirus replication have yet to be determined, but it is likely similar to picornaviruses, whose replication machinery involves RdRp in complex with or requiring host proteins and the viral-linked protein VPg, all of which interact with structural elements at the 5′ and 3′ ends of the dicistrovirus RNA genome (Paul and Wimmer, 2015). A recent study uses structural RNA probing approaches to begin modelling secondary structures at the 5′ end of the CrPV genome (Gross *et al.*, 2017). The importance of these secondary structures for replication has not been confirmed.

Packaging and egress

Encapsidation of the dicistrovirus RNA genome with capsid proteins has not been studied in detail; however, virion structures have been solved, which provide hints about this process (see section on virion structure). The release of the virus from infected cells also remains to be studied. Specifically, the mechanisms that underlie lytic versus non-lytic dicistroviruses are poorly understood, but the distinct dicistrovirus pathogenesis and transmission strategies probably dictate the route of viral egress.

Dicistroviruses are classified as nonenveloped viruses. An emerging theme is that classically defined nonenveloped viruses, such as Hepatitis A and C viruses and poliovirus, can hijack exosome-like or autophagic vesicles for virion release (Feng *et al.*, 2013; Ramakrishnaiah *et al.*, 2013; Bird *et al.*, 2014). A recent study shows that CrPV may also hijack the exosome pathway for viral release (Kerr *et al.*, 2018a). CrPV structural proteins and RNA are found in exosome-like vesicles (ELVs) and the ELVs from CrPV-infected cells are infectious. This suggests that the classically defined nonenveloped viruses of invertebrates and vertebrates may acquire an envelope as a fundamental viral strategy to evade host immune responses and/or facilitate transmission. In this context, it will be interesting to determine whether the use of ELVs extends to other dicistroviruses.

Unique translation mechanisms for dicistroviral protein synthesis

All viruses must use the host ribosome for viral protein synthesis. Some viruses have evolved elegant strategies to hijack the ribosome (Jan *et al.*, 2016). One distinguishing feature of *Dicistroviridae* is their use of distinct IRESs to drive translation of the two ORFs (Wilson *et al.*, 2000b). Translation of the upstream ORF1 is mediated by the 5′UTR IRES, whereas downstream ORF2 translation is mediated by the intergenic IRES (IGR IRES). The IRES arrangement allows for differential regulation of expression of the non-structural and structural proteins during infection, presumably as a viral strategy to optimally coordinate the different steps of the viral life cycle (Wilson *et al.*, 2000b; Khong *et al.*, 2016). While the mechanisms underlying the 5′UTR IRES are emerging, the IGR IRES mechanism has been extensively studied at the biochemical and structural level. This has yielded tremendous insights into both ribosome function and the translational control strategies viruses use to hijack the host ribosome. We provide a brief overview of canonical cap-dependent translation before delving into dicistrovirus IGR and 5′UTR IRES-mediated translation.

Canonical translation versus internal ribosome entry sites

The majority of eukaryotic mRNAs utilize a cap-dependent scanning mechanism to recruit the ribosome and initiate translation from an AUG codon. Typically, an mRNA contains a 5′ m7G cap and a 3′ poly A tail, which act synergistically to mediate translational initiation and to protect the mRNA from ribonucleases (Jackson *et al.*, 2010; Hinnebusch *et al.*, 2016). There are ≈ 12 core translation initiation factors that orchestrate a sequential process of 80S ribosome assembly at the AUG start codon. Briefly, the 5′ cap is bound to the complex eukaryotic initiation factor 4F complex, consisting of the cap-binding protein, eIF4E, the RNA helicase eIF4A, and the scaffold protein eIF4G. eIF4G interacts with several proteins, including the eIF3/40S subunit and the poly A binding protein, the latter of which promotes translation by mediating circularization of the 5′ and 3′ ends of the mRNA. The ternary complex eIF2:Met-tRNA$_i$:GTP along with eIF1 and eIF1A are recruited to the 43S preinitiation complex (PIC) at the 5′ end. Scanning of the PIC occurs until the appropriate AUG codon is recognized via anticodon-codon base pairing of the Met-tRNA$_i$. The GTPase activating protein eIF5 is recruited to stimulate GTP hydrolysis on eIF2, thereby leading to dissociation of factors and 60S subunit joining to assemble an elongation-competent 80S ribosome. At this step, the AUG codon and Met-tRNAi occupy the ribosomal P site, leaving the A site empty to allow delivery of the first aminoacyl tRNA by elongation factor 1A. This begins the elongation step of translation.

An alternate mechanism of translation initiation is through an IRES. IRESs, found in a subset of viral and cellular mRNAs, can direct translation under virus infection and cellular stress when cap-dependent translation is compromised (Leppek *et al.*, 2018; Mailliot and Martin, 2018). IRESs are structured RNAs that use limited initiation factors to recruit the ribosome. The limited factor requirement may allow IRES-containing mRNAs to be preferentially translated when global cap-dependent translation is inhibited. Viral IRESs have been studied extensively through reconstitution experiments and RNA structural probing analysis, revealing key conserved structural elements important for recruitment of the factors/ribosome. Viral IRESs are classified into four main groups based on these properties. The *Dicistroviridae* IGR IRES represents the most streamlined class of IRESs that can recruit the ribosome through an unprecedented mechanism.

Dicistrovirus IGR IRES – factorless IRES translation

The first hint that dicistrovirus IGR IRES is unique was the discovery that translation starts at a non-AUG codon (Sasaki and Nakashima, 1999; Wilson *et al.*, 2000a). For example, the CrPV IGR IRES initiates translation at a GCU alanine codon (Table 5.2 and Fig. 5.4). PSIV and CrPV IRESs have been modelled using extensive secondary structure modelling, RNA structural probing analysis and mutational analysis (Wilson *et al.*, 2000a; Kanamori and Nakashima, 2001; Jan and Sarnow, 2002). The core IGR IRES is approximately 180 nucleotides in length and adopts a triple pseudoknot (PK) RNA structure (Fig. 5.4). The RNA structure is separated into two domains that dictate function. PKII and PKIII domains are responsible for ribosomal recruitment, and PKI is responsible for ribosome positioning at the non-AUG start codon (Wilson *et al.*, 2000a; Kanamori and Nakashima, 2001; Jan and Sarnow, 2002; Costantino and Kieft, 2005; Kamoshita *et al.*, 2009) (Table 5.2). The IRES also contains unpaired nucleotides within loops and bulge regions (ex. L1.1, SLIV and SLV) that function to interact with specific regions of the ribosome (Fig. 5.4) (Pfingsten *et al.*, 2006; Schüler *et al.*, 2006; Jang *et al.*, 2009). Another unique aspect of dicistrovirus IGR IRES is that the IRES can directly

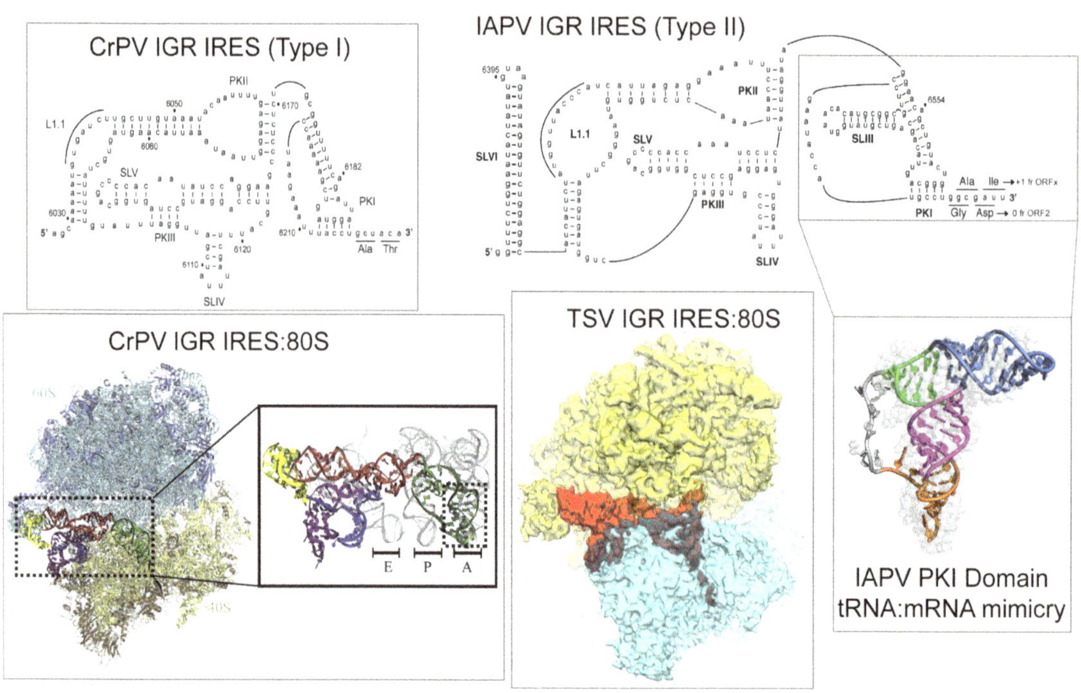

Figure 5.4 The dicistrovirus IGR IRES. (Top) The secondary structures of CrPV IGR IRES (Type I) and IAPV IGR IRES (Type II) are shown. (Bottom left) Cryo-EM structure of the Type I CrPV IGR IRES bound to the 80S ribosome (PDB ID 4V91). Figures republished from Fernandez *et al.* (2014), licensed under CC BY 3.0. (Bottom centre) Cryo EM structure of the Type 2 TSV IGR IRES bound to the 80S ribosome (PDB ID: 3J6Y). Figure republished with permission from Koh *et al.* (2014). (Bottom right) Global shape mimicry of tRNA by the IAPV IGR IRES PKI domain. The IAPV IGR IRES NMR/SAXS structure of the IAPV IRES PKI domain (PDB ID: 2N8V) is superimposed over $tRNA^{phe}$ (PDB ID:4TNA, in grey).

recruit the ribosome without initiation factors or initiator Met-tRNA$_i$ (Sasaki and Nakashima, 2000; Wilson *et al.*, 2000a). *In vitro* binding studies using purified ribosomal subunits show that the IRES binds to the purified 40S and 80S ribosomes tightly, with an affinity of 5–20 nM (Jan and Sarnow, 2002; Jan *et al.*, 2003). Reconstitution experiments using purified ribosomes, elongation factors, and aminoacyl-tRNAs showed that the IGR IRES can support translation in a minimal eukaryotic translation system (Jan *et al.*, 2003; Pestova and Hellen, 2003). Thus, the IGR IRES directly interacts with and manipulates the ribosomes to drive translation in *cis*, acting as its own mRNA template for protein synthesis.

The IGR IRESs are divided into Type I and II IRESs based on distinct elements. Type II IRESs contain a larger L1.1 domain, an extra stem–loop called SLIII within the PKI domain and lack a conserved UAC within the ribosome binding domain (Table 5.2, Fig. 5.4). Despite these differences, IRESs are functionally and structurally similar (Hatakeyama *et al.*, 2004; Cevallos and Sarnow, 2005; Pfingsten *et al.*, 2007; Nakashima and Uchiumi, 2009). Indeed, PKI domains can be functionally swapped between Type I and II IRESs (Jang and Jan, 2010; Hertz and Thompson, 2011).

Crystallography and cryo-EM structures of the IGR IRES have provided insights into the mechanism whereby the viral RNA actively manipulates the host ribosome to begin translation. The IGR IRES directly interacts with both the 40S and 60S subunits for 80S assembly, spanning all three tRNA ribosomal sites within the inter-subunit space of the 80S ribosome; PKII and PKIII occupy the E and P sites, while PKI occupies the A site (Spahn *et al.*, 2004; Schüler *et al.*, 2006; Fernández *et al.*, 2014; Koh *et al.*, 2014; Au *et al.*, 2015). Remarkably, the PKI domain, which positions the ribosome to start translation at a non-AUG codon, mimics a tRNA anticodon–codon interaction, thus explaining how the IRES occupies the ribosomal tRNA sites for initiation (Fig. 5.4) (Costantino *et al.*, 2008). Indeed, the PKI domain can be functionally swapped with a tRNA (Au and Jan, 2012). The PKII and PKIII domains adopt the compact globular core required for 40S and 60S recruitment to assemble the 80S ribosome (Pfingsten *et al.*, 2006; Schüler *et al.*, 2006). Specifically, SLIV and SLV contact ribosomal proteins rpS5 and eS25 of the 40S subunit, and the L1.1 domain interacts with the L1 stalk of the 60S subunit. Mutations within these domains block 40S or 60S subunit recruitment (Jan and Sarnow, 2002; Costantino and Kieft, 2005; Jang *et al.*, 2009).

The IRES also induces conformations within the ribosome that are important for IRES translation. For example, the ribosome bound to the IRES oscillates between canonical and rotated states, conformations that occur during translation normally (Fig. 5.5) (Fernández *et al.*, 2014; Petrov *et al.*, 2016). The rotated state involves a counter-clockwise rotation of the 40S subunit relative to the 60S subunit. This state involves the 40S domain swivelling, as well as the movement of the L1 stalk of the ribosome. Since the L1.1 loop of the IRES interacts with the L1 stalk, it couples the IRES to the movements of the ribosome and forces it to adopt conformations that depend on the rotational state of the ribosome.

Cryo-EM reconstructions of the IRES:ribosome complexes undergoing translocation by addition of elongation factors and aminoacyl-tRNAs have revealed additional conformations that facilitate IRES translation (Muhs *et al.*, 2015; Abeyrathne *et al.*, 2016; Murray *et al.*, 2016). When the IRES is in a rotated state, elongation factor eEF2-GTP can bind to the ribosome, interacting with the ribosomal A site and PKI of the IRES, thus promoting translocation of the PKI domain from the A to P sites (Fig. 5.5). Delivery of the next aminoacyl-tRNA by the action of eEF1A stabilizes the IRES:ribosome such that the PKI is committed to the P site. Interestingly, without delivery of the next aminoacyl-tRNA, the IRES will undergo back-translocation whereby the PKI domain will translocate back to the A site. This entire process mediated by the IRES, called pseudotranslocation, is unique because it occurs without peptide bond formation. A recent cryo-EM study has reconstituted translocation of the IRESs to the ribosomal E site, thus IRESs undergo two translocation steps on the ribosome. This revealed a surprising conformational change within the IRES: the PKI domain flips from the decoding E site in the 40S to the 60S subunit, resembling an acceptor tRNA (Pisareva *et al.*, 2018). This remarkable finding suggests that the viral IRES undergoes dynamic RNA gymnastics to manipulate the ribosome. Importantly, all of these structural studies complement and support the extensive functional, mutational and biochemical

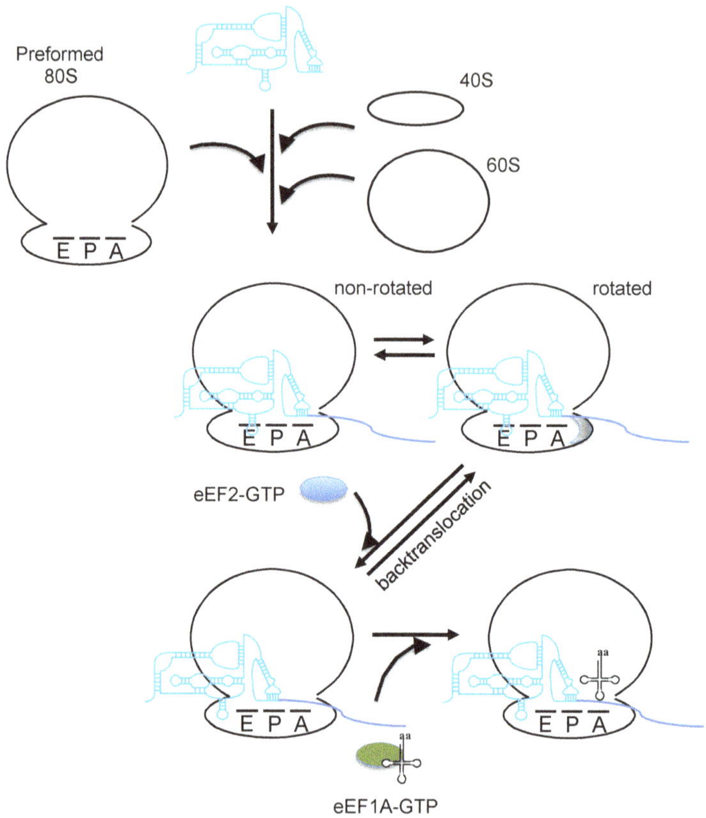

Figure 5.5 Model of IGR IRES mechanism. The 80S assembles on the IGR IRES through sequential recruitment of 40S and 60S or direct recruitment of a preformed 80S ribosome. Once in complex with the 80S ribosome, the ribosome alternates between a rotated and non-rotated state with the PKI tRNA anticodon:codon domain occupying the ribosomal A site. Pseudotranslocation occurs via the action of eEF2 with the PKI domain translocating from the P to A site of the ribosome. Back translocation of the PKI domain can occur unless there is delivery of the first aminoacyl-tRNA by eEF1A to lock the PKI domain in the ribosomal P site.

studies on the IRES (Yamamoto *et al.*, 2007). For example, mutations within the variable loop region (VLR) prevent pseudotranslocation (Ruehle *et al.*, 2015). In essence, the IRES hijacks the ribosome by 'tricking' the ribosome to start translation in the elongation mode of translation.

The use of single molecule fluorescence spectroscopy to capture the dynamics of IRES:ribosome interactions and conformational changes in real time have revealed that the IGR IRES can recruit 80S ribosomes through either sequential 40S then 60S binding or by recruitment of pre-formed 80S ribosomes (Petrov *et al.*, 2016) (Fig. 5.5). Monitoring the kinetic pathway of the initial steps of IGR IRES translation has revealed relatively slow initial translocation steps, which may suggest a quality control mechanism to ensure reading frame selection (Zhang *et al.*, 2016; Bugaud *et al.*, 2017). The future use of advanced biophysical approaches on IRES–ribosome complexes will provide new insights into this viral translational mechanism.

Although there have been studies on IGR IRES translation using reporter constructs in insect cell culture (Fernandez *et al.*, 2002; Wang *et al.*, 2013), the majority of studies of IGR IRESs have used *in vitro* approaches without the context provided by an *in vivo* virus infection. The development of the CrPV infectious clone allows reverse genetics approaches to test these *in vitro* models in a more physiologically realistic setting (Kerr *et al.*, 2015). Using this clone, mutations within the IGR IRES were tested for their effects on viral titre, translation and replication (Kerr *et al.*, 2016). Although most mutations tested that inhibited IRES translation *in vitro* correlated with a decrease in virus yield, some mutations did not correlate or could not be simply explained by a defect in IRES translation. This study highlights the importance of *in vivo* studies to monitor IRES translation mechanisms in the context of viral infection.

In summary, IGR IRESs directly interact with and manipulate the ribosome by inducing coordinated conformational changes on the ribosome that resemble tRNA–ribosome interactions during translation in general. As such, studies on this factorless translation initiation mechanism will continue to shed light into fundamental ribosome functions and viral translational control strategies.

A third overlapping ORF – IGR IRES-mediated translation of alternative reading frames

Phylogenetic analysis revealed an extra overlapping +1 frame ORF (termed ORFx) within ORF2 of a subset of dicistroviruses which includes the honey bee viruses IAPV, KBV and ABPV as well as the fire ant virus SINV-1 (Firth *et al.*, 2009; Sabath *et al.*, 2009) (Table 5.2 and Fig. 5.1B). Subsequent investigations demonstrated that these IGR IRESs can direct translation in two reading frames: the 0 frame structural protein ORF2 and the +1 frame ORFx (Ren *et al.*, 2012). Using an *in vitro* translation system, the IAPV IGR IRES directed +1 frame translation at ~20% frequency of 0 frame translation. Specific elements within the PKI domain and an adjacent U:G wobble base pair mediate reading frame selection. RNA structural probing analyses suggest that the IRES also adopts distinct conformations to direct reading frame selection (Ren *et al.*, 2014; Au *et al.*, 2015). Finally, NMR/SAXS has revealed that the PKI domain of IAPV IRES mimics an RNA structure that resembles a tRNA shape, with SLIII acting as an acceptor stem of a tRNA (Fig. 5.4) (Au *et al.*, 2015). Thus, IGR IRESs not only hijack the ribosome for viral protein synthesis but can also direct translation in two reading frames, constituting a novel recoding translation mechanism for increasing the viral genome's coding capacity.

An extra SLVI is also present immediately upstream of the IGR IRESs of the honey bee viruses IAPV, KBV and ABPV IGR (Fig. 5.4) (Firth *et al.*, 2009). In fact, the stop codon for ORF1 is within the SLVI domain, hinting that translation of ORF1 may affect translation of ORF2. Indeed, mutations that disrupt SLVI formation moderately inhibit IGR IRES translation. This suggests both that the integrity of SLVI contributes to IGR IRES translation and that ORF1 and ORF2 translation may be coupled (Au *et al.*, 2018; Ren *et al.*, 2012). A recent study using an IAPV/CrPV chimeric virus clone showed that IAPV SLVI is important for virus infection (Au *et al.*, 2018).

IRES-mediated recoding is not restricted to the honey bee and fire ant dicistroviruses. The CrPV IGR IRES also directs +1 frame translation to produce a small 41 amino acid ORFx protein (Fig. 5.1B) (Kerr *et al.*, 2018b). However, unlike honey bee IGR IRESs, the CrPV IRES directs ribosome bypass, where a subset of ribosomes recruited to the CrPV IRES reposition downstream by 37 nucleotides to initiate translation of the +1 frame ORFx. Blocking ORFx expression by inserting a stop codon within the ORFx reading frame attenuates virus infection in fruit flies, thus demonstrating the biological relevance of CrPV ORFx in virus infection.

The function and specific role of ORFx in virus infection have yet to be determined. Mass spectrometry analysis has shown that ORFx is expressed in IAPV-infected honey bees and in CrPV-infected *Drosophila* cells, a strong indication that ORFx has a role in virus infection and/or pathogenesis (Ren *et al.*, 2012; Kerr *et al.*, 2018b). The ORFx of CrPV and honey bee dicistroviruses do not have any obvious sequence similarity, nor do they have any obvious homology to known proteins (Fig. 5.1B). However, CrPV ORFx has a predicted transmembrane domain and can associate with the ER membrane (Kerr *et al.*, 2018b). Although the CrPV ORFx importance to virus infection has been established (Kerr *et al.*, 2018b), an infectious IAPV clone is not available to test IAPV ORFx functionality via reverse genetics approaches. However, the recent development of a chimeric IAPV/CrPV infectious clone with CrPV IGR IRES swapped with IAPV IGR IRES may improve future research into the role of the IAPV IGR IRES and ORFx protein (Au *et al.*, 2018).

Ribosome heterogeneity and IGR IRES translation

Because of its streamlined mechanism, the IGR IRES has become a model for understanding more complex IRES mechanisms, including those of clinically relevant viruses such as *Human immunodeficiency virus* (HIV) and *Hepatitis C virus* (HCV). Some of the conformational changes within the IGR IRES and ribosome resemble that of HCV IRES:ribosome interactions (Spahn *et al.*, 2004; Pisareva *et al.*, 2018), suggesting that these IRES-induced conformations are intrinsic to ribosome function. In a surprising twist, work with IGR IRES has led to investigations into insights of ribosome heterogeneity. As discussed, the IGR IRES contacts the ribosome directly; for example, SLV interacts with eS25 (Nishiyama *et al.*, 2003; Schüler *et al.*, 2006; Muhs *et al.*, 2011). Interestingly, depletion of some ribosomal proteins (e.g. eS25 and eS6) inhibits IGR IRES- but not cap-dependent translation, suggesting that distinct ribosomes can mediate translation of specific mRNAs (Landry *et al.*, 2009; Muhs *et al.*, 2011). These findings have led to research showing that the depletion of other ribosomal proteins not only impacts IGR IRES translation, but also that of *Hepatitis C virus* and cellular IRESs as well as ribosome shunting, indicating that ribosome heterogeneity may be an important factor in dictating translation of specific mRNAs (Cherry *et al.*, 2005; Hertz *et al.*, 2013; Carvajal *et al.*, 2016). Besides ribosomal proteins, IGR IRES has been linked to pseudouridylation in ribosomal RNA. Experimental depletion of pseudouridines led to ribosomes with reduced affinity for the IGR IRES that could not support IRES translation, despite having limited impact on overall cap-dependent translation (Jack *et al.*, 2011). These findings demonstrate that ribosome heterogeneity deriving from both post-translational and post-transcriptional modifications and from specialized functions of individual ribosomal components can affect the translation of specific mRNAs. The challenge is to identify the specific modifications and/or translation initiation complexes of importance. Clarification is still needed on whether specialized ribosomes are a major determinant in dicistrovirus infection.

The fact that dicistrovirus IGR IRES can bind directly to the conserved tRNA-binding core of the ribosome to direct translation could explain how this IRES can function across eukaryotes including yeast, insect and mammalian cells (Thompson *et al.*, 2001; Fernandez *et al.*, 2002). IRES can also bind to prokaryotic ribosomes and function in bacteria; however, the mechanism of 80S assembly is distinct from that in eukaryotic cells (Zhu *et al.*, 2011; Colussi *et al.*, 2014). The simplicity of the IGR IRES's direct recruitment of ribosomal machinery is a powerful model to understand ribosome function and dynamics. Furthermore, it provides insight into tRNA–ribosome interactions that mediate reading frame selection and maintenance, and highlights diverse strategies for directing viral protein synthesis.

5′UTR IRES

Despite the extensive characterization of the IGR IRES, studies of the 5′UTR IRES of dicistroviruses have been limited, partly due to the lack of obvious sequence and structural conservation (Jan, 2006). The 5′UTR IRESs range in length from 27 (SINV-1) to 964 (HiPV) nucleotides. This IRES mechanism has only been studied for the CrPV, RhPV and the dicistro-like Halastavi árva virus (HalV) 5′UTR IRESs. Biochemical reconstitution methods using purified translation factors and ribosomes assembled on the IRES have shown that HalV and RhPV IRESs use a similar subset of factors (eIF1, eIF1A, eIF2 and eIF3) to promote AUG recognition by the preinitiation 40S complex (Woolaway *et al.*, 2001; Roberts and Groppelli, 2009; Abaeva *et al.*, 2016). RhPV IRES is strongly dependent on the RNA helicase eIF4A, but the HalV IRES is not. Both HalV and RhPV use an unstructured region within the 5′UTR or near the AUG codon to recruit the 40S subunit. Once bound to the RhPV IRES, the 40S scans to locate the AUG codon. In contrast, binding of 40S to the HalV IRES results in retrograde scanning of the ribosome to locate the AUG codon (Abaeva *et al.*, 2016). Interestingly, both HalV and RhPV 5′UTR IRESs are functional across kingdoms; the HalV and RhPV 5′UTR IRESs are functional in plant, mammalian and insect cell lysates. Thus, there may be common properties between some dicistrovirus 5′UTR IRESs, despite the differences in their sequences and structures.

In contrast to RhPV and HalV, the 5′UTR IRES of CrPV does not use a scanning mechanism. Instead, it directly recruits the ribosome to the AUG codon (Gross *et al.*, 2017). RNA structural probing revealed several stem loops and a pseudoknot that are required for IRES translation. Based on these properties, the CrPV 5′UTR IRES mechanism may be similar to that of the HCV IRES. Interestingly, both HCV and the related cripavirus DCV require the ribosomal protein RACK1 for 5′UTR IRES translation and virus infection (Majzoub *et al.*, 2014). The finding that RACK1 is required for dicistrovirus 5′UTR IRES further supports the idea that ribosome heterogeneity plays a role in IRES-mediated translation in dicistrovirus and other virus infections.

The different 5′UTR IRES mechanisms of dicistroviruses may have evolved distinct translation mechanisms for each host species. The 5′UTR IRES directs translation of the dicistrovirus ORF1, which encodes the non-structural proteins RdRp, helicase and protease, all of which are required for viral replication. Thus, each dicistrovirus 5′UTR IRES mechanism is optimized for each host species during infection in order to promote translation of the non-structural proteins.

During dicistrovirus infection, the structural proteins are produced in supramolar excess compared to non-structural proteins (Moore *et al.*, 1980; Wilson *et al.*, 2000b; Garrey *et al.*, 2010), largely due to the distinct mechanisms of the 5′UTR and IGR IRESs. The 5′UTR IRES is relatively weak, while the IGR IRES is stimulated as infection progresses (Wilson *et al.*, 2000a; Khong *et al.*, 2016). Translation from the 5′UTR IRES and IGR IRES is also temporally regulated during infection; the 5′UTR IRES is active first, followed by the IGR IRES (Khong *et al.*, 2016). IGR IRES translation is presumably delayed on infection to allow for the 5′UTR IRES to express the non-structural proteins needed for replication of the viral genome. Once sufficient viral replication is achieved, IGR IRES translation is stimulated to drive expression of the structural proteins required for viral packaging. Thus, the CrPV uses a dual IRES translational control strategy to ensure optimal temporal expression of viral non-structural proteins prior to synthesis of viral structural proteins, which coordinates the timing of viral replication and assembly during the dicistrovirus life cycle. The mechanisms that regulate the dual IRESs in dicistrovirus-infected cells remain to be investigated. However, host translational shut-off and recycling of ribosomes by Pelo, a homologue of the yeast Dom34, may contribute to this temporal regulation (Garrey *et al.*, 2010; Wu *et al.*, 2014).

Host responses to viral infection

Shut off of host protein synthesis

Dicistrovirus infection leads to a rapid shut off of host protein synthesis concomitant with a switch to viral IRES-dependent translation (Moore *et al.*, 1980, 1981; Wilson *et al.*, 2000b; Garrey *et al.*, 2010) (Table 5.1). Although these observations are primarily from CrPV- and DCV-infected *Drosophila* cells, it is likely that other dicistrovirus infections lead to host translation shut-off. Viruses may employ host translational shut off (1) to block induction of antiviral innate immune protein expression and (2) to increase the pool of available ribosomes for viral protein synthesis. It is also possible that the shut off of translation is a host response akin to a cellular stress response. In poliovirus infections, host translation shutdown is a viral strategy; the poliovirus 2A protease cleaves translation factor eIF4G (Gradi *et al.*, 1998).

During dicistrovirus infection of S2 cells by CrPV, the translation factor eIF2α is phosphorylated and interactions between eIF4E, cap binding protein, and eIF4G are inhibited (Garrey *et al.*, 2010). eIF2 is a trimeric complex of α, β, and γ subunits, which brings the initiator Met-tRNA$_i$ to the 40S pre-initiation complex at the 5′cap of mRNAs. Phosphorylation of eIF2α reduces the pool of available eIF2, leading to a decrease in translation. Phosphorylation of eIF2α could be a viral strategy to inhibit host translation, as neither the viral IGR IRES mechanism nor the initiator Met-tRNA$_i$ requires translation factors to promote ORF2 protein synthesis. However, blocking eIF2α phosphorylation in CrPV infected S2 cells by either depleting both eIF2α kinases in *Drosophila* (dPERK, dGCN2) or inducing the phosphatase dGADD34 still results in host translation inhibition and viral protein synthesis, suggesting that another translation factor is targeted (Garrey *et al.*, 2010; Khong *et al.*, 2016). The disruption of eIF4E–eIF4G interactions in CrPV-infected S2 cells does not involve the 4E-binding protein or *Drosophila* Cup, both known inhibitors of eIF4E activity. Whether targeting eIF4E–eIF4G interactions is the main determinant for host translation shut off under dicistrovirus infection has yet to be established. Other mechanisms may contribute to translation shut off. For example, IAPV infection of honey bees results in ribosomal RNA fragmentation (Johnson *et al.*, 2009).

Antiviral responses

Dicistroviruses have served as excellent models to study fundamental virus–host interactions and antiviral responses in insects. We briefly review the main innate immune pathways in insects and the impact of these pathways on dicistrovirus infections using DCV and CrPV infections as a model. A more comprehensive and detailed overview of the insect immune responses is provided in Chapter 3.

Four main innate immune pathways exist in insects: Toll, IMD, JAK-STAT and RNA interference pathways (Kingsolver *et al.*, 2013). Initially identified from responses to fungal and bacterial infections, triggering of the Toll and IMD pathways activates NFκB-like transcription factors, Dif and Relish, respectively. In turn, activated Dif or Relish leads to transcription of immune genes for the antimicrobial peptides Drosomycin (Toll) and Diptericin (IMD). Activation of the JAK-STAT pathway responds to septic infection in *Drosophila* (Zeidler and Bausek, 2013). The JAK Kinase, Hopscotch protein and the STAT transcription factor (STAT92E), respond to activation of the Domeless receptor by three cytokines of the unpaired (UPD) family. Although these four pathways have all been implicated in response to dicistrovirus infection, the RNAi pathway is the main insect antiviral response (Chapter 3).

RNA interference

The RNA interference pathway is the major antiviral response in insects (Chapter 3). In virus-infected *Drosophila*, double stranded RNA (dsRNA) intermediates produced during replication are sensed by Dicer-2, a ribonuclease III enzyme. Dicer-2 processes the dsRNA into ≈21–23 nucleotide small interfering RNAs (siRNA). The siRNAs, in concert with the RNA-induced silencing complex (RISC), which contains Ago-2, surveils for target RNAs and induces cleavage (Wilson and Doudna, 2013). The importance of the RNAi pathway in antiviral insect responses has been established by multiple studies. *ago-2*- and *dcr-2*-deficient *Drosophila* are hypersensitive to CrPV and DCV infections (Galiana-Arnoux *et al.*, 2006; Kingsolver *et al.*, 2013; van Rij *et al.*, 2006). *ago2*-deficient flies showed a significant increase in viral RNA accumulation, a 1000-fold increase in viral titre, and an increased mortality rate.

Insects can mediate systemic immunity in response to virus infection. In *Drosophila*, mutant flies defective in the dsRNA update pathway (*egh*, *ninaC* or *CG4572*) are hypersensitive to DCV, suggesting that uninfected cells may participate in antiviral immunity (Saleh *et al.*, 2006, 2009). Systemic RNAi-based immunity may involve haemocytes, which take up dsRNA from infected cells to produce virus-derived complementary DNAs (vDNAs) through the action of endogenous transposon-encoded reverse transcriptase (Goic *et al.*, 2013; Lamiable *et al.*, 2016; Tassetto *et al.*, 2017). This vDNA is present in both linear and circular forms, with the circular vDNA (cVDNA) bearing homology to the defective viral genome (DVG) and serving as a template for vDNA and cVDNA (Poirier *et al.*, 2018). The vDNA serves as template for the synthesis of viral siRNA (vsRNA) which is then incorporated into and secreted in exosome-like vesicles that confer systemic RNAi antiviral immunity (Tassetto *et al.*, 2017). The systemic spread of dsRNA may be a virus-specific immune signal akin to the immune responses elicited by plants and nematodes.

Inhibitors of RNA interference

In support of the RNAi pathway as the major insect antiviral response, dicistroviruses encode a silencing suppressor (VSS) that dampens this immune response in *Drosophila*. For CrPV and DCV, these RNAi suppressors map to the 1A protein at the most amino terminus of ORF1 (Nayak *et al.*, 2010; van Rij *et al.*, 2006; Wang *et al.*, 2006) (Table 5.1). Expression of CrPV 1A and DCV 1A in *Drosophila* cells blocked double-stranded RNA induced gene silencing, resulting in cells that were more susceptible to virus infection (Nayak *et al.*, 2010; van Rij *et al.*, 2006). Interestingly, DCV 1A and CrPV 1A proteins inhibit RNA interference via distinct mechanisms.

DCV 1A, a 97-amino acid protein, inhibits RNAi by binding to dsRNA, thus blocking processing by Dcr-2 and siRNA loading into RISC (van Rij *et al.*, 2006). In contrast, CrPV 1A, a 166-amino-acid protein, shows no dsRNA binding affinity and does not prevent siRNA loading into RISC. Instead, CrPV 1A modulates RNAi by interacting directly with Ago-2 to suppress its cleavage activation (Nayak *et al.*, 2010). Using biochemical and single molecule imaging approaches, it has been shown that CrPV 1A blocks the Ago-2 RISC complex from targeting the RNA (Watanabe *et al.*, 2017). In a recent report by Andino and colleagues (Nayak *et al.*, 2018), CrPV 1A was shown to contain a BC box domain, which is responsible for mediating polyubiquitination of Ago-2 that results in proteosomal degradation. This divergence in the mechanisms of RNAi inhibition between DCV and CrPV may reflect differences in viral strategies for pathogenesis and/or the viral life cycle. Indeed, DCV and CrPV have distinct tissue tropisms in infected fruit flies, and CrPV but not DCV causes cytopathic effects in infected cells (Chtarbanova *et al.*, 2014). DCV 1A and CrPV 1A do not show any significant sequence similarity (9.27% similarity) and the proteins do not have obvious homology to known proteins. Moreover, it is not known if other dicistrovirus 1A proteins have similar VSS functions. Further structural and biochemical analysis should shed light into these mechanisms. Finally, CrPV 1A has recently been shown to have roles besides inhibiting the RNAi antiviral response (Khong *et al.*, 2017). How these multiple functions of 1A interplay to promote infection is a promising avenue for investigation.

JAK-STAT pathway

A global transcriptome analysis of flies injected with DCV revealed activation of the Jak-STAT pathway (Dostert *et al.*, 2005; Deddouche *et al.*, 2008; Kemp *et al.*, 2013). Virus injection into flies deficient in hopscotch, the main Jak kinase in *Drosophila*, resulted in higher viral loads and increased mortality to CrPV and DCV (Dostert *et al.*, 2005; Kemp *et al.*, 2013). The most induced genes under infection are *vir-1* (virus-induced RNA) and *vago*. Although vir-1 expression under virus infection is dependent on hopscotch via the Jak-STAT pathway, deleting vir-1 or overexpressing vir-1 does not affect virus infection, suggesting that vir-1 does not have a direct role in innate immunity (Dostert *et al.*, 2005). On the other hand, *vago* deficient flies are more susceptible to DCV infections. Interestingly, the induction of vago is dependent on Dcr-2 (Deddouche *et al.*, 2008). This finding suggests that Dcr-2 not only senses and activates RNAi immunity but also triggers an inducible antiviral pathway. How Dcr-2 signals to activate downstream genes is not clear. In a more recent study, flies deficient in the histone H3 lysine 9 methyltransferase, G9a, are more susceptible to DCV infection, which is associated with hyperactivation of the JAK-STAT pathway, thus linking an epigenetic mark to antiviral signalling (Merkling *et al.*, 2015). From these transcriptome analyses, there is little overlap between the genes induced from DCV infection versus bacteria or fungal infection (Dostert *et al.*, 2005). Moreover, following gene expression from a panel of different insect viruses, Jak-STAT signalling is virus specific (Kemp *et al.*, 2013).

Toll pathway

The Toll pathway is a major determinant for resistance to DCV oral infection of *Drosophila* (Ferreira *et al.*, 2014). *toll*-deficient flies are more susceptible to DCV oral infection, leading to increased mortality and higher viral loads. By contrast, this resistance is not observed with DCV injection of fruit flies, which is the most common method for infecting *Drosophila* with dicistrovirus. DCV oral infection leads to Toll pathway activation, resulting in translocation of the NFκB transcription factor, Dorsal, primarily in the fat body. Interestingly, this study found similar tissue tropism of DCV for both oral or injection-mediated infection (Ferreira *et al.*, 2014). In contrast, DCV oral infection resulted in infection of the midgut epithelial cells, which could be due to differences in feeding protocols (Xu *et al.*, 2013; Ferreira *et al.*, 2014). The latter study found that the extracellular signal-regulated kinase (ERK) signalling pathway was important for gut epithelium resistance to viral infection. Thus, *Drosophila* likely uses multiple antiviral pathways.

IMD pathway

The IMD pathway also contributes to antiviral activity against dicistrovirus infection. Mutant flies that are IMD pathway deficient are more susceptible to CrPV infection, with infection resulting in higher viral loads (Costa *et al.*, 2009). Although this suggests that the IMD pathway is protective, CrPV infection does not activate the IMD pathway based on monitoring the expression of downstream antimicrobial peptides. The IMD pathway may therefore depend on the gut microbiota to protect against virus infection. DCV oral infection in mutant flies deficient in the IMD pathway displayed an increase in viral replication in the intestine, supporting this assertion (Sansone *et al.*, 2015). Gut microbiota thus may activate the IMD signalling pathway to boost antiviral responses (Sansone *et al.*, 2015). A recent RNAi screen by Imler and colleagues (Goto *et al.*, 2018) showed that the kinase dIKKβ and the NFκB-like transcription factor Relish control virus infection through the induction of antiviral genes such as STING.

Other factors involved in antiviral response

Apart from the Toll, IMD, JAK-STAT and RNAi pathways, several other host factors play key roles in modulating virus infection. The gene *pastrel* (*pst*) has been shown to restrict CrPV and DCV infection in flies (Magwire *et al.*, 2012; Martins *et al.*, 2014). Its resistance activity seems to be dicistrovirus specific, as there is no significant association with susceptibility to other positive strand RNA viruses. The double-stranded RNA-binding protein DIP1 (Disconnected interacting protein), which participates in a wide range of cellular processes, plays an antiviral role in S2 cells (Zhang *et al.*, 2015). *dip1* mutant flies are hypersensitive to DCV infection. The exact role of DIP1 in innate immune response remains to be investigated.

Apoptosis

Evading or delaying apoptosis is a strategy used by many viruses to establish infection (Benedict *et al.*, 2002). DCV infection in fruit flies induces the expression of Damm, which is one of the seven caspases in *Drosophila*, suggesting a potential link between DCV infection and apoptosis (Dostert *et al.*, 2005). Furthermore, apoptosis-dependent phagocytosis that removes virus-infected cells is induced in DCV-injected flies (Nainu *et al.*, 2015). Apoptosis has been shown to plays only a minor role in CrPV and DCV infections (Lamiable *et al.*, 2016). Interestingly, an inhibitor of apoptosis protein (IAP) repeat domain has been identified at the N terminus of the TSV ORF1 replicase protein, which suggests that some dicistroviruses may modulate apoptotic pathways during infection (Mari *et al.*, 2002).

Stress granules

Stress granules (SG) are dynamic cytoplasmic aggregates of RNAs and proteins that assemble in response to translation inhibition (Beckham and Parker, 2008). The function and mechanism of SG formation remains poorly understood. During virus infections, SG assembly may act as a host antiviral response by sequestering either the translation components from the viral RNA or the viral RNA itself. Therefore, viruses have evolved distinct mechanisms to modulate stress granule formation to facilitate virus infection (Lloyd, 2013). For example, poliovirus inhibits stress granule formation via cleavage of G3BP, a key stress granule protein, by the virally encoded 3C protease (White *et al.*, 2007). In West Nile virus infected cells, the recruitment of key stress granule proteins TIA-1 and TIAR to the minus strand viral RNA for replication inhibit SG assembly (Emara and Brinton, 2007). In vertebrates, SGs can recruit and activate the antiviral factor, PKR, a kinase that phosphorylates eIF2α to block viral translation (Reineke and Lloyd, 2015). Host translation is shut off in CrPV- and DCV-infected *Drosophila* S2 cells which typically leads to SG formation; however, in these cells SG assembly is inhibited, suggesting that viral modulation of SG assembly is evolutionarily conserved and that SGs may play an antiviral role (Khong and Jan, 2011) (Table 5.1).

Besides inhibiting Ago-2 activity (Nayak *et al.*, 2010, 2018), the CrPV 1A protein can also inhibit stress granule formation, indicating that the 1A protein is multifunctional, modulating several host pathways to promote infection (Khong *et al.*, 2017) (Table 5.1). Expression of 1A in S2 cells can inhibit SG formation in cells challenged with a SG inducer (i.e. heat shock, arsenite). CrPV 1A expression also blocks transcription of a subset of genes. A specific arginine R146 within 1A is important for these functions, which are separate from the VSS RNAi silencing activity of 1A. This separation indicates that these functions of 1A can be uncoupled (Khong *et al.*, 2017). Expression of a mutant R146A CrPV 1A resulted in loss of its ability to block SG and transcription. Furthermore, a mutant CrPV virus containing R146A resulted in increased SG formation in infected cells and attenuation of mortality in fruit flies injected with virus (Khong *et al.*, 2017). Interestingly, the expression of wild-type CrPV 1A but not mutant R146A 1A in Hela cells blocked SG formation, suggesting that 1A interferes with a conserved pathway in SG formation (Khong *et al.*, 2017).

Heat shock response

Heat shock of cells is protective against CrPV and DCV infection (Cevallos and Sarnow, 2010; Merkling *et al.*, 2015). Both viral RNA and protein abundance are high in heat shocked infected cells, suggesting that heat shock does not affect viral protein synthesis or replication. In contrast, virion formation is blocked in heat shocked infected cells. Therefore, the heat shock response, a cellular mechanism preventing proteotoxicity, can act as an antiviral component in *Drosophila*. Analysis of *in vitro* and *in vivo* transcriptomes in *Drosophila* revealed a strong induction of heat shock response genes with DCV and CrPV infection (Merkling *et al.*, 2015). These responses may be activated by the accumulation of unfolded viral protein in the cytosol.

Wolbachia pipientis infection confers protection against RNA viruses

A significant proportion of the microbiota in animals can confer immunity to pathogen infection, either through direct competition or indirectly by stimulating innate immunity (Belkaid and Hand, 2014). *Wolbachia* are Gram-negative, vertically transmitted, obligatory, intracellular bacteria that infect a number of arthropods and nematodes. The endosymbiotic association of *Wolbachia pipientis* with *Drosophila* has been shown to delay CrPV and DCV accumulation and mortality by an unknown mechanism (Hedges *et al.*, 2008; Teixeira *et al.*, 2008). Many *Wolbachia* strains infect *Drosophila* species, but not all mediate antiviral protection (Osborne *et al.*, 2009). The evolutionary success of *Wolbachia* relies on the strong and efficient vertical maternal transfer, which involves localization to the posterior pole of the oocyte to ensure its incorporation in the next generation (Serbus and Sullivan, 2007). Infection with *Wolbachia* reduced CrPV and DCV replication and conferred protection against other RNA viruses; however, the induced resistance was not effective against DNA viruses. The mechanism by which *Wolbachia* limits RNA virus infection is unknown; however, it is likely that the antiviral action results from up-regulation of immune responses. Elevated oxidative stress through the activation of extracellular signal-regulated kinase pathway is potentially associated with antiviral protection against DCV in *Wolbachia*-infected flies (Wong *et al.*, 2015, 2016).

Pathogenesis

Dicistrovirus pathology ranges from subtle or no clear symptoms to highly pathogenic, resulting in rapid death. Dicistrovirus infections are often associated with paralysis of the hind legs. Dicistrovirus has a wide tissue tropism that may explain disease progression. DCV has been detected in the crop, midgut, fat body and testes of DCV-infected *Drosophila* (Ferreira *et al.*, 2014). In infected fruit flies, DCV infects the crop, a blind muscular food storage organ. This results in the inability of the crop to contract, leading to accumulation of food, which causes significant malnutrition as well as accumulation of toxic compounds and dehydration, eventually leading to death (Chtarbanova *et al.*, 2014). In contrast, CrPV infection leads to an increase in crop contractions in infected fruit flies, likely as a compensatory mechanism for the impaired function of muscle cells in the midgut (Chtarbanova *et al.*, 2014). CrPV and ALPV have been localized to insect brains and the central nervous system using immunostaining against CrPV (Chtarbanova *et al.*, 2014) and *in situ* hybridization against ALPV (Hatfill *et al.*, 1990). These findings may explain the paralytic symptoms that are common in many dicistrovirus infections (Hatfill *et al.*, 1990). Dicistrovirus infections induce not only neurological changes but also widespread behavioural changes. RhPV infection alters the sensitivity of aphids to chemical cues, affecting how these insects choose plants to feed on. Infected aphids were also found to be more sensitive to the alarm pheromone than uninfected insects (Ban *et al.*, 2008). These insects are more susceptible to attack by the predatory beetle *Coccinella septempunctata* and the parasitoid *Aphidius ervi*. ALPV and IAPV infections result in a range of behavioural defects affecting orientation and navigation (Williamson *et al.*, 1988). In addition, IAPV infections of honey bees result in a significant decrease in the homing ability of foragers and a lower response threshold to sucrose (Li *et al.*, 2013; Han *et al.*, 2015)

Some dicistrovirus infections can result in widespread persistent infections with little or no symptoms. DCV infections can persist in *Drosophila* cell lines (Wu *et al.*, 2010). DCV, among other viruses, can also persistently infect about one third of wild and laboratory *D. melanogaster* colonies (Plus *et al.*, 1975; Kapun *et al.*, 2010). These persistent infections can have unintentional effects on studies using this genetic model organism (Wu *et al.*, 2010). DCV infections also alter the viability of offspring; DCV-infected fruit flies produce more eggs and offspring than uninfected flies, and direct infection of larvae speeds up development (Gomariz-Zilber *et al.*, 1995). Conversely, KBV and IAPV infected honey bees result in slower colony startup and offspring production (Meeus *et al.*, 2014). Overall, the pathogenesis of this family of viruses varies dramatically. The details of the molecular mechanisms underlying the progression of these viral diseases remain to be examined.

Honey bee disease

The recent decline in honey bee populations worldwide is of critical agricultural importance. Crops that are predominantly pollinated by honey bees have an estimated value of more than US $215 billion dollars worldwide, and approximately a third of all food products consumed by humans are pollinated by honey bees (Gallai *et al.*, 2009). Honey bee losses resulting from colony collapse are multifactorial, but viral infections are significant contributors to bee disease (Cox-Foster *et al.*, 2007; Williams *et al.*, 2010). Honey bees are host to at least five dicistroviruses including ABPV, KBV, BQCV, IAPV, and an isolate of cricket paralysis virus. The latter has been shown to be asymptomatic (Christian and Scotti, 1998). Bee viruses spread horizontally via salivary secretions, food reserves, and faeces, as well through *Varroa* mite infestations (Chen and Siede, 2007).

Direct protection from and treatment of viral infections have been attempted for honey bees. Feeding bees with dsRNAs (encoding IAPV genome fragments) induced RNA interference protection of bees from IAPV infection (Maori *et al.*, 2009). Unfortunately, subsequent large-scale testing of 160 bee hives using RNAi approaches did not significantly affect bee mortality or infection by IAPV (Hunter *et al.*, 2010). A subset of dicistroviruses encode proteins that have anti-RNAi activity (van Rij *et al.*, 2006; Nayak *et al.*, 2010), which may reduce the efficacy of RNAi therapeutic approaches to treat virus infected bees.

Taura syndrome

The dicistrovirus *Taura syndrome virus* (TSV) is the causative agent of Taura syndrome in penaeid shrimp (Mari *et al.*, 2002). Taura syndrome outbreaks have had a devastating effect on the shrimp industry in the Americas, with a negative economic impact of over US $2 billion during the peak of the virus outbreak (Lightner, 1999). Originally limited to the Americas, Taura syndrome has spread to Asia, with outbreaks in Taiwan, Thailand, Myanmar, China, Korea and Indonesia (Kiatpathomchai *et al.*, 2008). Mortality of TSV is extremely high, with rates between 75% and 80% in *P. vannamei* (Wilson and Doudna, 2013). Intriguingly, infectious TSV has been found in seagull faeces, potentially providing a passive vector for virus transmission (Lightner, 1999; Garza *et al.*, 2011). Currently, resistant stocks of shrimps, selected through selective breeding for resistance along with the adoption of biosecurity practices, have reduced TSV to the point where it is no longer considered a major threat in many shrimp farming regions (Maori *et al.*, 2009; Moss *et al.*, 2012). Although direct antiviral strategies to combat Taura syndrome are lacking, a previous study showed that administering nonspecific dsRNA to shrimp resulted in resistance to TSV, suggesting that broad antiviral immunity can be achieved in arthropods (Robalino *et al.*, 2004, 2007).

Biopesticides

Management of insect pests is an ongoing global agricultural challenge. The use of pesticides results in a US$40 billion increase in agricultural production in the United States (Pimentel, 2005; Oerke, 2005; Popp *et al.*, 2012). Owing to the global concern that current pesticides can be detrimental to non-target species, there is emerging potential for the use of dicistroviruses as biopesticides. Dicistroviruses are attractive viral vectors for pest management for several reasons. Some dicistroviruses specifically infect host insect pests that negatively affect agriculture, such as RhPV infection of aphids. Viruses such as CrPV have a wide host range, and thus could target diverse hosts (Christian and Scotti, 1996). Since the establishment of the first dicistrovirus infectious clone (Kerr *et al.*, 2015), the potential use of dicistroviruses as biopesticides and the ability to use genetic engineering to maximize the effectiveness of this biopesticide presents an emerging opportunity.

Many dicistroviruses are pathogens of insects that cause substantial losses in crop production, resulting from insect feeding and from insect-mediated transmission of plant viruses. For example, the virus *Homalodisca coagulata virus* (HoCV-1) infects the glassy-winged sharpshooter (GWSS), an insect that feeds on over 100 plant species, and can devastate crops (Hunnicutt *et al.*, 2008). The GWSS also serves as a vector for *Xylella fastidiosa*, a bacterium that results in plant death and has caused several disease outbreaks over the years, including Pierce's disease of grapes (PD) and citrus variegated chlorosis (CVC) (Redak *et al.*, 2004). HoCV-1 could potentially be used as a control agent for GWSS populations to avoid these diseases without the use of chemical insecticides (Falk *et al.*, 1992). In another example, SINV-1 has been proposed for use as a biopesticide against the invasive red fire ant (*Solenopsis invicta*). The red fire ant has invaded over 138 million hectares in the USA, causing over US $6 billion in damage and control efforts annually (Valles, 2012). SINV-1 can only infect the genus *Solenopsis* and is vertically transmitted (Valles, 2012). However, a recent study has found that SINV-1 treatment results in higher survival rates than chemical treatment (Tufts *et al.*, 2014).

One of the most exciting possibilities for use of dicistroviruses as biopesticides is the ability of these viruses to potentially be expressed in plants, enabling the production of plants expressing these viruses as a means to control pests (Gordon and Waterhouse, 2006). However, attempts to express viruses in plants have been largely unsuccessful because of RNAi silencing of viruses in the plants (Gordon and Waterhouse, 2006). One method proposed to circumvent this problem is to use dicistrovirus-like particles to deliver either dsRNA against essential genes of the infected host or toxins against the host (Gordon and Waterhouse, 2006; Baum *et al.*, 2007; Bonning and Johnson, 2010; Bonning *et al.*, 2014). As proof of principal, a luteovirus plant virus coat protein can deliver a toxin to aphids as those aphids feed on plants expressing the viral coat protein-insect neurotoxin fusion (Bonning *et al.*, 2014). A similar strategy could be implemented using a dicistrovirus clone expressed in plants that would then be transmitted to insects feeding on the plant. Indeed, some dicistroviruses have been shown to use plants as vectors (RhPV) (Gildow and d'Arcy, 1988).

The lethality of some dicistroviruses also has potential for exploitation. Incapable of infecting mammals, these viruses could be used to control disease-carrying hosts. Chagas disease, a devastating illness that has infected between 8 to 11 million people in Latin America is caused by the parasite *Trypanosoma cruzi*, which is spread by the insect vector *Triatoma infestans*. *Triatoma virus* (TrV) could be deployed as a protective measure, as its infection of *T. cruzi* leads to >97% mortality (Muscio *et al.*, 1997). Moreover, TrV infection increases adherence of *T. cruzi* to intestinal cells, making the vector less likely to shed the pathogen (Marti *et al.*, 2017).

Conclusions and future perspectives

Dicistroviruses have provided a powerful model for studying fundamental viral IRES translation, ribosome functions, antiviral responses and innate immune responses. Indeed, dicistrovirus research has attracted researchers from a wide range of fields including ecologists, virologists, immunologists, biochemists and biophysicists, all of which have brought innovative approaches that have accelerated our understanding of dicistrovirus infection and associated mechanisms. Despite significant progress, there are still outstanding gaps in our knowledge.

Although dicistrovirus IRES translation has been extensively studied at the mechanistic level, the rest of the viral life cycle, including viral entry, replication and egress, is understudied. Many of the host proteins involved in mediating specific steps of the viral life cycle are unknown. For instance, the receptor for viral entry has not been identified, nor have the host proteins involved in replication steps such as membrane reorganization and formation of replication complexes at the 5′ and 3′ UTR. An initial direction for further research is to identify key elements/structures at the 5′ and 3′UTRs needed for replication. As discussed, the 5′UTR secondary structure has recently been modelled, which should further our understanding of replication and 5′UTR IRES translation (Gross *et al.*, 2017). There are still knowledge gaps even for the well-studied IGR IRES translation mechanism. The mechanism by which a subset of IGR IRESs mediate reading frame selection through RNA–ribosome interactions remains to be investigated. These details would shed light into fundamental tRNA-ribosome functions. Furthermore, the emerging idea that ribosome heterogeneity contributes to the translation of specific IRES-containing RNAs, warrants detailed investigation. Most studies in this field have focused on artificially depleting ribosomal proteins or specific RNA modifications that can impact dicistrovirus IRES translation. It will be important to determine which post-translational or post-transcriptional modifications of the ribosome are important for specifying viral IRES-mediated IRES translation. In particular, host translation shut off on dicistrovirus infection requires further study. Understanding the host targets for translational shut off on infection could reveal important insights into the cellular conditions that allow viral IGR IRES translation, as well as the temporal regulation of the two IRESs within dicistroviruses. Advances in whole genome-wide knockout screens using CRISPR-Cas9 approaches should help identify the functional proteins involved in these processes. From the virus side, the predicted functions of the non-structural and structural proteins are based on sequence alignments; however, it is likely that some have moonlighting functions that modulate cellular processes to promote infection. For example, the surprising finding that the CrPV 1A protein has multiple functions highlights the need to study other dicistrovirus proteins. The development of the CrPV infectious clone, which has already provided a framework for reverse genetics to explore the functions of 1A and IGR IRES translation, should accelerate our discovery of the functions of these viral proteins.

Understanding of these fundamental virus host interactions should allow insights into the pathogenesis of dicistroviruses. The use of the genetically tractable model organism *Drosophila*, has paved the way in studying dicistrovirus tropism, antiviral responses and pathogenesis. The use of the CrPV infectious clone has similar potential to provide insights into these processes in infected fruit flies. Transcriptomics has provided key insights into the antiviral responses in dicistrovirus infected *Drosophila*. Incorporating advanced translatome and proteomic approaches, such as ribosome profiling and siLAC labelling, should enable a global view of the pathways modulated during infection, potentially pinpointing key factors that can elucidate pathogenesis. Understanding of these mechanisms could allow for identification of new targets and strategies to combat diseases caused by dicistrovirus infections, such as virus-induced honey bee disease. For example, the function of the +1 frame overlapping ORFx in a subset of dicistroviruses remains elusive. The strong evolutionary selection of ORFx indicates an important function in infection and/or pathogenesis. Because the IGR IRES +1 frame translation and ORFx are found in honey bee dicistroviruses, it is possible that this mechanism/ORFx may be an attractive antiviral target. In a different vein, the use of dicistroviruses as biopesticides could have a large impact on serious agricultural pests. The diversity of dicistroviruses and the wide host range of some of these viruses makes them particularly attractive for this application. Harnessing and exploiting unique dicistrovirus mechanisms will be key in developing viral-based biopesticides. Finally, the recent identification of hundreds of dicistro-like viruses is exciting, highlighting the ubiquity and prevalence of dicistroviruses and should provide a treasure for researchers in identifying potentially new viral mechanisms.

Acknowledgements

We thank Jan lab members for discussions and critical reading of the chapter. This work was supported by the Canadian Institutes of Health Research (PJT-148761) to E.J. and a UBC-SERB scholarship to J.S.

References

Abaeva, I.S., Pestova, T.V., and Hellen, C.U. (2016). Attachment of ribosomal complexes and retrograde scanning during initiation on the Halastavi árva virus IRES. Nucleic Acids Res. *44*, 2362–2377. https://doi.org/10.1093/nar/gkw016

Abeyrathne, P.D., Koh, C.S., Grant, T., Grigorieff, N., and Korostelev, A.A. (2016). Ensemble cryo-EM uncovers inchworm-like translocation of a viral IRES through the ribosome. Elife *5*, e14874. https://doi.org/10.7554/eLife.14874

Agirre, J., Goret, G., LeGoff, M., Sánchez-Eugenia, R., Marti, G.A., Navaza, J., Guérin, D.M.A., and Neumann, E. (2013). Cryo-electron microscopy reconstructions of triatoma virus particles: a clue to unravel genome delivery and capsid disassembly. J. Gen. Virol. *94*, 1058–1068.

Au, H.H., and Jan, E. (2012). Insights into factorless translational initiation by the tRNA-like pseudoknot domain of a viral IRES. PLOS ONE *7*, e51477. https://doi.org/10.1371/journal.pone.0051477

Au, H.H., Cornilescu, G., Mouzakis, K.D., Ren, Q., Burke, J.E., Lee, S., Butcher, S.E., and Jan, E. (2015). Global shape mimicry of tRNA within a viral internal ribosome entry site mediates translational reading frame selection. Proc. Natl. Acad. Sci. U.S.A. *112*, E6446–55. https://doi.org/10.1073/pnas.1512088112

Au, H.H.T., Elspass, V.M., and Jan, E. (2018). Functional insights into the adjacent stem-loop in honey bee Dicistroviruses that promotes internal ribosome entry site-mediated translation and viral infection. J. Virol. *92*, e01725-17.

Bailey, L., and Woods, R.D. (1977). Two more small RNA viruses from honey bees and further observations on sacbrood and acute bee-paralysis viruses. J Gen. Virol. *37*, 175-182.

Bailey, L., Gibbs, A.J., and Woods, R.D. (1963). Two viruses from adult honey bees (*APIS mellifera linnaeus*). Virology *21*, 390–395.

Bailey, L., Gibbs, A.J., and Woods, R.D. (1964). Sacbrood virus of the larval honey bee (*APIS mellifera linnaeus*). Virology *23*, 425–429.

Ban, L., Didon, A., Jonsson, L.M., Glinwood, R., and Delp, G. (2007). An improved detection method for the Rhopalosiphum padi virus (RhPV) allows monitoring of its presence in aphids and movement within plants. J. Virol. Methods *142*, 136–142.

Ban, L., Ahmed, E., Ninkovic, V., Delp, G., and Glinwood, R. (2008). Infection with an insect virus affects olfactory behaviour and interactions with host plant and natural enemies in an aphid. Ent. Exp. Applicata *127*, 108-117.

Baum, J.A., Bogaert, T., Clinton, W., Heck, G.R., Feldmann, P., Ilagan, O., Johnson, S., Plaetinck, G., Munyikwa, T., Pleau, M., *et al.* (2007). Control of coleopteran insect pests through RNA interference. Nat. Biotechnol. *25*, 1322–1326.

Beckham, C.J., and Parker, R. (2008). P bodies, stress granules, and viral life cycles. Cell Host Microbe 3, 206–212. https://doi.org/10.1016/j.chom.2008.03.004

Belkaid, Y., and Hand, T.W. (2014). Role of the microbiota in immunity and inflammation. Cell *157*, 121–141. https://doi.org/10.1016/j.cell.2014.03.011

Benedict, C.A., Norris, P.S., and Ware, C.F. (2002). To kill or be killed: viral evasion of apoptosis. Nat. Immunol. 3, 1013–1018. https://doi.org/10.1038/ni1102-1013

Bird, S.W., Maynard, N.D., Covert, M.W., and Kirkegaard, K. (2014). Nonlytic viral spread enhanced by autophagy components. Proc. Natl. Acad. Sci. U.S.A. *111*, 13081–13086. https://doi.org/10.1073/pnas.1401437111

Bonning, B.C., and Johnson, K.N. (2010). Dicistroviruses. In Insect Virology, S. Asgari and K.N. Johnson, eds. (Caister Academic Press, Norfolk), pp. 201–229.

Bonning, B.C., Pal, N., Liu, S., Wang, Z., Sivakumar, S., Dixon, P.M., King, G.F., and Miller, W.A. (2014). Toxin delivery by the coat protein of an aphid-vectored plant virus provides plant resistance to aphids. Nat. Biotechnol. *32*, 102–105. https://doi.org/10.1038/nbt.2753

Brown, J.D., and Ryan, M.D. (2010). Ribosome 'Skipping': 'Stop-Carry On' or 'StopGo' translation. In Recoding: Expansion of Decoding Rules Enriches Gene Expression, J.F. Atkins, and R.F. Gesteland, eds. (Springer, New York), pp. 101–121.

Bugaud, O., Barbier, N., Chommy, H., Fiszman, N., Le Gall, A., Dulin, D., Saguy, M., Westbrook, N., Perronet, K., and Namy, O. (2017). Kinetics of CrPV and HCV IRES-mediated eukaryotic translation using single-molecule fluorescence microscopy. RNA *23*, 1626–1635. https://doi.org/10.1261/rna.061523.117

Carvajal, F., Vallejos, M., Walters, B., Contreras, N., Hertz, M.I., Olivares, E., Cáceres, C.J., Pino, K., Letelier, A., Thompson, S.R., *et al.* (2016). Structural domains within the HIV-1 mRNA and the ribosomal protein S25 influence cap-independent translation initiation. FEBS J. *283*, 2508–2527. https://doi.org/10.1111/febs.13756

Cevallos, R.C., and Sarnow, P. (2005). Factor-independent assembly of elongation-competent ribosomes by an internal ribosome entry site located in an RNA virus that infects penaeid shrimp. J. Virol. *79*, 677–683.

Cevallos, R.C., and Sarnow, P. (2010). Temperature protects insect cells from infection by cricket paralysis virus. J. Virol. *84*, 1652–1655. https://doi.org/10.1128/JVI.01730-09

Chen, Y.P., and Siede, R. (2007). Honey bee viruses. Adv. Virus Res. *70*, 33–80.

Cherry, S., and Perrimon, N. (2004). Entry is a rate-limiting step for viral infection in a *Drosophila* melanogaster model of pathogenesis. Nat. Immunol. *5*, 81–87. https://doi.org/10.1038/ni1019

Cherry, S., Doukas, T., Armknecht, S., Whelan, S., Wang, H., Sarnow, P., and Perrimon, N. (2005). Genome-wide RNAi screen reveals a specific sensitivity of IRES-containing RNA viruses to host translation inhibition. Genes Dev. *19*, 445–452.

Cherry, S., Kunte, A., Wang, H., Coyne, C., Rawson, R.B., and Perrimon, N. (2006). COPI activity coupled with fatty acid biosynthesis is required for viral replication. PLOS Pathog. *2*, e102.

Christian, P.D., and Scotti, P.D. (1996). Biopesticides from small RNA viruses on insects: aspects of their in vitro production. In Invertebrate Cell Culture: Looking Toward The Twenty First Century, K. Maramorosch, and M.J. Loeb, eds. (San Francisco: Proceedings of the IX International Conference on Invertebrate Cell Culture Society for In Vitro Biology), pp. 73–81.

Christian, P.D., and Scotti, P.D. (1998). Picornalike viruses of insects. In The Insect Viruses, L.K. Miller, and L.A. Ball, eds. (Springer, Boston, MA), pp. 301–336.

Chtarbanova, S., Lamiable, O., Lee, K.Z., Galiana, D., Troxler, L., Meignin, C., Hetru, C., Hoffmann, J.A., Daeffler, L., and Imler, J.L. (2014). *Drosophila* C virus systemic infection leads to intestinal obstruction. J. Virol. *88*, 14057–14069. https://doi.org/10.1128/JVI.02320-14

Colussi, T.M., Costantino, D.A., Hammond, J.A., Ruehle, G.M., Nix, J.C., and Kieft, J.S. (2014). The structural basis of transfer RNA mimicry and conformational plasticity by a viral RNA. Nature *511*, 366–369. https://doi.org/10.1038/nature13378

Costa, A., Jan, E., Sarnow, P., and Schneider, D. (2009). The Imd pathway is involved in antiviral immune responses in *Drosophila*. PLOS ONE *4*, e7436. https://doi.org/10.1371/journal.pone.0007436

Costantino, D., and Kieft, J.S. (2005). A preformed compact ribosome-binding domain in the cricket paralysis-like virus IRES RNAs. RNA *11*, 332–343.

Costantino, D.A., Pfingsten, J.S., Rambo, R.P., and Kieft, J.S. (2008). tRNA-mRNA mimicry drives translation initiation from a viral IRES. Nat. Struct. Mol. Biol. *15*, 57–64.

Cox-Foster, D.L., Conlan, S., Holmes, E.C., Palacios, G., Evans, J.D., Moran, N.A., Quan, P.L., Briese, T., Hornig, M., Geiser, D.M., *et al.* (2007). A metagenomic survey of microbes in honey bee colony collapse disorder. Science *318*, 283–287.

D'Arcy, C.J., Burnett, P.A., Hewings, A.D., and Goodman, R.M. (1981a). Purification and characterization of a virus from the aphid Rhopalosiphum padi. Virology *112*, 346–349.

D'Arcy, C.J., Burnett, P.A., and Hewings, A.D. (1981b). Detection, biological effects, and transmission of a virus of the aphid Rhopalosiphum padi. Virology *114*, 268–272.

Deddouche, S., Matt, N., Budd, A., Mueller, S., Kemp, C., Galiana-Arnoux, D., Dostert, C., Antoniewski, C., Hoffmann, J.A., and Imler, J.L. (2008). The DExD/H-box helicase Dicer-2 mediates the induction of antiviral activity in *drosophila*. Nat. Immunol. *9*, 1425–1432. https://doi.org/10.1038/ni.1664

Donnelly, M.L., Hughes, L.E., Luke, G., Mendoza, H., ten Dam, E., Gani, D., and Ryan, M.D. (2001). The 'cleavage' activities of foot-and-mouth disease virus 2A site-directed mutants and naturally occurring '2A-like' sequences. J. Gen. Virol. *82*, 1027–1041.

Dostert, C., Jouanguy, E., Irving, P., Troxler, L., Galiana-Arnoux, D., Hetru, C., Hoffmann, J.A., and Imler, J.L. (2005). The Jak-STAT signaling pathway is required but not sufficient for the antiviral response of *drosophila*. Nat. Immunol. *6*, 946–953.

Emara, M.M., and Brinton, M.A. (2007). Interaction of TIA-1/TIAR with West Nile and dengue virus products in infected cells interferes with stress granule formation and processing body assembly. Proc. Natl. Acad. Sci. U.S.A. *104*, 9041–9046.

Falk, M.M., Sobrino, F., and Beck, E. (1992). VPg gene amplification correlates with infective particle formation in foot-and-mouth disease virus. J. Virol. *66*, 2251–2260.

Feng, Z., Hensley, L., McKnight, K.L., Hu, F., Madden, V., Ping, L., Jeong, S.H., Walker, C., Lanford, R.E., and Lemon, S.M. (2013). A pathogenic picornavirus acquires an envelope by hijacking cellular membranes. Nature *496*, 367–371. https://doi.org/10.1038/nature12029

Fernández, I.S., Bai, X.C., Murshudov, G., Scheres, S.H., and Ramakrishnan, V. (2014). Initiation of translation by cricket paralysis virus IRES requires its translocation in the ribosome. Cell *157*, 823–831. https://doi.org/10.1016/j.cell.2014.04.015

Fernandez, J., Yaman, I., Sarnow, P., Snider, M.D., and Hatzoglou, M. (2002). Regulation of internal ribosomal entry site-mediated translation by phosphorylation of the translation initiation factor eIF2alpha. J. Biol. Chem. *277*, 19198–19205. https://doi.org/10.1074/jbc.M201052200

Ferreira, Á.G., Naylor, H., Esteves, S.S., Pais, I.S., Martins, N.E., and Teixeira, L. (2014). The Toll-dorsal pathway is required for resistance to viral oral infection in *Drosophila*. PLOS Pathog. *10*, e1004507. https://doi.org/10.1371/journal.ppat.1004507

Firth, A.E., Wang, Q.S., Jan, E., and Atkins, J.F. (2009). Bioinformatic evidence for a stem-loop structure 5'-adjacent to the IGR-IRES and for an overlapping gene in the bee paralysis dicistroviruses. Virol. J. *6*, 193. https://doi.org/10.1186/1743-422X-6-193

Galiana-Arnoux, D., Dostert, C., Schneemann, A., Hoffmann, J.A., and Imler, J.L. (2006). Essential function in vivo for Dicer-2 in host defense against RNA viruses in *Drosophila*. Nat. Immunol. *7*, 590–597.

Gallai, N., Salles, J.-M., Settele, J., and Vaissière, B.E. (2009). Economic valuation of the vulnerability of world agriculture confronted with pollinator decline. Ecol. Econ. *68*, 810–821.

Garrey, J.L., Lee, Y.Y., Au, H.H., Bushell, M., and Jan, E. (2010). Host and viral translational mechanisms during cricket paralysis virus infection. J. Virol. *84*, 1124–1138. https://doi.org/10.1128/JVI.02006-09

Garza, J.R., Hasson, K.W., Poulos, B.T., Redman, R.M., White, B.L., and Lightner, D.V. (2011). Demonstration of infectious taura syndrome virus in the feces of seagulls collected during an epizootic in Texas. J. Aquat. Anim. Health *9*, 156–159.

Gildow, F., and D'Arcy, C.J.P. (1988). Barley and oats as reservoirs for an aphid virus and the influence on barley yellow dwarf virus transmission. Phytopathology *78*, 811–816.

Goic, B., Vodovar, N., Mondotte, J.A., Monot, C., Frangeul, L., Blanc, H., Gausson, V., Vera-Otarola, J., Cristofari, G., and Saleh, M.C. (2013). RNA-mediated interference and reverse transcription control the persistence of RNA viruses in the insect model *Drosophila*. Nat. Immunol. *14*, 396–403. https://doi.org/10.1038/ni.2542

Gomariz-Zilber, E., Poras, M., and Thomas-Orillard, M. (1995). *Drosophila* C virus: experimental study of infectious yields and underlying pathology in *Drosophila* melanogaster laboratory populations. J. Invertebr. Pathol. *65*, 243–247.

Goodfellow, I. (2011). The genome-linked protein VPg of vertebrate viruses - a multifaceted protein. Curr. Opin. Virol. *1*, 355–362. https://doi.org/10.1016/j.coviro.2011.09.003

Gordon, K.H., and Waterhouse, P.M. (2006). Small RNA viruses of insects: expression in plants and RNA silencing. Adv. Virus Res. *68*, 459–502.

Goto, A., Okado, K., Martins, N., Cai, H., Barbier, V., Lamiable, O., Troxler, L., Santiago, E., Kuhn, L., Paik, D., *et al.* (2018). The Kinase IKKβ Regulates a STING- and NF-κB-Dependent Antiviral Response Pathway in *Drosophila*. Immunity *49*, 225–234.e4.

Gradi, A., Svitkin, Y.V., Imataka, H., and Sonenberg, N. (1998). Proteolysis of human eukaryotic translation initiation factor eIF4GII, but not eIF4GI, coincides with the shutoff of host protein synthesis after poliovirus infection. Proc. Natl. Acad. Sci. U.S.A. *95*, 11089–11094.

Gross, L., Vicens, Q., Einhorn, E., Noireterre, A., Schaeffer, L., Kuhn, L., Imler, J.L., Eriani, G., Meignin, C., and Martin, F. (2017). The IRES5'UTR of the dicistrovirus cricket paralysis virus is a type III IRES containing an essential pseudoknot structure. Nucleic Acids Res. *45*, 8993–9004. https://doi.org/10.1093/nar/gkx622

Han, Y., van Oers, M.M., van Houte, S., and Ros, V.I.D. (2015). Virus-induced behavioural changes in insects. In Host Manipulations by Parasites and Viruses, H. Mehlhorn, ed. (Springer International Publishing, Cham), pp. 149–174.

Hasson, K., Lightner, D., Poulos, B., Redman, R., White, B., Brock, J., and Bonami, J. (1995). Taura syndrome in Penaeus vannamei: demonstration of a viral etiology. J. Dis. Aquat. Organ. *23*, 115–126.

Hatakeyama, Y., Shibuya, N., Nishiyama, T., and Nakashima, N. (2004). Structural variant of the intergenic internal ribosome entry site elements in dicistroviruses and computational search for their counterparts. RNA *10*, 779–786.

Hatfill, S.J., Williamson, C., Kirby, R., and von Wechmar, M.B. (1990). Identification and localization of aphid lethal paralysis virus particles in thin tissue sections of the Rhopalosiphum padi aphid by in situ nucleic acid hybridization. J. Invertebr. Pathol. *55*, 265–271.

Hedges, L.M., Brownlie, J.C., O'Neill, S.L., and Johnson, K.N. (2008). Wolbachia and virus protection in insects. Science *322*, 702. https://doi.org/10.1126/science.1162418

Hertz, M.I., and Thompson, S.R. (2011). In vivo functional analysis of the Dicistroviridae intergenic region internal ribosome entry sites. Nucleic Acids Res. *39*, 7276–7288. https://doi.org/10.1093/nar/gkr427

Hertz, M.I., Landry, D.M., Willis, A.E., Luo, G., and Thompson, S.R. (2013). Ribosomal protein S25 dependency reveals a common mechanism for diverse internal ribosome entry sites and ribosome shunting. Mol. Cell. Biol. *33*, 1016–1026. https://doi.org/10.1128/MCB.00879-12

Hinnebusch, A.G., Ivanov, I.P., and Sonenberg, N. (2016). Translational control by 5'-untranslated regions of eukaryotic mRNAs. Science *352*, 1413–1416. https://doi.org/10.1126/science.aad9868

Hunnicutt, L.E., Hunter, W.B., Cave, R.D., Powell, C.A., and Mozoruk, J.J. (2006). Genome sequence and molecular characterization of Homalodisca coagulata virus-1, a novel virus discovered in the glassy-winged sharpshooter (Hemiptera: Cicadellidae). Virology *350*, 67–78.

Hunnicutt, L.E., Mozoruk, J., Hunter, W.B., Crosslin, J.M., Cave, R.D., and Powell, C.A. (2008). Prevalence and natural host range of Homalodisca coagulata virus-1 (HoCV-1). Arch. Virol. *153*, 61–67. https://doi.org/10.1007/s00705-007-1066-2

Hunter, W., Ellis, J., Vanengelsdorp, D., Hayes, J., Westervelt, D., Glick, E., Williams, M., Sela, I., Maori, E., Pettis, J., *et al.* (2010). Large-scale field application of RNAi technology reducing Israeli acute paralysis virus disease in honey bees (Apis mellifera, Hymenoptera: Apidae). PLOS Pathog. *6*, e1001160. https://doi.org/10.1371/journal.ppat.1001160

Jack, K., Bellodi, C., Landry, D.M., Niederer, R.O., Meskauskas, A., Musalgaonkar, S., Kopmar, N., Krasnykh, O., Dean, A.M., Thompson, S.R., *et al.* (2011). rRNA pseudouridylation defects affect ribosomal ligand binding and translational fidelity from yeast to human cells. Mol. Cell *44*, 660–666. https://doi.org/10.1016/j.molcel.2011.09.017

Jackson, R.J., Hellen, C.U., and Pestova, T.V. (2010). The mechanism of eukaryotic translation initiation and principles of its regulation. Nat. Rev. Mol. Cell Biol. *11*, 113–127. https://doi.org/10.1038/nrm2838

Jan, E. (2006). Divergent IRES elements in invertebrates. Virus Res. *119*, 16–28.

Jan, E., and Sarnow, P. (2002). Factorless ribosome assembly on the internal ribosome entry site of cricket paralysis virus. J. Mol. Biol. *324*, 889–902.

Jan, E., Kinzy, T.G., and Sarnow, P. (2003). Divergent tRNA-like element supports initiation, elongation, and termination of protein biosynthesis. Proc. Natl. Acad. Sci. U.S.A. *100*, 15410–15415. https://doi.org/10.1073/pnas.2535183100

Jan, E., Mohr, I., and Walsh, D. (2016). A cap-to-tail guide to mRNA translation strategies in virus-infected cells. Annu. Rev. Virol. 3, 283–307. https://doi.org/10.1146/annurev-virology-100114-055014

Jang, C.J., and Jan, E. (2010). Modular domains of the Dicistroviridae intergenic internal ribosome entry site. RNA *16*, 1182–1195. https://doi.org/10.1261/rna.2044610

Jang, C.J., Lo, M.C., and Jan, E. (2009). Conserved element of the dicistrovirus IGR IRES that mimics an E-site tRNA/ribosome interaction mediates multiple functions. J. Mol. Biol. *387*, 42–58. https://doi.org/10.1016/j.jmb.2009.01.042

Johnson, R.M., Evans, J.D., Robinson, G.E., and Berenbaum, M.R. (2009). Changes in transcript abundance relating to colony collapse disorder in honey bees (Apis mellifera). Proc. Natl. Acad. Sci. U.S.A. *106*, 14790–14795. https://doi.org/10.1073/pnas.0906970106

Jousset, F.X., Plus, N., Croizier, G., and Thomas, M. (1972). Existence chez *Drosophila* de deux groupes de Picornavirus de propriétés sérologiques et biologiques différentes. C. R. Acad. Sci. Hebd. Seances Acad. Sci. D. *275*, 3043–3046.

Kamoshita, N., Nomoto, A., and RajBhandary, U.L. (2009). Translation initiation from the ribosomal A site or the P site, dependent on the conformation of RNA pseudoknot I in dicistrovirus RNAs. Mol. Cell 35, 181–190. https://doi.org/10.1016/j.molcel.2009.05.024

Kanamori, Y., and Nakashima, N. (2001). A tertiary structure model of the internal ribosome entry site (IRES) for methionine-independent initiation of translation. RNA 7, 266–274.

Kapun, M., Nolte, V., Flatt, T., and Schlötterer, C. (2010). Host range and specificity of the *Drosophila* C virus. PLOS ONE *5*, e12421. https://doi.org/10.1371/journal.pone.0012421

Kemp, C., Mueller, S., Goto, A., Barbier, V., Paro, S., Bonnay, F., Dostert, C., Troxler, L., Hetru, C., Meignin, C., *et al.* (2013). Broad RNA interference-mediated antiviral immunity and virus-specific inducible responses in *Drosophila*. J. Immunol. *190*, 650–658. https://doi.org/10.4049/jimmunol.1102486

Kerr, C.H., Wang, Q.S., Keatings, K., Khong, A., Allan, D., Yip, C.K., Foster, L.J., and Jan, E. (2015). The 5' untranslated region of a novel infectious molecular clone of the dicistrovirus cricket paralysis virus modulates infection. J. Virol. *89*, 5919–5934. https://doi.org/10.1128/JVI.00463-15

Kerr, C.H., Ma, Z.W., Jang, C.J., Thompson, S.R., and Jan, E. (2016). Molecular analysis of the factorless internal ribosome entry site in Cricket Paralysis virus infection. Sci. Rep. *6*, 37319. https://doi.org/10.1038/srep37319

Kerr, C.H., Dalwadi, U., Scott, N.E., Yip, C.K., Foster, L.J., and Jan, E. (2018a). Transmission of Cricket paralysis virus via exosome-like vesicles during infection of

Drosophila cells. Sci. Rep. *8*, 17353. https://doi.org/10.1038/s41598-018-35717-5

Kerr, C.H., Wang, Q.S., Moon, K.M., Keatings, K., Allan, D.W., Foster, L.J., and Jan, E. (2018b). IRES-dependent ribosome repositioning directs translation of a +1 overlapping ORF that enhances viral infection. Nucleic Acids Res. *46*, 11952–11967. https://doi.org/10.1093/nar/gky1121

Khong, A., and Jan, E. (2011). Modulation of stress granules and P bodies during dicistrovirus infection. J. Virol. *85*, 1439–1451. https://doi.org/10.1128/JVI.02220-10

Khong, A., Bonderoff, J.M., Spriggs, R.V., Tammpere, E., Kerr, C.H., Jackson, T.J., Willis, A.E., and Jan, E. (2016). Temporal regulation of distinct internal ribosome entry sites of the Dicistroviridae cricket paralysis virus. Viruses *8*, E25. https://doi.org/10.3390/v8010025

Khong, A., Kerr, C.H., Yeung, C.H., Keatings, K., Nayak, A., Allan, D.W., and Jan, E. (2017). Disruption of stress granule formation by the multifunctional cricket paralysis virus 1A protein. J. Virol. *91*, e01779–16.

Kiatpathomchai, W., Jaroenram, W., Arunrut, N., Gangnonngiw, W., Boonyawiwat, V., and Sithigorngul, P. (2008). Experimental infections reveal that common Thai crustaceans are potential carriers for spread of exotic Taura syndrome virus. Dis. Aquat. Org. *79*, 183–190. https://doi.org/10.3354/dao01903

King, L.A., and Moore, N.F. (1988). Evidence for the presence of a genome-linked protein in two insect picornaviruses, cricket paralysis and *Drosophila* C viruses. FEMS Microbiol. Lett. *50*, 41–44.

Kingsolver, M.B., Huang, Z., and Hardy, R.W. (2013). Insect antiviral innate immunity: pathways, effectors, and connections. J. Mol. Biol. *425*, 4921–4936. https://doi.org/10.1016/j.jmb.2013.10.006

Koh, C.S., Brilot, A.F., Grigorieff, N., and Korostelev, A.A. (2014). Taura syndrome virus IRES initiates translation by binding its tRNA-mRNA-like structural element in the ribosomal decoding center. Proc. Natl. Acad. Sci. U.S.A. *111*, 9139–9144. https://doi.org/10.1073/pnas.1406335111

Koonin, E.V., Wolf, Y.I., Nagasaki, K., and Dolja, V.V. (2008). The Big Bang of picorna-like virus evolution antedates the radiation of eukaryotic supergroups. Nat. Rev. Microbiol. *6*, 925–939. https://doi.org/10.1038/nrmicro2030

Lamiable, O., Arnold, J., de Faria, I.J.D.S., Olmo, R.P., Bergami, F., Meignin, C., Hoffmann, J.A., Marques, J.T., and Imler, J.L. (2016). Analysis of the contribution of hemocytes and autophagy to *Drosophila* antiviral immunity. J. Virol. *90*, 5415–5426. https://doi.org/10.1128/JVI.00238-16

Landry, D.M., Hertz, M.I., and Thompson, S.R. (2009). RPS25 is essential for translation initiation by the Dicistroviridae and hepatitis C viral IRESs. Genes Dev. *23*, 2753–2764. https://doi.org/10.1101/gad.1832209

Leppek, K., Das, R., and Barna, M. (2018). Functional 5' UTR mRNA structures in eukaryotic translation regulation and how to find them. Nat. Rev. Mol. Cell Biol. *19*, 158–174. https://doi.org/10.1038/nrm.2017.103

Letunic, I., and Bork, P. (2016). Interactive tree of life (iTOL) v3: an online tool for the display and annotation of phylogenetic and other trees. Nucleic Acids Res. *44*, W242–5. https://doi.org/10.1093/nar/gkw290

Li, Z., Chen, Y., Zhang, S., Chen, S., Li, W., Yan, L., Shi, L., Wu, L., Sohr, A., and Su, S. (2013). Viral infection affects sucrose responsiveness and homing ability of forager honey bees, Apis mellifera L. PLOS ONE *8*, e77354. https://doi.org/10.1371/journal.pone.0077354

Lightner, D.V. (1999). The Penaeid shrimp viruses TSV, IHHNV, WSSV, and YHV. J. Appl. Aquaculture *9*, 27–52.

Lloyd, R.E. (2013). Regulation of stress granules and P-bodies during RNA virus infection. Wiley Interdiscip. Rev. RNA *4*, 317–331. https://doi.org/10.1002/wrna.1162

Luke, G.A., de Felipe, P., Lukashev, A., Kallioinen, S.E., Bruno, E.A., and Ryan, M.D. (2008). Occurrence, function and evolutionary origins of '2A-like' sequences in virus genomes. J. Gen. Virol. *89*, 1036–1042.

Magwire, M.M., Fabian, D.K., Schweyen, H., Cao, C., Longdon, B., Bayer, F., and Jiggins, F.M. (2012). Genome-wide association studies reveal a simple genetic basis of resistance to naturally coevolving viruses in *Drosophila* melanogaster. PLOS Genet. *8*, e1003057. https://doi.org/10.1371/journal.pgen.1003057

Mailliot, J., and Martin, F. (2018). Viral internal ribosomal entry sites: four classes for one goal. Wiley Interdiscip. Rev. RNA 9. https://doi.org/10.1002/wrna.1458

Majzoub, K., Hafirassou, M.L., Meignin, C., Goto, A., Marzi, S., Fedorova, A., Verdier, Y., Vinh, J., Hoffmann, J.A., Martin, F., *et al.* (2014). RACK1 controls IRES-mediated translation of viruses. Cell *159*, 1086–1095.

Maori, E., Lavi, S., Mozes-Koch, R., Gantman, Y., Peretz, Y., Edelbaum, O., Tanne, E., and Sela, I. (2007). Isolation and characterization of Israeli acute paralysis virus, a dicistrovirus affecting honeybees in Israel: evidence for diversity due to intra- and inter-species recombination. J. Gen. Virol. *88*, 3428–3438.

Maori, E., Paldi, N., Shafir, S., Kalev, H., Tsur, E., Glick, E., and Sela, I. (2009). IAPV, a bee-affecting virus associated with Colony Collapse Disorder can be silenced by dsRNA ingestion. Insect Mol. Biol. *18*, 55–60. https://doi.org/10.1111/j.1365-2583.2009.00847.x

Mari, J., Poulos, B.T., Lightner, D.V., and Bonami, J.R. (2002). Shrimp Taura syndrome virus: genomic characterization and similarity with members of the genus Cricket paralysis-like viruses. J. Gen. Virol. *83*, 915–926.

Marti, G.A., Ragone, P., Balsalobre, A., Ceccarelli, S., Susevich, M.L., Diosque, P., Echeverría, M.G., and Rabinovich, J.E. (2017). Can Triatoma virus inhibit infection of Trypanosoma cruzi (Chagas, 1909) in Triatoma infestans (Klug)? A cross infection and co-infection study. J. Invertebr. Pathol. *150*, 101–105.

Martins, N.E., Faria, V.G., Nolte, V., Schlötterer, C., Teixeira, L., Sucena, É., and Magalhães, S. (2014). Host adaptation to viruses relies on few genes with different cross-resistance properties. Proc. Natl. Acad. Sci. U.S.A. *111*, 5938–5943. https://doi.org/10.1073/pnas.1400378111

Masoumi, A., Hanzlik, T.N., and Christian, P.D. (2003). Functionality of the 5'- and intergenic IRES elements of cricket paralysis virus in a range of insect cell lines, and its relationship with viral activities. Virus Res. *94*, 113–120.

Meeus, I., de Miranda, J.R., de Graaf, D.C., Wäckers, F., and Smagghe, G. (2014). Effect of oral infection with Kashmir bee virus and Israeli acute paralysis virus on bumblebee (Bombus terrestris) reproductive success. J. Invertebr. Pathol. *121*, 64–69. https://doi.org/10.1016/j.jip.2014.06.011

Merkling, S.H., Bronkhorst, A.W., Kramer, J.M., Overheul, G.J., Schenck, A., and Van Rij, R.P. (2015). The epigenetic regulator G9a mediates tolerance to RNA virus infection in *Drosophila*. PLOS Pathog. *11*, e1004692. https://doi.org/10.1371/journal.ppat.1004692

Moore, N.F., Kearns, A., and Pullin, J.S. (1980). Characterization of cricket paralysis virus-induced polypeptides in *Drosophila* cells. J. Virol. *33*, 1–9.

Moore, N.F., Reavy, B., Pullin, J.S., and Plus, N. (1981). The polypeptides induced in *Drosophila* cells by *Drosophila C* virus (strain Ouarzazate). Virology *112*, 411–416.

Moss, S.M., Moss, D.R., Arce, S.M., Lightner, D.V., and Lotz, J.M. (2012). The role of selective breeding and biosecurity in the prevention of disease in penaeid shrimp aquaculture. J. Invertebr. Pathol. *110*, 247–250. https://doi.org/10.1016/j.jip.2012.01.013

Muhs, M., Yamamoto, H., Ismer, J., Takaku, H., Nashimoto, M., Uchiumi, T., Nakashima, N., Mielke, T., Hildebrand, P.W., Nierhaus, K.H., *et al.* (2011). Structural basis for the binding of IRES RNAs to the head of the ribosomal 40S subunit. Nucleic Acids Res. *39*, 5264–5275. https://doi.org/10.1093/nar/gkr114

Muhs, M., Hilal, T., Mielke, T., Skabkin, M.A., Sanbonmatsu, K.Y., Pestova, T.V., and Spahn, C.M. (2015). Cryo-EM of ribosomal 80S complexes with termination factors reveals the translocated cricket paralysis virus IRES. Mol. Cell *57*, 422–432. https://doi.org/10.1016/j.molcel.2014.12.016

Mullapudi, E., Přidal, A., Pálková, L., de Miranda, J.R., and Plevka, P. (2016). Virion structure of Israeli acute bee paralysis virus. J. Virol. *90*, 8150–8159. https://doi.org/10.1128/JVI.00854-16

Murray, J., Savva, C.G., Shin, B.S., Dever, T.E., Ramakrishnan, V., and Fernández, I.S. (2016). Structural characterization of ribosome recruitment and translocation by type IV IRES. Elife *5*, e13567. https://doi.org/10.7554/eLife.13567

Muscio, O.A., LaTorre, J.L., and Scodeller, E.A. (1987). Small nonoccluded viruses from triatomine bug Triatoma infestans (Hemiptera: Reduviidae). J. Invertebr. Pathol. *49*, 218–220.

Muscio, O.A., La Torre, J., Bonder, M.A., and Scodeller, E.A. (1997). Triatoma virus pathogenicity in laboratory colonies of Triatoma infestans (Hemiptera:Reduviidae). J. Med. Entomol. *34*, 253–256.

Muscio, O.A., Bonder, M.A., La Torre, J.L., and Scodeller, E.A. (2000). Horizontal transmission of triatoma virus through the fecal-oral route in Triatoma infestans (Hemiptera: Triatomidae). J. Med. Entomol. *37*, 271–275.

Nainu, F., Tanaka, Y., Shiratsuchi, A., and Nakanishi, Y. (2015). Protection of insects against viral infection by apoptosis-dependent phagocytosis. J. Immunol. *195*, 5696–5706. https://doi.org/10.4049/jimmunol.1500613

Nakashima, N., and Ishibashi, J. (2010). Identification of the 3C-protease-mediated 2A/2B and 2B/2C cleavage sites in the nonstructural polyprotein precursor of a Dicistrovirus lacking the NPGP motif. Arch. Virol. *155*, 1477–1482. https://doi.org/10.1007/s00705-010-0723-z

Nakashima, N., and Nakamura, Y. (2008). Cleavage sites of the 'P3 region' in the nonstructural polyprotein precursor of a dicistrovirus. Arch. Virol. *153*, 1955–1960. https://doi.org/10.1007/s00705-008-0208-5

Nakashima, N., and Shibuya, N. (2006). Multiple coding sequences for the genome-linked virus protein (VPg) in dicistroviruses. J. Invertebr. Pathol. *92*, 100–104.

Nakashima, N., and Uchiumi, T. (2009). Functional analysis of structural motifs in dicistroviruses. Virus Res. *139*, 137–147. https://doi.org/10.1016/j.virusres.2008.06.006

Nakashima, N., Sasaki, J., Tsuda, K., Yasunaga, C., and Noda, H. (1998). Properties of a new picorna-like virus of the brown-winged green bug, plautia stali J. Invertebr. Pathol. *71*, 151–158.

Nayak, A., Berry, B., Tassetto, M., Kunitomi, M., Acevedo, A., Deng, C., Krutchinsky, A., Gross, J., Antoniewski, C., and Andino, R. (2010). Cricket paralysis virus antagonizes Argonaute 2 to modulate antiviral defense in *Drosophila*. Nat. Struct. Mol. Biol. *17*, 547–554. https://doi.org/10.1038/nsmb.1810

Nayak, A., Kim, D.Y., Trnka, M.J., Kerr, C.H., Lidsky, P.V., Stanley, D.J., Rivera, B.M., Li, K.H., Burlingame, A.L., Jan, E., *et al.* (2018). A viral protein restricts *Drosophila* RNAi immunity by regulating argonaute activity and Stability. Cell Host Microbe *24*, 542–557.e9.

Nishiyama, T., Yamamoto, H., Shibuya, N., Hatakeyama, Y., Hachimori, A., Uchiumi, T., and Nakashima, N. (2003). Structural elements in the internal ribosome entry site of Plautia stali intestine virus responsible for binding with ribosomes. Nucleic Acids Res. *31*, 2434–2442.

Oerke, E. (2006). Crop losses to pests. J. Agric. Sci. *144*, 31–43.

Osborne, S.E., Leong, Y.S., O'Neill, S.L., and Johnson, K.N. (2009). Variation in antiviral protection mediated by different Wolbachia strains in *Drosophila* simulans. PLOS Pathog. *5*, e1000656. https://doi.org/10.1371/journal.ppat.1000656

Paul, A.V., and Wimmer, E. (2015). Initiation of protein-primed picornavirus RNA synthesis. Virus Res. *206*, 12–26. https://doi.org/10.1016/j.virusres.2014.12.028

Pestova, T.V., and Hellen, C.U. (2003). Translation elongation after assembly of ribosomes on the cricket paralysis virus internal ribosomal entry site without initiation factors or initiator tRNA. Genes Dev. *17*, 181–186.

Petrov, A., Grosely, R., Chen, J., O'Leary, S.E., and Puglisi, J.D. (2016). Multiple parallel pathways of translation initiation on the CrPV IRES. Mol. Cell *62*, 92–103. https://doi.org/10.1016/j.molcel.2016.03.020

Pfingsten, J.S., Costantino, D.A., and Kieft, J.S. (2006). Structural basis for ribosome recruitment and manipulation by a viral IRES RNA. Science *314*, 1450–1454.

Pfingsten, J.S., Costantino, D.A., and Kieft, J.S. (2007). Conservation and diversity among the three-dimensional folds of the Dicistroviridae intergenic region IRESes. J. Mol. Biol. *370*, 856–869.

Pimentel, D.J.E. (2005). Environmental and economic costs of the application of pesticides primarily in the United States. Development and Sustainability 7, 229–252.

Pisareva, V.P., Pisarev, A.V., and Fernández, I.S. (2018). Dual tRNA mimicry in the cricket paralysis virus IRES uncovers an unexpected similarity with the hepatitis C virus IRES. Elife *7*, e34062. https://doi.org/10.7554/eLife.34062

Plus, N., Croizier, G., Jousset, F.X., and David, J. (1975). Picornaviruses of laboratory and wild *Drosophila* melanogaster: geographical distribution and serotypic composition. Ann. Microbiol. *126*, 107–117.

Plus, N., Croizier, G., Reinganum, C., and Scott, P.D. (1978). Cricket paralysis virus and *drosophila* C virus: serological analysis and comparison of capsid polypeptides and host range. J. Invertebr. Pathol. *31*, 296–302.

Poirier, E.Z., Goic, B., Tomé-Poderti, L., Frangeul, L., Boussier, J., Gausson, V., Blanc, H., Vallet, T., Loyd, H., Levi, L.I., *et al.* (2018). Dicer-2-dependent generation of viral DNA from defective genomes of RNA viruses modulates antiviral immunity in insects. Cell Host Microbe *23*, 353–365.e8.

Popp, J., Peto, K., and Nagy, J. (2013). Pesticide productivity and food security. A review. Agron. Sustain. Dev. *33*, 243.

Ramakrishnaiah, V., Thumann, C., Fofana, I., Habersetzer, F., Pan, Q., de Ruiter, P.E., Willemsen, R., Demmers, J.A., Stalin Raj, V., Jenster, G., *et al.* (2013). Exosome-mediated transmission of hepatitis C virus between human hepatoma Huh7.5 cells. Proc. Natl. Acad. Sci. U.S.A. *110*, 13109–13113. https://doi.org/10.1073/pnas.1221899110

Redak, R.A., Purcell, A.H., Lopes, J.R., Blua, M.J., Mizell, R.F., and Andersen, P.C. (2004). The biology of xylem fluid-feeding insect vectors of Xylella fastidiosa and their relation to disease epidemiology. Annu. Rev. Entomol. *49*, 243–270. https://doi.org/10.1146/annurev.ento.49.061802.123403

Reineke, L.C., and Lloyd, R.E. (2015). The stress granule protein G3BP1 recruits protein kinase R to promote multiple innate immune antiviral responses. J. Virol. *89*, 2575–2589. https://doi.org/10.1128/JVI.02791-14

Reinganum, C. (1975). The isolation of cricket paralysis virus from the emperor gum moth, Antheraea eucalypti Scott, and its infectivity towards a range of insect species. Intervirology *5*, 97–102. https://doi.org/10.1159/000149886

Reinganum, C., O'Loughlin, G.T., and Hogan, T.W. (1970). A nonoccluded virus of the field crickets *Teleogryllus oceanicus* and *T. commodus* (Orthoptera: Gryllidae). J. Invertebr. Pathol. *16*, 214–220.

Ren, Q., Wang, Q.S., Firth, A.E., Chan, M.M., Gouw, J.W., Guarna, M.M., Foster, L.J., Atkins, J.F., and Jan, E. (2012). Alternative reading frame selection mediated by a tRNA-like domain of an internal ribosome entry site. Proc. Natl. Acad. Sci. U.S.A. *109*, E630–9. https://doi.org/10.1073/pnas.1111303109

Ren, Q., Au, H.H., Wang, Q.S., Lee, S., and Jan, E. (2014). Structural determinants of an internal ribosome entry site that direct translational reading frame selection. Nucleic Acids Res. *42*, 9366–9382. https://doi.org/10.1093/nar/gku622

Robalino, J., Browdy, C.L., Prior, S., Metz, A., Parnell, P., Gross, P., and Warr, G. (2004). Induction of antiviral immunity by double-stranded RNA in a marine invertebrate. J. Virol. *78*, 10442–10448. https://doi.org/10.1128/JVI.78.19.10442-10448.2004

Robalino, J., Bartlett, T.C., Chapman, R.W., Gross, P.S., Browdy, C.L., and Warr, G.W. (2007). Double-stranded RNA and antiviral immunity in marine shrimp: inducible host mechanisms and evidence for the evolution of viral counter-responses. Dev. Comp. Immunol. *31*, 539–547.

Roberts, L.O., and Groppelli, E. (2009). An atypical IRES within the 5' UTR of a dicistrovirus genome. Virus Res. *139*, 157–165. https://doi.org/10.1016/j.virusres.2008.07.017

Ruehle, M.D., Zhang, H., Sheridan, R.M., Mitra, S., Chen, Y., Gonzalez, R.L., Cooperman, B.S., and Kieft, J.S. (2015). A dynamic RNA loop in an IRES affects multiple steps of elongation factor-mediated translation initiation. Elife *4*, e08146. https://doi.org/10.7554/eLife.08146

Sabath, N., Price, N., and Graur, D. (2009). A potentially novel overlapping gene in the genomes of Israeli acute paralysis virus and its relatives. Virol. J. *6*, 144. https://doi.org/10.1186/1743-422X-6-144

Saleh, M.C., van Rij, R.P., Hekele, A., Gillis, A., Foley, E., O'Farrell, P.H., and Andino, R. (2006). The endocytic pathway mediates cell entry of dsRNA to induce RNAi silencing. Nat. Cell Biol. *8*, 793–802.

Saleh, M.C., Tassetto, M., van Rij, R.P., Goic, B., Gausson, V., Berry, B., Jacquier, C., Antoniewski, C., and Andino, R. (2009). Antiviral immunity in *Drosophila* requires systemic RNA interference spread. Nature *458*, 346–350. https://doi.org/10.1038/nature07712

Sánchez-Eugenia, R., Goikolea, J., Gil-Cartón, D., Sánchez-Magraner, L., and Guérin, D.M. (2015). Triatoma virus recombinant VP4 protein induces membrane permeability through dynamic pores. J. Virol. *89*, 4645–4654. https://doi.org/10.1128/JVI.00011-15

Sansone, C.L., Cohen, J., Yasunaga, A., Xu, J., Osborn, G., Subramanian, H., Gold, B., Buchon, N., and Cherry, S. (2015). Microbiota-dependent priming of antiviral intestinal immunity in *Drosophila*. Cell Host Microbe *18*, 571–581. https://doi.org/10.1016/j.chom.2015.10.010

Sasaki, J., and Nakashima, N. (1999). Translation initiation at the CUU codon is mediated by the internal ribosome entry site of an insect picorna-like virus in vitro. J. Virol. *73*, 1219–1226.

Sasaki, J., and Nakashima, N. (2000). Methionine-independent initiation of translation in the capsid protein of an insect RNA virus. Proc. Natl. Acad. Sci. U.S.A. *97*, 1512–1515. https://doi.org/10.1073/pnas.010426997

Schüler, M., Connell, S.R., Lescoute, A., Giesebrecht, J., Dabrowski, M., Schroeer, B., Mielke, T., Penczek, P.A., Westhof, E., and Spahn, C.M. (2006). Structure of the ribosome-bound cricket paralysis virus IRES RNA. Nat. Struct. Mol. Biol. *13*, 1092–1096.

Serbus, L.R., and Sullivan, W. (2007). A cellular basis for Wolbachia recruitment to the host germline. PLOS Pathog. *3*, e190.

Shi, M., Lin, X.D., Tian, J.H., Chen, L.J., Chen, X., Li, C.X., Qin, X.C., Li, J., Cao, J.P., Eden, J.S., *et al.* (2016). Redefining the invertebrate RNA virosphere. Nature [Epub ahead of print]. https://doi.org/10.1038/nature20167

Snijder, J., Uetrecht, C., Rose, R.J., Sanchez-Eugenia, R., Marti, G.A., Agirre, J., Guérin, D.M., Wuite, G.J., Heck, A.J., and Roos, W.H. (2013). Probing the biophysical interplay between a viral genome and its capsid. Nat. Chem. *5*, 502–509. https://doi.org/10.1038/nchem.1627

Spahn, C.M., Jan, E., Mulder, A., Grassucci, R.A., Sarnow, P., and Frank, J. (2004). Cryo-EM visualization of a viral internal ribosome entry site bound to human ribosomes: the IRES functions as an RNA-based translation factor. Cell *118*, 465–475. https://doi.org/10.1016/j.cell.2004.08.001

Spurny, R., Přidal, A., Pálková, L., Kiem, H.K., de Miranda, J.R., and Plevka, P. (2017). Virion structure of black queen cell virus, a common honeybee pathogen. J. Virol. *91*, e02100–16.

Squires, G., Pous, J., Agirre, J., Rozas-Dennis, G.S., Costabel, M.D., Marti, G.A., Navaza, J., Bressanelli, S., Guerin, D.M., and Rey, F.A. (2013). Structure of the Triatoma virus capsid. Acta Crystallogr. D Biol. Crystallogr. *69*, 1026–1037.

Talide, L., Imler, J.L., and Meignin, C. (2019). Sensing viral infections in insects: pathways looking for receptors. In Insect Molecular Virology: Advances and Emerging Trends, B.C. Bonning, ed. (Caister Academic Press, Norfolk, UK).

Tassetto, M., Kunitomi, M., and Andino, R. (2017). Circulating immune cells mediate a systemic RNAi-based adaptive antiviral response in *Drosophila*. Cell *169*, 314–325.e13.

Tate, J., Liljas, L., Scotti, P., Christian, P., Lin, T., and Johnson, J.E. (1999). The crystal structure of cricket paralysis virus: the first view of a new virus family. Nat. Struct. Biol. *6*, 765–774. https://doi.org/10.1038/11543

Teixeira, L., Ferreira, A., and Ashburner, M. (2008). The bacterial symbiont Wolbachia induces resistance to RNA viral infections in *Drosophila* melanogaster. PLOS Biol. *6*, e2. https://doi.org/10.1371/journal.pbio.1000002

Thompson, S.R., Gulyas, K.D., and Sarnow, P. (2001). Internal initiation in Saccharomyces cerevisiae mediated by an initiator tRNA/eIF2-independent internal ribosome entry site element. Proc. Natl. Acad. Sci. U.S.A. *98*, 12972–12977. https://doi.org/10.1073/pnas.241286698

Toriyama, S., Guy, P.L., Fuji, S., and Takahashi, M. (1992). Characterization of a new picorna-like virus, himetobi P virus, in planthoppers. J. Gen. Virol. 73, 1021–1023. https://doi.org/10.1099/0022-1317-73-4-1021

Tufts, D.M., Hunter, W.B., and Bextine, B. (2014). Solenopsis invicta virus (Sinv-1) infection and insecticide interactions in the red imported fire ant (Hymenoptera: Formicidae). Fla. Entomol. 97, 1251–1254.

Valles, S.M. (2012). Positive-strand RNA viruses infecting the red imported fire ant, *Solenopsis invicta*. Psyche *2012*, 14. https://doi.org/10.1155/2012/821591

Valles, S.M., and Hashimoto, Y. (2009). Isolation and characterization of Solenopsis invicta virus 3, a new positive-strand RNA virus infecting the red imported fire ant, Solenopsis invicta. Virology *388*, 354–361. https://doi.org/10.1016/j.virol.2009.03.028

Valles, S.M., Chen, Y., Firth, A.E., Guérin, D.M.A., Hashimoto, Y., Herrero, S., de Miranda, J.R., and Ryabov, E. (2017). ICTV virus taxonomy profile: Dicistroviridae. J. Gen. Virol. *98*, 355–356.

van Rij, R.P., Saleh, M.C., Berry, B., Foo, C., Houk, A., Antoniewski, C., and Andino, R. (2006). The RNA silencing endonuclease Argonaute 2 mediates specific antiviral immunity in *Drosophila* melanogaster. Genes Dev. *20*, 2985–2995.

Wamonje, F.O., Michuki, G.N., Braidwood, L.A., Njuguna, J.N., Musembi Mutuku, J., Djikeng, A., Harvey, J.J.W., and Carr, J.P. (2017). Viral metagenomics of aphids present in bean and maize plots on mixed-use farms in Kenya reveals the presence of three dicistroviruses including a novel Big Sioux River virus-like dicistrovirus. Virol. J. *14*, 188. https://doi.org/10.1186/s12985-017-0854-x

Wang, Q.S., Au, H.H., and Jan, E. (2013). Methods for studying IRES-mediated translation of positive-strand RNA viruses. Methods *59*, 167–179. https://doi.org/10.1016/j.ymeth.2012.09.004

Wang, X.H., Aliyari, R., Li, W.X., Li, H.W., Kim, K., Carthew, R., Atkinson, P., and Ding, S.W. (2006). RNA interference directs innate immunity against viruses in adult *Drosophila*. Science *312*, 452–454.

Watanabe, M., Iwakawa, H.O., Tadakuma, H., and Tomari, Y. (2017). Biochemical and single-molecule analyses of the RNA silencing suppressing activity of CrPV-1A. Nucleic Acids Res. *45*, 10837–10844. https://doi.org/10.1093/nar/gkx748

White, J.P., Cardenas, A.M., Marissen, W.E., and Lloyd, R.E. (2007). Inhibition of cytoplasmic mRNA stress granule formation by a viral proteinase. Cell Host Microbe *2*, 295–305.

Williams, G.R., Tarpy, D.R., vanEngelsdorp, D., Chauzat, M.P., Cox-Foster, D.L., Delaplane, K.S., Neumann, P., Pettis, J.S., Rogers, R.E., and Shutler, D. (2010). Colony Collapse Disorder in context. Bioessays *32*, 845–846. https://doi.org/10.1002/bies.201000075

Williamson, C., Rybicki, E.P., Kasdorf, G.G.F., and Von Wechmar, M.B. (1988). Characterization of a new Picorna-like virus isolated from aphids. J. Gen. Virol. *69*, 787–795.

Wilson, J.E., Pestova, T.V., Hellen, C.U., and Sarnow, P. (2000a). Initiation of protein synthesis from the A site of the ribosome. Cell *102*, 511–520.

Wilson, J.E., Powell, M.J., Hoover, S.E., and Sarnow, P. (2000b). Naturally occurring dicistronic cricket paralysis virus RNA is regulated by two internal ribosome entry sites. Mol. Cell. Biol. *20*, 4990–4999.

Wilson, R.C., and Doudna, J.A. (2013). Molecular mechanisms of RNA interference. Annu. Rev. Biophys. *42*, 217–239. https://doi.org/10.1146/annurev-biophys-083012-130404

Wong, Z.S., Brownlie, J.C., and Johnson, K.N. (2015). Oxidative stress correlates with Wolbachia-mediated antiviral protection in Wolbachia-*Drosophila* associations. Appl. Environ. Microbiol. *81*, 3001–3005. https://doi.org/10.1128/AEM.03847-14

Wong, Z.S., Brownlie, J.C., and Johnson, K.N. (2016). Impact of ERK activation on fly survival and Wolbachia-mediated protection during virus infection. J. Gen. Virol. 97, 1446–1452. https://doi.org/10.1099/jgv.0.000456

Woolaway, K.E., Lazaridis, K., Belsham, G.J., Carter, M.J., and Roberts, L.O. (2001). The 5' untranslated region of Rhopalosiphum padi virus contains an internal ribosome entry site which functions efficiently in mammalian, plant, and insect translation systems. J. Virol. *75*, 10244–10249. https://doi.org/10.1128/JVI.75.21.10244-10249.2001

Wu, Q., Luo, Y., Lu, R., Lau, N., Lai, E.C., Li, W.X., and Ding, S.W. (2010). Virus discovery by deep sequencing and assembly of virus-derived small silencing RNAs. Proc. Natl. Acad. Sci. U.S.A. *107*, 1606–1611. https://doi.org/10.1073/pnas.0911353107

Wu, X., He, W.T., Tian, S., Meng, D., Li, Y., Chen, W., Li, L., Tian, L., Zhong, C.Q., Han, F., *et al.* (2014). pelo is required for high efficiency viral replication. PLOS Pathog. *10*, e1004034. https://doi.org/10.1371/journal.ppat.1004034

Xu, J., Hopkins, K., Sabin, L., Yasunaga, A., Subramanian, H., Lamborn, I., Gordesky-Gold, B., and Cherry, S. (2013). ERK signaling couples nutrient status to antiviral defense in the insect gut. Proc. Natl. Acad. Sci. U.S.A. *110*, 15025–15030. https://doi.org/10.1073/pnas.1303193110

Yamamoto, H., Nakashima, N., Ikeda, Y., and Uchiumi, T. (2007). Binding mode of the first aminoacyl-tRNA in translation initiation mediated by Plautia stali intestine virus internal ribosome entry site. J. Biol. Chem. *282*, 7770–7776.

Yang, S., Yu, J., Fan, Z., Gong, S.T., Tang, H., and Pan, L. (2018). Bub1 facilitates virus entry through endocytosis in a model of *Drosophila* pathogenesis. J. Virol. *92*, e00254–18.

Zeidler, M.P., and Bausek, N. (2013). The *Drosophila* JAK-STAT pathway. JAKSTAT 2, e25353. https://doi.org/10.4161/jkst.25353

Zhang, H., Ng, M.Y., Chen, Y., and Cooperman, B.S. (2016). Kinetics of initiating polypeptide elongation in an IRES-dependent system. Elife 5, e13429. https://doi.org/10.7554/eLife.13429

Zhang, Q., Zhang, L., Gao, X., Qi, S., Chang, Z., and Wu, Q. (2015). DIP1 plays an antiviral role against DCV infection in *Drosophila* melanogaster. Biochem. Biophys. Res. Commun. *460*, 222–226. https://doi.org/10.1016/j.bbrc.2015.03.013

Zhu, J., Korostelev, A., Costantino, D.A., Donohue, J.P., Noller, H.F., and Kieft, J.S. (2011). Crystal structures of complexes containing domains from two viral internal ribosome entry site (IRES) RNAs bound to the 70S ribosome. Proc. Natl. Acad. Sci. U.S.A. *108*, 1839–1844. https://doi.org/10.1073/pnas.1018582108

Looking Through the Lens of 'Omics Technologies: Insights into the Transmission of Insect Vector-borne Plant Viruses

6

Jennifer R. Wilson[1,2], Stacy L. DeBlasio[1,2,3], Mariko M. Alexander[1,2] and Michelle Heck[1,2,3*]

[1]Plant Pathology and Plant Microbe Biology Section, School of Integrative Plant Sciences, Cornell University, Ithaca, NY, USA.
[2]Boyce Thompson Institute, Ithaca, NY, USA.
[3]Emerging Pests and Pathogens Research Unit, United States Department of Agriculture – Agricultural Research Service, Ithaca, NY, USA.

*Correspondence: mlc68@cornell.edu

https://doi.org/10.21775/9781912530083.06

Abstract

Insects in the orders Hemiptera and Thysanoptera transmit viruses and other pathogens associated with the most serious diseases of plants. Plant viruses transmitted by these insects target similar tissues, genes, and proteins within the insect to facilitate plant-to-plant transmission with some degree of specificity at the molecular level. 'Omics experiments are becoming increasingly important and practical for vector biologists to use towards better understanding the molecular mechanisms and biochemistry underlying transmission of these insect-borne diseases. These discoveries are being used to develop novel means to obstruct virus transmission into and between plants. In this chapter, we summarize 'omics technologies commonly applied in vector biology and the important discoveries that have been made using these methods, including virus and insect proteins involved in transmission, as well as the tri-trophic interactions involved in host and vector manipulation. Finally, we critically examine the limitations and new horizons in this area of research, including the role of endosymbionts and insect viruses in virus–vector interactions, and the development of novel control strategies.

Advances in whole genome sequencing of insect vectors of plant viruses fuels discovery

Major advances in next generation sequencing technologies and their increased affordability has led to an explosion in the availability of whole-genome sequence data for each biological player in the virus–vector–host relationship: virus, vector and plant (Fig. 6.1A). To date, draft reference genome sequences have been completed and/or published for nearly all the major types of agricultural vectors including whiteflies (Chen *et al.*, 2016), planthoppers (Zhu *et al.*, 2017), aphids (Consortium, 2010; Wenger *et al.*, 2017), and psyllids (Saha *et al.*, 2017). Whole genome studies have shed light on the gene families in these insects that are critical for plant adaptation, insecticide resistance, and virus transmission (Consortium, 2010; Chen *et al.*, 2016; Kaur *et al.*, 2016; Wenger *et al.*, 2017; Zhu *et al.*, 2017). Functional annotation of genes coded by

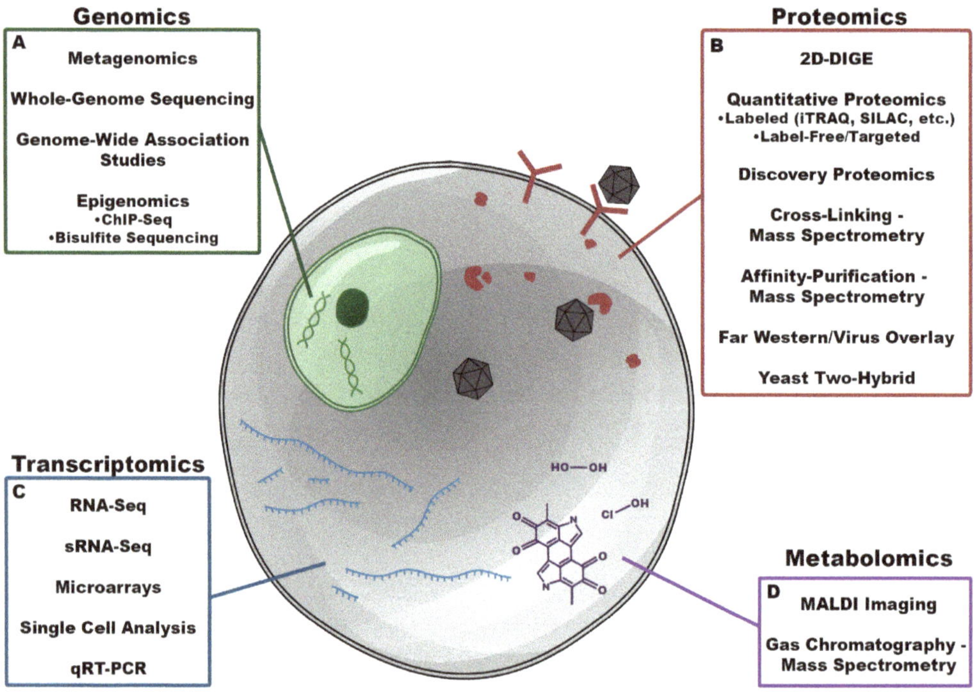

Figure 6.1 'Omics approaches to studying vector biology, including (A) genomics; (B) proteomics; (C) transcriptomics; and (D) metabolomics. 2D-DIGE, 2-dimensional difference gel electrophoresis; ChIP-seq, chromatin immunoprecipitation sequencing; iTRAQ, isobaric tags for relative and absolute quantification; MALDI, matrix-assisted laser desorption/ionization; qRT-PCR, quantitative reverse transcriptase polymerase chain reaction; RNA-seq, RNA sequencing; SILAC, stable isotope labelling by amino acids in cell culture; sRNA-seq, small RNA sequencing.

vector genomes is still highly reliant on homology-based methods to other arthropods (Oppenheim *et al.*, 2015), especially the model species *Drosophila melanogaster* (Saha *et al.*, 2017), which means many genes in non-model vectors lack a validated functional assignment. Structural annotations also need to be improved for these data to be more useful to vector biologists.

Genome sequence information is used as the foundation for the analysis of other types of 'omics data: transcriptomics, proteomics, and metabolomics (Fig. 6.1B–D). Transcriptomics (Fig. 6.1C), the most widely used 'omics technique in biology, involves the detection and quantification of gene expression at the RNA level that occurs during certain developmental stages and/or in response to changes in physiological conditions. There are several RNA species that can be measured including mRNA, small RNAs, long non-coding RNAs, and viral RNA, with each requiring slight variations in sample preparation and downstream bioinformatics analysis (Kukurba and Montgomery, 2015). These types of studies provide evidence for the putative involvement of those genes and/or their gene products with the insect's developmental stage (Wang *et al.*, 2010) or other conditions being studied. These can include changes in gene expression that occur during virus acquisition compared with non-viruliferous insects (Brault *et al.*, 2010), comparisons of the saliva of different aphid vector species (Thorpe *et al.*, 2016), the response of insects being reared on different host plants (Christodoulides *et al.*, 2017; Mathers *et al.*, 2017), or the molecular mechanisms associated with insecticide resistance (Yang *et al.*, 2013). Expressed sequence tags (EST) and microarrays were extensively used in the past for gene expression studies in vector species (Ramsey *et al.*, 2007; Brault *et al.*, 2010; Götz *et al.*, 2012). Now, next-generation sequencing techniques are emmployed. Proteome studies (Fig. 6.1B) often measure proteins using mass spectrometry (MS). These studies investigate gene expression at the translational level and identify post-translational modifications, which lead to differences in the levels of proteins being produced under a physiological condition or their localization within cells. Discovery-based proteomics techniques allow the researcher to identify the proteins present in a sample by matching peptide sequences to databases of known protein sequences (derived from the genome or transcriptome data) (Cilia *et al.*, 2011b) and to quantify their levels relative to other samples (Pinheiro *et al.*, 2014). Some techniques in proteomics allow one to measure the levels of a specific target protein (targeted experiments) (Cilia *et al.*, 2012a) or identify and quantify protein–protein interactions (co-immunoprecipitation-MS and crosslinking-MS technology) (Chavez *et al.*, 2012; DeBlasio *et al.*, 2015a). Metabolomics (Fig. 6.1D) is the study of the metabolome of an organism, that is, all the metabolites produced by an organism, tissue, or cell under a physiological condition. Metabolic outputs are downstream to gene transcription and protein expression. Methods may be targeted to measure individual metabolites with known standards, or non-targeted, enabling the discovery of new metabolites and metabolic signatures (Maag *et al.*, 2015). Combining 'omics techniques can provide a more complete picture of what is occurring on the molecular level. For instance, proteomics coupled to transcriptomics can give a more nuanced picture of how proteins are regulated on the post-transcriptional and post-translational level, independent of gene expression (Kruse *et al.*, 2017). Metabolomics can also be combined with transcriptomics and/or proteomics to understand how the regulation of enzymes gives rise to the differential expression of metabolites (Kersten *et al.*, 2013).

Plant virus transmission by insects involves a series of carefully orchestrated, temporally-regulated protein interactions among virus, vector, and the plant host, all of which can be captured and measured quantitatively using these various 'omics approaches. The first step in virus transmission is acquisition, or the retention of the virus in the vector. Acquisition may be as simple as virus transiently binding to insect mouthparts, or as complicated as virus transcytosis across the insect gut. Protein interaction studies have been used to identify viral, vector and host components involved (detailed in sections below). Next, there may be a latent period during which the virus must circulate through the vector, which, for the propagative viruses, includes virus replication. RNA sequencing and proteomics have been used to measure the various effects viruses and virus-infected plants have on their insect vectors. Finally, the virus must be inoculated into a new host, completing the transmission process. Protein interaction studies (such as in Fig. 6.1B), ultrastructural studies, classical molecular biological and entomological approaches (for example, electrical penetration graph) have been used to study the latter step. Modes of transmission are often defined by the length of these various steps of the process, as well as the vector tissues where the virus is retained following acquisition.

The terminology in the vector biology literature can be as confusing and full of jargon as the 'omics literature, so we attempt to clarify a few key points here when discussing insect vectors. In this chapter, we will follow the designations provided by Gray and Banerjee (1999): that the modes of transmission should be classified as 'circulative' or 'non-circulative', depending on whether the virus passed through insect vector cell membranes and was retained internally within the vector, which would constitute circulative transmission. Not all vector species transmit all viruses and not all individuals within a species are capable of transmitting a particular virus species. This natural variation in vectoring ability is found in many vector species and is genetically encoded (reviewed in Gray *et al.*, 2014). In these vector species, vector and non-vector are common terms to describe individual insects (or vector species), which are either capable or not capable, respectively, of efficient virus transmission. Finally, insects which have acquired circulative viruses are referred to as viruliferous. Insects which have acquired circulative, propagative viruses (which replicate in the insect vector as well as the plant host) may also be referred to as infected.

New knowledge in vector biology 'omics includes the identification of direct protein interactions facilitating virus transmission as well as indirect effects, such as host and vector manipulation (detailed in sections below). Whereas the viral components to transmission have largely been characterized by more traditional molecular biology techniques and microscopy, rapid advances in 'omics technologies have fuelled discovery on the vector side of vector–pathogen interactions (Heck, 2018) and have led to the development of novel strategies that block specific molecular events involved in the steps of virus transmission (Heck and Brault, 2018). Advances in RNA interference (RNAi) in non-model insects have also opened the door to functional validation of candidate genes and proteins involved in virus transmission by hemipteran insect vectors (Pitino *et al.*, 2011; Mulot *et al.*, 2018).

Virus–vector protein interactions involved in non-circulative transmission

Viruses transmitted in a non-circulative mode are characterized by rapid acquisition and retention in the vector mouthparts and foregut. There is no latency period between acquisition and inoculation, and limited virus persistence in the vector. To date, aphids are the only known vectors of viruses transmitted in this mode. Non-circulative virus transmission is intimately linked to aphid probing behaviour (Drucker and Then, 2015). In seeking a new plant host, aphids penetrate the cell epidermis with their stylets in a series of shallow probes that puncture cell membranes, allowing the aphid to 'taste' cellular components and to come into direct contact with virions. Virions bind the aphid mouthparts or foregut and can be acquired in a matter of seconds to minutes. The virions can later be dislodged from aphid mouthparts during the same probing process on a susceptible plant host. Therefore, aphids need not colonize a plant host to transmit virus in this mode. In fact, many non-circulative viruses are transmitted by non-colonizing aphids, and vector and virus host range do not entirely overlap (Berlandier *et al.*, 1997). While non-circulative viruses can be transmitted by many different aphid species, there is still some degree of specificity in virion binding to aphid mouthparts, which will be discussed in the subsections below.

Viruses using the capsid strategy

Transmission of non-circulative viruses can be broadly separated into two categories, depending on the viral proteins required: the capsid strategy or the helper component strategy. Early work on non-circulative viruses by Pirone and Megahed (1966) involved feeding gradient-purified virions to insects via a membrane sachet and testing if transmission occurs. For some viruses, this membrane feeding was successful, indicating that virion proteins alone were sufficient for transmission. Virions were thought to directly bind aphid mouthparts, in what has been termed the capsid strategy. *Cucumber mosaic virus* (CMV) is one of the best characterized among the non-circulative viruses to use the capsid strategy. CMV has a tripartite genome, and when reassortment of viral RNA segments occurs in mixed infections, the transmission phenotype always corresponds to the virus contributing RNA 3, which encodes the capsid coat protein (CP) (Mossop and Francki, 1977). Mutational analysis of natural and engineered variants of the CMV CP has identified many residues that are required for transmission by the aphid vectors, *Aphis gossypii* and *Myzus persicae* (Perry *et al.*, 1994, 1998; Liu *et al.*, 2002). Some of these mutations abolish aphid transmission by affecting virion stability, indicating that properly assembled virus particles are required for transmission (Ng *et al.*, 2000, 2005). The highly-conserved βH-βI loop of the capsid surface, which is comprised almost entirely of acidic residues, has also been found to be involved in transmission, as mutations that lead to substitutions of neutral or basic residues in this region abolish transmission (Liu *et al.*, 2002). Additionally, these virus mutants could regain transmission ability if compensatory residue substitutions occurred elsewhere in the capsid, which may have preserved virion stability or charge distribution (Liu *et al.*, 2002).

Progress has been slow in identifying insect proteins involved in the transmission of non-circulative viruses, largely due to the difficulty in identifying and extracting cuticle proteins as well as the small size of insect mouthparts. Only recently have vector proteins involved in transmission of non-circulative viruses been discovered. Liang and Gao (2017) performed yeast two-hybrid assays to detect the binding of *M. persicae* cuticle proteins with the capsid of CMV. The cuticle protein *M. persicae* cuticle protein 4 (MPCP4) was found to directly interact with the CMV CP (Fig. 6.2G). Knockdown of MPCP4 by double stranded RNA (dsRNA) feeding led to less acquisition of CMV CP in the aphids, as detected by quantitative polymerase chain reaction (qPCR), indicating that MPCP4 may be an aphid receptor of CMV (Liang and Gao, 2017). This same *M. persicae* protein was recently found to be involved in the transmission of *Cauliflower mosaic virus* (CaMV), a helper

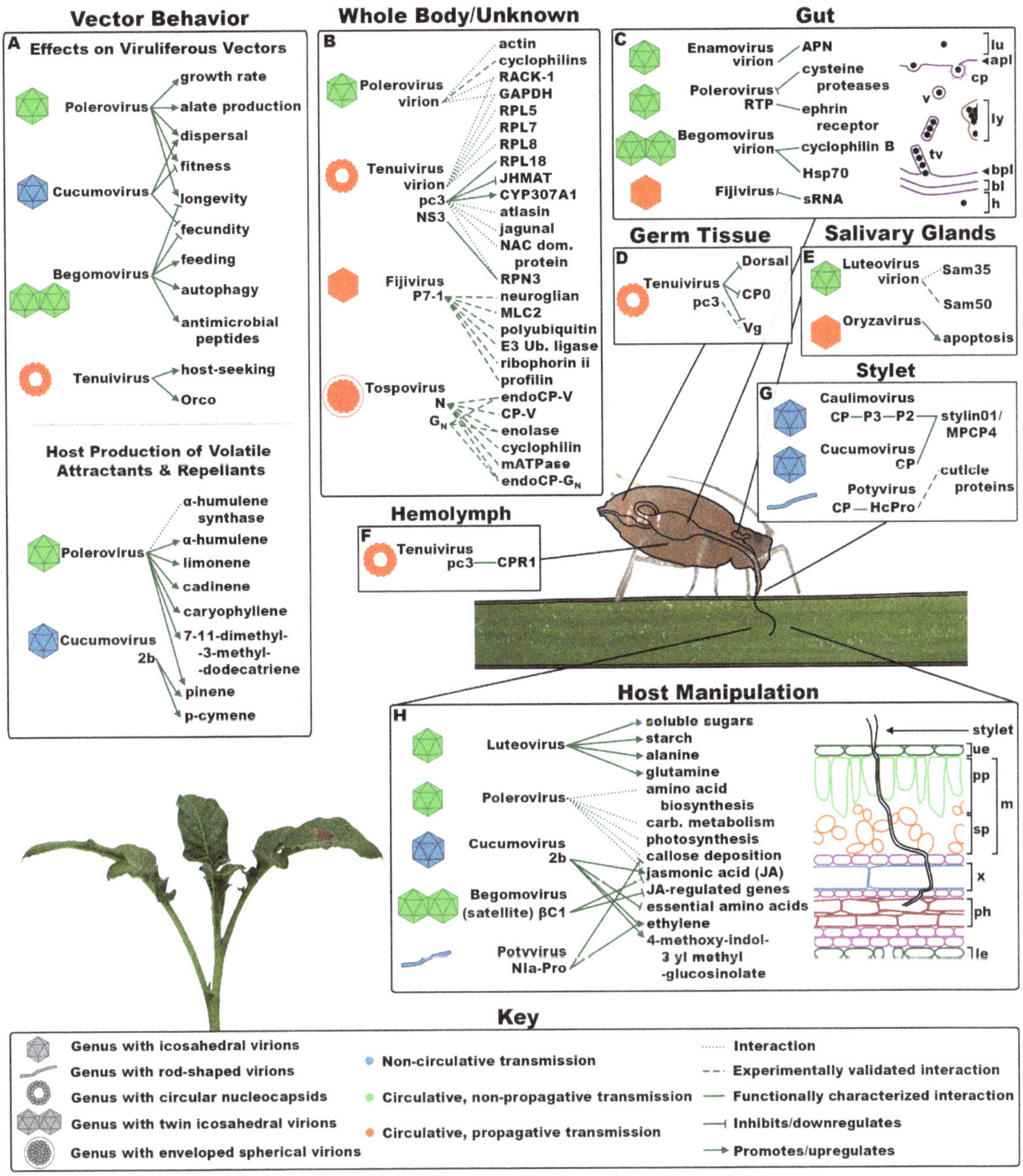

Figure 6.2 Overview of interactions between virus, vector and host involved in the transmission of plant viruses, as determined using 'omics technologies. (A) Interactions affecting vector behaviour, including the production of volatile attractants. (B) Interactions between virus and vector proteins with unknown or undetermined localization. (C) Interactions between virus and vector proteins and processes occurring in the insect gut. Diagram at right shows a model for virion movement across the gut. apl, apical plasmalemma; bl, basal lamina; bpl, basal plasmalemma; cp, coated pit; h, hemolymph; lu, lumen; ly, lysosome; tv, tubular vesicle; v, vesicle. (D) Interactions between virus and vector proteins occurring in the germ tissue. (E) Interactions between virus and vector proteins and processes occurring in the salivary glands. (F) Interactions between virus and vector proteins occurring in the hemolymph. (G) Interactions between viruses and vector proteins occurring in the stylet. (H) Interactions between virus and plant host factors affecting vector transmission. Diagram at right shows a cross-section of a hemipteran feeding on a leaf; le, lower epidermis; m, mesophyll; ph, phloem; pp, palisade parenchyma; sp, spongy parenchyma; ue, upper epidermis; x, xylem. Key for colour and symbol codes; left column: symbol for virion structure; centre column: transmission mode indicated by virion symbol colour; right column: edge types denoting different types of virus–host or virus–vector interactions.

component strategy virus (see below) (Webster *et al.*, 2018).

Viruses using the helper component strategy

Not all viruses can be transmitted as purified virions alone. Work by Kassanis and Govier (1971) identified a non-transmissible form of *Potato virus Y* (PVY) that could be conditionally transmitted if aphids were first given an acquisition period on wild-type PVY, but not if acquired in the reverse order. Therefore, they concluded that aphids captured a 'helper component' from the wild-type strain. This viral protein was later identified and named HC-Pro (for helper component-protease). Govier and Kassanis (1974) could reconstitute aphid transmission by purifying HC-Pro from plants and mixing it with purified virions in membrane feeding assays. Non-circulative viruses that require one or more non-structural viral proteins for transmission are said to use the helper strategy.

Since the discovery of helper components in the potyvirus PVY, there have been advances in understanding the mechanism of transmission for the *Potyviridae*. Potyviruses have filamentous rod-shaped capsids composed of a single CP (Zamora *et al.*, 2017). During transmission, the CP is bound by HC-Pro, which is a multifunctional viral protein with protease activity (Ivanov *et al.*, 2016). HC-Pro forms a 'molecular bridge' between the capsid and the aphid stylet (Ammar *et al.*, 1994). A surface-exposed domain in the N-terminus of the CP and the amino acid sequence motif DAG have been found to be highly conserved across potyviruses and necessary for aphid transmission (Allison *et al.*, 1985; Harrison and Robinson, 1988; Shukla *et al.*, 1988; Atreya *et al.*, 1991, 1995). The DAG motif is thought to bind the PTK amino acid motif near the C-terminus of HC-Pro (Peng *et al.*, 1998). KITC, near the N-terminus of HC-Pro, is another highly-conserved amino acid motif that has been tied to aphid transmission and is thought to bind the aphid mouthparts (Peng *et al.*, 1998).

CaMV, an icosahedral virus in the *Caulimoviridae* family, also binds insect mouthparts using the helper strategy. However, it requires not one, but two viral proteins in addition to the capsid: P2 and P3 (Woolston *et al.*, 1983). Using far western blot analysis, the P3 protein was found to directly bind virion (Leh *et al.*, 2001; Plisson *et al.*, 2005) and P2 (Leh *et al.*, 1999), while the P2 protein is thought to coordinate the interaction between virions. This viral protein forms aggregates termed 'transmission bodies' (Espinoza *et al.*, 1991; Drucker *et al.*, 2002; Khelifa *et al.*, 2007) that are responsive to wounding, such as during aphid probing (Bak *et al.*, 2013; Martinière *et al.*, 2013). When an infected leaf is wounded, the transmission bodies dissociate and P2 disperses onto microtubules throughout the cell (Martinière *et al.*, 2013). Virions of CaMV are also shown to change localization in response to aphid probing and colocalize with P2 along the microtubules, purportedly to increase the probability of transmission (Bak *et al.*, 2013).

To date, few vector proteins have been identified to be involved in the transmission of non-circulative viruses using the helper strategy. Dombrovsky *et al.* (2007) extracted proteins from *M. persicae* and tested binding of these proteins to the HC-Pro of the potyvirus *Zucchini yellow mosaic virus* using protein membrane overlay. The authors identified nine proteins binding HC-Pro, four of which were identified to be cuticle proteins via analysis by MS (Fig. 6.2G). Their binding to HC-Pro, but not to the virus capsid, supports the role of these proteins in transmission, as potyviruses use the helper strategy, and the virus capsid alone is not sufficient to directly bind the aphid stylet. Additionally, these proteins failed to bind a mutant form of HC-Pro where the lysine residue in the KLSC domain was changed to glutamic acid, a residue substitution that also abolishes aphid transmission (Dombrovsky *et al.*, 2007). For CaMV, the *M. persicae* protein stylin-01 (formerly known as MPCP4) was identified as the receptor for the virus in the aphid stylet by competitive binding between a stylin-01 antibody and the CaMV P2 protein (Webster *et al.*, 2018) (Fig. 6.2G). Additionally, knockdown of the *stylin-01* gene by feeding aphids on short interfering RNAs (siRNAs) specific to *stylin-01* led to diminished transmission of CaMV by *M. persicae*. Considering this same protein was previously implicated in the transmission of CMV (Liang and Gao, 2017), it seems the same aphid cuticle protein can serve as a receptor for two non-circulative viruses using different transmission strategies. Further identification of receptors for non-circulative viruses in aphid mouthparts as well as characterization of protein interaction topologies using breakthrough proteomics technologies involving cross-linking coupled to MS, such as Protein Interaction Reporter (Chavez *et al.*, 2012; DeBlasio *et al.*, 2015a), may lead to the development of interdiction strategies to block non-circulative virus transmission.

Virus-vector protein interactions involved in circulative, non-propagative transmission

Circulative, non-propagative transmission is characterized by a longer virus acquisition time of hours to days, a latent period during which the virus circulates within the insect, and no replication of the virus within the insect vector tissues. Insects often remain viruliferous for their entire lifespan after acquisition. Viruses from three families are transmitted in this mode: the *Luteoviridae*, *Geminiviridae*, and *Nanoviridae*. Viruses in the *Luteoviridae* (referred to in this chapter as luteovirids) and nanoviruses are transmitted by aphids. Historically, viruses in the *Geminiviridae* have been known to be transmitted by whitefly, leafhopper or treehopper vectors, but recently a report described the cowpea aphid, *Aphis craccivora*, as a vector for *Alfalfa leaf curl virus* (Roumagnac *et al.*, 2015). Circulative viruses are often only acquired when vector insects initiate phloem feeding (Fig. 6.2H). Phloem limitation of circulative, non-propagative viruses is a hallmark feature shared among viruses in these families, which is thought to help promote transmission by their sap-sucking hemipteran vectors (Peter *et al.*, 2009; Gray *et al.*, 2014). Circulative transmission depends upon many species-specific protein interactions. Most circulative virus species are only transmitted by one or a few vector species with varying degrees of transmission efficiency (Rochow, 1969, 1970; Gray *et al.*, 2002; Gray, 2007;).

Much of what is known about this mode of transmission comes from the luteovirid pathosystems. Several ultrastructural studies using electron microscopy have detailed the transmission process of luteovirids on the cellular level (Gildow, 1987; Gildow, 1993; Garret *et al.*, 1993, 1996; Peiffer *et al.*, 1997) (Fig. 6.2C). After feeding on a source of virus, virions enter the gut lumen of the insect and are acquired through the gut epithelial cells. Different luteovirids are acquired through different parts of the gut; either the posterior midgut or hindgut (Garret *et al.*, 1993; Gildow, 1993). In contrast, viruses in the *Geminiviridae* are acquired across the filter chamber or midgut (Ghanim and Medina, 2007; Czosnek *et al.*, 2017), whereas nanoviruses have been observed to passage through the foregut and anterior midgut (Ghanim and Medina, 2007; Bressan and Watanabe, 2011). Virions travel through the endomembrane system of gut epithelial cells in the form of coated pits or tubular vesicles (Gildow, 1993; Garret *et al.*, 1993, 1996). Once the virions are trafficked through the cell, they are released at the basal plasmalemma via exocytosis and pass through the basal lamina into the hemocoel, where they diffuse through the open circulatory system of the vector (Garret *et al.*, 1993, 1996) until they reach the salivary glands. The route used by these viruses in the hemolymph from the gut to the salivary tissues is a largely unexplored and understudied research area. At the primary or accessory salivary glands, virions are again endocytosed and trafficked through vesicles until they reach the posterior cell membranes where they are released into the salivary canal. Luteovirids enter the accessory salivary gland (Peiffer *et al.*, 1997), whereas viruses in the *Geminiviridae* and nanoviruses pass into the primary salivary glands (Ghanim and Medina, 2007; Bressan and Watanabe, 2011). Finally, virions are injected into the next host with salivary secretions for feeding (Peiffer *et al.*, 1997), completing the process of transmission.

Protein interactions regulating acquisition and transmission of luteovirids

Only the structural proteins of the luteovirid capsid have been shown to be involved in directly facilitating the circulative movement of virions through their aphid vectors. Luteovirid icosahedral virions are comprised of two proteins, a CP that makes up the majority of the 180 subunits of the capsid, and the readthrough protein (RTP), which comprises a minor but unknown number of monomer units. The RTP is generated by sporadic ribosomal readthrough of an amber stop codon at the end of the CP open reading frame, leading to the generation of a protein extension known as the readthrough domain (RTD) that protrudes from the surface of the virion. Domains in both the luteovirid CP and N-terminal region of the RTD have been implicated in insect transmission (Chay *et al.*, 1996; Brault *et al.*, 2000, 2003, 2005; Reinbold *et al.*, 2001; Lee *et al.*, 2005; Kaplan *et al.*, 2007; Peter *et al.*, 2008;

Boissinot *et al.*, 2014). The roles for the luteovirid capsid proteins in insect transmission and plant infection have been extensively reviewed elsewhere (Gray *et al.*, 2014; Oppenheim *et al.*, 2015; Whitfield *et al.*, 2015; Heck and Brault, 2018). Protein interaction work using Protein Interaction Reporter (PIR) technology, a chemical crosslinking MS approach to measure protein interaction topologies, has shown that capsid proteins from different luteovirid species share common structural features in their virions (Chavez *et al.*, 2012; DeBlasio *et al.*, 2015a; Alexander *et al.*, 2017).

Virus overlay assays coupled to MS analysis have led to the identification of aphid proteins that can directly bind the virions of luteovirids (van den Heuvel *et al.*, 1994). These include five proteins from *M. persicae* homogenate found to bind virions of *Potato leafroll virus* (PLRV). One of these proteins, symbionin (a homologue of GroEL), is produced by the aphid obligate endosymbiont *Buchnera aphidicola*. However, recent work carried out by Bouvaine and colleagues show that symbionin is unlikely to come into contact with, nor would the predicted binding site be structurally accessible to, luteovirids in aphids (Bouvaine *et al.*, 2011), underscoring the importance of validating the role of protein interactions in virus transmission identified using *in vitro* approaches. For *Barley yellow dwarf virus-MAV* (BYDV-MAV), two aphid proteins found to bind virion, SaM35 and SaM50, were detected in the head of the aphid vector *Sitobion avenae* (Li *et al.*, 2001) (Fig. 6.2E). Yet, proteins isolated from the heads of the aphid *Rhopalosiphum maidis*, a non-vector species of this virus, were found not to bind MAV (Li *et al.*, 2001). SaM50 from *S. avenae* and *Schizaphis graminum* was also found to directly interact with BYDV-GDV, which is only transmitted by these two aphid species (Wang, 2003). Immunogold labelling of ultrathin sections of aphid heads showed that SaM50 localized to the plasma membrane of the accessory salivary gland of *S. graminum* but not in the non-vector species *Rhopalosiphum padi*. In addition, aphid feeding on artificial diets containing antibodies towards this protein reduced transmission of BYDV-GAV by both *S. avenae* and *S. graminum*. Using the same far western blot technique, Seddas *et al.* (2004) identified several proteins binding to *Beet western yellows virus*, including symbionin, Rack-1, GAPDH, and actin (Fig. 6.2B), although their function in aphid transmission has yet to be determined. Microarray analysis of gene expression in the gut of pea aphids after acquisition of *Pea enation mosaic virus 1* (PEMV1) identified several differentially expressed transcripts with limited fold change of up- or down-regulation, including genes involved in endocytosis, intracellular trafficking and signal transduction, all important processes in virus acquisition (Brault *et al.*, 2010). Only 20% of the genome was represented in this transcriptomic analysis due to the limited availability of aphid genome sequences at that time and the low enrichment of gut-specific transcripts in the microarray (Brault *et al.*, 2010). Further transcriptomic analysis is needed to determine if this low level of differential expression is due to the limitations of the study or if aphid gut physiology is not greatly perturbed by luteovirid acquisition.

A powerful way to identify proteins involved in transmission is proteomic phenotyping by comparing the proteomes of vector and non-vector aphids within a species. Papura *et al.* (2002) first performed this method using 2-D gel electrophoresis on an F1 population from a selfed *S. avenae* clone differing in BYDV-PAV transmission efficiency. They identified two differentially expressed protein spots with dissimilar isoelectric points but identical molecular weights between vector and non-vector aphids, indicating that these differential spots were isoforms of the same protein most likely derived from one biallelic locus (Papura *et al.*, 2002). Advances in 2-D proteomics methods in the early 2000s, namely in the development of Difference In Gel Electrophoresis (DIGE), which relies on the use of an internal standard included in every gel for relative quantification of all samples in the study, enabled highly precise, quantitative comparison of proteomes from across a large number of samples. Comparative proteomic profiling by 2-D DIGE of an F2 population of *S. graminum* differing in transmission ability of the luteovirid *Cereal yellow dwarf virus-RPV* (CYDV-RPV) also identified differentially expressed protein isoforms from biallelic loci that correlated with transmission phenotype and could even pinpoint which proteins were involved in passage across the gut as compared to the accessory salivary gland (Yang *et al.*, 2008; Cilia *et al.*, 2011a). One such protein, cyclophilin B, was present in two isoforms, S28 and S29, differing in a single amino acid change (Tamborindeguy *et al.*, 2013). Only the S29 isoform was found to be present in genotypes and field-collected aphids that were efficient transmitters of CYDV-RPV (Tamborindeguy *et al.*, 2013). Additionally, both S28 and S29 were found to bind CYDV-RPV *in vitro* and in co-immunoprecipitations from aphid tissue (Fig. 6.2B), but could not bind PLRV *in vitro*, for which *S. graminum* is a non-vector (Tamborindeguy *et al.*, 2013). Co-immunoprecipitations from aphid tissue coupled to MS also detected another cyclophilin protein binding CYDV-RPV, cyclophilin A, indicating a larger role for cyclophilins in the transmission of this virus.

Recent work in luteovirid systems has discovered the first putative receptors for circulative virus transmission within the gut of aphids. In a hallmark paper by Linz *et al.* (2015), aminopeptidase-N (APN) from pea aphid, *Acyrthosiphon pisum*, was identified to bind PEMV1 using a 2-D far western blot method coupled to MS (Fig. 6.2C). This interaction was confirmed via co-immunoprecipitation, immunofluorescence binding assays, and surface plasmon resonance (Linz *et al.*, 2015). The role of APN as a receptor was confirmed *in vitro* by showing the internalization of green fluorescent protein (GFP)-labelled PEMV1 CP by Sf9 insect cells expressing APN from pea aphid (Linz *et al.*, 2015). Additionally, Linz *et al.* (2015) showed that PEMV1 competes with a peptide that also binds APN, GBP3.1, which was previously shown to inhibit the uptake of this virus by aphids (Liu *et al.*, 2010). Recently, a second luteovirid receptor, the ephrin receptor from *M. persicae*, was found to unambiguously interact, via yeast two-hybrid screening, with the RTP of *Turnip yellows virus* (TuYV) (Mulot *et al.*, 2018) (Fig. 6.2C). Knockdown of ephrin in aphids by dsRNA feeding reduced transmission of TuYV by 43–77%, providing *in vivo* evidence of its role in transmission (Mulot *et al.*, 2018). Additionally, the authors found evidence of this receptor binding to the *Cucurbit aphid borne yellows virus* (CABYV) CP and RTP via yeast two-hybrid screening. The knockdown of the ephrin receptor in *M. persicae* reduced acquisition of CABYV as well as *Beet mild yellowing virus*, two other closely related luteovirids primarily transmitted by *M. persicae*, suggesting its broader role as a general receptor for luteovirids in this widespread vector (Mulot *et al.*, 2018).

Protein interactions regulating acquisition and transmission of begomoviruses

Begomoviruses, in the *Geminiviridae* family, are circulatively transmitted by whitefly vectors and may have mono- or bipartite genomes. While historically referred to as non-propagative, there is some possible evidence for virus replication in vector tissues (Pakkianathan *et al.*, 2015). The CP is thought to be the sole viral determinant of plant-to-plant transmission in begomoviruses, supported by changes in transmission specificity when CP sequences are swapped between geminiviruses. For instance, replacing the CP of *African cassava mosaic virus* (ACMV) with the CP of the leafhopper-transmissible *Beet curly top virus* resulted in a leafhopper-transmissible ACMV (Briddon *et al.*, 1990). Furthermore, exchanging the CP of the non-transmissible *Abutilon mosaic virus* (AbMV) with that of the transmissible *Sida golden mosaic virus* rendered AbMV transmissible (Höfer *et al.*, 1997). The CP of geminiviruses have been shown to bind the midgut of their respective insect vectors, the site of virus acquisition (Wang, Y. *et al.*, 2014). Feeding on antibodies raised against the CP alone or those specific for assembled virion inhibited transmission of *Wheat dwarf virus* by its leafhopper vector (Wang, Y. *et al.*, 2014). Detailed mutational analysis of begomovirus CPs has identified many residues that are important for transmission, especially the amino acids between residues 129 and 152 (Azzam *et al.*, 1994; Noris *et al.*, 1998; Liu *et al.*, 1999; Höhnle *et al.*, 2001; Soto *et al.*, 2005; Caciagli *et al.*, 2009). As for other plant viruses, proper virion assembly and stability is important for transmission: the CP of *Tomato yellow leaf curl Sardinia virus* with an asparagine to aspartate mutation at position 130 was unable to assemble and was not acquired by its whitefly vector, *Bemisia tabaci* (Caciagli *et al.*, 2009).

Two vector proteins have been implicated in regulating transmission of begomoviruses by *B. tabaci*. Microarray, qPCR, and western blot analysis showed an up-regulation of heat shock protein 70 (Hsp70) in response to *Tomato yellow leaf curl virus* (TYLCV) and *Squash leaf curl virus* (Götz *et al.*, 2012). Immunocapture PCR, virus overlay assays, and co-immunoprecipitation showed that Hsp70 interacts with TYLCV (Fig. 6.2C). Immunostaining showed co-localization of Hsp70 with TYLCV and *Watermelon chlorotic stunt virus* in whitefly midgut epithelial cells (Götz *et al.*, 2012). Finally, the role of Hsp70 in virus transmission was confirmed functionally by feeding whiteflies on antibody raised against Hsp70 prior to virus acquisition. This antibody feeding assay resulted in increased transmission of TYLCV, indicating that Hsp70 may be a negative regulator of transmission and serves to protect whiteflies against begomoviruses (Götz *et al.*, 2012). Cyclophilin B was identified as a positive regulator of TYLCV transmission by *B. tabaci* (Kanakala and Ghanim, 2016) (Fig. 6.2C). Similar feeding studies using a cyclophilin B antibody resulted in a decrease in TYLCV transmission. Delivery of cyclosporin A, a cyclophilin B inhibitor from the fungus *Tolypocladium inflatum*, to whiteflies via artificial diet decreased co-localization of cyclophilin B and TYLCV in whitefly midguts (Kanakala and Ghanim, 2016).

Virus–vector protein interactions regulating nanovirus acquisition and transmission

Viruses in the family *Nanoviridae* are multipartite, with genomes consisting of six to eight single-stranded DNA segments, each individually encapsidated into icosahedral particles comprised of a single coat protein. Progress in genetically dissecting these viruses has been slowed by the difficulty in generating infectious clones of the many genome segments. Unlike the other viruses transmitted in the circulative, non-propagative mode, early studies showing the inability of purified nanovirus particles to be transmitted by aphids indicated that a helper component could be required (Franz *et al.*, 1999). This component was recently identified in *Faba bean necrotic stunt virus* (FBNSV) to be the nuclear shuttle protein (NSP) encoded by DNA-N (Grigoras *et al.*, 2018). The authors found that FBNSV was no longer aphid-transmissible when the infectious clone containing the DNA-N viral segment, which encodes the NSP, was excluded from plant inoculation (Grigoras *et al.*, 2018). How the NSP protein facilitates transmission is unknown. However, the function of this viral protein is conserved, as the NSP of FBNSV could complement that of the closely related *Faba bean necrotic yellows virus* and the distantly related *Pea necrotic yellow dwarf virus* (Grigoras *et al.*, 2018). Moreover, the mechanism may be distinct from the role of helper components in non-circulative viruses, as NSP was found to not directly bind the CP in yeast two-hybrid and bimolecular fluorescence complementation assays (Krenz *et al.*, 2017). To date, aphid proteins involved in nanovirus transmission have not been reported. However, a recent study has shown that different genome segments of nanoviruses exhibit unequal relative frequencies in their aphid vectors compared with in plants (Gallet *et al.*, 2018). Further understanding of virus and vector proteins that regulate the transmission of nanoviruses is an exciting frontier, considering the unique challenges presented by their extreme multipartite virus lifestyle.

Virus–vector interactions regulating circulative, propagative transmission

Plant viruses transmitted in the circulative, propagative mode are characterized by acquisition times of hours to days, a latent period of days to weeks during which the virus replicates and disseminates within the vector, and long-term persistence in the vectors, which remain viruliferous for their entire lives. The route taken by circulative, propagative viruses through their vector often mirrors that taken by circulative, non-propagative viruses: virions pass through the gut epithelial cells into the hemolymph, then enter the salivary glands, with the key distinction that propagative viruses replicate in nearly every tissue encountered, notably the gut and salivary glands. Propagative viruses also replicate in other vector tissues not directly involved in the circulative pathway for non-propagative viruses, such as muscle cells adjacent to the gut, from which virions can pass into the hemolymph. These viruses can also directly enter the salivary glands in vectors and developmental stages where these tissues are in contact, such as in thrips nymphs (Kritzman *et al.*, 2002; Moritz *et al.*, 2004). Viruses transmitted in a propagative mode often infect the ovarioles and can be vertically transmitted between generations of the vector (Huo *et al.*, 2014). Some viruses transmitted in this mode can produce tubules that facilitate escape from the midgut (Liu *et al.*, 2011; Chen *et al.*, 2012; Mar *et al.*, 2014; Wang, H. *et al.*, 2014).

Virus–vector protein interactions regulating rhabdovirus acquisition and transmission

Plant-infecting viruses in the *Rhabdoviridae* family (genera *Nucleorhabdovirus* and *Cytorhabdovirus*) are transmitted in a circulative, propagative mode by planthopper, leafhopper, and aphid species. Considering this mode of transmission, rhabdoviruses replicate in both their plant hosts and insect vectors (reviewed in Jackson *et al.*, 2005). Almost all viral proteins in this family are thought to act similarly in the two hosts, except for the movement proteins, which are used exclusively in the plant to transverse plasmodesmata (Huang *et al.*, 2005). Rhabdovirus capsids are enveloped in a membrane derived from the insect or plant host with viral-encoded tripartite glycoprotein (G) spikes protruding from these membranes. For rhabdoviruses that infect vertebrates, such as *Vesicular stomatitis virus*, the G protein is involved in facilitating the adhesion and entry into cells (Schlegel and Wade, 1983). A seminal paper by Gaedigk *et al.* (1986) showed that antibodies raised against the G protein of *Potato yellow dwarf virus* (PYDV) neutralized the infectivity of PYDV by preventing entry of the virus into vector cells. Therefore, the G protein of plant-infecting rhabdoviruses likely functions similarly to the G protein in vertebrate-infecting viruses of the same family.

A few transcriptomic studies have begun to investigate vector pathways involved in the transmission of plant rhabdoviruses. Comparing gene expression between *Maize mosaic virus* (MMV)-infected *Peregrinus maidis* and uninfected planthoppers revealed down-regulation of four genes involved in innate immune responses: the autophagocytosis associated protein ATG3, phosphoinositide 3-kinase (PI3K), c-JUN NH2-terminal kinase (Jnk), and tripeptidyl-peptidase II (TPPii) (Whitfield *et al.*, 2011). In a comparative transcriptomic study of *Graminella nigrifrons*, the vector of *Maize fine streak virus* (MFSV), insects were separated according to whether they were vectors (based on the results of a virus transmission assay to plants), and infected or uninfected. Vectors could be distinguished by the down-regulation of three peptidoglycan recognition proteins (PGRPs), which are involved in innate immune responses to bacteria and fungi (Chen *et al.*, 2012). A later study also implicated innate immunity genes in MFSV transmission by *G. nigrifrons* (Cassone *et al.*, 2014). After just four hours, the authors observed an up-regulation of many immune response genes, including genes involved in the Toll pathway, peroxidases, a superoxide dismutase, and PGRPs, in addition to genes related to cytoskeleton organization, hemopoiesis, glycosylation, and phagocytosis. However, this immune response was transient, as, by the seven-day time point, transcript levels had returned to the same levels as in the healthy controls (Cassone *et al.*, 2014). Emerging transcriptome data for vector species reveal homologues to proteins known to interact with vertebrate-infecting rhabdoviruses. For instance, the *P. maidis* genome contains a homologue of the nicotinic acetylcholine receptor, which is thought to bind and facilitate entry of *Rabies virus* (Whitfield *et al.*, 2011). No rhabdovirus receptor has been identified in insects.

Virus-vector protein interactions regulating tospovirus acquisition and transmission

Tospoviruses (in the newly-designated family *Tospoviridae*) are enveloped, negative-sense RNA viruses with a tripartite genome that are transmitted by thrips in a circulative, propagative mode. The tospovirus genome is protected by the nucleoprotein (N), which is surrounded by a host-derived membrane embedded with two tospovirus-encoded glycoproteins, G_N and G_C, that are involved in transmission. In reassortments, virus transmission determinants always map to RNA M, which encodes the glycoproteins (Sin *et al.*, 2005). Additionally, mutations in the G_N/G_C open reading frame abolish thrips transmission without altering virus assembly (Zheng *et al.*, 2011). G_N plays a role in virus attachment to the thrips midgut as evidenced by immunolocalization and virus overlay assays (Ullman *et al.*, 1995; Bandla *et al.*, 1998; Kikkert *et al.*, 1998). G_C is hypothesized to be a viral fusion protein based on structure (Garry and Garry, 2004) and the fact that it undergoes pH-induced conformational changes typical of type II fusion proteins (Whitfield *et al.*, 2005). Functional analysis confirmed that G_C mediates fusion of host and virus membranes (Plassmeyer *et al.*, 2005, 2007).

There have been several transcriptomic and one proteomic analysis characterizing the effects that tospoviruses have on their thrips vectors. Badillo-Vargas *et al.* (2012) used transcriptome data to help identify proteins resolved using 2-D gel electrophoresis in uninfected and *Tomato spotted wilt virus* (TSWV)-infected *Frankliniella occidentalis.* The transcriptome was particularly useful for protein identification in this study as thrips genes have remarkably low similarity to genes found in other sequenced insect genomes. The authors identified 37 differentially expressed proteins between virus treatments, 62% of which were down-regulated. Many of these proteins were identified to be components of innate immune responses. Another *F. occidentalis* transcriptome analysis looking at the thrips response to TSWV revealed up-regulation of genes involved in defense, signal transduction, and endocytosis (Zhang *et al.*, 2013). Two later 'omics studies identified genes differentially expressed among the various developmental stages of *F. occidentalis* and *Frankliniella fusca* in response to TSWV. Attaining resolution at the level of developmental stage is crucial for tospoviruses, as they are only acquired during the larval stages (Ullman *et al.*, 1992; Nagata *et al.*, 1999). The authors found that in *F. occidentalis*, there was a global down-regulation of TSWV-responsive transcripts in the first instar larvae, including those involved in proteolysis and detoxification, which were up-regulated in the prepupal stage (Schneweis *et al.*, 2017). Cuticle proteins were one of the largest groups of TSWV-responsive transcripts, largely down-regulated and unique to larval and prepupal stages (Schneweis *et al.*, 2017). In *F. fusca*, transcripts of proteins involved in immune response, intracellular transport, and virus replication were found to be up-regulated (Shrestha *et al.*, 2017). Collectively, these studies have identified vector proteins potentially involved in tospovirus transmission and may serve to launch further investigation.

In an exciting preprint by Badillo-Vargas *et al.* (2018), the first ever tospovirus-binding thrips proteins were identified. Using 2-D gel overlay assays coupled to MS, the authors identified six thrips proteins from the first instar larvae that bind TSWV: the endocuticle structural glycoprotein endoCP-V, the cuticular protein CP-V, cyclophilin, enolase, mitochondrial ATP synthase α (mATPase), and the endocuticle structural glycoprotein endoCP-G_N (Fig. 6.2B). They confirmed binding of all six proteins to the TWSV N protein using yeast two-hybrid assays, and binding of endoCP-G_N to the TWSV glycoprotein G_N. These binding interactions were also validated via biomolecular fluorescence complementation. Furthermore, via immunolocalization, the authors confirmed the expression of these proteins in both the midgut and salivary glands of larval thrips, key tissues involved in tospovirus transmission. Co-localization of virus with cyclophilin and endoCP-G_N in the midgut, the site of tospovirus acquisition, was also observed. Considering the consistent and direct interaction between endoCP-G_N and TSWV G_N, the authors hypothesize that endoCP-G_N may serve some receptor-like role in the acquisition of tospoviruses, and future work should focus on pinpointing and verifying its function *in vivo*.

Virus–vector protein interactions regulating tenuivirus acquisition and transmission

Viruses in the genus *Tenuivirus* are transmitted by planthopper or leafhopper vectors in a circulative, propagative mode. The tenuivirus nucleocapsid protein pc3 has been found to be at the interface of interactions with the vector (described below). The non-structural viral protein NS4 forms inclusion bodies in insect tissues and directly interacts with pc3 and virus ribonucleoprotein complexes (Wu *et al.*, 2014a). RNAi targeting of NS4 shows that it is important for movement in the vector, but has no effect on replication (Wu *et al.*, 2014a).

Most of the research on tenuiviruses has focused on the type species, *Rice stripe virus* (RSV) transmitted by *Laodelphax striatellus*, the small brown planthopper. Several vector proteins have been found to interact with RSV. Using a dot immunobinding assay, *L. striatellus* proteins RACK1, GAPDH3, RPL5, RPL7, and RPL8 were found to interact with virus (Li *et al.*, 2011) (Fig. 6.2B). A Gal4 yeast two-hybrid screen was also used on the RSV system to look for interacting proteins localized to the nucleus, where they found binding between the RSV nucleocapsid protein pc3 and RPL18 (Li *et al.*, 2018) (Fig. 6.2B). Knockdown of RPL18 inhibited RSV translation and replication in the vector, indicating this vector protein is crucial for virus replication (Li *et al.*, 2018).

Recently, a split-ubiquitin yeast two-hybrid

screen between the RSV pc3 nucleocapsid protein and a library of small brown planthopper proteins identified atlasin, jagunal, NAC domain protein, a novel cuticular protein (CPR1), and vitellogenin (Vg) as interacting with pc3 (Liu *et al.*, 2015) (Fig. 6.2B). CPR1 was also found to co-localize with and bind virus *in vivo*. Knockdown of CPR1 using RNAi led to decreased RSV accumulation in the hemolymph and salivary gland as well as overall transmission (Liu *et al.*, 2015). The authors proposed that CPR1 binds to RSV and stabilizes it in the hemolymph (Liu *et al.*, 2015) (Fig. 6.2F). Vg was also identified as being down-regulated in RSV-viruliferous planthoppers in a separate transcriptomics study (Zhang *et al.*, 2010). On the protein level, Vg was found to co-localize with RSV *in vivo* (Huo *et al.*, 2014). RSV binding to Vg is thought to play a crucial role in vertical transmission by allowing RSV to enter nurse cells via endocytosis and ovarioles through the nutritive cords (Huo *et al.*, 2014) (Fig. 6.2D). Vg proteins may play a broader role in transmission of propagative plant pathogens through the manipulation of vector fecundity, as it has been found to be up-regulated in the hemolymph of *Diaphorina citri* (the Asian citrus psyllid) infected with '*Candidatus* Liberibacter asiaticus', (CLas) the Gram-negative, circulative, propagative bacterium associated with citrus greening disease (Kruse *et al.*, 2018). *D. citri* infection with CLas has a positive benefit on vector fitness for lab-reared colonies (Pelz-Stelinski and Killiny, 2016).

Small RNA sequencing of *L. striatellus* has shown that the vector produces viral-derived small interfering RNAs against RSV (Xu *et al.*, 2012a). Knockdown of the *Ago2* gene in planthoppers resulted in higher accumulation of RSV in the vector, indicating that RNAi-mediated antiviral immunity is active in the planthopper (Xu *et al.*, 2012a,b). RSV encodes a viral silencing suppressor protein, NS3, which is the most abundant viral transcript produced in the vector (Zhang *et al.*, 2010). Yeast two-hybrid experiments showed the RPN3 subunit of the 26S proteasome in the vector interacts with NS3 (Fig. 6.2B), and knockdown of NS3 results in increased RSV infection (Xu *et al.*, 2015). The authors hypothesize that NS3 hijacks the 26S proteasome by interacting with RPN3 to counter host defenses (Xu *et al.*, 2015).

Virus–vector protein interactions regulating reovirus acquisition and transmission

Three genera in the family *Reoviridae* infect plants (*Fijivirus*, *Phytoreovirus*, and *Oryzavirus*), and all are transmitted by planthoppers or leafhoppers in a circulative, propagative mode. Plant reoviruses are non-enveloped and icosahedral with an inner and outer capsid (Artimo *et al.*, 2012). The outer minor capsid protein of *Rice dwarf virus* (RDV), P2, contains features similar to the viral fusion proteins of other enveloped viruses, as indicated by syncytium formation when expressed in insect cells (Zhou *et al.*, 2007). Therefore, P2 is proposed to bind receptors in the midgut of the vector and facilitate virus entry by inducing fusion of the insect epithelial cell and viral membranes (Omura *et al.*, 1998). Interestingly, several plant reoviruses have been found to form tubule-like structures for movement in the vector. Tubules formed by RDV have been observed moving along microvilli in the midgut of its vector (Chen *et al.*, 2012) and crossing the basal lamina to the midgut visceral muscles, in the case of *Southern rice black-streaked dwarf virus* (SRBSDV) (Wang, H. *et al.*, 2014). The role of these tubules in virus movement was confirmed by knockdown of the tubule-forming protein P7–1 from SRBSDV, which inhibited tubule formation and virus spread in the vector (Liu *et al.*, 2011; Mar *et al.*, 2014).

SRBSDV is one of the most well-studied plant-infecting reoviruses. It is transmitted by the white backed planthopper, *Sogatella furcifera*. Transcriptomic analysis of gene expression between viruliferous and non-viruliferous insects showed that genes involved in primary metabolism, the ubiquitin-proteosome system, cytoskeletal organization, and immune pathways were responsive to SRBSDV (Xu *et al.*, 2012b). Another transcriptome study comparing insects varying in viral load identified the differential expression of transcripts involved in the induction of vector defense responses, with the RNA interference pathway being the most up-regulated (Wang *et al.*, 2016). *S. furcifera* proteins binding to the SRBSDV P7–1 tubule-forming protein were initially identified via yeast two-hybrid screening and 18 of the interactions were confirmed via chemiluminescent co-immunoprecipitation (Mar *et al.*, 2014). Organ-specific expression data gathered via quantitative reverse transcriptase PCR (qRT-PCR) for six of these vector proteins (neuroglian, myosin light chain 2, polyubiquitin, E3 ubiquitin ligase, ribophorin ii, and profilin; Fig. 6.2B) were used to resolve the spatial organization of these virus–protein interaction networks (Mar *et al.*, 2014). Follow up studies should focus on determining the function of these proteins interacting with SRBSDV.

During transmission of circulative propagative viruses, infection in the vector advances from the midgut to midgut visceral muscles. Recent work by Lan *et al.* (2016a) shows that the RNA interference pathway is sufficient to prevent escape of SRBSDV from the midgut of the non-vector *L. striatellus* in cultured cells (Fig. 6.2C). However, knockdown of a key enzyme in this pathway, Dicer-2, allows virus accumulation in the midgut to reach the threshold necessary for dissemination to the visceral muscles (Lan *et al.*, 2016a). Interestingly, knockdown of Dicer-2 in whole insects even allowed the non-vector *L. striatellus* to transmit SRBSDV (Lan *et al.*, 2016a). A similar study on another reovirus system, *Rice gall dwarf virus* (RGDV) and its vector *Recilia dorsalis*, showed that the insect also uses the RNAi pathway to modulate virus infection (Lan *et al.*, 2016b). Silencing of Dicer-2 in this vector insect caused the virus to reach such high titres that the infection was lethal rather than persistent, which would preclude virus transmission (Lan *et al.*, 2016b). Therefore, a balance of virus infection in the vector must be maintained to allow for transmission.

Escape from the salivary glands is another crucial step in the transmission of circulative, propagative viruses. In *Nilaparvata lugens*, the vector of *Rice ragged stunt virus* (RRSV), virus infection appeared to induce apoptosis in the salivary gland of the vector (Fig. 6.2E), as visualized using a dUTP nick-end labelling assay (Huang *et al.*, 2015). Caspases are known to mediate apoptosis in animals, including insects. The authors searched the newly assembled *N. lugens* genome to identify caspases. Silencing of all copies of *Nlcaspase-1* prevented apoptosis in the salivary gland and diminished transmission of RRSV (Huang *et al.*, 2015). Therefore, inducing apoptosis in the vector salivary glands may be an important means for the virus to complete the transmission process (Huang *et al.*, 2015). Apoptosis may also be involved in transmission of CLas from *D. citri*, as enhanced levels of apoptosis are observed in adults insects reared on CLas-infected citrus trees (Ghanim *et al.*, 2016; Mann *et al.*, 2018).

Host manipulation by plant viruses: an indirect strategy for promoting vector transmission

Beyond the direct vector-pathogen protein interactions described above, successful virus transmission also relies on indirect effects on the vector and these effects have also been studied using 'omics methods. Over 50 years ago, Holmes and Bethel (1972) introduced the paradigm known as the 'host manipulation hypothesis' to explain the 'suicidal' behaviour displayed by parasitized animals that increased their risk of predation. From these observations, they proposed that pathogens evolve ways to control aspects of their host's behaviour to enhance their rate of transmission (reviewed in Heil, 2016). A most dramatic example of host manipulation is perhaps the *Ophiocordyceps unilateralis* fungus that infects *Camponotus leonardi* ants living in tropical rainforest trees. The infected ants, referred to as 'zombie' ants, adhere to very specific regions of vegetation using their mouth parts, which is critical for parasite fitness and dispersal (Andersen *et al.*, 2009).

It is well documented that insect-borne plant viruses and pathogens induce a myriad of adaptive visual and biochemical changes in host plants as infection progresses from the initial site of inoculation to full invasion of distal plant tissues (Pallas and Garcia, 2011). It was originally believed that disease symptoms such as chlorosis (yellowing of leaves) and inhibition of growth were simply deleterious side effects due to host resources being commandeered to support virus replication (Pallas and Garcia, 2011). However, numerous studies correlating plant pathology with insect vector performance and behaviour have since shown that most plant pathogens modify host physiology in adaptive ways that facilitate the type of host–vector relationship that favours their specific mode of transmission (Bosque-Pérez and Eigenbrode, 2011; Mann *et al.*, 2012; Mauck *et al.*, 2012; Gray *et al.*, 2014). Work with the citrus greening system has shown that effects of host manipulation extend beyond the ecology of plant virus–vector interactions: plant infection with the citrus greening bacterium induce symptoms in plants which attract parasitoids of the insect vector (Martini *et al.*, 2014).

In general, beneficial effects on fitness and behaviour are observed when vectors are reared on plants infected with circulative pathogens whose dissemination is highly dependent on the prolonged feeding of one or a few vector species (Ajayi and Dewar, 1983; Castle and Berger, 1993; Castle *et al.*, 1998; Jiménez-Martíneza *et al.*, 2004; Maris *et al.*, 2004; Kotzampigikis *et al.*, 2010; Pelz-Stelinski and Killiny, 2016). For example, compared to virus-free hosts, the growth rate, longevity, and fecundity of the aphid *M. persicae* were enhanced when aphids were caged on potato plants infected with *Potato leafroll virus* (PLRV) (Fig. 6.2A), a circulative luteovirid that is primarily transmitted by this aphid species (Castle and Berger, 1993). For luteovirids, extension of feeding times increases the likelihood of vectors acquiring virus (Kotzampigikis *et al.*, 2010) while enhanced reproduction and increased alatae production creates a large reservoir of viruliferous insects that can carry and transmit virions to other plants (Gildow, 1980, 1983). In contrast, intermediate to no effects were observed when the same aphid species was reared on potato plants infected with the non-circulative virus, PVY, (Revers and Garcia, 2015), which requires only a brief interaction with aphids for transmission, or *Potato virus X* (PVX, *Potexvirus*), a vector-independent virus (Castle and Berger, 1993).

Vector manipulation by plant viruses promotes feeding behaviours aligned with the transmission mode that maximizes virus acquisition and transmission. When given a choice, non-viruliferous *M. persicae* prefer to immigrate and to settle on PLRV-infected leaves compared to those infected with PVX or PVY, or virus-free leaves (Castle *et al.*, 1998; Eigenbrode *et al.*, 2002). Over time, aphid emigration from PLRV-infected leaves is also reduced compared to the three other infection conditions (Eigenbrode *et al.*, 2002) and aphid-feeding behaviour becomes increased on older, symptomatic leaves (Alvarez *et al.*, 2007). However, as luteovirids are acquired, host preference switches and viruliferous aphids become more attracted to healthy plants (Medina-Ortega *et al.*, 2009) (Fig. 6.2A). For non-circulative viruses like CMV where acquisition occurs in seconds and retention is short-lived (Jacquemond, 2012), aphids are initially attracted to diseased plants, but dispersal occurs rapidly (Fig. 6.2A), indicating that these infected plants are quickly being perceived as poor-quality hosts (Mauck *et al.*, 2010a,b). Collectively, these studies along with other works testing additional plant-virus-vector systems (Ajayi and Dewar, 1983; Blua and Perring, 1992; Jiménez-Martíneza *et al.*, 2004; Maris *et al.*, 2004; Donaldson and Gratton, 2007; McMenemy *et al.*, 2012; Chen *et al.*, 2013; Maluta *et al.*, 2014; Wu *et al.*, 2014b; Claudel *et al.*, 2018) support Holmes and Bethel's hypothesis. However, exceptions in the literature can be found where effects on vectors are host specific (Castle *et al.*, 1998; Claudel *et al.*, 2018; DeBlasio *et al.*, 2018), neutral (Hodge and Powell, 2008), nuanced due to mixed infections (Salvaudon *et al.*, 2013; Lightle and Lee, 2014) or apparently contradictory to the host manipulation hypothesis (Blua *et al.*, 1994; Fiebig *et al.*, 2004; Boquel *et al.*, 2010; Casteel *et al.*, 2014). The application of 'omics technologies to identify the mechanisms and genetic features mediating these virus-induced effects on host–vector interactions has been instrumental in formulating a deeper understanding of plant virus epidemiology. Potential mechanisms proposed by these studies and details of the supporting data are discussed below.

The role of volatile cues in vector attraction

Vector attraction to infected plants has mainly been attributed to two factors: visual symptoms of disease such as leaf chlorosis, which play a general role in attracting insect vectors (Macias and Mink, 1969; Ajayi and Dewar, 1983; Holopainen *et al.*, 2009; Webster, 2012) and virus-induced changes to host volatile emissions that specifically influence insect responses (Eigenbrode *et al.*, 2002; Jiménez-Martíneza *et al.*, 2004; Ngumbi *et al.*, 2007; Medina-Ortega *et al.*, 2009; Werner *et al.*, 2009; Mauck *et al.*, 2010a; Rajabaskar *et al.*, 2013, 2014; Claudel *et al.*, 2018). Eigenbrode *et al.* (2002) were the first to show that non-viruliferous *M. persicae* were attracted to and preferentially arrested on filter paper models treated with headspace volatiles collected from PLRV-infected potatoes compared to those collected from healthy, PVY-, or PVX-infected plants, a result similar to what was observed when aphids were allowed to come in contact with infected leaves. In the context of a real infection, virus-induced changes to volatile cues are dynamic, as aphids were more attracted to plants that were inoculated at a younger age compared to leaves that were more mature (Alvarez *et al.*, 2007; Werner *et al.*, 2009; Rajabaskar *et al.*, 2013). Metabolic profiling of headspace volatiles using quantitative gas chromatography coupled to MS (GC–MS) showed that six host compounds (limonene, pinene, cadinene, caryophyllene, α-humulene, and 7,11-dimethyl-3-methyl dodecatriene) of the 21 detected were elevated by PLRV-infection compared to the other virus infection treatments (Eigenbrode *et al.*, 2002) (Fig. 6.2A). Using a bead-free, quantitative affinity purification (AP)–MS workflow, which enables the identification of host protein interaction networks complexing with viruses, DeBlasio and colleagues have since shown that PLRV forms complexes *in planta* with the *N. benthamiana* protein α-humulene/(–)-(E)-b-caryophyllene synthase (DeBlasio *et al.*, 2017) (Fig. 6.2A). This enzyme is required for the biosynthesis of volatile sesquiterpenes including α-humulene (Tholl *et al.*, 2005), suggesting that PLRV influences the production of host volatile emissions at the protein level. Interestingly, only synthetic blends mimicking the exact concentration and composition of the naturally occurring PLRV-infected blend could elicit significant attraction/arrest of aphids compared to single applications of each compound and volatile blends from uninfected plants (Ngumbi *et al.*, 2007) indicating a more complex molecular mechanism may be at play. However, results from analysing other luteovirid-host-aphid systems favour the hypothesis that it may be the concentration of host volatiles emitted rather than metabolite composition that influences vector behaviour and that these changes are host-dependent (Jiménez-Martíneza *et al.*, 2004; Claudel *et al.*, 2018). Further experimentation looking at the effects that altering host and luteovirid protein expression/activity have on vector attraction as well as identifying the biochemical changes occurring within the insect in response to host volatile cues is needed to refine these predictions.

Selection has also favoured non-circulative viruses to have complex, multi-trophic mechanisms to ensure their plant-to-plant spread. For the non-circulative virus, CMV, Mauck *et al.* (2010a) demonstrated that, like circulative viruses, aphid vectors were initially attracted to the emissions of squash plants infected with the FNY strain compared to healthy hosts. However, aphid fitness was reduced, and insects quickly migrated away from these infected plants (Fig. 6.2A). Furthermore, volatile metabolites from infected plants were elevated in concentration but similar in composition to those produced by healthy plants. Thus, authors proposed that this strategy promotes CMV transmission by deceptively luring vectors to infected hosts with chemical cues that make the plant appear more appetizing from afar, while changes to plant quality (discussed below) facilitate immediate dispersal once virus is acquired by probing. A recent study has shown that volatiles emitted by CMV-infected tomato and *Arabidopsis thaliana*, such as pinene and p-cymene (Fig. 6.2A) allow for an additional evolutionary advantage by attracting host pollinators (Groen *et al.*, 2016). In choice experiments, bumble bees preferred CMV-infected plants to uninfected plants. Although CMV infection decreased seed yield in these hosts, enhanced fertilization due to increased interactions with bumble bees produced yields equivalent to uninfected plants. In addition, the bees could distinguish between volatiles emitted by plants infected with a CMV mutant unable to express the 2b RNA silencing suppressor and *A. thaliana* silencing mutants, implicating a role for host small RNA pathways in the production of virus-induced volatile emissions. Mathematical modelling showed that pollinator preference for virus-infected plants in the field could impart a selective advantage for the virus by allowing genes for disease susceptibility to persist in the population over host pathogen resistance (Groen *et al.*, 2016)

Virus-induced changes to the nutritional quality of hosts

Although volatile cues have been shown to play a major part in vector attraction and preference for infected hosts, data generated from metabolomic and proteomic profiling experiments have revealed that the nutritional quality of plant sap, the main food source for most insect vectors, is also significantly altered in ways that potentially make plants superior hosts when infected with circulative viruses (Bosque-Pérez and Eigenbrode, 2011) and poor ones when infected with non-circulative viruses (Mauck *et al.*, 2010a, 2014). Experiments with diet sachets have demonstrated that feeding behaviour and aphid fitness are positively influenced when the levels of soluble sugars are higher in diet relative to amount of amino acids (Mittler, 1967; Puterka *et al.*, 2017). In wheat infected with the circulative BYDV, levels of soluble sugars, starch (Jensen, 1972) and the amino acids alanine and glutamine (Ajayi, 1986) were found to be elevated in symptomatic leaf tissue where chloroplast function was compromised (Fig. 6.2H). Indeed, positive effects on aphid survival and fitness can be observed when chloroplast function is severely disrupted through the down-regulation of *phytoene desaturase* (*PDS*), an essential enzyme in the biosynthesis of plastid-associated pigments (DeBlasio *et al.*, 2018). However, this effect was negated when aphids were reared for an extended time on *PDS*-silenced plants co-infected with PLRV, demonstrating that plants infected with circulative viruses eventually become poor-quality hosts, which would favour insect dispersal after virus is acquired. Characterization of host-virus protein interaction networks using quantitative AP-MS revealed that within host plants, PLRV proteins form complexes with host proteins that function in amino acid biosynthesis, carbohydrate metabolism, and photosynthesis (DeBlasio *et al.*, 2015b) (Fig. 6.2H). Interestingly, interactions between PLRV and a subset of these proteins are lost or weakened in the absence of the RTP, a PLRV structural protein known to facilitate symptom development (DeBlasio *et al.*, 2015c). Collectively these data suggest that, like the regulation of host volatile emissions, PLRV capsid proteins may function to change the composition of plant sap by binding to plant proteins involved in nutrient metabolism and altering or redirecting their activity.

Studies evaluating the nutritional quality of host plants infected with the non-circulative virus CMV-FNY showed the ratio of carbohydrates to amino acids to be reduced in phloem and non-vascular cells of infected squash compared to healthy plants (Mauck *et al.*, 2014). These infected plants also had lower levels of the essential amino acids that are required for aphid survival (Mauck *et al.*, 2014). Thus, CMV-infection effectively creates an antagonistic feeding environment that could potentially promote vector dispersal. Electrical penetration graph measurements show that aphids do ingest less phloem sap on CMV-infected *A. thaliana* plants (Westwood *et al.*, 2013). CMV-induced deterrence to aphid feeding in *A. thaliana* is regulated by the interplay between CMV 2b and two other viral proteins (1a and 2a) and their effects on host Argonaute 1 (Westwood *et al.*, 2013). However, these effects are most likely host-dependent as CMV-FNY infection leads to an increase of soluble sugars in melon phloem sap (Shalitin and Wolf, 2000) and positive effects on aphid performance in tobacco plants, which has also been shown to be regulated by the CMV 2b-silencing suppressor (Ziebell *et al.*, 2011).

Alteration of plant defenses to insects

Manipulation of host defense against the insect is yet another way plant viruses can indirectly influence vector performance and behaviour. In the absence of virus, plants respond to herbivore attack through the up-regulation of an array of host defensive molecules that decrease insect fitness and deter feeding (War *et al.*, 2012). Global profiling of mRNA and protein expression in infected hosts show that plant viruses work to either suppress or activate host pathways regulating these responses (Whitham *et al.*, 2006; Di Carli *et al.*, 2012). In general, the phytohormones jasmonic acid (JA) and ethylene (ET) act as anti-herbivore signalling molecules (Morkunas *et al.*, 2011). In CMV-infected squash plants where negative impacts on aphid performance have been observed, significantly higher levels of JA are induced by aphid feeding compared to observations on healthy plants (Fig. 6.2H). Ethylene emissions were also increased in infected plants (Mauck *et al.*, 2014) (Fig. 6.2H). Microarray analysis of CMV-infected *A. thaliana* showed up-regulation of a gene coding for an enzyme involved in the biosynthesis of 4-methoxy-indol-3-yl-methylglucosinolate, a known aphid deterrent (Westwood *et al.*, 2013) (Fig. 6.2H). These effects are consistent with favouring the non-circulative mode of transmission where insect deterrence would be advantageous. In contrast, ET signalling is required for *Turnip mosaic virus* (TuMV, *Potyvirus*) suppression of callose deposition against *M. persicae*, which leads to a positive effect on aphid performance (Casteel *et al.*, 2014, 2015). In an elegant demonstration of the tritrophic interaction between plant, virus, and insect, this process was found to be mediated by the aphid-induced localization of the potyvirus Nuclear Inclusion a-Protease (NIa-Pro) to the plant vacuole showing that potyviruses manipulate their host plant into promoting vector performance only when their vector is present and the response is needed (Bak *et al.*, 2017).

Circulative viruses manipulate plant host physiology in ways that impact plant defense against the vector. For the begomovirus *Tomato yellow leaf curl China virus* (TYLCCNV), which is transmitted in a circulative manner by whiteflies, co-infection of tobacco with its betasatellite represses host JA-mediated responses against the insect (Zhang *et al.*, 2012). Specifically, the betasatellite encoded βC1 protein represses the expression of host JA-biosynthesis and JA-regulated genes (Fig. 6.2H), which leads to a decrease in JA production and an increase in vector fitness. Although information is lacking on the global effects luteovirid infection has on plant defense and how that relates to vector performance, AP-MS experiments show that host proteins with functions in host defense, including callose deposition and JA biosynthesis (i.e. lipoxygenases), complex *in planta* with PLRV proteins (DeBlasio *et al.*, 2015b,c, 2017).

Changes induced within the vector

Although an extensive amount of work has been done to understand how viruses modify plants, scientists are now starting to focus on characterizing the changes within the insect that occur due to ingestion/perception of virus-induced host plant signals. There have been some studies comparing the transcription profiles of insect vectors fed on virus-infected plants compared to healthy hosts (Brault *et al.*, 2010; Zhang *et al.*, 2010; Cassone *et al.*, 2014). However, there are few reports where the effects of infected host on vector biochemistry can be clearly separated from the direct effects of virus (Brizard *et al.*, 2006; Bencharki *et al.*, 2010; Cilia *et al.*, 2012b; Pinheiro *et al.*, 2017). Proteomics has been used to identify interactions between host and virus that are required for acquisition and transmission (Brizard *et al.*, 2006; Bencharki *et al.*, 2010; Cilia *et al.*, 2012b). Treatment of the purified luteovirid CYDV with sodium sulfite eliminates its ability to be transmitted by its aphid vector *R. padi*, even though virion morphology was unaffected (Cilia *et al.*, 2012b). Analysis of virus preparations by nanoscale liquid chromatography tandem MS (nLC-MS/MS) revealed host plant proteins co-purifying with transmissible virion that were lost when CYDV was treated with sodium sulfite, suggesting that these host–virus interactions may be critical for virus uptake. Targeted MS analysis of aphids fed on CYDV-infected plants showed that several of these CYDV-associated host proteins accumulated to higher levels in the insect compared to those fed on healthy plants (Cilia *et al.*, 2012b), consistent with a previous observation that addition of high quantities of soluble host proteins to diet increases the transmission of a related polerovirus (Bencharki *et al.*, 2010). Yet, in both of these reports, the cellular and biochemical effects these plant proteins have on the vector were never assessed. Recently, Pinheiro *et al.* (2017) found that the circulative transmission of PLRV is reduced when *M. persicae* is reared on turnip plants, a host for the aphid but a non-host for the virus. Using a combination of proteomic, biochemical, and microscopic approaches, they demonstrated that signals derived from the non-host plant inhibited PLRV transmission by altering the activity and localization of the aphid cysteine protease cathepsin B within gut cells (Fig. 6.2C). It is possible that up-regulation of cathepsin B in the aphid gut degrades host proteins needed for successful virus acquisition, thus, having an inhibitory effect on virus transmission.

Vector manipulation by plant viruses: a direct strategy for promoting vector transmission

The vector manipulation hypothesis was described as recently as 2012 in a landmark paper by Ingwell *et al.* (2012). The authors observed that while non-viruliferous *R. padi* aphids were attracted to plants infected with the luteovirid BYDV, once the aphids acquired virus, their preference switched to healthy, uninfected plants, a behaviour change which would promote transmission (Ingwell *et al.*, 2012). Therefore, the authors described the vector manipulation hypothesis as ‘the evolution of strategies in plant pathogens to enhance their spread to new hosts.’ This hypothesis describes direct effects of the virus on the vector and is distinguished from the host manipulation hypothesis, which deals with indirect effects on the vector mediated by an infected plant host. To observe direct effects of a plant virus on its insect vector, insects must be removed from the infected plant to allow gut clearing of non-acquired virus and transient signals from the infected plant.

This is accomplished either by having insects acquire purified virus in a membrane feeding setup as in the BYDV study (Ingwell *et al.*, 2012), and/or by moving insects to a non-host plant after acquisition (Pinheiro *et al.*, 2017). This section describes novel insights gained into the evolutionary interactions involved in transmission by observing the direct effects of virus on vector behaviour/performance and leveraging 'omics techniques.

Insights from viruses transmitted in a circulative, non-propagative mode

In a follow-up to the 2012 study with BYDV (Ingwell *et al.*, 2012), the same research group examined the preferences of *M. persicae* in response to PLRV, another luteovirid transmitted in the circulative, non-propagative mode (Rajabaskar *et al.*, 2014). Consistent with their previous findings, the authors found that non-viruliferous aphids preferentially settle on PLRV-infected plants, whereas viruliferous *M. persicae* preferred mock-inoculated potato plants (Rajabaskar *et al.*, 2014). Using GC-MS, the authors identified many of the volatile organic compounds produced by healthy and PLRV-infected plants. Similar to the work of Eigenbrode *et al.* (2002) with non-viruliferous *M. persicae*, Rajabaskar *et al.* (2014) repeated the choice assays with non-viruliferous and viruliferous aphids using only trapped plant headspace volatiles or synthetic volatile blends mimicking healthy and infected plants. Even in the absence of actual potato plants, the authors still observed the same aphid preferences, showing that these olfactory cues alone were enough to determine aphid host selection.

The interactions between begomoviruses and their whitefly vectors are often more akin to a pathogen–host relationship rather than a non-propagative virus–vector one. TYLCV reduces the life expectancy and fecundity of its vector *B. tabaci* (Rubinstein and Czosnek, 1997). In transcriptome analysis of TYLCCNV-viruliferous whiteflies, the virus was shown to alter genes related to the cell cycle and primary metabolism, which may explain the reductions in insect longevity and fecundity (Luan *et al.*, 2011). Additionally, whitefly immune responses such as autophagy and antimicrobial peptide production were activated in response to TYLCCNV (Luan *et al.*, 2011). Transcriptome profiling of dissected guts of TYLCV-viruliferous whiteflies also showed perturbation of the cell cycle and induction of defense, including antimicrobial peptides (Geng *et al.*, 2018). In electrical penetration graph feeding experiments, TYLCV-viruliferous whiteflies fed more readily than their non-viruliferous counterparts and spent more time salivating into the phloem, which is key for virus transmission (Liu *et al.*, 2013). A similar study with TYLCCNV found distinct differences in feeding behaviour of viruliferous *B. tabaci* when feeding on a host or non-host of the virus (He *et al.*, 2015). The virus inhibited whitefly feeding on cotton, a non-host, and on a resistant cultivar of tobacco, but improved whitefly feeding on TYLCCNV-infected tobacco (He *et al.*, 2015). Therefore, this improvement in feeding may be an indirect effect mediated by the infected plant rather than a direct effect.

Insights from viruses transmitted in a circulative, propagative mode

Recent work in the RSV-small brown planthopper system, guided by transcriptomics data, delved into the two common phenotypes of vector manipulation: attraction to infected plant volatiles and alteration of vector fecundity. While several studies have shown that volatiles produced by virus-infected plants may help attract insect vector species (reviewed above) little work has been done on the vector side of this interaction. In a functional study from the RSV system, a key olfactory receptor was identified in the small brown planthopper (Li *et al.*, 2019). Using the *L. striatellus* transcriptome (Zhang *et al.*, 2010) and a known sequence of the olfactory co-receptor Orco from the closely related *Nilaparvata lugens*, the authors confirmed the existence of a homologue of Orco in the small brown planthopper. Following orally delivered dsRNA-mediated knock-down of the *Orco* gene in *L. striatellus*, the insects were given the choice between a healthy rice plant or open air. After silencing of *Orco*, the response time of the insects was greatly increased, and a larger proportion presented no response or chose the air treatment. This finding confirms the role of Orco in olfactory host-seeking behaviour. The authors also found that RSV-viruliferous planthoppers had higher *Orco* expression, and present stronger host-seeking behaviour as indicated by a lower proportion of insects with no response in the choice assay. These findings indicate that the spread of RSV may be increased by the up-regulation of an olfactory receptor in the RSV vector, which would aid the planthopper in seeking out host plants.

There is a lack of consensus in the literature regarding how RSV affects the fecundity of its vector, but fecundity manipulation may be important for virus epidemiology. Li *et al.* (2015) observed a decrease in hatchability of planthopper eggs when one or more parent was viruliferous with RSV, consistent with a previous study (Nasu, 1963). The authors assessed the expression of 115 genes related to embryonic development, which they chose by harnessing the power of *D. melanogaster* as a model for insect developmental biology and mining homologous genes from *L. striatellus* (Zhang *et al.*, 2010). They attributed the decrease in hatchability and correlated egg defects to significant down-regulation of two important embryonic development genes: *Ls-Dorsal*, a transcription factor regulating tissue differentiation and *Ls-CPO*, an enzyme responsible for crosslinking and hardening of the protective chorion surrounding eggs (Fig. 6.2D). In contrast, Wan *et al.* observed no significant difference in hatching between RSV-viruliferous and non-viruliferous planthoppers but instead observed a decreased number of eggs laid by viruliferous females (Wan *et al.*, 2015). These authors tied this observation to decreased expression of *Vg* in viruliferous-females, which is a key yolk protein necessary for both egg development and the passage of RSV into nurse cells for transovarial transmission (Huo *et al.*, 2014). Furthermore, Wan *et al.* observed accelerated nymphal development, attributed to down-regulation of *JHMAT* in the juvenile hormone (JH) pathway, and up-regulation of *CYP307A1* in the ecdysteroid pathway (Fig. 6.2B). This expression pattern is consistent with immunosuppression by RSV, as JH serves as an immune activator and 20-hydroxy-ecdysone (an ecdysteroid) acts as an inhibitor in other insect systems (Beckstead *et al.*, 2005; Tian *et al.*, 2010). Despite these conflicting findings, RSV has a major role in altering embryogenesis and development of its insect vector. Work by He and colleagues suggests that fecundity manipulation may be a beneficial strategy for the virus to spread within a crop, as the proportion of viruliferous insects in the population rather than the absolute population size of the small brown plant-hopper vector is positively correlated with disease severity in rice fields in China (He *et al.*, 2016).

Vector manipulation has also been observed in plant-infecting reoviruses. In a compelling parallel to the studies performed on the luteovirid system, Wang, H. *et al.* (2014) showed that while non-viruliferous *S. furcifera* prefer SRBSDV-infected plants, SRBSDV-viruliferous insects prefer uninfected plants, which would promote virus dispersal by viruliferous individuals seeking new susceptible host plants (Wang, H. *et al.*, 2014). These authors also looked at the preferences of planthoppers given an acquisition access period on SRBSDV-infected plants, which did not become viruliferous. Intriguingly, these exposed, nonviruliferous *S. furcifera* preferred uninfected rice plants over SRBSDV-infected ones, which is inconsistent with the vector manipulation hypothesis. As SRBSDV-infected plants have adverse effects on vector fecundity, development, and longevity (Tu *et al.*, 2013), the authors propose that the planthoppers 'remember' the unfavourable quality of these infected plants, as well as their odour, and are subsequently deterred when exposed to the same volatile cues, although these same insects preferred the odour of an infected plant to no plant (air). In a final twist of this experiment, the authors looked at the host preference of a non-vector of SRBSDV, *N. lugens*, the brown planthopper, and vector of RRSV, another reovirus that commonly co-infects plants with SRBSDV. *N. lugens* preferred uninfected rice plants to SRBSDV-infected plants, consistent with previous studies showing a virus-infected plant is often unattractive to non-vector species (van Molken *et al.*, 2012; Mauck *et al.*, 2014). However, after *N. lugens* acquired RRSV, its preference switched from healthy rice plants to SRBSDV-infected ones. The authors hypothesize that either RRSV causes this change in its vector's behaviour, or SRBSDV is altering the behaviour of a non-vector species. Either way, this manipulation favours co-infection of the two viruses, which are commonly found in mixed infections.

One relatively unexplored area of vector manipulation is response to abiotic stress. Xu *et al.* (2016) looked at the thermal tolerance of SRBSDV-viruliferous planthoppers under extreme heat and cold conditions, as well as performing transcriptomics on insects exposed to virus and/or extreme temperature. They found that viruliferous insects were better able to tolerate extreme heat stress, which has important implications for the spread of

SRBSDV epidemics in the summer. Additionally, viruliferous insects experienced higher mortality under cold stress, which may affect the ability of planthoppers to overwinter. The transcriptomics data revealed general up-regulation of expression in viruliferous insects exposed to cold stress and down-regulation of transcription in those exposed to extreme heat. The authors observed nuanced expression of heat shock proteins in insects exposed to SRBSDV and heat stress, which may be due to the crosstalk between heat stress and antiviral pathways, both of which would be activated under this dual stress and potentially repressed by viral counter-defense strategies. In a follow-up metabolomics study of *S. furcifera* exposed to the same virus and environmental stressors, there was up-regulation of polyols and sugars in viruliferous planthoppers under both heat and cold stress, and up-regulation of certain amino acids, such as methionine, proline, threonine, ornithine, L-homoserine and β-alanine, under heat stress (Zhang *et al.*, 2018). These same amino acids were down-regulated in response to cold stress. Therefore, the virus may alter the insect's abiotic stress response, which could impact virus and vector seasonal migration, especially in the face of climate change.

Future directions

Rapid advances in 'omics technologies include instruments used for detection with increased sensitivity being built more quickly and cheaply. The advent of recent technologies such as single-cell transcriptomics (Ziegenhain *et al.*, 2018), laser-capture microdissection (Zhu *et al.*, 2016) and MALDI-imaging (Vrkoslav *et al.*, 2010), which can be used to detect RNA, protein, or metabolite species (respectively) within a small population of cells or an individual cell, has allowed for the generation of expression atlas resources for model insect species that detail tissue specific expression patterns/responses that would have been lost in studies using whole insects.

As more genomic, proteomic, and metabolomic information becomes available for non-model vector species, systems-wide studies comparing 'omics datasets across different virus-vector systems is needed to determine what is generally involved in plant pathogen transmission by insects and what is specific. In the literature, many of the same vector proteins are reported to be involved in the transmission process for diverse virus groups, indicating potential conservation or co-evolution of viruses to co-opt these conserved proteins during transmission. These same proteins may also be involved in the ability of insect vectors to rapidly adapt and colonize new host plants, as expanded gene families for specific sets of proteins in aphids, including cuticular proteins and cathepsin B, are also involved in host adaptation (Mathers *et al.*, 2017). However, considering genome annotations are largely derived from homology to distant species (Saha *et al.*, 2017), it is difficult to know to what extent 'omics methods are biased towards these conserved proteins and are currently unable to identify species-specific genes and proteins that may play crucial roles in transmission or other aspects of vector biology. Future genome annotation efforts should seek to overcome these limitations. Improving RNAi techniques in non-model insects can aid functional validation of annotations.

Many of the 'omics techniques discussed can be applied to understanding the biology and transmission of insect-infecting viruses, which have the potential for use as RNAi delivery systems (Heck, 2018). RNA sequencing techniques are being used to discover novel insect-infecting viruses (reviewed in Nouri *et al.*, 2018; see also Chapter 1), and the development of infectious clones can aid reverse genetic approaches to understand the function of viral proteins. Transcriptomics and proteomics can be used to discover insect proteins involved in transmission of these insect viruses, as described in this chapter for plant viruses, and functional genomics, such as dsRNA feeding, can be used to validate the function of insect proteins in some cases.

Insect-infecting viruses may be an undescribed partner in the transmission of plant viruses. A compelling paper by Pinheiro *et al.* (2019) provides a fascinating example. Using small RNA sequencing, the authors discovered that plants infected with the luteovirid PLRV alter the production of small RNAs (sRNAs) in the aphid vector, *M. persicae*, producing unusually large-sized RNAs matching to *Myzus persicae densovirus* (MpDNV). Aphids viruliferous with PLRV displayed higher titres of MpDNV, suggesting these aberrant sRNA sizes are a reflection of an altered anti-virus defense response in the aphid. Densoviruses are ubiquitous among arthropods (Fédière, 2000), and have even been shown to promote wing morph development in aphids, which are polyphenic and produce both winged and non-winged individuals (Ryabov *et al.*, 2009). Pinheiro *et al.* (2019) hypothesize that PLRV suppresses the immune system of its aphid vector to allow for the proliferation of MpDNV, an insect-infecting virus that promotes the development of more winged individuals, which in turn would increase plant-to-plant spread of PLRV.

In addition to the impact of insect-infecting viruses on transmission, vector endosymbionts are other crucial biological players that may affect virus transmission indirectly. Many economically important insect vectors, such aphids and whiteflies, harbour obligate and secondary endosymbionts. The role of endosymbionts in the transmission of plant viruses has been debated and reviewed elsewhere (Pinheiro *et al.*, 2015). Manipulation of endosymbionts may prove to be an important frontier in developing novel control strategies against insect-transmitted plant viruses, as successfully shown in animal-infecting arboviruses (Durvasula *et al.*, 2003; Hoffmann *et al.*, 2011).

Finally, these compelling insights into the mechanisms of transmission in vector biology in recent years need to be channelled into the development of control strategies. As more receptors for virus acquisition in insects become known, strategies can be developed to block or reduce virus binding to these crucial proteins, either through down-regulation of the receptor in the insect by RNAi (Liang and Gao, 2017; Mulot *et al.*, 2018; Webster *et al.*, 2018), or through the application of chemical or peptide inhibitors (Liu *et al.*, 2010). Other control strategies tailored to the biology of the system in question, such as decreasing the susceptibility of vectors of circulative, propagative viruses so that virus titres do not reach levels conducive for transmission in the vector, an approach that has been deployed for mosquito-borne arboviruses (Hoffmann *et al.*, 2011), should also be explored. Conversely, considering the finding that insect sRNA-mediated antiviral immunity can keep circulative, propagative virus populations in check (Lan *et al.*, 2016a,b), an alternate strategy may be to suppress the antiviral defenses of the vector, such that virus titres reach a level that kill vectors before transmission can occur. As newer and more powerful technologies are developed, our understanding of the biology of these insect-transmitted plant viruses will only increase, as should development and deployment of more effective control strategies.

Funding

The authors are funded by NSF grant 1354309 (to MH), USDA-ARS CRIS Project 8062-22410-006-00-D (to MH) and an NSF Graduate Research Fellowship DGE-1650441 (to JRW).

References

Ajayi, O. (1986). The effect of *Barley yellow dwarf virus* on the amino acid composition of spring wheat. Ann. Appl. Biol. *108*, 145–149.

Ajayi, O., and Dewar, A.M. (1983). The effect of *Barley yellow dwarf virus* on field populations of the cereal aphids, *Sitobion avenae* and *Metopolophium dirhodum*. Ann. Appl. Biol. *103*, 1–11.

Alexander, M.M., Mohr, J.P., DeBlasio, S.L., Chavez, J.D., Ziegler-Graff, V., Brault, V., Bruce, J.E., and Heck, M.C. (2017). Insights in luteovirid structural biology guided by chemical cross-linking and high resolution mass spectrometry. Virus Res. *241*, 42–52. https://doi.org/10.1016/j.virusres.2017.05.005

Allison, R.F., Dougherty, W.G., Parks, T.D., Willis, L., Johnston, R.E., Kelly, M., and Armstrong, F.B. (1985). Biochemical analysis of the capsid protein gene and capsid protein of *Tobacco etch virus*: N-terminal amino acids are located on the virion's surface. J. Virol. *147*, 309–316. https://doi.org/10.1016/0042-6822(85)90133-3

Alvarez, A.E., Garzo, E., Verbeek, M., Vosman, B., Dicke, M., and Tjallingii, W.F. (2007). Infection of potato plants with *Potato leafroll virus* changes attraction and feeding behaviour of *Myzus persicae*. Entomol. Exp. Appl. *125*, 135–144. https://doi.org/10.1111/j.1570-7458.2007.00607.x

Ammar, E.D., Jarlfors, U., and Pirone, T.P. (1994). Association of potyvirus helper component protein with virions and the cuticle lining the maxillary food canal and foregut of an aphid vector. Phytopathology *84*, 1054–1060.

Andersen, S.B., Gerritsma, S., Yusah, K.M., Mayntz, D., Hywel-Jones, N.L., Billen, J., Boomsma, J.J., and Hughes, D.P. (2009). The life of a dead ant: the expression of an adaptive extended phenotype. Am. Nat. *174*, 424–433. https://doi.org/10.1086/603640

Artimo, P., Jonnalagedda, M., Arnold, K., Baratin, D., Csardi, G., de Castro, E., Duvaud, S., Flegel, V., Fortier, A., Gasteiger, E., *et al.* (2012). ExPASy: SIB bioinformatics resource portal. Nucleic Acids Res. *40*, W597–603. https://doi.org/10.1093/nar/gks400

Atreya, P.L., Atreya, C.D., and Pirone, T.P. (1991). Amino acid substitutions in the coat protein result in loss of insect transmissibility of a plant virus. Proc. Natl. Acad. Sci. U.S.A. *88*, 7887–7891.

Atreya, P.L., Lopez-Moya, J.J., Chu, M., Atreya, C.D., and Pirone, T.P. (1995). Mutational analysis of the coat protein N-terminal amino acids involved in potyvirus transmission by aphids. J. Gen. Virol. *76*, 265–270. https://doi.org/10.1099/0022-1317-76-2-265

Azzam, O., Frazer, J., de la Rosa, D., Beaver, J.S., Ahlquist, P., and Maxwell, D.P. (1994). Whitefly transmission and efficient ssDNA accumulation of *Bean golden mosaic* geminivirus require functional coat protein. Virology *204*, 289–296.

Badillo-Vargas, I.E., Rotenberg, D., Schneweis, D.J., Hiromasa, Y., Tomich, J.M., and Whitfield, A.E. (2012). Proteomic analysis of *Frankliniella occidentalis* and differentially expressed proteins in response to *Tomato spotted wilt virus* infection. J. Virol. *86*, 8793–8809. https://doi.org/10.1128/JVI.00285-12

Badillo-Vargas, I.E., Chen, Y., Martin, K.M., Rotenberg, D., and Whitfield, A.E. (2018). Discovery of novel thrips vector proteins that bind to the plant bunyavirus, *Tomato spotted wilt virus*. bioRxiv 416560 (preprint). https://doi.org/10.1101/416560

Bak, A., Gargani, D., Macia, J.L., Malouvet, E., Vernerey, M.S., Blanc, S., and Drucker, M. (2013). Virus factories of *Cauliflower mosaic virus* are virion reservoirs that engage actively in vector transmission. J. Virol. *87*, 12207–12215. https://doi.org/10.1128/JVI.01883-13

Bak, A., Cheung, A.L., Yang, C., Whitham, S.A., and Casteel, C.L. (2017). A viral protease relocalizes in the presence of the vector to promote vector performance. Nat. Commun. *8*, 14493. https://doi.org/10.1038/ncomms14493

Bandla, M.D., Campbell, L.R., Ullman, D.E., and Sherwood, J.L. (1998). Interaction of *Tomato spotted wilt* tospovirus (TSWV) glycoproteins with a Thrips midgut protein, a potential cellular receptor for TSWV. Phytopathology *88*, 98–104. https://doi.org/10.1094/PHYTO.1998.88.2.98

Beckstead, R.B., Lam, G., and Thummel, C.S. (2005). The genomic response to 20-hydroxyecdysone at the onset of *Drosophila* metamorphosis. Genome Biol. *6*, R99.

Bencharki, B., Boissinot, S., Revollon, S., Ziegler-Graff, V., Erdinger, M., Wiss, L., Dinant, S., Renard, D., Beuve, M., Lemaitre-Guillier, C., *et al.* (2010). Phloem protein partners of *Cucurbit aphid borne yellows virus*: possible involvement of phloem proteins in virus transmission by aphids. Mol. Plant Microbe Interact. *23*, 799–810. https://doi.org/10.1094/MPMI-23-6-0799

Berlandier, F., Thackray, D., Jones, R., Latham, L., and Cartwright, L. (1997). Determining the relative roles of different aphid species as vectors of *Cucumber mosaic* and *Bean yellow mosaic* viruses in lupins. Ann. Appl. Biol. *131*, 297–314. https://doi.org/10.1111/j.1744-7348.1997.tb05158.x

Blua, M.J., and Perring, T.M. (1992). Effects of *Zucchini yellow mosaic virus* on coloization and feeding behavior of *Aphis gossypii* (Homoptera: Aphididae) alatae. Environ. Entomol. *21*, 578–585. https://doi.org/10.1093/ee/21.3.578

Blua, M.J., Perring, T.M., and Madore, M.A. (1994). Plant virus-induced changes in aphid population development and temporal fluctuations in plant nutrients. J. Chem. Ecol. *20*, 691–707. https://doi.org/10.1007/BF02059607

Boissinot, S., Erdinger, M., Monsion, B., Ziegler-Graff, V., and Brault, V. (2014). Both structural and non-structural forms of the readthrough protein of *Cucurbit aphid-borne yellows virus* are essential for efficient systemic infection of plants. PLOS ONE *9*, e93448. https://doi.org/10.1371/journal.pone.0093448

Boquel, S., Giordanengo, P., and Ameline, A. (2010). Divergent effects of PVY-infected potato plant on aphids. Eur. J. Plant Pathol. *129*, 507–510. https://doi.org/10.1007/s10658-010-9732-8

Bosque-Pérez, N.A., and Eigenbrode, S.D. (2011). The influence of virus-induced changes in plants on aphid vectors: insights from luteovirus pathosystems. Virus Res. *159*, 201–205. https://doi.org/10.1016/j.virusres.2011.04.020

Bouvaine, S., Boonham, N., and Douglas, A.E. (2011). Interactions between a luteovirus and the GroEL chaperonin protein of the symbiotic bacterium *Buchnera aphidicola* of aphids. J. Gen. Virol. *92*, 1467–1474. https://doi.org/10.1099/vir.0.029355-0

Brault, V., Mutterer, J., Scheidecker, D., Simonis, M.T., Herrbach, E., Richards, K., and Ziegler-Graff, V. (2000). Effects of point mutations in the readthrough domain of the *Beet western yellows virus* minor capsid protein on virus accumulation in planta and on transmission by aphids. J. Virol. *74*, 1140–1148. https://doi.org/10.1128/JVI.74.3.1140-1148.2000

Brault, V., Bergdoll, M., Mutterer, J., Prasad, V., Pfeffer, S., Erdinger, M., Richards, K.E., and Ziegler-Graff, V. (2003). Effects of point mutations in the major capsid protein of *Beet western yellows virus* on capsid formation, virus accumulation, and aphid transmission. J. Virol. *77*, 3247–3256. https://doi.org/10.1128/JVI.77.5.3247-3256.2003

Brault, V., Périgon, S., Reinbold, C., Erdinger, M., Scheidecker, D., Herrbach, E., Richards, K., and Ziegler-Graff, V. (2005). The polerovirus minor capsid protein determines vector specificity and intestinal tropism in the aphid. J. Virol. *79*, 9685–9693. https://doi.org/10.1128/JVI.79.15.9685-9693.2005

Brault, V., Tanguy, S., Reinbold, C., Le Trionnaire, G., Arneodo, J., Jaubert-Possamai, S., Guernec, G., and Tagu, D. (2010). Transcriptomic analysis of intestinal genes following acquisition of *Pea enation mosaic virus* by the pea aphid *Acyrthosiphon pisum*. J. Gen. Virol. *91*, 802–808. https://doi.org/10.1099/vir.0.012856-0

Bressan, A., and Watanabe, S. (2011). Immunofluorescence localisation of *Banana bunchy top virus* (family *Nanoviridae*) within the aphid vector, *Pentalonia nigronervosa*, suggests a virus tropism distinct from aphid-transmitted luteoviruses. Virus Res. *155*, 520–525. https://doi.org/10.1016/j.virusres.2010.12.005

Briddon, R.W., Pinner, M.S., Stanley, J., and Markham, P.G. (1990). Geminivirus coat protein gene replacement alters insect specificity. Virology *177*, 85–94. https://doi.org/10.1016/0042-6822(90)90462-z

Brizard, J.P., Carapito, C., Delalande, F., Van Dorsselaer, A., and Brugidou, C. (2006). Proteome analysis of plant-virus interactome: comprehensive data for virus multiplication inside their hosts. Mol. Cell Proteomics *5*, 2279–2297. https://doi.org/10.1074/mcp.M600173-MCP200

Caciagli, P., Medina Piles, V., Marian, D., Vecchiati, M., Masenga, V., Mason, G., Falcioni, T., and Noris, E. (2009). Virion stability is important for the circulative transmission of *Tomato yellow leaf curl sardinia virus* by

Bemisia tabaci, but virion access to salivary glands does not guarantee transmissibility. J. Virol. *83*, 5784–5795. https://doi.org/10.1128/JVI.02267-08

Cassone, B.J., Wijeratne, S., Michel, A.P., Stewart, L.R., Chen, Y., Yan, P., and Redinbaugh, M.G. (2014). Virus-independent and common transcriptome responses of leafhopper vectors feeding on maize infected with semi-persistently and persistent propagatively transmitted viruses. BMC Genomics *15*, 133. https://doi.org/10.1186/1471-2164-15-133

Casteel, C.L., Yang, C., Nanduri, A.C., De Jong, H.N., Whitham, S.A., and Jander, G. (2014). The NIa-Pro protein of *Turnip mosaic virus* improves growth and reproduction of the aphid vector, *Myzus persicae* (green peach aphid). Plant J. *77*, 653–663. https://doi.org/10.1111/tpj.12417

Casteel, C.L., De Alwis, M., Bak, A., Dong, H., Whitham, S.A., and Jander, G. (2015). Disruption of ethylene responses by *Turnip mosaic virus* mediates suppression of plant defense against the green peach aphid vector. Plant Physiol. *169*, 209–218. https://doi.org/10.1104/pp.15.00332

Castle, S.J., and Berger, P.H. (1993). Rates of growth and increase of *Myzus persicae* on virus-infected potatoes according to type of virus-vector relationship. Entomol. Exp. Appl. *69*, 51–60. https://doi.org/10.1111/j.1570-7458.1993.tb01727.x

Castle, S.J., Mowry, T.M., and Berger, P.H. (1998). Differential settling by *Myzus persicae* (Homoptera: Aphididae) on various virus infected host plants. Ann. Entomol. Soc. Am. *91*, 661–667. https://doi.org/10.1093/aesa/91.5.661

Chavez, J.D., Cilia, M., Weisbrod, C.R., Ju, H.J., Eng, J.K., Gray, S.M., and Bruce, J.E. (2012). Cross-linking measurements of the *Potato leafroll virus* reveal protein interaction topologies required for virion stability, aphid transmission, and virus–plant interactions. J. Proteome Res. *11*, 2968–2981. https://doi.org/10.1021/pr300041t

Chay, C.A., Gunasinge, U.B., Dinesh-Kumar, S.P., Miller, W.A., and Gray, S.M. (1996). Aphid transmission and systemic plant infection determinants of *Barley yellow dwarf luteovirus-PAV* are contained in the coat protein readthrough domain and 17-kDa protein, respectively. Virology *219*, 57–65. https://doi.org/10.1006/viro.1996.0222

Chen, G., Pan, H., Xie, W., Wang, S., Wu, Q., Fang, Y., Shi, X., and Zhang, Y. (2013). Virus infection of a weed increases vector attraction to and vector fitness on the weed. Sci. Rep. *3*, 2253. https://doi.org/10.1038/srep02253

Chen, W., Hasegawa, D.K., Kaur, N., Kliot, A., Pinheiro, P.V., Luan, J., Stensmyr, M.C., Zheng, Y., Liu, W., Sun, H., *et al.* (2016). The draft genome of whitefly *Bemisia tabaci* MEAM1, a global crop pest, provides novel insights into virus transmission, host adaptation, and insecticide resistance. BMC Biol. *14*, 110. https://doi.org/10.1186/s12915-016-0321-y

Chen, Y., Cassone, B.J., Bai, X., Redinbaugh, M.G., and Michel, A.P. (2012). Transcriptome of the plant virus vector *Graminella nigrifrons*, and the molecular interactions of *Maize fine streak* rhabdovirus transmission. PLOS ONE *7*, e40613. https://doi.org/10.1371/journal.pone.0040613

Christodoulides, N., Van Dam, A.R., Peterson, D.A., Frandsen, R.J.N., Mortensen, U.H., Petersen, B., Rasmussen, S., Normark, B.B., and Hardy, N.B. (2017). Gene expression plasticity across hosts of an invasive scale insect species. PLOS ONE *12*, e0176956. https://doi.org/10.1371/journal.pone.0176956

Cilia, M., Tamborindeguy, C., Fish, T., Howe, K., Thannhauser, T.W., and Gray, S. (2011a). Genetics coupled to quantitative intact proteomics links heritable aphid and endosymbiont protein expression to circulative polerovirus transmission. J. Virol. *85*, 2148–2166. https://doi.org/10.1128/JVI.01504-10

Cilia, M., Tamborindeguy, C., Rolland, M., Howe, K., Thannhauser, T.W., and Gray, S. (2011b). Tangible benefits of the aphid *Acyrthosiphon pisum* genome sequencing for aphid proteomics: Enhancements in protein identification and data validation for homology-based proteomics. J. Insect Physiol. *57*, 179–190. https://doi.org/10.1016/j.jinsphys.2010.11.001

Cilia, M., Bereman, M., Fish, T., MacCoss, M.J., and Gray, S. (2012a). Homopteran vector biomarkers for efficient circulative plant virus transmission are conserved in multiple aphid species and the whitefly *Bemisia tabaci*. J. Integr. Agr. *11*, 249–262. https://doi.org/10.1016/S2095-3119(12)60009-4

Cilia, M., Peter, K.A., Bereman, M.S., Howe, K., Fish, T., Smith, D., Gildow, F., MacCoss, M.J., Thannhauser, T.W., and Gray, S.M. (2012b). Discovery and targeted LC-MS/MS of purified polerovirus reveals differences in the virus-host interactome associated with altered aphid transmission. PLOS ONE *7*, e48177. https://doi.org/10.1371/journal.pone.0048177

Claudel, P., Chesnais, Q., Fouche, Q., Krieger, C., Halter, D., Bogaert, F., Meyer, S., Boissinot, S., Hugueney, P., Ziegler-Graff, V., *et al.* (2018). The aphid-transmitted *Turnip yellows virus* differentially affects volatiles emission and subsequent vector behavior in two Brassicaceae plants. Int. J. Mol. Sci. *19*. https://doi.org/10.3390/ijms19082316

Consortium, I.A.G. (2010). Genome sequence of the pea aphid *Acyrthosiphon pisum*. PLOS Biol. *8*, e1000313. https://doi.org/10.1371/journal.pbio.1000313

Czosnek, H., Hariton-Shalev, A., Sobol, I., Gorovits, R., and Ghanim, M. (2017). The incredible journey of begomoviruses in their whitefly vector. Viruses 9, E273.

DeBlasio, S.L., Chavez, J.D., Alexander, M.M., Ramsey, J., Eng, J.K., Mahoney, J., Gray, S.M., Bruce, J.E., and Cilia, M. (2015a). Visualization of host-polerovirus interaction topologies using protein interaction reporter technology. J. Virol. *90*, 1973–1987. https://doi.org/10.1128/JVI.01706-15

DeBlasio, S.L., Johnson, R., Mahoney, J., Karasev, A., Gray, S.M., MacCoss, M.J., and Cilia, M. (2015b). Insights into the polerovirus-plant interactome revealed by coimmunoprecipitation and mass spectrometry. Mol. Plant Microbe Interact. *28*, 467–481. https://doi.org/10.1094/MPMI-11-14-0363-R

DeBlasio, S.L., Johnson, R., Sweeney, M.M., Karasev, A., Gray, S.M., MacCoss, M.J., and Cilia, M. (2015c). *Potato leafroll virus* structural proteins manipulate overlapping, yet distinct protein interaction networks during infection. Proteomics *15*, 2098–2112. https://doi.org/10.1002/pmic.201400594

DeBlasio, S.L., Bereman, M.S., Mahoney, J., Thannhauser, T.W., Gray, S.M., MacCoss, M.J., and Cilia Heck, M. (2017). Evaluation of a bead-free coimmunoprecipitation technique for identification of virus-host protein interactions using high-resolution mass spectrometry. J. Biomol. Tech. *28*, 111–121. https://doi.org/10.7171/jbt.17-2803-002

DeBlasio, S.L., Rebelo, A.R., Parks, K., Gray, S.M., and Heck, M.C. (2018). Disruption of chloroplast function through downregulation of phytoene desaturase enhances the systemic accumulation of an aphid-borne, phloem-restricted virus. Mol. Plant Microbe Interact. *31*, 1095–1110. https://doi.org/10.1094/MPMI-03-18-0057-R

Di Carli, M., Benvenuto, E., and Donini, M. (2012). Recent insights into plant-virus interactions through proteomic analysis. J. Proteome Res. *11*, 4765–4780. https://doi.org/10.1021/pr300494e

Dombrovsky, A., Gollop, N., Chen, S., Chejanovsky, N., and Raccah, B. (2007). *In vitro* association between the helper component-proteinase of *Zucchini yellow mosaic virus* and cuticle proteins of *Myzus persicae*. J. Gen. Virol. *88*, 1602–1610. https://doi.org/10.1099/vir.0.82769-0

Donaldson, J.R., and Gratton, C. (2007). Antagonistic effects of soybean viruses on soybean aphid performance. Environ. Entomol. *36*, 918–925.

Drucker, M., and Then, C. (2015). Transmission activation in non-circulative virus transmission: a general concept? Curr. Opin. Virol. *15*, 63–68. https://doi.org/10.1016/j.coviro.2015.08.006

Drucker, M., Froissart, R., Hébrard, E., Uzest, M., Ravallec, M., Espérandieu, P., Mani, J.C., Pugnière, M., Roquet, F., Fereres, A., *et al.* (2002). Intracellular distribution of viral gene products regulates a complex mechanism of *Cauliflower mosaic virus* acquisition by its aphid vector. Proc. Natl. Acad. Sci. U.S.A. *99*, 2422–2427. https://doi.org/10.1073/pnas.042587799

Durvasula, R.V., Sundaram, R.K., Cordon-Rosales, C., Pennington, P., and Beard, B. (2003). *Rhodnius prolixus* and its symbiont, *Rhodococcus rhodnii*: a model for paratransgenic control of disease transmission. In Insect Symbiosis, Miller, T.A., ed. (CRC Press, Boca Raton, FL, USA), pp. 83–95.

Eigenbrode, S.D., Ding, H., Shiel, P., and Berger, P.H. (2002). Volatiles from potato plants infected with *Potato leafroll virus* attract and arrest the virus vector, *Myzus persicae* (Homoptera: Aphididae). Proc. Biol. Sci. *269*, 455–460. https://doi.org/10.1098/rspb.2001.1909

Espinoza, A.M., Medina, V., Hull, R., and Markham, P.G. (1991). *Cauliflower mosaic virus* gene II product forms distinct inclusion bodies in infected plant cells. Virology *185*, 337–344.

Fédière, G. (2000). Epidemiology and pathology of *Densovirinae*. Contrib. Microbiol. *4*, 1–11.

Fiebig, M., Poehling, H.M., and Borgemeister, C. (2004). *Barley yellow dwarf virus*, wheat, and *Sitobion avenae*: a case of trilateral interactions. Entomol. Exp. Appl. *110*, 11–21.

Franz, A.W., van der Wilk, F., Verbeek, M., Dullemans, A.M., and van den Heuvel, J.F. (1999). *Faba bean necrotic yellows virus* (genus *Nanovirus*) requires a helper factor for its aphid transmission. Virology *262*, 210–219. https://doi.org/10.1006/viro.1999.9904

Gaedigk, K., Adam, G., and Mundry, K.-W. (1986). The spike protein of *Potato yellow dwarf virus* and its functional role in the infection of insect vector cells. J. Gen. Virol. *67*, 2763–2773. https://doi.org/10.1099/0022-1317-67-12-2763

Garret, A., Kerlan, C., and Thomas, D. (1993). The intestine is a site of passage for *Potato leafroll virus* from the gut lumen into the haemocoel in the aphid vector, *Myzus persicae* Sulz. Arch. Virol. *131*, 377–392. https://doi.org/10.1007/bf01378639

Garret, A., Kerlan, C., and Thomas, D. (1996). Ultrastructural study of acquisition and retention of *Potato leafroll* luteovirus in the alimentary canal of its aphid vector, *Myzus persicae* Sulz. Arch. Virol. *141*, 1279–1292. https://doi.org/10.1007/bf01718830

Gallet, R., Fabre, F., Thébaud, G., Sofonea, M.T., Sicard, A., Blanc, S., and Michalakis, Y. (2018). Small bottleneck size in a highly multipartite virus during a complete infection cycle. J. Virol. *92*, e00139–00118. https://doi.org/10.1128/JVI.00139

Garry, C.E., and Garry, R.F. (2004). Proteomics computational analyses suggest that the carboxyl terminal glycoproteins of bunyaviruses are class II viral fusion protein (beta-penetrenes). Theor. Biol. Med. Model. *1*, 10. https://doi.org/10.1186/1742-4682-1-10

Geng, L., Qian, L.X., Shao, R.X., Liu, Y.Q., Liu, S.S., and Wang, X.W. (2018). Transcriptome profiling of whitefly guts in response to *Tomato yellow leaf curl virus* infection. Virol. J. *15*, 14. https://doi.org/10.1186/s12985-018-0926-6

Ghanim, M., and Medina, V. (2007). Localization of *Tomato yellow leaf curl virus* in its whitefly vector *Bemisia tabaci*. In Tomato Yellow Leaf Curl Virus Disease: Management, Molecular Biology, Breeding for Resistance, H. Czosnek, ed. (Springer Netherlands, Dordrecht), pp. 171–183.

Ghanim, M., Fattah-Hosseini, S., Levy, A., and Cilia, M. (2016). Morphological abnormalities and cell death in the Asian citrus psyllid (*Diaphorina citri*) midgut associated with Candidatus *Liberibacter asiaticus*. Sci. Rep. *6*, 33418. https://doi.org/10.1038/srep33418

Gildow, F.E. (1980). Increased production of alatae by aphids reared on oats infected with *Barley yellow dwarf virus*. Ann. Entomol. Soc. Am. *73*, 343347.

Gildow, F.E. (1983). Influence of *Barley yellow dwarf virus*-infected oats and barley on morphology of aphid vectors. Phytopathology *73*, 1196–1199. https://doi.org/10.1094/Phyto-73-1196

Gildow, F.E. (1987). Virus-membrane interactions involved in circulative transmission of luteoviruses by aphids. In Current Topics in Vector Research, K.F. Harris, ed. (Springer New York, New York, NY), pp. 93–120.

Gildow, F.E. (1993). Evidence for receptor-mediated endocytosis regulating luteovirus acquisition by aphids. Phytopathology *83*, 270–277. https://doi.org/10.1094/Phyto-83-270

Götz, M., Popovski, S., Kollenberg, M., Gorovits, R., Brown, J.K., Cicero, J.M., Czosnek, H., Winter, S., and Ghanim, M. (2012). Implication of *Bemisia tabaci* heat shock protein 70 in begomovirus-whitefly interactions. J. Virol. *86*, 13241–13252. https://doi.org/10.1128/JVI.00880-12

Govier, D.A., and Kassanis, B. (1974). A virus-induced component of plant sap needed when aphids acquire

Potato virus Y from purified preparations. Virology *61*, 420–426. https://doi.org/10.1016/0042-6822(74)90278-5

Gray, S.M., Smith, D.M., Barbierri, L., and Burd, J. (2002). Virus transmission phenotype is correlated with host adaptation among genetically diverse populations of the aphid *Schizaphis graminum*. Phytopathology *92*, 970–975. https://doi.org/10.1094/PHYTO.2002.92.9.970

Gray, S.M., Caillaud, M.C., Burrows, M., and Smith, D.M. (2007). Transmission of two viruses that cause Barley Yellow Dwarf is controlled by different loci in the aphid, *Schizaphis graminum*. J. Insect Sci. *7*, 1–15. https://doi.org/10.1673/031.007.2501

Gray, S.M., Cilia, M., and Ghanim, M. (2014). Circulative, 'nonpropagative' virus transmission: an orchestra of virus-, insect-, and plant-derived instruments. Adv. Virus Res. *89*, 141–199. https://doi.org/10.1016/B978-0-12-800172-1.00004-5

Grigoras, I., Vetten, H.J., Commandeur, U., Ziebell, H., Gronenborn, B., and Timchenko, T. (2018). Nanovirus DNA-N encodes a protein mandatory for aphid transmission. Virology *522*, 281–291.

Groen, S.C., Jiang, S., Murphy, A.M., Cunniffe, N.J., Westwood, J.H., Davey, M.P., Bruce, T.J., Caulfield, J.C., Furzer, O.J., Reed, A., *et al.* (2016). Virus infection of plants alters pollinator preference: A payback for susceptible hosts? PLOS Pathog. *12*, e1005790. https://doi.org/10.1371/journal.ppat.1005790

Harrison, B.D., and Robinson, D.J. (1988). Molecular variation in vector-borne plant viruses: epidemiological significance. Philos. Trans. R. Soc. Lond., B, Biol. Sci. *321*, 447–462. https://doi.org/10.1098/rstb.1988.0102

He, D.C., Zhan, J., Cheng, Z.B., and Xie, L.H. (2016). Viruliferous rate of small brown planthopper is a good indicator of rice stripe disease epidemics. Sci. Rep. *6*, 21376. https://doi.org/10.1038/srep21376

He, W.B., Li, J., and Liu, S.S. (2015). Differential profiles of direct and indirect modification of vector feeding behaviour by a plant virus. Sci. Rep. *5*, 7682. https://doi.org/10.1038/srep07682

Heck, M. (2018). Insect transmission of plant pathogens: a systems biology perspective. mSystems 3, e00168–17. https://doi.org/10.1128/mSystems.00168-17

Heck, M., and Brault, V. (2018). Targeted disruption of aphid transmission: a vision for the management of crop diseases caused by *Luteoviridae* members. Curr. Opin. Virol. 33, 24–32.

Heil, M. (2016). Host manipulation by parasites: cases, patterns, and remaining doubts. Front. Ecol. Evol. 4. https://doi.org/10.3389/fevo.2016.00080

Hodge, S., and Powell, G. (2008). Do plant viruses facilitate their aphid vectors by inducing symptoms that alter behavior and performance? Environ. Entomol. 37, 1573–1581.

Höfer, P., Bedford, I.D., Markham, P.G., Jeske, H., and Frischmuth, T. (1997). Coat protein gene replacement results in whitefly transmission of an insect nontransmissible geminivirus isolate. Virology *236*, 288–295.

Hoffmann, A.A., Montgomery, B.L., Popovici, J., Iturbe-Ormaetxe, I., Johnson, P.H., Muzzi, F., Greenfield, M., Durkan, M., Leong, Y.S., Dong, Y., *et al.* (2011). Successful establishment of *Wolbachia* in *Aedes* populations to suppress dengue transmission. Nature *476*, 454–457. https://doi.org/10.1038/nature10356

Höhnle, M., Höfer, P., Bedford, I.D., Briddon, R.W., Markham, P.G., and Frischmuth, T. (2001). Exchange of three amino acids in the coat protein results in efficient whitefly transmission of a nontransmissible *Abutilon mosaic virus* isolate. Virology *290*, 164–171. https://doi.org/10.1006/viro.2001.1140

Holmes, J.C., and Bethel, W.M. (1972). Modification of intermediate host behaviour by parasites. In Behavioural Aspects of Parasite Transmission, E.U. Canning, and C.A. Wright, eds. (Academic Press, London and New York), pp. 123–149.

Holopainen, J.K., Semiz, G., and Blande, J.D. (2009). Life-history strategies affect aphid preference for yellowing leaves. Biol. Lett. *5*, 603–605. https://doi.org/10.1098/rsbl.2009.0372

Huang, H.J., Bao, Y.Y., Lao, S.H., Huang, X.H., Ye, Y.Z., Wu, J.X., Xu, H.J., Zhou, X.P., and Zhang, C.X. (2015). *Rice ragged stunt virus*-induced apoptosis affects virus transmission from its insect vector, the brown planthopper to the rice plant. Sci. Rep. *5*, 11413. https://doi.org/10.1038/srep11413

Huang, Y.W., Geng, Y.F., Ying, X.B., Chen, X.Y., and Fang, R.X. (2005). Identification of a movement protein of *Rice yellow stunt* rhabdovirus. J. Virol. *79*, 2108–2114. https://doi.org/10.1128/JVI.79.4.2108-2114.2005

Huo, Y., Liu, W., Zhang, F., Chen, X., Li, L., Liu, Q., Zhou, Y., Wei, T., Fang, R., and Wang, X. (2014). Transovarial transmission of a plant virus is mediated by vitellogenin of its insect vector. PLOS Pathog. *10*, e1003949. https://doi.org/10.1371/journal.ppat.1003949

Ingwell, L.L., Eigenbrode, S.D., and Bosque-Pérez, N.A. (2012). Plant viruses alter insect behavior to enhance their spread. Sci. Rep. *2*, 578. https://doi.org/10.1038/srep00578

Ivanov, K.I., Eskelin, K., Bašić, M., De, S., Lõhmus, A., Varjosalo, M., and Mäkinen, K. (2016). Molecular insights into the function of the viral RNA silencing suppressor HCPro. Plant J. *85*, 30–45. https://doi.org/10.1111/tpj.13088

Jackson, A.O., Dietzgen, R.G., Goodin, M.M., Bragg, J.N., and Deng, M. (2005). Biology of plant rhabdoviruses. Annu. Rev. Phytopathol. *43*, 623–660. https://doi.org/10.1146/annurev.phyto.43.011205.141136

Jacquemond, M. (2012). *Cucumber mosaic virus*. Adv. Virus Res. *84*. https://doi.org/10.1016/S0065-3527(08)60039-1

Jensen, S.G. (1972). Metabolism and carbohydrate composition in *Barley yellow dwarf virus*-infected wheat. Phytopathology *62*, 587–592.

Jiménez-Martínez, E.S., Bosque-Pérez, S.A., Berger, P.H., Zemetrad, R.S., Ding, H., and Eigenbrode, S.D. (2004). Volatile cues influence the response of *Rhopalosiphum padi* (Homoptera: Aphididae) to *Barley yellow dwarf virus*–infected transgenic and untransformed wheat. Environ. Entomol. 33, 1207–1216. https://doi.org/10.1603/0046-225X-33.5.1207

Kanakala, S., and Ghanim, M. (2016). Implication of the whitefly *Bemisia tabaci* Cyclophilin B protein in the transmission of *Tomato yellow leaf curl virus*. Front. Plant Sci. 7, 1702. https://doi.org/10.3389/fpls.2016.01702

Kaplan, I.B., Lee, L., Ripoll, D.R., Palukaitis, P., Gildow, F., and Gray, S.M. (2007). Point mutations in the *Potato leafroll virus* major capsid protein alter virion stability and aphid transmission. J. Gen. Virol. *88*, 1821–1830. https://doi.org/10.1099/vir.0.82837-0

Kassanis, B., and Govier, D.A. (1971). New evidence on the mechanism of aphid transmission of *Potato C* and *Potato aucuba mosaic* viruses. J. Gen. Virol. *10*, 99–101. https://doi.org/10.1099/0022-1317-10-1-99

Kaur, N., Hasegawa, D.K., Ling, K.S., and Wintermantel, W.M. (2016). Application of genomics for understanding plant virus-insect vector interactions and insect vector control. Phytopathology *106*, 1213–1222. https://doi.org/10.1094/PHYTO-02-16-0111-FI

Kersten, B., Ghirardo, A., Schnitzler, J.P., Kanawati, B., Schmitt-Kopplin, P., Fladung, M., and Schroeder, H. (2013). Integrated transcriptomics and metabolomics decipher differences in the resistance of pedunculate oak to the herbivore *Tortrix viridana* L. BMC Genomics *14*, 737. https://doi.org/10.1186/1471-2164-14-737

Khelifa, M., Journou, S., Krishnan, K., Gargani, D., Esperandieu, P., Blanc, S., and Drucker, M. (2007). Electron-lucent inclusion bodies are structures specialized for aphid transmission of *Cauliflower mosaic virus*. J. Gen. Virol. *88*, 2872–2880. https://doi.org/10.1099/vir.0.83009-0

Kikkert, M., Meurs, C., van de Wetering, F., Dorfmüller, S., Peters, D., Kormelink, R., and Goldbach, R. (1998). Binding of *Tomato spotted wilt virus* to a 94-kDa thrips protein. Phytopathology *88*, 63–69. https://doi.org/10.1094/PHYTO.1998.88.1.63

Kotzampigikis, A., Hristova, D., and Tasheva-Terzieva, E. (2010). Virus-vector relationship between *Potato leafroll virus* PLRV and *Myzus persicae* Sulzer. Bulg. J. Agri. Sci. *16*, 412–421.

Krenz, B., Schießl, I., Greiner, E., and Krapp, S. (2017). Analyses of *Pea necrotic yellow dwarf virus*-encoded proteins. Virus Genes *53*, 454–463. https://doi.org/10.1007/s11262-017-1439-x

Kritzman, A., Gera, A., Raccah, B., van Lent, J.W., and Peters, D. (2002). The route of *Tomato spotted wilt virus* inside the thrips body in relation to transmission efficiency. Arch. Virol. *147*, 2143–2156. https://doi.org/10.1007/s00705-002-0871-x

Kruse, A., Fattah-Hosseini, S., Saha, S., Johnson, R., Warwick, E., Sturgeon, K., Mueller, L., MacCoss, M.J., Shatters, R.G., and Cilia Heck, M. (2017). Combining 'omics and microscopy to visualize interactions between the Asian citrus psyllid vector and the Huanglongbing pathogen Candidatus *Liberibacter asiaticus* in the insect gut. PLOS ONE *12*, e0179531. https://doi.org/10.1371/journal.pone.0179531

Kruse, A., Ramsey, J.S., Johnson, R., Hall, D.G., MacCoss, M.J., and Heck, M. (2018). Candidatus *Liberibacter asiaticus* minimally alters expression of immunity and metabolism proteins in hemolymph of *Diaphorina citri*, the insect vector of Huanglongbing. J. Proteome Res. *17*, 2995–3011. https://doi.org/10.1021/acs.jproteome.8b00183

Kukurba, K.R., and Montgomery, S.B. (2015). RNA sequencing and analysis. Cold Spring Harb. Protoc. *2015*, 951–969. https://doi.org/10.1101/pdb.top084970

Lan, H., Chen, H., Liu, Y., Jiang, C., Mao, Q., Jia, D., Chen, Q., and Wei, T. (2016a). Small interfering RNA pathway modulates initial viral infection in midgut epithelium of insect after ingestion of virus. J. Virol. *90*, 917–929. https://doi.org/10.1128/JVI.01835-15

Lan, H., Wang, H., Chen, Q., Chen, H., Jia, D., Mao, Q., and Wei, T. (2016b). Small interfering RNA pathway modulates persistent infection of a plant virus in its insect vector. Sci. Rep. *6*, 20699. https://doi.org/10.1038/srep20699

Lee, L., Kaplan, I.B., Ripoll, D.R., Liang, D., Palukaitis, P., and Gray, S.M. (2005). A surface loop of the *Potato leafroll virus* coat protein is involved in virion assembly, systemic movement, and aphid transmission. J. Virol. *79*, 1207–1214.

Leh, V., Jacquot, E., Geldreich, A., Hermann, T., Leclerc, D., Cerutti, M., Yot, P., Keller, M., and Blanc, S. (1999). Aphid transmission of *Cauliflower mosaic virus* requires the viral PIII protein. EMBO J. *18*, 7077–7085. https://doi.org/10.1093/emboj/18.24.7077

Leh, V., Jacquot, E., Geldreich, A., Haas, M., Blanc, S., Keller, M., and Yot, P. (2001). Interaction between the open reading frame III product and the coat protein is required for transmission of *Cauliflower mosaic virus* by aphids. J. Virol. 75, 100–106. https://doi.org/10.1128/JVI.75.1.100-106.2001

Li, C., Cox-Foster, D., Gray, S.M., and Gildow, F. (2001). Vector specificity of *Barley yellow dwarf virus* (BYDV) transmission: identification of potential cellular receptors binding BYDV-MAV in the aphid, *Sitobion avenae*. Virology *286*, 125–133. https://doi.org/10.1006/viro.2001.0929

Li, S., Xiong, R., Wang, X., and Zhou, Y. (2011). Five proteins of *Laodelphax striatellus* are potentially involved in the interactions between *Rice stripe virus* and vector. PLOS ONE 6, e26585. https://doi.org/10.1371/journal.pone.0026585

Li, S., Wang, S., Wang, X., Li, X., Zi, J., Ge, S., Cheng, Z., Zhou, T., Ji, Y., Deng, J., *et al.* (2015). *Rice stripe virus* affects the viability of its vector offspring by changing developmental gene expression in embryos. Sci. Rep. 5, 7883. https://doi.org/10.1038/srep07883

Li, S., Li, X., and Zhou, Y. (2018). Ribosomal protein L18 is an essential factor that promote *Rice stripe virus* accumulation in small brown planthopper. Virus Res. *247*, 15–20.

Li, S., Zhou, C., and Zhou, Y. (2019). Olfactory co-receptor Orco stimulated by *Rice stripe virus* is essential for host seeking behavior in small brown planthopper. Pest Manag. Sci. 75, 187–194. https://doi.org/10.1002/ps.5086

Liang, Y., and Gao, X.W. (2017). The cuticle protein gene MPCP4 of *Myzus persicae* (Homoptera: Aphididae) plays a critical role in *Cucumber mosaic virus* acquisition. J. Econ. Entomol. *110*, 848–853. https://doi.org/10.1093/jee/tox025

Lightle, D., and Lee, J. (2014). Raspberry viruses affect the behavior and performance of *Amphorophora agathonica* in single and mixed infections. Entomol. Exp. Appl. *151*, 57–64. https://doi.org/10.1111/eea.12170

Linz, L.B., Liu, S., Chougule, N.P., and Bonning, B.C. (2015). *In vitro* evidence supports membrane Alanyl Aminopeptidase N as a receptor for a plant virus in the pea aphid vector. J. Virol. *89*, 11203–11212. https://doi.org/10.1128/JVI.01479-15

Liu, B., Preisser, E.L., Chu, D., Pan, H., Xie, W., Wang, S., Wu, Q., Zhou, X., and Zhang, Y. (2013). Multiple forms of vector manipulation by a plant-infecting virus: *Bemisia tabaci* and *Tomato yellow leaf curl virus*. J. Virol. *87*, 4929–4937. https://doi.org/10.1128/JVI.03571-12

Liu, S., Briddon, R.W., Bedford, I.D., Pinner, M.S., and Markham, P.G. (1999). Identification of genes directly and indirectly involved in the insect transmission of *African cassava mosaic* geminivirus by *Bemisia tabaci*. Virus Genes *18*, 5–11. https://doi.org/10.1023/A:1008017017020

Liu, S., He, X., Park, G., Josefsson, C., and Perry, K.L. (2002). A conserved capsid protein surface domain of *Cucumber mosaic virus* is essential for efficient aphid vector transmission. J. Virol. *76*, 9756–9762. https://doi.org/10.1128/JVI.76.19.9756-9762.2002

Liu, S., Sivakumar, S., Sparks, W.O., Miller, W.A., and Bonning, B.C. (2010). A peptide that binds the pea aphid gut impedes entry of *Pea enation mosaic virus* into the aphid hemocoel. Virology *401*, 107–116. https://doi.org/10.1016/j.virol.2010.02.009

Liu, W., Gray, S., Huo, Y., Li, L., Wei, T., and Wang, X. (2015). Proteomic analysis of interaction between a plant virus and its vector insect reveals new functions of hemipteran cuticular protein. Mol. Cell Proteomics *14*, 2229–2242. https://doi.org/10.1074/mcp.M114.046763

Liu, Y., Jia, D., Chen, H., Chen, Q., Xie, L., Wu, Z., and Wei, T. (2011). The P7-1 protein of *Southern rice black-streaked dwarf virus*, a fijivirus, induces the formation of tubular structures in insect cells. Arch. Virol. *156*, 1729–1736. https://doi.org/10.1007/s00705-011-1041-9

Luan, J.B., Li, J.M., Varela, N., Wang, Y.L., Li, F.F., Bao, Y.Y., Zhang, C.X., Liu, S.S., and Wang, X.W. (2011). Global analysis of the transcriptional response of whitefly to *Tomato yellow leaf curl China virus* reveals the relationship of coevolved adaptations. J. Virol. *85*, 3330–3340. https://doi.org/10.1128/JVI.02507-10

Maag, D., Erb, M., and Glauser, G. (2015). Metabolomics in plant-herbivore interactions: challenges and applications. Entomol. Exp. Appl. *157*, 18–29. https://doi.org/10.1111/eea.12336

Macias, W., and Mink, G.I. (1969). Preference of green peach aphid for virus-infected sugarbeet leaves. J. Econ. Entomol. *62*, 28–29. https://doi.org/10.1093/jee/62.1.28

Maluta, N.K., Garzo, E., Moreno, A., Lopes, J.R., and Fereres, A. (2014). *Tomato yellow leaf curl virus* benefits population growth of the Q Biotype of *Bemisia tabaci* (Gennadius) (Hemiptera: Aleyrodidae). Neotrop. Entomol. *43*, 385–392. https://doi.org/10.1007/s13744-014-0223-z

Mann, M., Fattah-Hosseini, S., Ammar, E.D., Stange, R., Warrick, E., Sturgeon, K., Shatters, R., and Heck, M. (2018). *Diaphorina citri* nymphs are resistant to morphological changes induced by 'Candidatus *Liberibacter asiaticus*' in midgut epithelial cells. Infect. Immun. *86*, e00889–17.

Mann, R.S., Ali, J.G., Hermann, S.L., Tiwari, S., Pelz-Stelinski, K.S., Alborn, H.T., and Stelinski, L.L. (2012). Induced release of a plant-defense volatile 'deceptively' attracts insect vectors to plants infected with a bacterial pathogen. PLOS Pathog. *8*, e1002610. https://doi.org/10.1371/journal.ppat.1002610

Mar, T., Liu, W., and Wang, X. (2014). Proteomic analysis of interaction between P7-1 of *Southern rice black-streaked dwarf virus* and the insect vector reveals diverse insect proteins involved in successful transmission. J. Proteomics *102*, 83–97. https://doi.org/10.1016/j.jprot.2014.03.004

Maris, P.C., Joosten, N.N., Goldbach, R.W., and Peters, D. (2004). *Tomato spotted wilt virus* infection improves host suitability for its vector *Frankliniella occidentalis*. Phytopathology *94*, 706–711. https://doi.org/10.1094/PHYTO.2004.94.7.706

Martini, X., Pelz-Stelinski, K.S., and Stelinski, L.L. (2014). Plant pathogen-induced volatiles attract parasitoids to increase parasitism of an insect vector. Front. Ecol. Evol. *2*. https://doi.org/10.3389/fevo.2014.00008

Martinière, A., Bak, A., Macia, J.L., Lautredou, N., Gargani, D., Doumayrou, J., Garzo, E., Moreno, A., Fereres, A., Blanc, S., *et al.* (2013). A virus responds instantly to the presence of the vector on the host and forms transmission morphs. Elife *2*, e00183. https://doi.org/10.7554/eLife.00183

Mathers, T.C., Chen, Y., Kaithakottil, G., Legeai, F., Mugford, S.T., Baa-Puyoulet, P., Bretaudeau, A., Clavijo, B., Colella, S., Collin, O., *et al.* (2017). Rapid transcriptional plasticity of duplicated gene clusters enables a clonally reproducing aphid to colonise diverse plant species. Genome Biol. *18*, 27. https://doi.org/10.1186/s13059-016-1145-3

Mauck, K.E., De Moraes, C.M., and Mescher, M.C. (2010a). Deceptive chemical signals induced by a plant virus attract insect vectors to inferior hosts. Proc. Natl. Acad. Sci. U.S.A. *107*, 3600–3605. https://doi.org/10.1073/pnas.0907191107

Mauck, K.E., De Moraes, C.M., and Mescher, M.C. (2010b). Effects of *Cucumber mosaic virus* infection on vector and non-vector herbivores of squash. Commun. Integr. Biol. *3*, 579–582. https://doi.org/10.4161/cib.3.6.13094

Mauck, K.E., Bosque-Perez, N.A., Eigenbrode, S.D., De Moraes, C.M., and Mescher, M.C. (2012). Transmission mechanisms shape pathogen effects on host-vector interactions: evidence from plant viruses. Funct. Ecol. *26*, 1162–1175. https://doi.org/10.1111/j.1365-2435.2012.02026.x

Mauck, K.E., De Moraes, C.M., and Mescher, M.C. (2014). Biochemical and physiological mechanisms underlying effects of *Cucumber mosaic virus* on host-plant traits that mediate transmission by aphid vectors. Plant Cell Environ. *37*, 1427–1439. https://doi.org/10.1111/pce.12249

McMenemy, L.S., Hartley, S.E., MacFarlane, S.A., Karley, A.J., Shepherd, T., and Johnson, S.N. (2012). Raspberry viruses manipulate the behaviour of their insect vectors. Entomol. Exp. Appl. *144*, 56–68. https://doi.org/10.1111/j.1570-7458.2012.01248.x

Medina-Ortega, K.J., Bosque-Pérez, N.A., Ngumbi, E., Jiménez-Martínez, E.S., and Eigenbrode, S.D. (2009). *Rhopalosiphum padi* (Hemiptera: Aphididae) responses to volatile cues from *Barley yellow dwarf virus*-infected wheat. Environ. Entomol. *38*, 836–845. https://doi.org/10.1603/022.038.0337

Mittler, T.E. (1967). Effect of amino acid and sugar concentrations on the food uptake of the aphid *Myzus persicae*. Entomol. Exp. Appl. *10*, 39–51. https://doi.org/10.1111/j.1570-7458.1967.tb00043.x

Moritz, G., Kumm, S., and Mound, L. (2004). Tospovirus transmission depends on thrips ontogeny. Virus Res. *100*, 143–149. https://doi.org/10.1016/j.virusres.2003.12.022

Morkunas, I., Mai, V.C., and Gabryś, B. (2011). Phytohormonal signaling in plant responses to aphid feeding. Acta Physiol. Plant. 33, 2057–2073. https://doi.org/10.1007/s11738-011-0751-7

Mossop, D.W., and Francki, R.I. (1977). Association of RNA 3 with aphid transmission of *Cucumber mosaic virus*. Virology *81*, 177–181. https://doi.org/10.1016/0042-6822(77)90071-X

Mulot, M., Monsion, B., Boissinot, S., Rastegar, M., Meyer, S., Bochet, N., and Brault, V. (2018). Transmission of *Turnip yellows virus* by *Myzus persicae* is reduced by feeding aphids on double-stranded RNA targeting the ephrin receptor protein. Front. Microbiol. *9*, 457. https://doi.org/10.3389/fmicb.2018.00457

Nagata, T., Inoue-Nagata, A.K., Smid, H.M., Goldbach, R., and Peters, D. (1999). Tissue tropism related to vector competence of *Frankliniella occidentalis* for *Tomato spotted wilt* tospovirus. J. Gen. Virol. *80*, 507–515. https://doi.org/10.1099/0022-1317-80-2-507

Nasu, S. (1963). Studies on some leafhoppers and planthoppers which transmit virus diseases of rice plant in Japan. Bull. Kyushu Agric. Exp. Stn. *8*, 153–349.

Ng, J.C., Liu, S., and Perry, K.L. (2000). *Cucumber mosaic virus* mutants with altered physical properties and defective in aphid vector transmission. Virology *276*, 395–403. https://doi.org/10.1006/viro.2000.0569

Ng, J.C., Josefsson, C., Clark, A.J., Franz, A.W., and Perry, K.L. (2005). Virion stability and aphid vector transmissibility of *Cucumber mosaic virus* mutants. Virology 332, 397–405.

Ngumbi, E., Eigenbrode, S.D., Bosque-Pérez, N.A., Ding, H., and Rodriguez, A. (2007). *Myzus persicae* is arrested more by blends than by individual compounds elevated in headspace of PLRV-infected potato. J. Chem. Ecol. 33, 1733–1747. https://doi.org/10.1007/s10886-007-9340-z

Noris, E., Vaira, A.M., Caciagli, P., Masenga, V., Gronenborn, B., and Accotto, G.P. (1998). Amino acids in the capsid protein of *Tomato yellow leaf curl virus* that are crucial for systemic infection, particle formation, and insect transmission. J. Virol. 72, 10050–10057.

Nouri, S., Matsumura, E.E., Kuo, Y.W., and Falk, B.W. (2018). Insect-specific viruses: from discovery to potential translational applications. Curr. Opin. Virol. *33*, 33–41.

Omura, T., Yan, J., Zhong, B., Wada, M., Zhu, Y., Tomaru, M., Maruyama, W., Kikuchi, A., Watanabe, Y., Kimura, I., *et al.* (1998). The P2 protein of *Rice dwarf* phytoreovirus is required for adsorption of the virus to cells of the insect vector. J. Virol. 72, 9370–9373.

Oppenheim, S.J., Baker, R.H., Simon, S., and DeSalle, R. (2015). We can't all be supermodels: the value of comparative transcriptomics to the study of non-model insects. Insect Mol. Biol. *24*, 139–154. https://doi.org/10.1111/imb.12154

Pakkianathan, B.C., Kontsedalov, S., Lebedev, G., Mahadav, A., Zeidan, M., Czosnek, H., and Ghanim, M. (2015). Replication of *Tomato yellow leaf curl virus* in its whitefly vector, *Bemisia tabaci*. J. Virol. *89*, 9791–9803. https://doi.org/10.1128/JVI.00779-15

Pallas, V., and Garcia, J.A. (2011). How do plant viruses induce disease? Interactions and interference with host components. J. Gen. Virol. *92*, 2691–2705. https://doi.org/10.1099/vir.0.034603-0

Papura, D., Jacquot, E., Dedryver, C.A., Luche, S., Riault, G., Bossis, M., and Rabilloud, T. (2002). Two-dimensional electrophoresis of proteins discriminates aphid clones of *Sitobion avenae* differing in BYDV-PAV transmission. Arch. Virol. *147*, 1881–1898. https://doi.org/10.1007/s00705-002-0859-6

Peiffer, M.L., Gildow, F.E., and Gray, S.M. (1997). Two distinct mechanisms regulate luteovirus transmission efficiency and specificity at the aphid salivary gland. J. Gen. Virol. *78*, 495–503. https://doi.org/10.1099/0022-1317-78-3-495

Pelz-Stelinski, K.S., and Killiny, N. (2016). Better together: association with '*Candidatus Liberibacter* Asiaticus' increases the reproductive fitness of its insect vector, *Diaphorina citri* (Hemiptera: Liviidae). Ann. Entomol. Soc. Am. *109*, 371–376. https://doi.org/10.1093/aesa/saw007

Peng, Y.H., Kadoury, D., Gal-On, A., Huet, H., Wang, Y., and Raccah, B. (1998). Mutations in the HC-Pro gene of *Zucchini yellow mosaic* potyvirus: effects on aphid transmission and binding to purified virions. J. Gen. Virol. *79*, 897–904. https://doi.org/10.1099/0022-1317-79-4-897

Perry, K.L., Zhang, L., Shintaku, M.H., and Palukaitis, P. (1994). Mapping determinants in *Cucumber mosaic virus* for transmission by *Aphis gossypii*. Virology *205*, 591–595. https://doi.org/10.1006/viro.1994.1686

Perry, K.L., Zhang, L., and Palukaitis, P. (1998). Amino acid changes in the coat protein of *Cucumber mosaic virus* differentially affect transmission by the aphids *Myzus persicae* and *Aphis gossypii*. Virology *242*, 204–210. https://doi.org/10.1006/viro.1998.8991

Peter, K.A., Liang, D., Palukaitis, P., and Gray, S.M. (2008). Small deletions in the *Potato leafroll virus* readthrough protein affect particle morphology, aphid transmission, virus movement and accumulation. J. Gen. Virol. *89*, 2037–2045. https://doi.org/10.1099/vir.0.83625-0

Peter, K.A., Gildow, F., Palukaitis, P., and Gray, S.M. (2009). The C terminus of the polerovirus p5 readthrough domain limits virus infection to the phloem. J. Virol. *83*, 5419–5429. https://doi.org/10.1128/JVI.02312-08

Pinheiro, P., Bereman, M.S., Burd, J., Pals, M., Armstrong, S., Howe, K.J., Thannhauser, T.W., MacCoss, M.J., Gray, S.M., and Cilia, M. (2014). Evidence of the biochemical basis of host virulence in the greenbug aphid, *Schizaphis graminum* (Homoptera: Aphididae). J. Proteome Res. *13*, 2094–2108. https://doi.org/10.1021/pr4012415

Pinheiro, P.V., Kliot, A., Ghanim, M., and Cilia, M. (2015). Is there a role for symbiotic bacteria in plant virus transmission by insects? Curr. Opin. Insect Sci. *8*, 69–78. https://doi.org/10.1016/j.cois.2015.01.010

Pinheiro, P.V., Ghanim, M., Alexander, M., Rebelo, A.R., Santos, R.S., Orsburn, B.C., Gray, S., and Cilia, M. (2017). Host plants indirectly influence plant virus transmission by altering gut cysteine protease activity of aphid vectors. Mol. Cell Proteomics 16 (Suppl. 4), S230–S243. https://doi.org/10.1074/mcp.M116.063495

Pinheiro, P.V., Wilson, J.R., Xu, Y., Zheng, Y., Rebelo, A.R., Fattah-Hosseini, S., Kruse, A., Santos Dos Silva, R., Xu, Y., Kramer, M., *et al.* (2019). Plant viruses transmitted in two different modes produce differing effects on small RNA-mediated processes in their insect vector. Phytobiomes J. [Epub ahead of print] https://doi.org/10.1094/PBIOMES-10-18-0045-R.

Pirone, T.P., and Megahed, S. (1966). Aphid transmissibility of some purified viruses and viral RNAs. Virology *30*, 631–637.

Pitino, M., Coleman, A.D., Maffei, M.E., Ridout, C.J., and Hogenhout, S.A. (2011). Silencing of aphid genes by dsRNA feeding from plants. PLOS ONE *6*, e25709. https://doi.org/10.1371/journal.pone.0025709

Plassmeyer, M.L., Soldan, S.S., Stachelek, K.M., Martín-García, J., and González-Scarano, F. (2005). California serogroup Gc (G1) glycoprotein is the principal determinant of pH-dependent cell fusion and entry. Virology *338*, 121–132.

Plassmeyer, M.L., Soldan, S.S., Stachelek, K.M., Roth, S.M., Martín-García, J., and González-Scarano, F. (2007). Mutagenesis of the *La Crosse virus* glycoprotein supports a role for Gc (1066-1087) as the fusion peptide. Virology *358*, 273–282.

Plisson, C., Uzest, M., Drucker, M., Froissart, R., Dumas, C., Conway, J., Thomas, D., Blanc, S., and Bron, P. (2005). Structure of the mature P3-virus particle complex of *Cauliflower mosaic virus* revealed by cryo-electron microscopy. J. Mol. Biol. *346*, 267–277.

Puterka, G.J., Nicholson, S.J., and Cooper, W.R. (2017). Survival and feeding rates of four aphid species (Hemiptera: Aphididae) on various sucrose concentrations in diets. J. Econ. Entomol. *110*, 1518–1524. https://doi.org/10.1093/jee/tox168

Rajabaskar, D., Wu, Y., Bosque-Pérez, N.A., and Eigenbrode, S.D. (2013). Dynamics of *Myzus persicae* arrestment by volatiles from *Potato leafroll virus*-infected potato plants during disease progression. Entomol. Exp. Appl. *148*, 172–181. https://doi.org/10.1111/eea.12087

Rajabaskar, D., Bosque-Pérez, N.A., and Eigenbrode, S.D. (2014). Preference by a virus vector for infected plants is reversed after virus acquisition. Virus Res. *186*, 32–37. https://doi.org/10.1016/j.virusres.2013.11.005

Ramsey, J.S., Wilson, A.C., de Vos, M., Sun, Q., Tamborindeguy, C., Winfield, A., Malloch, G., Smith, D.M., Fenton, B., Gray, S.M., *et al.* (2007). Genomic resources for *Myzus persicae*: EST sequencing, SNP identification, and microarray design. BMC Genomics *8*, 423.

Reinbold, C., Gildow, F.E., Herrbach, E., Ziegler-Graff, V., Goncalves, M.C., van Den Heuvel, J.F., and Brault, V. (2001). Studies on the role of the minor capsid protein in transport of *Beet western yellows virus* through *Myzus persicae*. J. Gen. Virol. *82*, 1995–2007. https://doi.org/10.1099/0022-1317-82-8-1995

Rochow, W.F. (1969). Biological properties of four isolates of *Barley yellow dwarf virus*. Phytopathology *59*, 1580–1589.

Rochow, W.F. (1970). *Barley yellow dwarf virus*: phenotypic mixing and vector specificity. Science *167*, 875–878. https://doi.org/10.1126/science.167.3919.875

Roumagnac, P., Granier, M., Bernardo, P., Deshoux, M., Ferdinand, R., Galzi, S., Fernandez, E., Julian, C., Abt, I., Filloux, D., *et al.* (2015). *Alfalfa leaf curl virus*: an aphid-transmitted geminivirus. J. Virol. *89*, 9683–9688. https://doi.org/10.1128/JVI.00453-15

Rubinstein, G., and Czosnek, H. (1997). Long-term association of *Tomato yellow leaf curl virus* with its whitefly vector *Bemisia tabaci*: effect on the insect transmission capacity, longevity and fecundity. J. Gen. Virol. *78*, 2683–2689. https://doi.org/10.1099/0022-1317-78-10-2683

Ryabov, E.V., Keane, G., Naish, N., Evered, C., and Winstanley, D. (2009). Densovirus induces winged morphs in asexual clones of the rosy apple aphid, *Dysaphis plantaginea*. Proc. Natl. Acad. Sci. U.S.A. *106*, 8465–8470. https://doi.org/10.1073/pnas.0901389106

Saha, S., Hosmani, P.S., Villalobos-Ayala, K., Miller, S., Shippy, T., Flores, M., Rosendale, A., Cordola, C., Bell, T., Mann, H., *et al.* (2017). Improved annotation of the insect vector of citrus greening disease: biocuration by a diverse genomics community. Database *2017*. https://doi.org/10.1093/database/bax032

Salvaudon, L., De Moraes, C.M., and Mescher, M.C. (2013). Outcomes of co-infection by two potyviruses: implications for the evolution of manipulative strategies. Proc. Biol. Sci. *280*, 20122959. https://doi.org/10.1098/rspb.2012.2959

Schlegel, R., and Wade, M. (1983). Neutralized *Vesicular stomatitis virus* binds to host cells by a different 'receptor'. Biochem. Biophys. Res. Commun. *114*, 774–778.

Schneweis, D.J., Whitfield, A.E., and Rotenberg, D. (2017). Thrips developmental stage-specific transcriptome response to *Tomato spotted wilt virus* during the virus infection cycle in *Frankliniella occidentalis*, the primary vector. Virology *500*, 226–237.

Shalitin, D., and Wolf, S. (2000). *Cucumber mosaic virus* infection affects sugar transport in melon plants. Plant Physiol. *123*, 597–604.

Shrestha, A., Champagne, D.E., Culbreath, A.K., Rotenberg, D., Whitfield, A.E., and Srinivasan, R. (2017). Transcriptome changes associated with *Tomato spotted wilt virus* infection in various life stages of its thrips vector, *Frankliniella fusca* (Hinds). J. Gen. Virol. *98*, 2156–2170. https://doi.org/10.1099/jgv.0.000874

Shukla, D.D., Strike, P.M., Tracy, S.L., Gough, K.H., and Ward, C.W. (1988). The N and C termini of the coat proteins of potyviruses are surface-located and the N terminus contains the major virus-specific epitopes. J. Gen. Virol. *69*, 1497–1508. https://doi.org/10.1099/0022-1317-69-7-1497

Sin, S.H., McNulty, B.C., Kennedy, G.G., and Moyer, J.W. (2005). Viral genetic determinants for thrips transmission of *Tomato spotted wilt virus*. Proc. Natl. Acad. Sci. U.S.A. *102*, 5168–5173.

Soto, M.J., Chen, L.F., Seo, Y.S., and Gilbertson, R.L. (2005). Identification of regions of the *Beet mild curly top virus* (family *Geminiviridae*) capsid protein involved in systemic infection, virion formation and leafhopper transmission. Virology *341*, 257–270.

Tamborindeguy, C., Bereman, M.S., DeBlasio, S., Igwe, D., Smith, D.M., White, F., MacCoss, M.J., Gray, S.M., and Cilia, M. (2013). Genomic and proteomic analysis of *Schizaphis graminum* reveals cyclophilin proteins are involved in the transmission of *Cereal yellow dwarf virus*. PLOS ONE *8*, e71620. https://doi.org/10.1371/journal.pone.0071620

Tholl, D., Chen, F., Petri, J., Gershenzon, J., and Pichersky, E. (2005). Two sesquiterpene synthases are responsible for the complex mixture of sesquiterpenes emitted from *Arabidopsis* flowers. Plant J. *42*, 757–771.

Thorpe, P., Cock, P.J., and Bos, J. (2016). Comparative transcriptomics and proteomics of three different aphid species identifies core and diverse effector sets. BMC Genomics *17*, 172. https://doi.org/10.1186/s12864-016-2496-6

Tian, L., Guo, E., Diao, Y., Zhou, S., Peng, Q., Cao, Y., Ling, E., and Li, S. (2010). Genome-wide regulation of innate immunity by juvenile hormone and 20-hydroxyecdysone in the *Bombyx* fat body. BMC Genomics *11*, 549. https://doi.org/10.1186/1471-2164-11-549

Tu, Z., Ling, B., Xu, D., Zhang, M., and Zhou, G. (2013). Effects of *Southern rice black-streaked dwarf virus* on the development and fecundity of its vector, *Sogatella furcifera*. Virol. J. *10*, 145. https://doi.org/10.1186/1743-422X-10-145

Ullman, D.E., Cho, J.J., Mau, R.F., Westcot, D.M., and Custer, D.M. (1992). A midgut barrier to *Tomato spotted wilt virus* acquisition by adult western flower thrips. Phytopathology *82*, 1333–1333.

Ullman, D.E., Westcot, D.M., Chenault, K.D., Sherwood, J.L., German, T.L., Bandla, M.D., Cantone, F., and Duer, H. (1995). Compartmentalization, intracellular transport, and autophagy of *Tomato spotted wilt* tospovirus proteins in infected thrips cells. Phytopathology *85*, 644–654.

van den Heuvel, J.F., Verbeek, M., and van der Wilk, F. (1994). Endosymbiotic bacteria associated with circulative transmission of *Potato leafroll virus* by *Myzus persicae*. J. Gen. Virol. *75*, 2559–2565. https://doi.org/10.1099/0022-1317-75-10-2559

van Molken, T., de Caluwe, H., Hordijk, C.A., Leon-Reyes, A., Snoeren, T.A., van Dam, N.M., and Stuefer, J.F. (2012). Virus infection decreases the attractiveness of white clover plants for a non-vectoring herbivore. Oecologia *170*, 433–444. https://doi.org/10.1007/s00442-012-2322-z

Vrkoslav, V., Muck, A., Cvacka, J., and Svatos, A. (2010). MALDI imaging of neutral cuticular lipids in insects and plants. J. Am. Soc. Mass Spectrom. *21*, 220–231. https://doi.org/10.1016/j.jasms.2009.10.003

Wan, G., Jiang, S., Wang, W., Li, G., Tao, X., Pan, W., Sword, G.A., and Chen, F. (2015). *Rice stripe virus* counters reduced fecundity in its insect vector by modifying insect physiology, primary endosymbionts and feeding behavior. Sci. Rep. *5*, 12527. https://doi.org/10.1038/srep12527

Wang, H., Xu, D., Pu, L., and Zhou, G. (2014). *Southern rice black-streaked dwarf virus* alters insect vectors' host orientation preferences to enhance spread and increase *Rice ragged stunt virus* co-infection. Phytopathology *104*, 196–201. https://doi.org/10.1094/PHYTO-08-13-0227-R

Wang, L., Tang, N., Gao, X., Guo, D., Chang, Z., Fu, Y., Akinyemi, I.A., and Wu, Q. (2016). Understanding the immune system architecture and transcriptome responses to *Southern rice black-streaked dwarf virus* in *Sogatella furcifera*. Sci. Rep. *6*, 36254. https://doi.org/10.1038/srep36254

Wang, X. (2003). Identification of a protein associated with circulative transmission of *Barley yellow dwarf virus* from cereal aphids, *Schizaphis graminum* and *Sitobion avenae*. Chin. Sci. Bull. *48*, 2083–2087. https://doi.org/10.1360/03wc0153

Wang, X.W., Luan, J.B., Li, J.M., Bao, Y.Y., Zhang, C.X., and Liu, S.S. (2010). *De novo* characterization of a whitefly transcriptome and analysis of its gene expression during development. BMC Genomics *11*, 400. https://doi.org/10.1186/1471-2164-11-400

Wang, Y., Mao, Q., Liu, W., Mar, T., Wei, T., Liu, Y., and Wang, X. (2014). Localization and distribution of *Wheat dwarf virus* in its vector leafhopper, *Psammotettix alienus*. Phytopathology *104*, 897–904. https://doi.org/10.1094/PHYTO-09-13-0251-R

War, A.R., Paulraj, M.G., Ahmad, T., Buhroo, A.A., Hussain, B., Ignacimuthu, S., and Sharma, H.C. (2012). Mechanisms of plant defense against insect herbivores. Plant Signal. Behav. *7*, 1306–1320. https://doi.org/10.4161/psb.21663

Webster, B.E.N. (2012). The role of olfaction in aphid host location. Physiol. Entomol. *37*, 10–18. https://doi.org/10.1111/j.1365-3032.2011.00791.x

Webster, C.G., Pichon, E., van Munster, M., Monsion, B., Deshoux, M., Gargani, D., Calevro, F., Jimenez, J., Moreno, A., Krenz, B., *et al.* (2018). Identification of plant virus receptor candidates in the stylets of their aphid vectors. J. Virol. e00432–00418. Epub ahead of print]. https://doi.org/10.1128/JVI.00432-18

Wenger, J.A., Cassone, B.J., Legeai, F., Johnston, J.S., Bansal, R., Yates, A.D., Coates, B.S., Pavinato, V.A., and Michel, A. (2017). Whole genome sequence of the soybean aphid, *Aphis glycines*. Insect Biochem. Mol. Biol. [Epub ahead of print]. https://doi.org/10.1016/j.ibmb.2017.01.005

Werner, B.J., Mowry, T.M., Bosque-Pérez, N.A., Ding, H., and Eigenbrode, S.D. (2009). Changes in green peach aphid responses to *Potato leafroll virus*-induced volatiles emitted during disease progression. Environ. Entomol. *38*, 1429–1438. https://doi.org/10.1603/022.038.0511

Westwood, J.H., Groen, S.C., Du, Z., Murphy, A.M., Anggoro, D.T., Tungadi, T., Luang-In, V., Lewsey, M.G., Rossiter, J.T., Powell, G., *et al.* (2013). A trio of viral proteins tunes aphid-plant interactions in *Arabidopsis thaliana*. PLOS ONE *8*, e83066. https://doi.org/10.1371/journal.pone.0083066

Whitfield, A.E., Ullman, D.E., and German, T.L. (2005). *Tomato spotted wilt virus* glycoprotein G(C) is cleaved at acidic pH. Virus Res. *110*, 183–186. https://doi.org/10.1016/j.virusres.2005.01.007

Whitfield, A.E., Rotenberg, D., Aritua, V., and Hogenhout, S.A. (2011). Analysis of expressed sequence tags from *Maize mosaic* rhabdovirus-infected gut tissues of *Peregrinus maidis* reveals the presence of key components of insect innate immunity. Insect Mol. Biol. *20*, 225–242. https://doi.org/10.1111/j.1365-2583.2010.01060.x

Whitfield, A.E., Falk, B.W., and Rotenberg, D. (2015). Insect vector-mediated transmission of plant viruses. Virology 479-480, 278–289. https://doi.org/10.1016/j.virol.2015.03.026

Whitham, S.A., Yang, C., and Goodin, M.M. (2006). Global impact: elucidating plant responses to viral infection. Mol. Plant Microbe Interact. *19*, 1207–1215. https://doi.org/10.1094/MPMI-19-1207

Woolston, C.J., Covey, S.N., Penswick, J.R., and Davies, J.W. (1983). Aphid transmission and a polypeptide are specified by a defined region of the *Cauliflower mosaic virus* genome. Gene 23, 15–23.

Wu, W., Zheng, L., Chen, H., Jia, D., Li, F., and Wei, T. (2014a). Nonstructural protein NS4 of *Rice stripe virus* plays a critical role in viral spread in the body of vector insects. PLOS ONE *9*, e88636. https://doi.org/10.1371/journal.pone.0088636

Wu, Y., Davis, T.S., and Eigenbrode, S.D. (2014b). Aphid behavioral responses to virus-infected plants are similar despite divergent fitness effects. Entomol. Exp. Appl. *153*, 246–255. https://doi.org/10.1111/eea.12246

Xu, D., Zhong, T., Feng, W., and Zhou, G. (2016). Tolerance and responsive gene expression of *Sogatella furcifera* under extreme temperature stresses are altered by its vectored plant virus. Sci. Rep. *6*, 31521. https://doi.org/10.1038/srep31521

Xu, Y., Huang, L., Fu, S., Wu, J., and Zhou, X. (2012a). Population diversity of *Rice stripe virus*-derived siRNAs in three different hosts and RNAi-based antiviral immunity in *Laodelphgax striatellus*. PLOS ONE *7*, e46238. https://doi.org/10.1371/journal.pone.0046238

Xu, Y., Zhou, W., Zhou, Y., Wu, J., and Zhou, X. (2012b). Transcriptome and comparative gene expression analysis of *Sogatella furcifera* (Horváth) in response to *Southern rice black-streaked dwarf virus*. PLOS ONE *7*, e36238. https://doi.org/10.1371/journal.pone.0036238

Xu, Y., Wu, J., Fu, S., Li, C., Zhu, Z.R., and Zhou, X. (2015). *Rice stripe tenuivirus* nonstructural protein 3 hijacks the 26S proteasome of the small brown planthopper via direct interaction with regulatory particle non-ATPase subunit 3. J. Virol. *89*, 4296–4310. https://doi.org/10.1128/JVI.03055-14

Yang, N., Xie, W., Jones, C.M., Bass, C., Jiao, X., Yang, X., Liu, B., Li, R., and Zhang, Y. (2013). Transcriptome profiling of the whitefly *Bemisia tabaci* reveals stage-specific gene expression signatures for thiamethoxam resistance. Insect Mol. Biol. *22*, 485–496. https://doi.org/10.1111/imb.12038

Yang, X., Thannhauser, T.W., Burrows, M., Cox-Foster, D., Gildow, F.E., and Gray, S.M. (2008). Coupling genetics and proteomics to identify aphid proteins associated with vector-specific transmission of polerovirus (luteoviridae). J. Virol. *82*, 291–299.

Zamora, M., Méndez-López, E., Agirrezabala, X., Cuesta, R., Lavín, J.L., Sánchez-Pina, M.A., Aranda, M.A., and Valle, M. (2017). Potyvirus virion structure shows conserved protein fold and RNA binding site in ssRNA viruses. Sci. Adv. *3*, eaao2182. https://doi.org/10.1126/sciadv.aao2182

Zhang, F., Guo, H., Zheng, H., Zhou, T., Zhou, Y., Wang, S., Fang, R., Qian, W., and Chen, X. (2010). Massively parallel pyrosequencing-based transcriptome analyses of small brown planthopper (*Laodelphax striatellus*), a vector insect transmitting *Rice stripe virus* (RSV). BMC Genomics *11*, 303. https://doi.org/10.1186/1471-2164-11-303

Zhang, T., Luan, J.B., Qi, J.F., Huang, C.J., Li, M., Zhou, X.P., and Liu, S.S. (2012). Begomovirus-whitefly mutualism is achieved through repression of plant defences by a virus pathogenicity factor. Mol. Ecol. *21*, 1294–1304. https://doi.org/10.1111/j.1365-294X.2012.05457.x

Zhang, T., Feng, W., Ye, J., Li, Z., and Zhou, G. (2018). Metabolomic changes in *Sogatella furcifera* under *Southern rice black-streaked dwarf virus* infection and temperature stress. Viruses *10*, 344. https://doi.org/10.3390/v10070344

Zhang, Z., Zhang, P., Li, W., Zhang, J., Huang, F., Yang, J., Bei, Y., and Lu, Y. (2013). *De novo* transcriptome sequencing in *Frankliniella occidentalis* to identify genes involved in plant virus transmission and insecticide resistance. Genomics *101*, 296–305. https://doi.org/10.1016/j.ygeno.2013.02.005

Zheng, Y.X., Chen, C.C., and Jan, F.J. (2011). Complete nucleotide sequence of *Capsicum chlorosis virus* isolated from *Phalaenopsis* orchid and the prediction of the unexplored genetic information of tospoviruses. Arch. Virol. *156*, 421–432. https://doi.org/10.1007/s00705-010-0874-y

Zhou, F., Pu, Y., Wei, T., Wei, T., Liu, H., Deng, W., Wei, C., Ding, B., Omura, T., and Li, Y. (2007). The P2 capsid protein of the nonenveloped *Rice dwarf* phytoreovirus induces membrane fusion in insect host cells. Proc. Natl. Acad. Sci. U.S.A. *104*, 19547–19552. https://doi.org/10.1073/pnas.0708946104

Zhu, J., Jiang, F., Wang, X., Yang, P., Bao, Y., Zhao, W., Wang, W., Lu, H., Wang, Q., Cui, N., *et al.* (2017). Genome sequence of the small brown planthopper, *Laodelphax striatellus*. Gigascience 6, 1–12. https://doi.org/10.1093/gigascience/gix109

Zhu, Y., Li, H., Bhatti, S., Zhou, S., Yang, Y., Fish, T., and Thannhauser, T.W. (2016). Development of a laser capture microscope-based single-cell-type proteomics tool for studying proteomes of individual cell layers of plant roots. Hortic. Res. *3*, 16026. https://doi.org/10.1038/hortres.2016.26

Ziebell, H., Murphy, A.M., Groen, S.C., Tungadi, T., Westwood, J.H., Lewsey, M.G., Moulin, M., Kleczkowski, A., Smith, A.G., Stevens, M., *et al.* (2011). *Cucumber mosaic virus* and its 2b RNA silencing suppressor modify plant-aphid interactions in tobacco. Sci. Rep. *1*, 187. https://doi.org/10.1038/srep00187

Ziegenhain, C., Vieth, B., Parekh, S., Hellmann, I., and Enard, W. (2018). Quantitative single-cell transcriptomics. Brief. Funct. Genomics *17*, 220–232. https://doi.org/10.1093/bfgp/ely009

Advances in Tetravirus Research: New Insight into the Infectious Virus Life Cycle and an Expanding Host Range

7

Rosemary Ann Dorrington*, Meesbah Jiwaji, Janet Awino Awando and Mart-Mari de Bruyn

Department of Biochemistry and Microbiology, Rhodes University, Grahamstown, South Africa.

*Correspondence: r.dorrington@ru.ac.za

https://doi.org/10.21775/9781912530083.07

Abstract

Tetraviruses are a group of relatively unknown small RNA viruses with particles that display a characteristic $T=4$ capsid architecture. Tetraviruses are classified into three families, the *Alphatetraviridae, Permutotetraviridae* and *Carmotetraviridae,* according to the divergent characteristics of their respective viral replicases. Tetraviruses generally infect the larvae of lepidopteran insect species, many of which are important agricultural pests and, until recently, were thought to have an unusually narrow host range and tissue tropism. The development of experimental systems for studying the viral infectious life cycle in tissue culture has permitted the extension of the virus host range to mammalian cells and plants. This chapter will review recent advances in the understanding of the biology of tetraviruses, highlighting new information on the expression and functional characterization of viral proteins and the development of biological systems for elucidating the molecular mechanisms of infection, viral replication and host range.

Introduction

Tetraviruses are small, RNA viruses that infect lepidopteran (butterflies and moths) insects, notably Saturniid, Limacodid and Noctuid moths, many of which are important agricultural pests. Tetravirus particles are characterized by the unique $T=4$ quasi-symmetry of their capsids (the name derived from the Greek *téttares*, which means 'four') encapsidating one or two single-stranded (ss), positive sense (+) genomic RNAs. The viral genome may encode up to four genes translated directly into proteins that may be processed further into smaller, functionally distinct peptides (Dorrington *et al.*, 2011; Mendes *et al.*, 2015).

The first described tetravirus was isolated from diseased larvae of the pine emperor moth (*Nudaurelia cytherea capensis*) in South Africa (Hendry *et al.*, 1968). The larvae were infected with at least five structurally distinct small RNA viruses including *Nudaurelia capensis β virus* (NβV). NβV was the most plentiful and extensively characterized of these viruses and was found to have a single coat protein of approximately 60 kDa and a (+) ssRNA genome (Struthers and Hendry, 1974; Juckes, 1979). Structural analysis of the NβV particles subsequently revealed a new type of virus capsid with unusual $T=4$ icosahedral symmetry (Finch *et al.*, 1974) and this virus became the type-member of a new group of viruses named the Nudaurelia β-like viruses (Moore *et al.*, 1981; Matthews, 1982), later classified as the *Tetraviridae* family (Francki *et al.*, 1991).

For more than 35 years, tetravirus research focused on the structure and assembly of their unique capsids, the molecular characterization of their genomes and on developing their potential as biocontrol agents for important agricultural pests (Christian *et al.*, 1993, 2005; Hanzlik *et al.*, 1999). However, research on tetravirus pathogenicity and replication biology lagged behind due to a lack of tissue culture cell lines susceptible to infection and difficulties in establishing laboratory colonies of most of the host species. The discovery of *Providence virus* (PrV) as a persistent infection in a *Helicoverpa zea* midgut cell line (Pringle *et al.*, 2003) provided new opportunities for studying tetravirus replication and virus–host interactions *in vivo*. This chapter will review recent advances in the understanding of the biology of tetraviruses, highlighting new information on the expression and functional characterization of viral proteins and the development of biological systems for elucidating the molecular mechanisms of infection, viral replication and host range.

Taxonomy

The distinguishing feature of all tetraviruses is the characteristic $T=4$ icosahedral symmetry of their capsids. Tetraviruses are generally identified by the size of their capsids (approximately 40 nm in diameter) and the presence of major and minor capsid proteins of 58–64 kDa and 6–8 kDa in size, respectively (Dorrington *et al.*, 2011). These common structural features, together with serological relationships led to their classification within a single family, the *Tetraviridae*, (Francki *et al.*, 1991) comprising two genera according to their genome organization, namely *Betatetravirus* with monopartite genomes and *Omegatetravirus*, which have bipartite genomes (Fauquet *et al.*, 2005). Most of the betatetraviruses, with the exception of *Pseudoplusia includens virus* (PiV), were serologically related to some degree (Reinganum and Scotti, 1976; Reinganum *et al.*, 1978; Greenwood and Moore, 1981; Chao *et al.*, 1983; Hendry and Agrawal, 1994). However, this was not the case for the omegatetraviruses (Hanzlik *et al.*, 1993), despite significant sequence conservation between their capsid proteins (Hanzlik *et al.*, 1995; Yi *et al.*, 2005), highlighting the limitations of using serological criteria for distinguishing between tetravirus species.

Phylogenetic analyses revealed that the RNA-dependent RNA polymerase (RdRp) domains of tetravirus replicases are clustered into three distinct groups, each belonging to a different superfamily of RNA viral replicases (Walter *et al.*, 2010). The permuted RdRp domains of *Thosea asigna virus* (TaV) and *Euprosterna elaeasa virus* (EeV) form a group distantly related to the picorna-like birnaviruses (Gorbalenya *et al.*, 2002; Zeddam *et al.*, 2010), while the alpha-like RdRps of NβV and *Helicoverpa armigera stunt virus* (HaSV) form a second, distinct cluster. The PrV replicase is isolated from other tetraviruses, being more closely related to the tombusvirus, *Pelargonium chlorotic ring pattern virus* (PCRPV), and other carmo-like plant viruses (Walter *et al.*, 2010; Dorrington *et al.*, 2011). Therefore, based upon the diverse characteristics of their viral replicases, the tetraviruses were reclassified into three new families, namely the *Permutotetraviridae*, the *Alphatetraviridae* and the *Carmotetraviridae* (ICTV, 2011). There are currently 14 known tetravirus species, the majority of which are classified within the *Alphatetraviridae* family (Table 7.1). With the exception of NβV, the remaining viruses within the *Betatetravirus* genus of this family have been classified based upon their monopartite genome organization and not on the characteristics of their viral replicase. Thus, some of these viruses may well be reclassified into the *Permutotetraviridae or Carmotetraviridae* families in the future, when their genome sequence becomes available.

The tetravirus capsid

Tetravirus virions have non-enveloped, spherical particles exhibiting $T=4$ icosahedral shell quasi-symmetry, ranging in diameter from 38–41 nm (Dorrington *et al.*, 2011). The particles assemble as procapsids composed of 240 identical capsid protein precursor (CP) subunits that adopt different conformations arranged in four quasi-equivalent positions in the mature virus particle (Munshi *et al.*, 1996). During maturation, autoproteolytic cleavage of the CP (α subunit) at its C-terminus results in the production of the β and γ subunits (Agrawal and Johnson, 1995; Hanzlik *et al.*, 1995; Gordon *et al.*, 1999; Pringle *et al.*, 1999, 2003). The internal helical domain (N-terminus and γ subunit) of the CP interacts with viral RNA forming the molecular switch that determines the $T=4$ icosahedral symmetry of the virus particle. The

Table 7.1 Taxonomic classification of known tetraviruses

Virus species	Acronym	Host family	Geographical location	GenBank Accession number	References
Family: *Alphatetraviridae*					
Genus: *Omegatetravirus*					
Nudaurelia capensis ω virus[1]	NωV	*Saturniidae*	South Africa	[S43937.1]	Hendry *et al.* (1985); du Plessis *et al.* (2005)
Helicoverpa armigera stunt virus	HaSV	*Noctuidae*	Australia	[KX423453.1]; [L37299]	Hanzlik *et al.* (1993); Hanzlik *et al.* (1995); Gordon *et al.* (1995)
Dendrolimus punctatus Tetravirus	DpTV	*Lasiocampidae*	China	[AY594352]; [AY594353]	Yi *et al.* (2005)
Genus: *Betatetravirus*					
Nudaurelia capensis β virus[1]	NβV	*Saturniidae*	South Africa	[AF102884]	Juckes *et al.* (1973); Struthers and Hendry (1974); Gordon *et al.* (1999)
Antheraea eucalypti virus[2]	AeV	*Saturniidae*	Australia	NA	Grace and Mercer (1976)
Darna trima virus	DtV	*Lymacodidae*	Malaysia	NA	Reinganum *et al.* (1978); King and Moore (1984)
Dasychira pudibunda virus *Calliteara pudibunda virus*[3]	DpV CpV	*Lymantriidae*	UK	NA	Reinganum *et al.* (1978); King and Moore (1984)
Philosamia cynthia × ricini virus	PxV	*Saturniidae*	UK	NA	Reinganum *et al.* (1978)
Pseudoplusia includens virus	PiV	*Noctuidae*	USA	NA	Chao *et al.* (1983)
Trichoplusia ni virus	TnV	*Noctuidae*	USA	NA	Morris *et al.* (1979)
Family: *Permutotetraviridae*					
Genus: *Alphapermutotetravirus*					
Thosea asigna virus[1]	TaV	*Lymacodidae*	Malaysia, USA	[AF282930], [AF062037]	Pringle *et al.* (1999)
Euprosterna elaeasa virus	EeV	*Zygaenidae*	Peru	[AF461742]	Gorbalenya *et al.* (2002); Zeddam *et al.* (2010)
Family: *Carmotetraviridae*					
Genus *Alphacarmotetravirus*					
Providence virus[1]	PrV	Noctuidae	USA	[AF062037]	Pringle *et al.* (2003); Walter *et al.* (2010)

[1]Type strain for the genus
[2]Serologically indistinguishable from NβV (Grace and Mercer, 1976)
[3]Host renamed *Calliteara pudibunda*.

capsid shell comprises a β-barrel and the external immunoglobulin-like domain believed to function in receptor binding during infection of host cells (Munshi *et al.*, 1996; Helgstrand *et al.*, 2004). In the alphatetraviruses, autoproteolytic cleavage releases the γ peptide, resulting in stabilization of the mature virus particle, which is important for infectivity (Munshi *et al.*, 1996; Taylor *et al.*, 2002; Helgstrand *et al.*, 2004; Domitrovic *et al.*, 2012). In TaV and EeV (*Permutotetraviridae*) and PrV (*Carmotetraviridae*) the CP undergoes additional co-translational processing at its amino terminus prior to assembly

(Pringle *et al.*, 1999, 2003; Gorbalenya *et al.*, 2002; Zeddam *et al.*, 2010).

The molecular mechanism of the assembly and maturation of alphatetravirus capsids is well characterized (Canady *et al.*, 2000; Banerjee *et al.*, 2010; Tang *et al.*, 2014). At neutral pH, NωV and HaSV virus-like particles (VLPs) purified as immature procapsids undergo maturation *in vitro* at pH 5.0 (Canady *et al.*, 2000; Tomasicchio *et al.*, 2007). Maturation of NωV capsids involves conformational changes promoting autoproteolytic cleavage of CP, releasing the γ peptide to stabilize the molecular switch that supports the flat contacts of the virus particle (Canady *et al.*, 2001). Not all subunits are cleaved at the same rate and the γ peptides of the A and D subunits positioned at the five-fold and three-fold axes, respectively, are stabilized and cleaved within 30 minutes of incubation at pH 5.0. Cleavage of the B and C subunits follows more slowly (Matsui *et al.*, 2010). Lowering the pH is proposed to result in protonation of glutamate and aspartate side chains that facilitate a conformational change and shrinking of the capsid shell enabling autoproteolytic cleavage of the γ peptide. The conformational change is reversible, provided autoproteolytic cleavage of the γ peptide is inhibited (Taylor *et al.*, 2002). How the conformational changes observed *in vitro* relate to the infectious life cycle of tetraviruses will be addressed below.

Genome organisation and regulation of viral gene expression

Alphatetraviridae

Alphatetraviruses may have either monopartite (Genus: *Betatetravirus*) or bipartite (Genus: *Omegatetravirus*) genomes (Fig. 7.1). The genomic RNAs are capped at their 5′ ends and not polyadenylated at their 3′ ends, which instead, terminate at their 3′ ends with tRNA-like structures (Gordon *et al.*, 1995; Hanzlik *et al.*, 1995, du Plessis *et al.*, 2005; Yi *et al.*, 2005). In plant viruses, tRNA-like structures play a role in promoting replication of the minus strand RNAs, encapsidation of viral RNAs and maintaining the integrity of the 3′ end of the viral RNA (Dreher, 2009). They are also involved in the recoding of host translational machinery and evading the host antiviral response (Ariza-Mateos and Gómez, 2017).

The type member of the *Betatetravirus* genus, NβV, encodes a large replicase (REP) open reading frame (ORF) that almost entirely overlaps the CP coding sequence located at the 3′ end of the gRNA (Fig. 7.1). The CP is translated from a subgenomic (sg) RNA originating just upstream of the CP coding sequence (Gordon *et al.*, 1999). In omegatetraviruses the smaller RNA2 encodes two ORFs. At the 5′ end is the non-structural p17 ORF, which overlaps with the start of the second ORF encoding the CP, which ranges between 70 and 71 kDa in size (Hanzlik *et al.*, 1995; Mendes *et al.*, 2015).

The omegatetravirus REP is translated from RNA1, which also encodes three additional small ORFs (p11, p15 and p8), all three of which overlap with the 3′ end of the REP ORF in HaSV (Fig. 7.1). This arrangement is not conserved in the other sequenced omegatetravirus, *Dendrolimus punctatus tetravirus* (DpTV) (Yi *et al.*, 2005). In both HaSV and DpTV the p11 stop codon is immediately upstream of and in-frame with the AUG start of p15. In HaSV the p8 AUG is three nucleotides downstream of the p15 stop codon (de Bruyn *et al.*, 2016), while the DpTV p8 is in a different reading frame and is 99 nucleotides downstream of the stop codon of p15 (Yi *et al.*, 2005).

The alphatetravirus replicases include three conserved functional domains, namely methytransferase (MT), helicase (HEL) and RdRp domains, in an N- to C-terminal modular arrangement (Fig. 7.1) that are characteristic of all members of the alpha-like virus superfamily (van der Heijden and Bol, 2002). The RdRp domain of the DpTV replicase has been expressed in *E. coli* and shown to initiate primer-independent synthesis of viral RNA *in vitro* (Zhou *et al.*, 2008). Although there are a number of insect-borne animal (alphaviruses) and plant viruses (marafiviruses) within the alpha-like virus family, the alphatetraviruses represent the only insect-specific family within the group (van der Heijden and Bol, 2002). Another feature of the alphatetravirus REPs that is unusual among the alpha-like viruses, is the apparent lack of co- or post-translational processing, a characteristic shared with a small group of plant viruses that includes the *Foveavirus* and *Potexvirus* genera (van der Heijden and Bol, 2002).

Of the alphatetraviruses, the best studied is HaSV, which employs diverse strategies to modulate expression of viral genes including

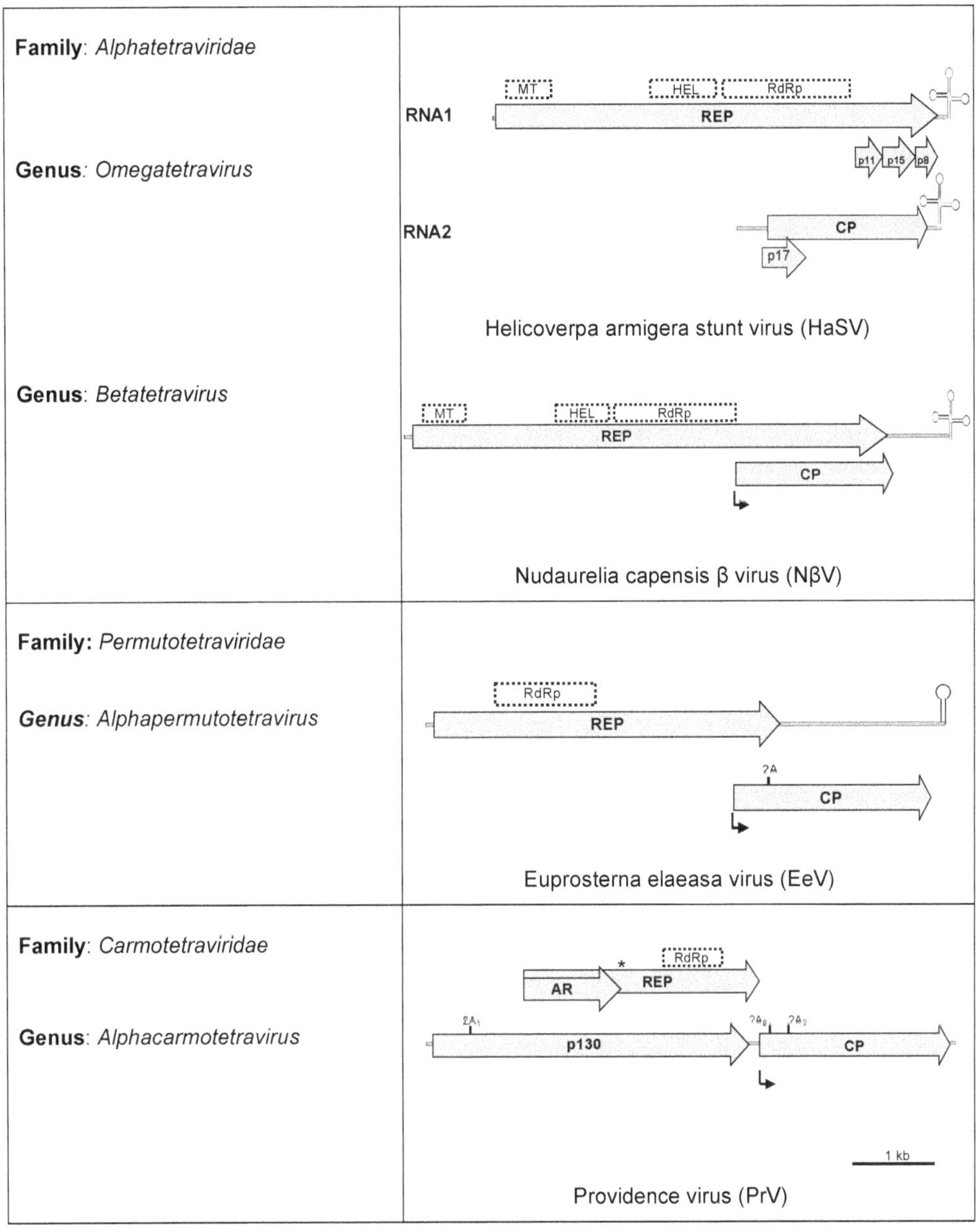

Figure 7.1 Genome organization of tetraviruses representative of the *Alphatetraviridae*, *Permutotetraviridae* and *Carmotetraviridae* families. Viral genes encoding the replicase (REP), Capsid Protein precursor (CP) and replication accessory protein of PrV (p40) are indicated. Putative tRNA-like (HaSV and NβV) and pseudoknot-like (EeV) structures are shown at the 3′ ends of the genomic RNAs. The positions of 2A-like processing sequences (2A) and read-through stop codon (*) are indicated as are the functional domains of the viral replicases (HEL, helicase; MT, methyltransferase; RdRp, RNA-dependent RNA polymerase). The bent arrow shows the start of the subgenomic RNAs of the monopartite viruses (NβV, EeV and PrV) from which the viral CP is translated. GenBank Accession numbers: KX423453.1 (HaSV RNA1), L37299 (HaSV RNA2), AF102884 (NβV), AF461742 (EeV) and AF062037 (PrV).

temporal regulation of the replication of the two viral genomic RNAs, translational control and regulation of protein turnover. In HaSV-infected *Helicoverpa armigera* larvae, RNA1 can be detected three days after infection, while small amounts of RNA2 appear after a further two days (Bawden *et al.*, 1999). This implies that RNA1, encoding the REP, is replicated during the early stages of infection and that replication of RNA2 occurs during the later stages of infection. This would ensure that expression of the CP and p17, encoded by RNA2, would commence later in the virus life cycle. This phenomenon has been well documented in the nodavirus, *Flock House virus* (FHV). Nodaviruses are small, non-enveloped insect RNA viruses that are structurally related to tetraviruses and also have a bipartite genome. Nodavirus RNA1 encodes the viral REP and two small ORFs (B1 and B2) that overlap with the 3′ end of the REP ORF (Thiéry *et al.*, 2011). B1 and B2 are translated from a sgRNA (RNA3) originating from the 3′ end of RNA1 (Eckerle *et al.*, 2003). Mutations in B2 suppress the replication of *Nodamura virus* (Johnson *et al.*, 2004) and the FHV B2 protein has been shown to inhibit Dicer function in infected cells (Galiana-Arnoux *et al.*, 2006). FHV RNA3 is required for promoting replication of RNA2 during the later stages of infection (Eckerle *et al.*, 2003).

Alphatetraviruses also encode small ORFs that overlap with the 3′ end of REP on RNA1 that could play a similar role to nodavirus B2 in the infectious virus life cycle (discussed in detail below). The three ORFs are expressed in the midgut cells of HaSV-infected *H. armigera* larvae as 11 kDa (p11) and 8 kDa (p8) proteins, together with a 34 kDa polyprotein comprising p11, p15 and p8 (de Bruyn *et al.*, 2016). Given the organization of the three ORFs in a tandem in-frame arrangement, separated by a UGA stop (p11 and p15) and UAA stop followed by GGG (p15 and p8), it is likely that the translation of these three ORFs involves an as yet uncharacterised read-through stop mechanism (de Bruyn *et al.*, 2016).

In HaSV the expression of the non-structural p17 is down-regulated relative to CP via translation regulation and the control of protein turnover (Mendes *et al.*, 2015). The p17 ORF is found upstream of and overlapping with the 5′ end of the CP ORF of RNA2 in all three omegatetraviruses for which genome sequence is available (Hanzlik *et al.*, 1995; du Plessis *et al.*, 2005; Yi *et al.*, 2005). In HaSV-infected *H. armigera* larval midgut tissues p17 is expressed at levels several orders of magnitude lower than that of the viral CP (Mendes *et al.*, 2015). The most likely explanation for low levels of expression of p17 is the poor context of the translational start (AUUUAUGA in HaSV and conserved in DpTV and NωV). This is in contrast to the translational context of the CP ORF downstream of p17 as compared with the consensus for invertebrates (A/CAAA/CAUGG, Cavener, 1987). In addition, high level of PEST amino acids predicted to be present in all three omegatetravirus p17 translation products are thought to promote rapid turnover of the p17 protein *in vivo* (Mendes *et al.*, 2015).

Permutotetraviridae

The monopartite genomes of EeV and TaV encode two overlapping ORFs, with the viral replicase located at the 5′ end and the CP at the 3′ end (Gordon *et al.*, 1999; Gorbalenya *et al.*, 2002; Zeddam *et al.*, 2010). Bioinformatic analysis suggests the presence of a pseudoknot at the 3′ ends of the EeV and TaV genomic RNAs, which may be involved in the regulation of genome replication in these two permutotetraviruses or they may play a role in preventing degradation of the genomic RNA by nucleases (Zeddam *et al.*, 2010). The replicase ORF of EeV and TaV overlaps CP by 536 nt and 529 nt, respectively in contrast to the NβV replicase ORF, which is significantly larger (Fig. 7.1).

As with the alphatetravirus, NβV, the permutotetravirus CP is translated from a sgRNA originating just upstream of the CP coding sequence (Gordon *et al.*, 1999; Pringle *et al.*, 1999). The TaV and EeV CP ORFs encode proteins with a predicted molecular mass of 82.5 kDa (Pringle *et al.*, 1999; Zeddam *et al.*, 2010). There is a 2A-like processing site located near the N-terminus of the TaV CP, the activity of which, in TaV, results in the production of a 17 kDa amino-terminal peptide (p17) as well as the CP of 65 kDa (Pringle *et al.*, 1999). The CP is then autoproteolytically cleaved at its C-terminus during particle maturation to produce the 58 kDa and 7 kDa proteins found in mature capsids (Pringle *et al.*, 1999). This 2A-like site is conserved in the same position in the EeV capsid protein precursor ORF (Zeddam *et al.*, 2010).

The replicases of TaV and EeV are structurally related to those of two members of the double-stranded (ds)RNA birnaviruses and *Drosophila A virus* (DAV), and are distantly related to members of the picorna-like virus superfamily (Gorbalenya *et al.*, 2002; Ambrose, *et al.*, 2009; Zeddam *et al.*, 2010). This distinct cluster of replicases encode an internally-permuted C-A-B arrangement of the canonical A, B and C motifs found in the palm subdomain of all template-dependent polynucleotide polymerases (Poch *et al.*, 1989; Gorbalenya *et al.*, 2002; Ambrose *et al.*, 2009). A putative nucleotidylation (VPg) signal is present upstream of the TaV and EeV RdRp domains, a feature shared with the *Birnaviridae* and absent from other tetravirus replicases (Zeddam *et al.*, 2010). The TaV and EeV replicases both lack discernible MT and HEL domains (Gorbalenya *et al.*, 2002), which is unusual for picorna-like replicases that usually encode a HEL domain (van der Heijden and Bol, 2002).

The structure of the TaV RdRp domain encoded within an N-terminal peptide was solved at 2.1 Å resolution (Ferrero *et al.*, 2012) and is to date the only tetravirus replicase RdRp domain for which the structure is known. Data obtained from the crystal structure of the TaV RdRp domain implies that the TaV REP likely dimerizes via a region within the RdRp domain (Ferrero *et al.*, 2015). These data confirm the bioinformatic predictions of structural conservation with the RdRp domains of the *Birnaviridae* but interestingly also revealed similarities with an extended loop known to be involved in the initiation of RNA replication in the RdRp domains of flaviviruses and in bacteriophage ϕ6 suggesting an evolutionary relationship between these distantly related virus groups (Ferrero *et al.*, 2015).

Carmotetraviridae

PrV has a monopartite genome encoding three viral genes. The CP ORF is located at the 3′ end of the genome and upstream of CP is an ORF encoding the viral REP, which ends with a stop codon (UGA) four nucleotides upstream of the AUG start of the CP coding sequence. As in the other monopartite tetraviruses, the (p81) is translated from an sgRNA originating just upstream of the CP AUG (Pringle *et al.*, 2003). There are two 2A-like processing sites (PrV-2A$_2$ and PrV-2A$_3$) located near the amino terminus of the CP coding sequence, the activity of which result in the translation of two small peptides of 7 kDa and 8 kDa along with the capsid α protein of 68 kDa (Pringle *et al.*, 2003). As with all other tetraviruses, p68 also undergoes assembly-dependent autoproteolytic cleavage at its carboxyl terminus during particle maturation to produce the β (p62) and γ (p6) subunits, respectively (Pringle *et al.*, 2003). In addition to REP and CP, PrV encodes a third, large ORF (p130) at its 5′ end and upstream of the replicase ORF, which is not found in any other tetraviruses (Fig. 7.1). A third 2A-like processing site is present in the N-terminus of p130, the activity of which is predicted to result in the translation of two products of 17 kDa and 113 kDa (Walter *et al.*, 2010).

The PrV REP ORF is distinguished from other tetravirus replicases in that it encodes a read-through stop codon that results in the translation of a smaller product (p40) in addition to REP, designated p104 (Walter *et al.*, 2010). The only similarity between the replicases of PrV and those of the other tetraviruses is the absence of MT and HEL domains, as is the case with TaV and EeV. This along with the sequence similarity of the RdRp domain with the replicases of the tombus- and umbraviruses [both (+)ssRNA plant viruses] places PrV within the carmo-like supergroup. The PrV RdRp domain carries yet another unusual arrangement of the canonical A-B-C palm subdomain, with an additional A and (poorly conserved) B motif spatially arranged in a similar manner to the permuted RdRps of TaV and EeV (Walter *et al.*, 2010).

The PrV REP ORF contains an in-frame Group 1 read-through stop codon following amino acid 359. This results in leaky read-through by the ribosome (for a review on viral ribosomal recoding, see Dabrowski *et al.*, 2015) and the production of a shorter translation product (p40) in addition to the full-length REP (p104) at a ratio of approximately 10:1 (Walter *et al.*, 2010). The PrV signal, which shares little homology with read-through stop signals found in insect and animal viruses, was the first example of a Group 1 read-through signal that is active in an animal virus, raising interesting questions about the origin of PrV.

The use of co-translational modification via 2A-like processing is a strategy shared by both permutotetraviruses and PrV with biologically active 2A-like sites at the amino-termini of the CP of TaV (TrV-2A), EeV (EeV-2A), two in the PrV (PrV-2A$_2$, PrV-2A$_3$) CP as well as in the PrV p130 (PrV-2A$_1$) (Pringle *et al.*, 2001, Luke *et al.*, 2008; Walter *et al.*, 2010). In contrast to PrV-2A$_1$ and PrV-2A$_3$, processing at the PrV-2A$_2$ site is less efficient *in vitro* (Luke *et al.*, 2008). These data are consistent with *in vitro* translation experiments that show translation products of approximately 15 kDa and 68 kDa, corresponding to a peptide with a predicted molecular mass of 13.2 kDa (p13) resulting from processing only at PrV-2A$_3$ and the capsid protein precursor p68 (Walter *et al.*, 2010). The 13 kDa peptide does not cross-react with antibodies raised against PrV virus particles nor does it co-purify with wild-type PrV virus particles, as is the case with the N-terminal 2A-like processing product of TaV, p17 (Pringle *et al.*, 1999, 2001, 2003).

2A sequences were first identified in the *Picornaviridae*, but they are present in other mammalian viruses (e.g. human type C rotaviruses) as well as a wide range of insect viruses (Luke *et al.*, 2008) providing the viruses with yet another mechanism to translate multiple peptides with potentially different functions, from a single ORF (Roulston *et al.*, 2016). The mechanism of 2A processing has been well characterized *in vitro* and a *Saccharomyces cerevisiae* experimental system has been developed to identify host-cell (ribosomal translation) factors that interact with these 2A sequences (Doronina *et al.*, 2008a,b).

Elucidating the infectious virus life cycle

For many years, the absence of tissue culture cell lines permissive to tetravirus replication and the difficulties associated with studies on whole animals (Bawden *et al.*, 1999; Brooks *et al.*, 2002) meant that most research on the tetraviruses was focused on the structure, assembly and maturation of the viral capsid, molecular characterization of their genomes and their potential uses as biocontrol agents. Tetraviruses were believed to infect only lepidopteran insects, with host ranges confined to a few closely related species replicating only in larval midgut cells (Fauquet *et al.*, 2005). The lack of insect tissue culture cell lines supporting tetravirus replication was attributed to the absence of host factors required for replication of HaSV and the lack of suitable receptors to mediate virus binding and cell entry (Bawden *et al.*, 1999). The discovery of PrV as a persistent infection in a *H. zea* midgut cell line (Pringle *et al.*, 2003) and recent breakthroughs in establishing alphatetravirus replication in tissue culture (Penkler *et al* 2016), have dispelled this notion and resulted in significant advances in our understanding of the infectious viral life cycle.

Binding and entry

The capsids of alphatetraviruses are unique amongst non-enveloped viruses because they are stabilized under acidic conditions and release their membrane-lytic peptides under alkaline conditions (Canady *et al.*, 2002; Domitrovic *et al.*, 2012). Precisely how the unusual biophysical properties relate to the infectious viral life cycle was revealed in a ground-breaking study in which conformational changes in the HaSV particles induced by alkaline pH promote binding to *Spodoptera frugiperda* (Sf9) cells. Binding was followed by the initiation of infection and RNA replication (Fig. 7.2) and the production of infectious HaSV particles in these cells (Penkler *et al.*, 2016). These results were confirmed *in vivo*, showing that an alkaline midgut pH was required for HaSV infection of *H. armigera* larvae. This study provided for the first time, an experimental system for studying the alphatetravirus infectious life cycle in cell culture.

Not all tetraviruses use an alkaline pH to induce the conformational changes required for binding and entry. The carmotetravirus, PrV, readily infects tissue culture cells in the neutral to slightly acidic conditions of tissue culture medium (Jiwaji *et al.*, 2016). The CP termini of PrV are structurally and functionally distinct from those of the alphatetraviruses: in contrast to the alphatetraviruses, the N-terminal and not the γ peptides of the PrV CP serves as the molecular switch. The arrangement of peptide termini in the PrV capsid shares an evolutionary relationship with nodaviruses (Speir *et al.*, 2010), in which infectivity is triggered by acidic pH conditions (Odegard *et al.*, 2010). HaSV and PrV both infect and replicate in the non-host *S. frugiperda* (Sf9) cell line (Fig. 7.2), which was derived from *S. frugiperda* ovarian tissue.

Virus replication

Tetraviruses replicate in the host cell cytoplasm in association with host cell membranes. Transient expression of HaSV REP fused to EGFP at its C terminus, in *S. frugiperda* Sf9 and mammalian (HeLa and BHK-21) cells, results in the subcellular localization of fluorescence to punctate cytoplasmic structures (Short *et al.*, 2010) that are associated with detergent-resistant membranes (Short and Dorrington, 2012). The punctate pattern is abolished when EGFP is fused to the N-terminus of the

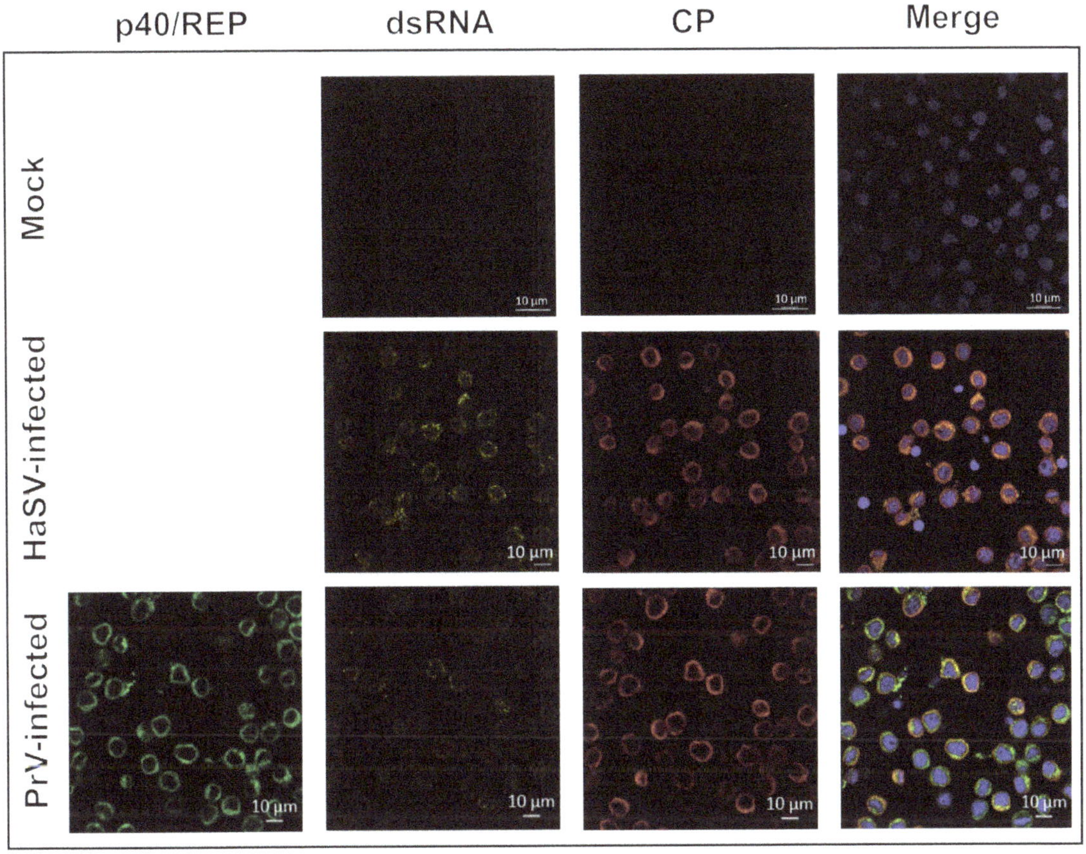

Figure 7.2 Subcellular localization of HaSV and PrV replication complexes in *Spodoptera frugiperda* (Sf9) cells. HaSV infection was carried out as described by Penkler *et al.* (2016) and the cells collected for immunofluorescence staining 8-hour post infection. PrV infection was carried out by incubating confluent Sf9 cells with filtered supernatant from *H. zea* MG8 cultures for 4 hours. Thereafter the supernatant was replaced with fresh growth medium and the cells incubated for 8h. Cells were probed with mouse monoclonal anti-dsRNA antibodies (English and Scientific Consulting; 10020500) and anti-mouse AF546 (Invitrogen; A11003) as described in Short *et al.* (2013). Viral replicase was stained with biotin-conjugated anti-PrV p40 and streptavidin AF488 (Invitrogen; S11223), while rabbit polyclonal anti-HaSV CP or anti-PrV CP antiserum and anti-rabbit AF647 (Invitrogen; A20170) were used to detect the CP. All images represent 1 μm optical slices taken using a Zeiss LSM 780 laser scanning confocal microscope using a 63, 0.75 NA objective. Scale bar: 10 μm.

replicase, implying a function for the N-terminus in subcellular localization of the replicase (Short *et al.*, 2010). Two domains, located at the N-terminus and within the C-terminal half of the HaSV replicase are both required for targeting the HaSV REP–EGFP fusion to punctate structures within the cytoplasm (Short and Dorrington, 2012).

Immunofluorescence microscopy studies have shown that PrV REP localizes to similar punctate structures in the cytoplasm of persistently infected *H. zea* MG8 cells (Short and Dorrington, 2012) and also in *S. frugiperda* cells expressing EGFP–REP fusion proteins (Short *et al.*, 2013). Co-localization of the PrV REP with replicating viral RNA confirmed that these punctate structures are the site of viral replication in PrV-infected infected *H. zea* MG8 (Short *et al.*, 2013) and *S. frugiperda* cells (Fig. 7.2) and that as with HaSV, virus replication occurs in association with detergent-resistant membranes (Short *et al.*, 2012).

Initially, studies with HeLa cells expressing HaSV REP-EGFP showed that EGFP fluorescence partially co-localized with CD63, a marker that is cycled between the endocytic and secretory pathways (Kobayashi *et al.*, 2000; Short *et al.*, 2010). Subsequently, recombinant organellar markers for insect cells (Maroniche *et al.*, 2011) were used to show that both the PrV and HaSV REP co-localize with vesicles originating from the Golgi and secretory vesicles (Short *et al.*, 2013; de Bruyn *et al.*, 2016). Thus, both PrV and HaSV share the same site of replication and replicate in association with detergent-resistant membranes of vesicles derived from the Golgi and secretory pathway. Given the differences between their replicases and the presence of additional virus-specific non-structural proteins, it is likely that each virus uses a distinct mechanism to initiate infection and assemble the virus replication factory.

As mentioned previously, the PrV REP ORF encodes a read-through stop codon, the activity of which results in the production of a truncated accessory replication protein, p40 and the full-length REP (p104). p40 and REP co-localize with dsRNA and are involved in the establishment of the replication complex (Short *et al.*, 2013). Located within the N-terminal region of REP and within p40 are membrane spanning domains that anchor both proteins in the target membrane. A protein–protein interaction domain facilitates self-interaction between p40 proteins and interaction with REP, which encodes the RdRp domain at its C-terminal end (Nakayinga, 2013). p40 and REP are trafficked to the site of replication where they interact with the target membranes and each other to form the replication complex (Short *et al.*, 2013). The function of the non-structural protein, p130, is unclear. However, preliminary data suggest that expression of p130 is required for translation of the PrV REP, implying that p130 is likely involved in initiation of viral replication (J.A. Awando and R.A. Dorrington, unpublished results).

There is biochemical evidence that suggests that the HaSV REP also self-interacts, via a region located near to or within the RdRp domain (Short and Dorrington, 2012). The assembly of a replication complex in HaSV-infected cells likely also involves accessory proteins in addition to the REP. The translation products of two of the three small ORFs located at the 3′ end of HaSV RNA1 co-localize with REP and are trafficked to secretory vesicles providing strong evidence that they play a role in the assembly of the viral replication complex (de Bruyn *et al.*, 2016). Precisely what their function is in the viral life cycle is unclear but over-expression of all three ORFs in a recombinant baculovirus appears to be toxic to *S. frugiperda* cells (de Bruyn *et al.*, 2016).

Packaging of tetraviral RNAs

There have been several apparently conflicting reports whether tetravirus VLPs are produced in the absence of viral replication package viral RNAs. Agrawal and Johnson (1995) demonstrated the assembly of VLPs in insect cells with evidence that these VLPs packaged viral capsid (VCAP) mRNA. However, in other studies, omegatetravirus VLPs produced by recombinant baculovirus expression of VCAP do not specifically encapsidate viral RNAs (Taylor, 2003; Gordon and Waterhouse, 2006; Routh *et al.*, 2012). This is also the case with VLPs produced in the yeast, *Saccharomyces cerevisiae* (Tomasicchio, 2007). These VLPs package mostly cellular RNAs as well as some mRNA expressing the capsid protein precursor. Gordon *et al.* (2001) reported the replication independent assembly of infectious HaSV particles in tobacco protoplasts. Their experimental system was based on the co-expression of HaSV RNA1 and RNA2 cDNAs fused at their 5′ ends to the (–1) nucleotide of the Cauliflower mosaic virus promoter and flanked at their 3′ ends by a *cis*-acting ribozyme, producing precise copies of viral RNA. In addition, the plant protoplasts were co-transfected with a third plasmid expressing HaSV VCAP message after which the protoplasts were fed to neonate *H. armigera* larvae. Evidence for assembly of infectious particles was the recovery of infectious HaSV from these larvae and detection of virus-like particles by immuno-electron microscopy, but the yields of virus particles were too low to confirm successful encapsidation of viral RNAs (Gordon *et al.*, 2001).

The same experimental system was used to demonstrate delivery of non-viral RNA encoding EGFP to larval midgut cells by recombinant HaSV VLPs assembled in tobacco protoplasts (Gordon and Waterhouse, 2006). This was achieved by transfecting protoplasts with plasmids expressing RNA1, VCAP and a third with RNA2 in which VCAP had been replaced with EGFP just downstream of the p17 stop codon. The presence of EGFP fluorescence in larval midgut cells up to 10 days after ingestion suggested replication of recombinant RNA2-EGFP transcripts in the midgut cells. However, the low level of VLP assembly in the system precluded analysis to show specific encapsidation of the viral RNAs (Gordon and Waterhouse, 2006).

The observation that p17 is expressed in infected insect cells and that the protein is present at low concentrations in wild-type virus particles purified form HaSV infected insects, has led to the proposal that p17 is involved in packaging of viral RNA during capsid assembly (Mendes *et al.*, 2015). To test this hypothesis, p17 was co-expressed with VCAP in *S. frugiperda* cells, the resulting VLPs were purified and analysed for the presence of viral RNA2. VLPs assembled in cells co-expressing VCAP and p17 contained viral mRNA, but when p17 was absent, no viral RNA could be detected in purified capsids. VLPs produced by recombinant baculoviruses package vRNA2 at detectable levels even in the absence of p17 and apparently excluded baculoviral transcripts. Deletion analysis showed that the 5′ and 3′ UTR sequences of HaSV RNA2 were not required for packaging (Mendes *et al.*, 2015). These data, together with those of Gordon and Waterhouse (2006), suggest that a sequence within the p17 ORF or immediately downstream of the p17 ORF may be required for packaging HaSV RNA2.

Are p17-like homologues present in other tetraviruses? Analysis of the amino acid sequence of p13, produced by 2A-like processing of the PrV VCAP, reveals a high incidence of PEST amino acid residues, raising the possibility that this peptide might have a similar function in facilitating packaging of viral RNAs by PrV particles. Similarly, a peptide of approximately 17 kDa is also produced by 2A-like processing of the TaV (and probably EeV) capsid protein precursor. The observation by Pringle *et al.* (1999) that the presence of TaV p17 appears to stabilize TaV VLPs suggests that this peptide may serve a similar function as the omegatetraviral p17 in the betatetraviruses TaV and EeV. There is no evidence for the presence of a p17-like ORF on the NβV genome sequence (Gordon *et al.*, 1999), which may either indicate an alternative packaging mechanism in this virus, or mutations in cDNA fragments used to assemble the NβV genome sequence.

Redefining the host range of tetraviruses

For many years, the most striking feature of the tetraviruses other than their distinctive capsid structure, was their narrow host range, often confined to a single species or selected species within one family (Dorrington *et al.*, 2011). This was a characteristic that made tetraviruses prime candidates for the development of biopesticides to control agricultural pests (Christian *et al.*, 1993; Hanzlik *et al.*, 1999). However, there are several exceptions to this rule. Trichoplusia ni virus and Dasychira pudibunda virus infect species in families other than that of their normal host (Greenwood and Moore, 1981). The host range of the omegatetravirus HaSV has been extended to several Noctuid species, including *H. armigera, H. zea, Heliothis virescens, Heliocheilus pallida,* (Subfamily *Heliothenae*) as well as *Earias huegeliana* and *Chrysodeixis argentifera* (Christian *et al.*, 2001) and recently, *S. frugiperda* (Penkler *et al.*, 2016).

HaSV infectivity requires the induction of conformational changes within the virus particle by an alkaline pH (Penkler *et al.*, 2016) suggesting that the requirement for an alkaline gut pH might be the barrier to infection that restricts HaSV and other alphatetraviruses such as NωV, with their conserved capsid structures and the shared biophysical properties of their VLPs to noctuid insects. The alkaline environment of the midgut of herbivorous insects is believed to be an adaptation to the chemical defence mechanisms of plants (Appel and Martin, 1990), which in turn has required their pathogens to evolve strategies to overcome some of the highest pH environments known in living systems.

There appear to be no such constraints on the host range of the carmotetravirus, PrV, which does not require an alkaline environment for infectivity. The distinct biophysical characteristics of its capsids allow PrV to infect and replicate in mammalian tissue culture cells (Fig. 7.3), including baby hamster kidney and human cancer cell lines (Jiwaji *et al.*, 2016; M. Jiwaji, unpublished data). Also unique to PrV is the ability to establish asymptomatic, persistent infections in both insect and mammalian cells (Pringle *et al.*, 2003; Jiwaji *et al.*, 2016). This implies conservation of the host cell receptor(s) used by PrV for binding and entry as well as host factors required to support virus replication in the invertebrate and vertebrate host cells. A PrV-like virus with 85% sequence identity was recently discovered in the guano of a female western barbastelle bat (*Barbastella barbastellus*) in Hungary (Kemenesi *et al.*, 2016). Western barbastelle bats are insectivorous with a predominantly lepidopteran diet (Zeale *et al.*, 2010), so the presence of PrV was most likely due to ingestion of PrV-infected prey (Kemenesi *et*

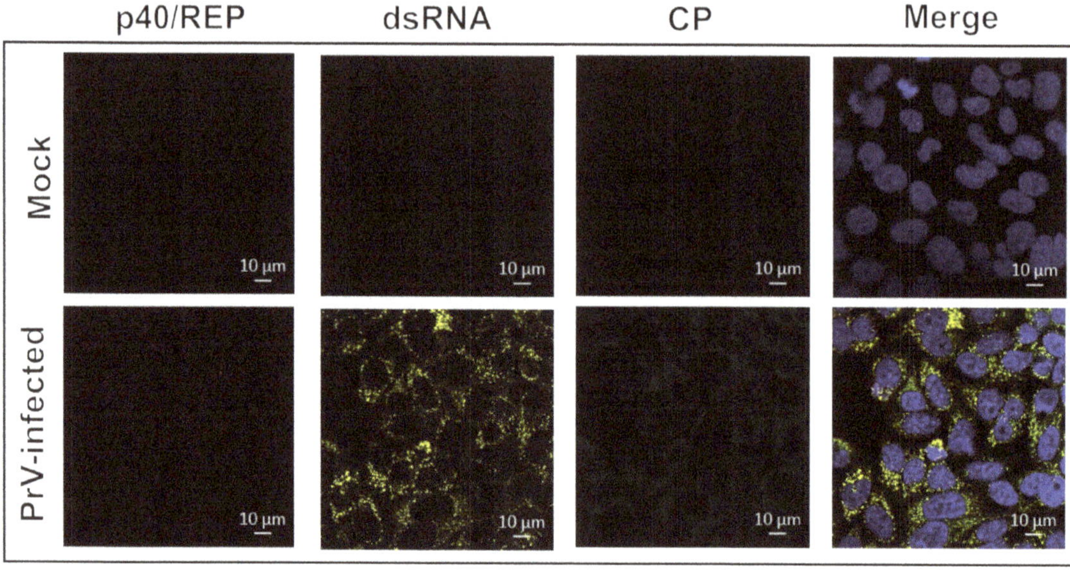

Figure 7.3 Replication of PrV in mammalian cells. PrV infections were carried out as described in Jiwaji *et al.* (2016). HeLa cells were probed with mouse monoclonal anti-dsRNA antibodies (English and Scientific Consulting; 10020500) and anti-mouse AF546 (Invitrogen; A11003) as described in Short *et al.* (2013). Viral replicase was stained with biotin-conjugated anti-PrV p40 and streptavidin AF488 (Invitrogen; S11223), while rabbit polyclonal anti-CP and anti-rabbit AF647 (Invitrogen; A20170) were used to detect PrV CP. All images represent 1 µm optical slices taken using a Zeiss LSM 780 laser scanning confocal microscope using a 63, 0.75 NA objective. Bar, 10 µm.

al., 2016). However, it is possible that the bat could have become infected with PrV.

PrV is the only tetravirus with a replicase that is more closely related to those of plant-infecting tombus- and umbraviruses, (Family: *Tombusviridae*). Not only is there conservation between their RdRp domains, but the translation strategy of PrV and tombusviruses is also conserved, with the presence of a read-through stop that results in the expression of a truncated accessory replication protein (Walter *et al.*, 2010). This has given rise to the hypothesis that PrV may have arisen via horizontal transmission from a plant to a phytophagous insect and recombination with a tetravirus in the insect gut. The observation that PrV replicates in plant cell-free extracts (Jiwaji *et al.*, 2016) raised the question of whether the virus may still be able to replicate in plants. PrV particles isolated from persistently infected *H. zea* MG8 cells were used to mechanically infect the leaves of *Vigna unguiculata* (cowpea) plants. After two weeks, *V. unguiculata* plants infected with PrV showed lesions and yellowing compared to uninfected control plants (Fig. 7.4A) and PrV virus purified from plants (Fig. 7.4B) was able to infect mammalian

Figure 7.4 PrV infection of cowpea plants. (A) Comparison of leaves from mock infected and PrV-infected cowpea plants. (B) Transmission electron microscopic visualization of the virus particles purified from *H. zea* MG8 cells and cowpea plants as described in Jiwaji *et al.* (2018). Scale bar: 50 nm.

tissue culture cells. Remarkably, the virus had moved from the site of infection to the new leaves at the top of the plant (Figure 7.4A, M. Jiwaji, unpublished data), demonstrating that not only has PrV retained the ability to infect and replicate in plants, but it has also retained the ability to move from cell to cell, like umbraviruses. These findings provide strong evidence that PrV has indeed evolved from an ancestral plant virus via horizontal transmission to an insect. Thus, not only is PrV able to infect and replicate in invertebrate and vertebrate animal cells, but it can also initiate a productive infection in plants, implying that the host factors required to support the PrV infectious life cycle are conserved across the plant and animal Kingdoms.

Are there new tetraviruses yet to be discovered? A study by Water *et al.* (2008) suggests that this is likely to be the case. A survey of the viruses in a wild population of *Nudaurelia cytherea capensis* (pine emperor moths) over a 3-year period resulted in the isolation of three distinct small RNA viruses, including NβV, which had not been found in wild South African *Nudaurelia* populations for almost 30 years. The other two viruses, designated NζV and NψV, were serologically related to NβV and NωV, respectively, but were distinct in terms of virion diameter, surface morphology and the sizes of their major capsid proteins.

Metagenomic studies have reported the discovery of tetravirus-like genomes, but with the exception of PrV, discovered in the virome of an insectivorous bat (Kemenesi *et al.*, 2016) no bona fide tetraviruses have been discovered. One of the first metagenomic studies reported the discovery of a tetravirus-like sequence, encoding a 984 aa protein, assembled from virus-derived siRNAs isolated from the *Drosophila melanogaster* S2 cell line (Wu *et al.*, 2010). The encoded protein shared 29% identity with the RdRp of EeV and TaV, leading the authors to propose that this sequence represented a new permutotetravirus that was named *Drosophila* tetravirus. Subsequently, it was determined that that this sequence likely originated from a strain of the unclassified ssRNA virus, DAV (Bronkhorst and van Rij, 2014), the RdRp of which shares homology with that of the permutotetraviruses (Ambrose *et al.*, 2009). Metagenomic sequencing of mosquito viromes resulted in the assembly of several ssRNA virus genomes including Sarawak virus, which is proposed to be a new tetravirus, due to sequence homology to the RdRp of the DpTV replicase (Sadeghi *et al.*, 2017). The classification of Sarawak virus as a tetravirus, however, will require further investigation into the biophysical characteristics of the virus capsid.

Concluding remarks

Tetraviruses have long been considered to have relatively simple genomes, encoding just the viral replicase and a single CP sequence (Hanzlik *et al.*, 1999). However, the apparent complexity of their genomes together with the identification of additional non-structural viral proteins has challenged the notion of genomic simplicity and focused attention on the mechanisms by which tetraviruses express multiple viral proteins from their relatively compact genomic RNAs.

The tetraviruses are a diverse group of small non-enveloped RNA viruses unified by the distinct $T=4$ icosahedral architecture of their capsids. However, the distinguishing characteristics of the RdRp domains of their replicases led to their classification in three separate families. It is reasonable to propose that the three tetravirus lineages are the result of genome re-assortment with (+)ssRNA viruses contributing the replication machinery and ancestral tetraviruses contributing the structural genes, on co-infection of the lepidopteran host gut. This hypothesis is supported by observations that (1) tetravirus VLPs package non-viral RNAs; (2) viruses belonging to two of the three families (*Alphatetraviridae* and *Carmotetraviridae*) replicate in association with secretory vesicles; and (3) PrV as an extant example of such a cross-Kingdom re-assortment.

In recent years, there have been significant breakthroughs in tetravirus research, most notably the establishment of virus replication in tissue culture that have provided the tools to elucidate the infectious virus life cycle and to develop a fundamental understanding of the interactions between virus and host that determine host range and infectivity. It is becoming clear that the biology of these viruses is very different, despite structural conservation of their capsid architecture and a shared site of replication in the host cell. To our knowledge, PrV is the only (+)ssRNA virus that infects and productively replicates in invertebrate and vertebrate cells as well as in plants, providing an unprecedented opportunity to study the fundamental mechanisms of RNA virus replication in eukaryotic cells.

Acknowledgements

All unpublished results cited were funded by grants awarded to R.A.D. by the South African National Research Foundation (GUN: 65512) and an SIR Research grant from the South African Medical Research Council. M.-M.d.B. was supported by a Rhodes University Henderson fellowship, while J.A.A. was supported by an Organisation for Women in Science in the Developed World (OWSD) PhD Fellowship. Opinions expressed and conclusions drawn, are those of the authors and are not necessarily to be attributed to any of the above-mentioned donors.

References

Agrawal, D.K., and Johnson, J.E. (1995). Assembly of the $T = 4$ *Nudaurelia capensis* omega virus capsid protein, post-translational cleavage, and specific encapsidation of its mRNA in a baculovirus expression system. Virology *207*, 89–97. https://doi.org/10.1006/viro.1995.1054

Ariza-Mateos, A., and Gómez, J. (2017). Viral tRNA mimicry from a biocommunicative perspective. Front. Microbiol. *8*, 2395. https://doi.org/10.3389/fmicb.2017.02395

Ambrose, R.L., Lander, G.C., Maaty, W.S., Bothner, B., Johnson, J.E., and Johnson, K.N. (2009). *Drosophila A virus* is an unusual RNA virus with a $T = 3$ icosahedral core and permuted RNA-dependent RNA polymerase. J. Gen. Virol. *90*, 2191–2200. https://doi.org/10.1099/vir.0.012104-0

Appel, H.M., and Martin, M.M. (1990). Gut redox conditions in herbivorous lepidopteran larvae. J. Chem. Ecol. *16*, 3277–3290. https://doi.org/10.1007/BF00982098

Banerjee, M., Speir, J.A., and Johnson, J.E. (2010). Structural comparison of insect RNA viruses. In Insect Virology, Asgari, S., Johnson, K., eds. (Caister Academic Press, Norfolk), pp. 327–346.

Bawden, A.L., Gordon, K.H., and Hanzlik, T.N. (1999). The specificity of Helicoverpa armigera stunt virus infectivity. J. Invertebr. Pathol. *74*, 156–163. https://doi.org/10.1006/jipa.1999.4858

Bronkhorst, A.W., and van Rij, R.P. (2014). The long and short of antiviral defense: small RNA-based immunity in insects. Curr. Opin. Virol. *7*, 19–28. https://doi.org/10.1016/j.coviro.2014.03.010

Brooks, E.M., Gordon, K.H., Dorrian, S.J., Hines, E.R., and Hanzlik, T.N. (2002). Infection of its lepidopteran host by the Helicoverpa armigera stunt virus (*Tetraviridae*). J. Invertebr. Pathol. *80*, 97–111.

Canady, M.A., Tihova, M., Hanzlik, T.N., Johnson, J.E., and Yeager, M. (2000). Large conformational changes in the maturation of a simple RNA virus, Nudaurelia capensis omega virus (NωV). J. Mol. Biol. *299*, 573–584. https://doi.org/10.1006/jmbi.2000.3723

Canady, M.A., Tsuruta, H., and Johnson, J.E. (2001). Analysis of rapid, large-scale protein quaternary structural changes: time-resolved X-ray solution scattering of Nudaurelia capensis omega virus (NωV) maturation. J. Mol. Biol. *311*, 803–814. https://doi.org/10.1006/jmbi.2001.4896

Cavener, D.R. (1987). Comparison of the consensus sequence flanking translational start sites in *Drosophila* and vertebrates. Nucleic Acids Res. *15*, 1353–1361. https://doi.org/10.1093/nar/15.4.1353

Christian, P.D., Dorian, S.J., Gordon, K.H.J., and Hanzlik, T.N. (2001). Pathology and properties of the tetravirus Helicoverpa armigera stunt virus. Biol. Control *20*, 65–75. https://doi.org/10.1006/bcon.2000.0887

Christian, P.D., Murray, D., Powell, R., Hopkinson, J., Gibb, N.N., and Hanzlik, T.N. (2005). Effective control of a field population of *Helicoverpa armigera* by using the small RNA virus *Helicoverpa armigera* stunt virus (*Tetraviridae*: *Omegatetravirus*). J. Econ. Entomol. *98*, 1839–1847.

Chao, Y.C., Scott, H.A., and Young, S.Y. (1983). An icosahedral RNA virus of the soybean looper (*Pseudoplusia includens*). J. Gen. Virol. *64*, 1835–1838. https://doi.org/10.1099/0022-1317-64-8-1835

de Bruyn, M. (2016). Investigating the expression of three small open reading frames encoded on *Helicoverpa armigera* stunt virus RNA 1. MSc thesis. Rhodes University, Grahamstown, South Africa. Available online: http://hdl.handle.net/10962/59168. Accessed 13 March 2019.

Domitrovic, T., Matsui, T., and Johnson, J.E. (2012). Dissecting quasi-equivalence in non-enveloped viruses: membrane disruption is promoted by lytic peptides released from subunit pentamers, not hexamers. J. Virol. *86*, 9976–9982. https://doi.org/10.1128/JVI.01089-12

Doronina, V.A. de Felipe, P., Wu, C., Sharma, P., Sachs, M.S., Ryan, M.D., and Brown, J.D. (2008a). Dissection of a co-translational nascent chain separation event. Biochem. Soc. Trans. *36*, 712–716. https://doi.org/10.1042/BST0360712

Doronina, V.A., Wu, C., de Felipe, P., Sachs, M.S., Ryan, M.D., and Brown, J.D. (2008b). Site-specific release of nascent chains from ribosomes at a sense codon. Mol. Cell. Biol. *28*, 4227–4239. https://doi.org/10.1128/MCB.00421-08

Dreher, T.W. (2009). Role of tRNA-like structures in controlling plant virus replication. Virus Res. *139*, 217–229. https://doi.org/10.1016/j.virusres.2008.06.010

Dorrington, R.A., Gorbalenya, A.E., Gordon, K.H.J., Lauber, C., and Ward, V.K. (2011). Tetraviridae. In Virus Taxonomy, King, A.M.Q., Adams, M.J., Carstens, E.B., and Lefkowitz, E.J., eds. (Elsevier, Oxford), pp. 1091–1102.

du Plessis, L., Hendry, D.A., Dorrington, R.A., Hanzlik, T.N., Johnson, J.E., and Appel, M. (2005). Revised RNA2 sequence of the tetravirus, Nudaurelia capensis ω virus (NωV). Arch. Virol. *150*, 2397–2402. https://doi.org/10.1007/s00705-005-0575-0

Eckerle, L.D., Albariño, C.G., and Ball, L.A. (2003). Flock House virus subgenomic RNA3 is replicated and its replication correlates with transactivation of RNA2. Virology *317*, 95–108. https://doi.org/10.1016/j.virol.2003.08.029

Fauquet, C.M., Mayo, M.A., Maniloff, J., Desselberger, U., and Ball, L.A. (2005). Virus taxonomy. Eighth Report of the International Committee on Taxonomy of Viruses (Academic Press, London).

Ferrero, D., Buxaderas, M., Rodriguez, J.F., and Verdaguer, N. (2012). Purification, crystallization and preliminary X-ray diffraction analysis of the RNA-dependent RNA polymerase from Thosea asigna virus. Acta Crystallogr. Sect. F Struct. Biol. Cryst. Commun. *68*, 1263–1266. https://doi.org/10.1107/S1744309112037529

Ferrero, D.S., Buxaderas, M., Rodríguez, J.F., and Verdaguer, N. (2015). The structure of the RNA-dependent RNA polymerase of a permutotetravirus suggests a link between primer-dependent and primer-independent polymerases. PLOS Pathog. *11*, e1005265. https://doi.org/10.1371/journal.ppat.1005265

Finch, J.T., Crowther, R.A., Hendry, D.A., and Struthers, J.K. (1974). The structure of Nudaurelia capensis beta virus: the first example of a capsid with icosahedral surface symmetry T-4. J. Gen. Virol. *24*, 191–200. https://doi.org/10.1099/0022-1317-24-1-191

Francki, R.I.B., Fauquet, C.M., Knudson, D.L., and Brown, F. (1991). Classification and nomenclature of viruses. Fifth Report of the International Committee on Taxonomy of Viruses (Springer-Verlag, Vienna). Available online: https://doi.org/10.1007/978-3-7091-9163-7

Galiana-Arnoux, D., Dostert, C., Schneemann, A., Hoffmann, J.A., and Imler, J.L. (2006). Essential function *in vivo* for Dicer-2 in host defence against RNA viruses in *Drosophila*. Nat. Immunol. *7*, 590–597. https://doi.org/10.1038/ni1335

Gorbalenya, A.E., Pringle, F.M., Zeddam, J.L., Luke, B.T., Cameron, C.E., Kalmakoff, J., Hanzlik, T.N., Gordon, K.H., and Ward, V.K. (2002). The palm subdomain-based active site is internally permuted in viral RNA-dependent RNA polymerases of an ancient lineage. J. Mol. Biol. *324*, 47–62.

Gordon, K.H., and Waterhouse, P.M. (2006). Small RNA viruses of insects: expression in plants and RNA silencing. Adv. Virus Res. *68*, 459–502.

Gordon, K.H.J., Johnson, K.N., and Hanzlik, T.N. (1995). The larger genomic RNA of Helicoverpa armigera stunt virus encodes the viral RNA polymerase and has a novel 3'-terminal tRNA-like structure. Virology *208*, 84–98. https://doi.org/10.1006/viro.1995.1132

Gordon, K.H., Williams, M.R., Hendry, D.A., and Hanzlik, T.N. (1999). Sequence of the genomic RNA of nudaurelia beta virus (*Tetraviridae*) defines a novel virus genome organization. Virology *258*, 42–53.

Gordon, K.H., Williams, M.R., Baker, J.S., Gibson, J.M., Bawden, A.L., Millgate, A.G., Larkin, P.J., and Hanzlik, T.N. (2001). Replication-independent assembly of an insect virus (*Tetraviridae*) in plant cells. Virology *288*, 36–50. https://doi.org/10.1006/viro.2001.1049

Grace, T.D.C., and Mercer, E.H. (1965). A new virus of the Saturniid *Antherea eucalypti* Scott. J. Invertebr. Pathol *7*, 241–244. https://doi.org/10.1016/0022-2011(65)90044-3

Greenwood, L.K., and Moore, N.F. (1981). The *Nudaurelia* β group of small RNA viruses of insects: serological comparison of four members. Microbiologica *4*, 271–280. https://doi.org/10.1016/0022-2011(82)90068-4

Hanzlik, T.N., Dorrian, S.J., Gordon, K.H., and Christian, P.D. (1993). A novel small RNA virus isolated from the cotton bollworm, *Helicoverpa armigera*. J. Gen. Virol. *74*, 1805–1810. https://doi.org/10.1099/0022-1317-74-9-1805

Hanzlik, T.N., Dorrian, S.J., Johnson, K.N., Brooks, E.M., and Gordon, K.H.J. (1995). Sequence of RNA 2 of the Helicoverpa armigera stunt virus (*Tetraviridae*) and bacterial expression of its genes. J. Gen. Virol. *76*, 799–811. https://doi.org/10.1099/0022-1317-76-4-799

Hanzlik, T.N., Zeddam, J.-L., Gordon, K.H.J., and Christian, P.D. (1999). A new view of small RNA viruses of insects. Pestic. Outlook *10*, 22–26.

Helgstrand, C., Munshi, S., Johnson, J.E., and Liljas, L. (2004). The refined structure of Nudaurelia capensis ω virus reveals control elements for a $T = 4$ capsid maturation. Virology *318*, 192–203. https://doi.org/10.1016/j.virol.2003.08.045

Hendry, D.A., and Agrawal, D.K. (1994). Tetraviruses. In Encyclopaedia of Virology, Granoff, A., and Webster, R.G., eds. (Academic Press, San Diego, CA), pp. 1416–1422.

Hendry, D.A., Becker, M.F., and van Regenmortel M.V.H. (1968). A non-inclusion virus of the pine emperor moth *Nudaurelia cytherea capensis* Stoll. S. Afr. Med. J. *42*, 117.

Hendry, D.A., Hodgson, V., Clark, R., and Newman, J. (1985). Small RNA viruses co-infecting the pine emperor moth (*Nudaurelia cytherea capensis*). J. Gen. Virol. *66*, 627–632. https://doi.org/10.1099/0022-1317-66-3-627

ICTV (International Committee on Taxonomy of Viruses) (2011). Taxonomy Release. Available online: http://www.ictvonline.org/virusTaxonomy.asp?version=2011. Accessed 14 December 2018.

Jiwaji, M., Short, J.R., and Dorrington, R.A. (2016). Expanding the host range of small insect RNA viruses: Providence virus (*Carmotetraviridae*) infects and replicates in a human tissue culture cell line. J. Gen. Virol. *97*, 2763–2768. https://doi.org/10.1099/jgv.0.000578

Johnson, K.L., Price, B.D., Eckerle, L.D., and Ball, L.A. (2004). Nodamura virus nonstructural protein B2 can enhance viral RNA accumulation in both mammalian and insect cells. J. Virol. *78*, 6698–6704. https://doi.org/10.1128/JVI.78.12.6698-6704.2004

Juckes, I.R.M. (1979). Comparison of some biophysical properties of the *Nudaurelia* β and ε viruses. J. Gen. Virol. *42*, 89–94. https://doi.org/10.1099/0022-1317-42-1-89

Kemenesi, G., Földes, F., Zana, B., Kurucz, K., Estók, P., Boldogh, S., Görföl, T., Bányai, K., Oldal, M., and Jakab, F. (2016). Genetic characterization of providence virus isolated from bat guano in Hungary. Genome Announc. *4*, e00403–16. https://doi.org/10.1128/genomeA.00403-16

King, L.A., and Moore, N.F. (1984). The RNAs of two viruses of the Nudaurelia β family share little homology and have no terminal poly(A) tracts. FEMS Microbiol. Lett. *26*, 41–43. https://doi.org/10.1111/j.1574-6968.1985.tb01562.x

Kobayashi, T., Vischer, U.M., Rosnoblet, C., Lebrand, C., Lindsay, M., Parton, R.G., Kruithof, E.K., and Gruenberg, J. (2000). The tetraspanin CD63/lamp3 cycles between endocytic and secretory compartments in human endothelial cells. Mol. Biol. Cell *11*, 1829–1843. https://doi.org/10.1091/mbc.11.5.1829

Luke, G.A., de Felipe, P., Lukashev, A., Kallioinen, S.E., Bruno, E.A., and Ryan, M.D. (2008). Occurrence, function and evolutionary origins of '2A-like' sequences in virus genomes. J. Gen. Virol. *89*, 1036–1042. https://doi.org/10.1099/vir.0.83428-0

Maroniche, G.A., Mongelli, V.C., Alfonso, V., Llauger, G., Taboga, O., and del Vas, M. (2011). Development of a novel set of Gateway-compatible vectors for live imaging in insect cells. Insect Mol. Biol. *20*, 675–685. https://doi.org/10.1111/j.1365-2583.2011.01100.x

Matsui, T., Tsuruta, H., and Johnson, J.E. (2010). Balanced electrostatic and structural forces guide the large conformational change associated with maturation of T = 4 virus. Biophys. J. *98*, 1337–1343. https://doi.org/10.1016/j.bpj.2009.12.4283

Matthews, R.E.F. (1982). Classification and nomenclature of viruses. Fourth Report of the International Committee on Taxonomy of Viruses. Intervirology *17*, 1–160. https://doi.org/10.1159/000149278

Mendes, A., Vlok, M., Short, J.R., Matsui, T., and Dorrington, R.A. (2015). An encapsidated viral protein and its role in RNA packaging by a non-enveloped animal RNA virus. Virology *476*, 323–333. https://doi.org/10.1016/j.virol.2014.12.026

Moore, N.F., Greenwood, L.K., and Rixon, K.R. (1981). Studies on the capsid protein of several members of the *Nudaurelia* β group of small RNA viruses of insects. Microbiologica *4*, 59–71.

Morris, T.J., Hess, R.T., and Pinnock, D.E. (1979). Physicochemical characterization of a small RNA virus associated with baculovirus infection in Trichoplusia ni. Intervirology *11*, 238–247. https://doi.org/10.1159/000149040

Munshi, S., Liljas, L., Cavarelli, J., Bomu, W., McKinney, B., Reddy, V., and Johnson, J.E. (1996). The 2.8 A structure of a *T* = 4 animal virus and its implications for membrane translocation of RNA. J. Mol. Biol. *261*, 1–10. https://doi.org/10.1006/jmbi.1996.0437

Nakayinga, R. (2013). Understanding the replication biology of Providence virus: elucidating the function of non-structural proteins. PhD thesis, Rhodes University, Grahamstown, South Africa.

Odegard, A., Banerjee, M., and Johnson, J.E. (2010). Flock house virus: a model system for understanding non-enveloped virus entry and membrane penetration. Curr. Top. Microbiol. Immunol. *343*, 1–22. https://doi.org/10.1007/82_2010_35

Penkler, D.L., Jiwaji, M., Domitrovic, T., Short, J.R., Johnson, J.E., and Dorrington, R.A. (2016). Binding and entry of a non-enveloped T=4 insect RNA virus is triggered by alkaline pH. Virology *498*, 277–287. https://doi.org/10.1016/j.virol.2016.08.028

Poch, O., Sauvaget, I., Delarue, M., and Tordo, N. (1989). Identification of four conserved motifs among the RNA-dependent polymerase encoding elements. EMBO J. *8*, 3867–3874.

Pringle, F.M., Gordon, K.H., Hanzlik, T.N., Kalmakoff, J., Scotti, P.D., and Ward, V.K. (1999). A novel capsid expression strategy for Thosea asigna virus (*Tetraviridae*). J. Gen. Virol. *80*, 1855–1863. https://doi.org/10.1099/0022-1317-80-7-1855

Pringle, F.M., Kalmakoff, J., and Ward, V.K. (2001). Analysis of the capsid processing strategy of Thosea asigna virus using baculovirus expression of virus-like particles. J. Gen. Virol. *82*, 259–266. https://doi.org/10.1099/0022-1317-82-1-259

Pringle, F.M., Johnson, K.N., Goodman, C.L., McIntosh, A.H., and Ball, L.A. (2003). *Providence virus*: a new member of the *Tetraviridae* that infects cultured insect cells. Virology *306*, 359–370.

Reinganum, C., and Scotti, P.D. (1976). Serological relations between twelve small RNA viruses of insects. J. Gen. Virol. *31*, 131–134. https://doi.org/10.1099/0022-1317-31-1-131

Reinganum, C., Robertson, J.S., and Tinsley, T.W. (1978). A new group of RNA viruses from insects. J. Gen. Virol. *40*, 195–202. https://doi.org/10.1099/0022-1317-40-1-195

Roulston, C., Luke, G.A., de Felipe, P., Ruan, L., Cope, J., Nicholson, J., Sukhodub, A., Tilsner, J., and Ryan, M.D. (2016). '2A-Like' signal sequences mediating translational recoding: a novel form of dual protein targeting. Traffic *17*, 923–939. https://doi.org/10.1111/tra.12411

Routh, A., Domitrovic, T., and Johnson, J.E. (2012). Host RNAs, including transposons, are encapsidated by a eukaryotic single-stranded RNA virus. Proc. Natl. Acad. Sci. U.S.A. *109*, 1907–1912. https://doi.org/10.1073/pnas.1116168109

Sadeghi, M., Popov, V., Guzman, H., Phan, T.G., Vasilakis, N., Tesh, R., and Delwart, E. (2017). Genomes of viral isolates derived from different mosquitos species. Virus Res. *242*, 49–57. https://doi.org/10.1016/j.virusres.2017.08.012

Short, J.R., and Dorrington, R.A. (2012). Membrane targeting of an alpha-like tetravirus replicase is directed by a region within the RNA-dependent RNA polymerase domain. J. Gen. Virol. *93*, 1706–1716. https://doi.org/10.1099/vir.0.038992-0

Short, J.R., Knox, C., and Dorrington, R.A. (2010). Subcellular localisation and live cell imaging of the Helicoverpa armigera stunt virus replicase in mammalian and *Spodpoptera frugiperda* cells. J. Gen. Virol. *91*, 1514–1523. https://doi.org/10.1099/vir.0.020156-0

Short, J.R., Nakayinga, R., Hughes, G.E., Walter, C.T., and Dorrington, R.A. (2013). Providence virus (family: *Carmotetraviridae*) replicates vRNA in association with the Golgi apparatus and secretory vesicles. J. Gen. Virol. *94*, 1073–1078. https://doi.org/10.1099/vir.0.047647-0

Speir, J.A., Taylor, D.J., Natarajan, P., Pringle, F.M., Ball, L.A., and Johnson, J.E. (2010). Evolution in action: N and C termini of subunits in related T = 4 viruses exchange roles as molecular switches. Structure *18*, 700–709. https://doi.org/10.1016/j.str.2010.03.010

Struthers, J.K., and Hendry, D.A. (1974). Studies of the protein and nucleic acid components of *Nudaurelia capensis β virus*. J. Gen. Virol. *22*, 355–362. https://doi.org/10.1099/0022-1317-22-3-355

Tang, J., Kearney, B.M., Wang, Q., Doerschuk, P.C., Baker, T.S., and Johnson, J.E. (2014). Dynamic and geometric analyses of Nudaurelia capensis ω virus maturation reveal the energy landscape of particle transitions. J.

Mol. Recognit. *27*, 230–237. https://doi.org/10.1002/jmr.2354

Taylor, T.J. (2003). Molecular and biophysical analysis of a conformational change in a nonenveloped RNA virus. PhD thesis, University of California, San Diego, CA.

Taylor, D.J., Krishna, N.K., Canady, M.A., Schneemann, A., and Johnson, J.E. (2002). Large-scale, pH-dependent, quaternary structure changes in an RNA virus capsid are reversible in the absence of subunit autoproteolysis. J. Virol. *76*, 9972–9980. https://doi.org/10.1128/JVI.76.19.9972-9980.2002

Thiéry, R., Johnson, K.L., Nakai, T., Schneemann, A., Bonami, J.R., and Lightner, D.V. (2011). *Nodaviridae*. In Virus Taxonomy. King, A.M.Q., Adams, M.J., Carstens, E.B., and Lefkowitz, E.J., eds. (Elsevier, Oxford), pp. 1061–1067. https://doi.org/10.1016/B978-0-12-384684-6.00092-6

Tomasicchio, M. (2007). Assembly of *Omegatetravirus* virus-like particles in the yeast *Saccharomyces cerevisiae*. PhD Thesis, Rhodes University, Grahamstown, South Africa.

Tomasicchio, M., Venter, P.A., Gordon, K.H.J., Hanzlik, T.N., and Dorrington, R.A. (2007). Induction of apoptosis in *Saccharomyces cerevisiae* results in the spontaneous maturation of tetravirus procapsids *in vivo*. J. Gen. Virol. *88*, 1576–1582. https://doi.org/10.1099/vir.0.82250-0

van der Heijden, M.W., and Bol, J.F. (2002). Composition of alphavirus-like replication complexes: involvement of virus and host encoded proteins. Arch. Virol. *147*, 875–898. https://doi.org/10.1007/s00705-001-0773-3

Walter, C.T., Tomasicchio, M., Hodgson, V., Hendry, D.A., Hill, M.P., and Dorrington, R.A. (2008). Characterisation of a succession of small insect viruses in a wild South African population of *Nudaurelia cytherea capensis*. S. Afr. J. Sci. *104*, 147–152.

Walter, C.T., Pringle, F.M., Nakayinga, R., de Felipe, P., Ryan, M.D., Ball, L.A., and Dorrington, R.A. (2010). Genome organization and translation products of Providence virus: insight into a unique tetravirus. J. Gen. Virol. *91*, 2826–2735. https://doi.org/10.1099/vir.0.023796-0

Wu, Q., Luo, Y., Lu, R., Lau, N., Lai, E.C., Li, W.X., and Ding, S.W. (2010). Virus discovery by deep sequencing and assembly of virus-derived small silencing RNAs. Proc. Natl. Acad. Sci. U.S.A. *107*, 1606–1611. https://doi.org/10.1073/pnas.0911353107

Yi, F., Zhang, J., Yu, H., Liu, C., Wang, J., and Hu, Y. (2005). Isolation and identification of a new tetravirus from *Dendrolimus punctatus* larvae collected from Yunnan province, China. J. Gen. Virol. *86*, 789–796. https://doi.org/10.1099/vir.0.80543-0

Zeale, M.R., Butlin, R.K., Barker, G.L., Lees, D.C., and Jones, G. (2011). Taxon-specific PCR for DNA barcoding arthropod prey in bat faeces. Mol. Ecol. Resour. *11*, 236–244. https://doi.org/10.1111/j.1755-0998.2010.02920.x

Zeddam, J.L., Gordon, K.H., Lauber, C., Alves, C.A., Luke, B.T., Hanzlik, T.N., Ward, V.K., and Gorbalenya, A.E. (2010). Euprosterna elaeasa virus genome sequence and evolution of the *Tetraviridae* family: emergence of bipartite genomes and conservation of the VPg signal with the dsRNA *Birnaviridae* family. Virology *397*, 145–154. https://doi.org/10.1016/j.virol.2009.10.042

Zhou, L., Zheng, Y., Jiang, H., Zhou, W., Lin, M., Han, Y., Cao, X., Zhang, J., and Hu, Y. (2008). RNA-binding properties of Dendrolimus punctatus tetravirus p17 protein. Virus Res. *138*, 1–6. https://doi.org/10.1016/j.virusres.2008.05.006

Polydnaviruses: Evolution and Function

8

Michael R. Strand* and Gaelen R. Burke

Department of Entomology, University of Georgia, Athens, GA, USA.

*Correspondence: mrstrand@uga.edu

https://doi.org/10.21775/9781912530083.08

Abstract

Polydnaviruses (PDVs) were originally viewed as large DNA viruses that are beneficial symbionts of parasitoid wasps. Two groups of PDVs were also recognized: bracoviruses (BVs), which are associated with wasps in the family Braconidae, and ichnoviruses (IVs), which are associated with wasps in the family Ichneumonidae. Results to date indicate that BVs are endogenous virus elements (EVEs) that evolved from an ancient betanudivirus. IVs are also likely EVEs but are unrelated to BVs. BVs and IVs are very unusual relative to most known EVEs because they retain many viral functions that benefit wasps in parasitizing hosts. However, BVs and IVs cannot be considered beneficial symbionts because all components of their genomes are fixed in wasps. Recent studies indicate that other nudiviruses have endogenized in insects. Each exhibits a different functional fate from BVs but shares certain architectural features. We discuss options for classifying BVs and other endogenized nudiviruses. We also discuss future directions.

Introduction

Parasitoids are defined as free living insects as adults that feed parasitically on other arthropods as larvae (Pennacchio and Strand, 2006). Adult females usually lay their eggs in (endoparasitoids) or on (ectoparasitoids) a host, which is usually another insect species, and hosts usually die after offspring complete their development. The vast majority of parasitoids are wasps (order Hymenoptera) that have specialized to parasitize only one or a few host species. Host range specialization in turn is associated with high speciation rates; estimates suggest more than 25% of all insect species are parasitic Hymenoptera (Godfray, 1994; Forbes, *et al.*, 2018).

The *Polydnaviridae* was recognized in 1991 as a family of large double-stranded DNA viruses associated with endoparasitoid wasps in the Braconidae and Ichneumonidae families (Francki, 1991). Braconid-associated polydnaviruses (PDVs) were assigned to the genus *Bracovirus* (BV) and ichneumonid-associated PDVs were assigned to the genus *Ichnovirus* (IV). Subsequent studies indicated that all BV-carrying braconids reside in six subfamilies that form a monophyletic and hyperdiverse (≈ 50,000 species) assemblage named the microgastroid complex, while all IV-carrying ichneumonids reside in two subfamilies named the Campopleginae and Banchinae (≈ 14,000 species) (Whitfield, 2002; Quicke *et al.*, 2009). Both BV and IV-carrying wasps primarily parasitize larval stage Lepidoptera, with several also being used as biological control agents for management of pest species. However, the primary interest in these insects stemmed from the finding that each wasp species relies on its PDV to successfully parasitize hosts, while each PDV relies on its wasp for transmission. Thus, the association between PDVs and wasps was viewed as the first example of viruses evolving into beneficial symbionts (Edson *et al.*, 1981).

We wrote a contribution for the first edition of *Insect Virology* that summarized the PDV literature through 2009 (Strand, 2010). Several findings, reviews and essays have since been published that substantially advanced understanding of PDVs (Beckage and Drezen, 2012; Strand and Burke, 2013, 2014, 2015; Gundersen-Rindal *et al.*, 2013;

Herniou *et al.*, 2013; Drezen *et al.*, 2017; Gauthier *et al.*, 2018). In this chapter, we first summarize the early literature that led to the view of PDVs as beneficial symbionts of wasps. We next discuss results that shed important light on PDV evolution and function. We conclude by summarizing recent results showing that viruses related to the ancestor of BVs have independently associated themselves with other insects, including parasitoids outside of the microgastrine complex.

The early literature concluded that PDVs are beneficial symbionts of wasps

PDVs were first discovered in the late 1960s and early 1970s through electron microscopy (EM) studies that identified large numbers of particles in the ovaries and lateral oviducts of certain braconid and ichneumonid wasps (Rotheram, 1967; Vinson and Scott, 1975; Stoltz *et al.*, 1976). Mature particles in some braconid wasp species had one cylindrical capsid of similar width (≈40 nm) but variable length (25–100 nm) surrounded by a single envelope, while in others multiple capsids were surrounded by a single envelope (Stoltz and Vinson, 1979). Mature particles in ichneumonid wasps had biconvex capsids (≈85–330 nm) that were individually surrounded by two envelopes (Stoltz and Vinson, 1979). While morphologically distinct, braconid and ichneumonid particles both exhibited morphological features that suggested they were viruses. Molecular data further indicated that particles from both braconids and ichneumonids contained multiple, circular, dsDNAs that were non-equimolar in abundance (Webb and Strand, 2005). The estimated number of DNA segments in particles (20 to more than 100), their individual sizes (2 to more than 30 kb) and their estimated aggregate sizes (150 to more than 700 kb) varied between wasp species, with particles from braconids usually contain a smaller number of larger segments than ichneumonids (Krell and Stoltz, 1979; Krell *et al.*, 1980, 1982; Espagne *et al.*, 2004; Webb *et al.*, 2006; Lapointe *et al.*, 2007). Each capsid in mature particles from braconids further contained only one segment (Albrecht *et al.*, 1994; Beck *et al.*, 2007). This indicated that the total number of DNA segments detected in particles from a given wasp species are distributed across a population of particles and that the non-equimolar abundance of segments is due to particles containing certain segments being more abundant than others (Beck *et al.*, 2007).

Early studies also identified several shared life cycle features. First, sequences homologous to the DNA segments in particles were detected in the genomes of both male and female wasps across all life stages and cell types including the germ line (Stoltz, 1990; Fleming and Summers, 1991). These results suggested the DNAs in braconid and ichneumonid particles are integrated into the genome of wasps. They further suggested these entities persist as proviruses that can be transmitted vertically through the germline of wasps (Fleming, 1992). EM studies also provided strong, visual evidence that braconid and ichneumonid particles replicate in pupal and adult stage females in specialized cells of the proximal ovaries named calyx cells (Stoltz and Vinson, 1979; Wyler and Lanzrein, 2003). Braconid and ichneumonid-associated particles both assembled in calyx cell nuclei. Braconid particles were released by lysis of calyx cells. In contrast, ichneumonid particles were released by budding, which also resulted in particles acquiring a second envelope (Stoltz and Vinson, 1979; Norton and Vinson, 1983). Mature particles accumulated to high densities in the lumen of the proximal ovaries and lateral oviducts of female wasps, where they formed a thick paste called calyx fluid (Stoltz and Vinson, 1979). Female wasps were also shown to inject calyx fluid into the hosts they parasitized, and that particles rapidly infect haemocytes, fat bodies and other tissues (Stoltz and Vinson, 1979; Strand *et al.*, 1992; Strand, 1994; Li and Webb, 1994). After infecting a host cell, capsids migrated to nuclear pores followed by discharge of encapsidated DNAs into the nucleus and expression of genes encoded on encapsidated DNAs (Stoltz and Vinson, 1979; Fleming *et al.*, 1983; Blissard *et al.*, 1986). Functional assays further indicated that infection of hosts disabled immune defences that would otherwise kill wasp offspring and altered growth and metabolism in ways that prevented hosts from surviving (Edson *et al.*,1981; Stoltz and Cook, 1983; Davies *et al.*, 1987; Dover *et al.*, 1987). In contrast, no replication of braconid and ichneumonid particles occurred in infected hosts (Theilmann and Summers, 1986).

Altogether, particle morphology, the packaging

of gene-encoding dsDNAs into particles, and the ability of particles to infect and express genes in the hosts of wasps were identified as key virus-like features. The life cycle also showed parallels with known viruses that can persist as proviruses (*Retroviridae*, several bacteriophages) or that can be transmitted by insects to other hosts (*Togaviridae, Bunyaviridae, Reoviridae*). However, no other known viruses had large segmented circular dsDNA genomes or exhibited traits consistent with being beneficial symbionts of one host (wasps) and pathogens of another (hosts of wasps). Furthermore, no other viruses were known to only be transmitted vertically through one host (wasps), replicate in the same host to produce particles that infect but never replicate in another host (lepidopteran larvae). Thus, the family *Polydnaviridae*, genus *Bracovirus* (BV) and genus *Ichnovirus* (IV) were proposed (Stoltz *et al.*, 1984) and later recognized (Francki, 1991) as mutualistic viruses in braconid and ichneumonid wasps that have large 'polyDNA' genomes.

Subsequent phylogenetic studies further refined the boundaries of BV and IV wasp associations by showing that: (1) BVs are restricted to species in six subfamilies that form the microgastroid complex, (2) the microgastroid complex diverged approximately 100 million years ago from other subfamilies of braconids that lack BVs, and (3) IVs are restricted to species in two subfamilies of the Ichneumonidae: the Campopleginae and Banchinae (Whitfield, 2002; Murphy *et al.*, 2008; Thézé *et al.*, 2011; Quicke *et al.*, 2009) (Fig. 8.1). Higher order data also indicated that BV-carrying braconids and IV-carrying ichneumonids belong to the same superfamily (Ichneumonoidea) but are distantly related (Quicke *et al.*, 2009; Heraty *et al.*, 2011) (Fig. 8.1). Taken together, these results strongly suggested that all BV-carrying species evolved from a common ancestor, the BV–microgastroid association is ancient, and BVs likely evolved independently from IVs (Fig. 8.1). In contrast, these data do not address what BVs and IVs evolved from, or whether the ancestors of BVs

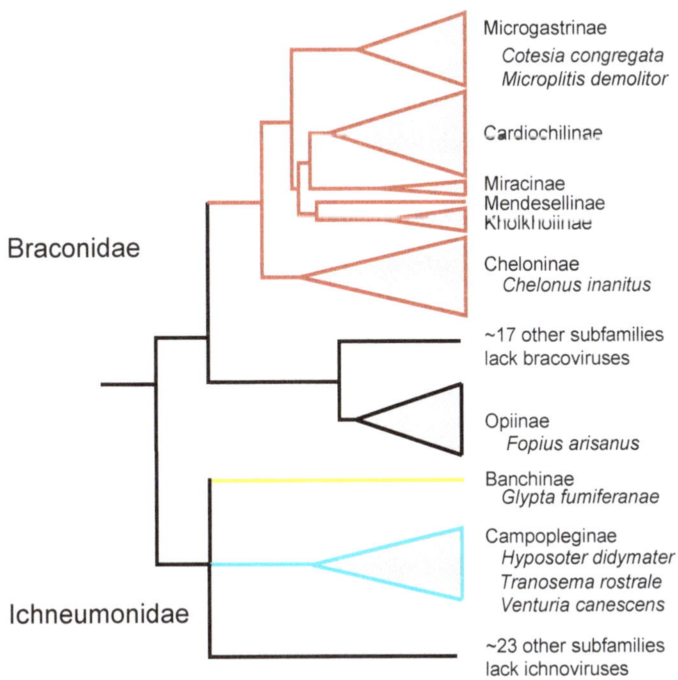

Figure 8.1 Evolutionary relationships in the superfamily Ichneumonoidea, which contains the families Braconidae and Ichneumonidae. Within the Braconidae, the monophyletic microgastroid complex consists of six subfamilies in which all species are thought to produce BVs. Other branches identify subfamilies of braconids that lack BVs. Within the Ichneumonidae, species in two subfamilies, the Campoleginae and Banchinae, produce IVs while other subfamilies lack IVs. Subfamily affiliations of select wasp species are also indicated.

and IVs are related. The uncertain phylogenetic relationship between campoplegine and banchine ichneumonids also made it unclear whether all IVs share a common ancestor.

Key insights into PDV evolution and function required genomic data from wasps

Microbes within the Eubacteria, Archaea, and Eukarya each share molecular machinery and genes that can be used as 'barcodes' for discerning ancestry and assigning species/isolates to particular taxa. In contrast, multiple evolutionary origins result in no molecular machinery or genes that are universally shared among viruses (Forterre, 2006; Holmes, 2011a). Different lineages of viruses can be classified into families and sometimes higher order associations through core genes that are shared among certain isolates. Most RNA and DNA viruses with small genomes (<20 kb) have gene inventories consisting entirely or mostly of core genes (Payne, 2017). Large DNA viruses, on the other hand, encode a small proportion of core genes of ancient and usually uncertain origin and a much larger proportion of non-core genes that are present in only some isolates or are unique to particular isolates (McGeoch and Davison, 1999; Iyer *et al.*, 2006; Holmes, 2011a; Garavaglia *et al.*, 2012). This is because large DNA viruses, like prokaryotes, have over evolutionary time acquired a diversity of genes from hosts and other organisms by horizontal gene transfer (HGT) (Shackelton and Holmes, 2004). In addition, large DNA viruses have sufficient coding capacity to maintain genes acquired by HGT through selection if they confer advantageous functions (Holmes, 2011a). Core genes in large DNA viruses often include structural and other genes with essential functions in replication, while non-core genes are often virulence-related, enabling isolates to infect and successfully propagate in particular hosts (McGeoch and Davison, 1999; Iyer *et al.*, 2006; Garavaglia *et al.*, 2012). Very little sequence data existed for any PDV when they were first classified as mutualistic viruses due to several technical challenges (Strand and Burke, 2015). Furthermore, the data needed to discern ancestry and key functions of PDVs took many years to generate. Acquisition of this information occurred in three major steps.

The DNA segments in BV and IV virions contain virulence genes with functions in parasitizing the hosts of wasps

The first step was accomplished in the 2000s when methodology progressed to the point that the DNA segments in BV virions from five species of braconids (Espagne *et al.*, 2004, Webb *et al.*, 2006; Desjardins *et al.*, 2008) and IV virions from four species of ichneumonids (Webb *et al.*, 2006; Tanaka *et al.*, 2007; Lapointe *et al.*, 2007) could be sequenced and assembled. The DNAs in select other PDVs were also later partially or fully sequenced (Chen *et al.*, 2011; Yu *et al.*, 2016). These results showed that coding densities for the DNAs in BV and IV virions were low and were more similar to eukaryotic genomic regions than to the genomes of other known large DNA viruses. Identified genes were either single copy or formed multimember families; some contained small introns, while others were intronless. Several of these genes shared homology with known eukaryotic genes but no genes in either BV or IV virions share homology with any known viral core gene. BVs from closely related wasps shared several genes with one another as did IVs from closely related wasps. BVs and IVs from more distantly related wasps shared fewer genes with one another (Webb *et al.*, 2006; Tanaka *et al.*, 2007; Lapointe *et al.*, 2007; Pruijssers and Strand, 2007; Desjardins *et al.*, 2008; Serbielle *et al.*, 2012). However, only a few types of genes were shared between BVs and campoplegine IVs [ankyrin repeat genes (*anks*)] or banchine ichneumonids [*ank* and protein tyrosine phosphatase genes (*ptps*)]. Analysis of the genes in BVs and IVs further support historical acquisition at different times and from different sources (Huguet *et al.*, 2012). For example, some genes unique to particular BVs show strong evidence of being wasp genes (Desjardins *et al.*, 2008; Burke and Strand, 2014). Other genes present in some or most isolates of BVs or IVs show evidence of acquisition from other viruses or other types of organisms or are of ancient enough lineage to be of indeterminate origin (Huguet *et al.*, 2012; Burke and Strand, 2012a; Serbielle *et al.*, 2012; Herniou *et al.*, 2013). Genes like the *ank* families in BVs and campoplegine IVs, and *ptps* in BVs and banchine IVs show evidence of independent acquisition rather than common ancestry (Lapointe *et al.*, 2007; Huguet *et al.*, 2012).

From a functional perspective these results indicate that: (1) BVs and IVs do not replicate in the hosts of wasps because no genes with discernible functions in replication are present on the DNAs in virions, and (2) a number of genes likely function as virulence factors with roles in parasitism. Further support for the latter came from studies showing that most genes on DNAs in virions are transcribed in hosts but are not transcribed in wasps (Rasoolizadeh *et al.*, 2009; Bitra *et al.*, 2011; Provost *et al.*, 2011; Burke and Strand, 2012b; Chevignon *et al.*, 2014; Dorémus *et al.*, 2014; Bitra *et al.*, 2015). Experimental studies also showed that genes/microRNAs in BV or IV virions disrupt specific host immune defences against wasp offspring (Beck and Strand, 2005; Thoetkiattikul *et al.*,, 2005; Beck and Strand, 2007; Pruijssers and Strand, 2007; Ibrahim and Kim, 2008; Labropoulou *et al.*,, 2008; Kwon and Kim, 2008; Cooper *et al.*, 2011; Magkrioti *et al.*, 2011; Bitra *et al.*, 2012; Gueguen *et al.*, 2013), alter host growth (Provost *et al.*, 2004; Falabella *et al.*, 2006; Pruijssers *et al.*, 2009; Kim *et al.*, 2013; Prasad *et al.*, 2014; Ignesti *et al.*, 2018; Wang *et al.*, 2018), and potentially affect host interactions with food plants and other organisms (Zhu *et al.*, 2018; Tan *et al.*, 2018).

From an evolutionary perspective, the paucity of shared genes between BVs and IVs was consistent with phylogenetic studies of wasps that suggested BVs and IVs arose independently. The diverse origins of genes in BV and IV virions were also similar to the variation in non-core genes that other families of large DNA viruses exhibit. In contrast, the absence of any recognizable core genes suggested that BVs and IVs are either not of viral origin or that core genes with functions in producing virions reside in the genomes of wasps (Stoltz, 1993; Federici and Bigot, 2003).

PDV core genes reside in the genomes of wasps

The second step in characterizing PDV genomes was analysing proteins in virions and determining whether the corresponding genes were present in the genomes of wasps and if they were related to core genes in known viruses. The first studies were conducted in the campoplegine ichneumonid wasp *Campoletis sonorensis* (Fig. 8.1), which produces *C. sonorensis ichnovirus* (CsIV). N-terminal sequencing of a protein (p44) in CsIV virions enabled identification of the corresponding cDNA, which was detected in calyx cells and hybridized to wasp genomic DNA but did not hybridize to DNAs in CsIV virions (Deng *et al.*, 2000). These results indicated that *p44* resides in the *C. sonorensis* genome but did not provide any evidence for viral ancestry because the genes share no homology with any known gene.

Scaling by proteomic analysis of virions, Sanger sequencing of ovary cDNAs, and sequencing bacterial artificial chromosome (BAC) genomic libraries yielded the first evidence that viral core genes reside in the genomes of two microgastrine braconids, *Cotesia congregata* and *Chelonus inanitus* (Fig. 8.1), that produce the BVs CcBV and CiBV (Bézier *et al.*, 2009; Wetterwald *et al.*, 2010). Proteomic analysis of MdBV virions from *Microplitis demolitor* (Fig. 8.1) together with deep sequencing of wasp ovaries using the Illumina platform further identified the full set of viral core genes identifiable by homology (Burke and Strand, 2012b; Burke *et al.*, 2013). Altogether, these studies identified 42 genes that are specifically upregulated in wasp ovaries when BV virions are being produced and that share significant homology with core genes in two families of large DNA viruses: the *Nudiviridae* and *Baculoviridae* (see Chapter 9). This finding was intuitively appealing because nudiviruses and baculoviruses both infect insects (Rohrmann, 2013) and because nudiviruses produce virions that are morphologically similar to BVs (Stoltz and Vinson, 1979).

All baculoviruses and nudiviruses have genomes that consist of a single large circular dsDNA. Six nudiviruses have been sequenced which share 33 genes with one another, forming a candidate core gene set (Wang and Jehle, 2009; Jehle *et al.*, 2013; Bézier *et al.*, 2015). Results further place these nudiviruses into two groups: the *alpha*- and *betanudivirus* genera (Bézier *et al.*, 2015) (Fig. 8.2). Of the 33 candidate nudivirus core genes, 21 are also shared with all baculoviruses (Garavaglia *et al.*, 2012; Rohrmann, 2013), which supports that the *Nudiviridae* and *Baculoviridae* are sister families (Jehle *et al.*, 2013; Bézier *et al.*, 2015; Burke *et al.*, 2018a) (Fig. 8.2). None of the candidate core genes in nudiviruses have been functionally studied, but those shared with baculoviruses have

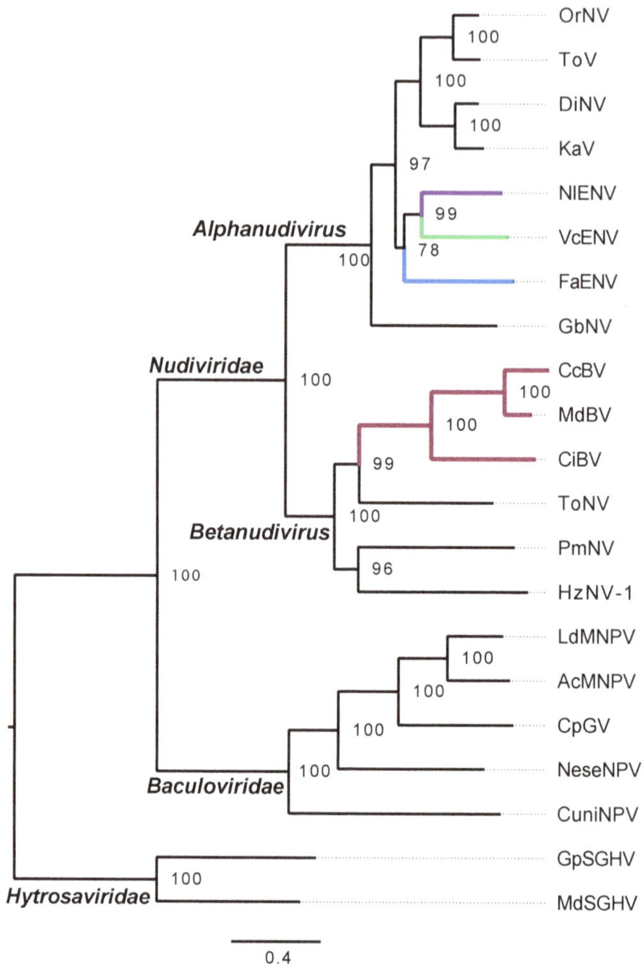

Figure 8.2 A maximum likelihood phylogeny for the Baculoviridae and Nudiviridae with the Hytrosaviridae as the outgroup. Numbers on nodes indicate support from 1000 bootstrap replicates. Black branches indicate exogenous taxa in each family while coloured branches identify endogenous taxa. Representative species for exogenous taxa include: Tipula oleracea nudivirus (ToNV), Heliothis zea nudivirus 1 (HzNV-1), Penaeus monodon nudivirus (PmNV), *Drosophila* innubila nudivirus (DiNV), Kallithea virus (KaV), Tomelloso virus (ToV), Oryctes rhinoceros nudivirus (OrNV), Gryllus bimaculatus nudivirus (GbNV), Autographa californica multiple nucleopolyhedrovirus (AcMNPV), Lymantria dispar multiple nucleopolyhedrovirus (LdMNPV), Cydia pomonella granulovirus (CpGV), Neodiprion sertifer nucleopolyhedrovirus (NeseMNPV), Culex nigripalpus nucleopolyhedrosis virus (CuniNPV), Glossina pallidipes salivary gland hytrosavirus (GpSGHV), and Musca domestica salivary gland hytrosavirus (MdSGHV). Representative species for endogenous taxa includes: Microplitis demolitor bracovirus (MdBV), Cotesia congregata bracovirus (CcBV), and Chelonus insularis bracovirus (CiBV) (magenta); Fopius arisanus endogenous nudivirus (FaENV) (blue); Venturia canescens endogenous nudivirus (VcENV) (green); and Nilaparvata lugens endogenous nudivirus (NlENV) (purple).

been functionally studied in model isolates like *Autographa californica multiple nucleopolyhedrovirus* (AcMNPV) (Rohrmann, 2013) (Fig. 8.2). These include a DNA polymerase (*dnapol*) and helicase (*helicase*) required for replicating the viral genome, four subunits (*lef-4, lef-8, lef-9, p47*) of a DNA dependent RNA polymerase, and structural genes such as *vp39, vlf-1, p74* and *pif-1* present in virions and containing promoter sequences specifically recognized by viral RNA polymerase (Rohrmann, 2013).

Orthologues to 21 of the core genes shared between nudiviruses and baculoviruses were identified in the *M. demolitor* ovary transcriptome data set and partially identified in *C. congregata* (Bézier *et al.*, 2009; Burke and Strand, 2012b). The only omission in *M. demolitor* was *dnapol*, which suggests that amplification of the DNAs packaged into BV virions requires wasp replication machinery (Burke and Strand, 2012b). In contrast, the presence of all of the RNA polymerase subunits and structural genes suggested that BV virion assembly is conserved with nudiviruses and baculoviruses. Further support for this latter conclusion comes from: (1) detection of all nudivirus-like structural gene products in MdBV virions and partial identification of these products in CcBV and CiBV (Bézier *et al.*, 2009; Wetterwald *et al.*, 2010; Burke *et al.*, 2013), and (2) RNAi experiments which show that the RNA polymerase subunit genes produce a functional enzyme in *M. demolitor* that transcribes BV structural genes but not wasp genes (Burke *et al.*, 2013). RNAi knockdown of *vp39*, *p74*, and *pif-1* also indicates each is required for producing normal virions (Burke *et al.*, 2013). A tyrosine recombinase named *integrase* (*int*) related to baculovirus *vlf-1* is present in all nudiviruses and BV-carrying wasps, which suggests *int* genes are also part of a BV core gene set. Functional studies further implicate *int-1* and *vlf-1* in *M. demolitor* in processing of the DNAs that are packaged into MdBV virions (Burke *et al.*, 2013, 2015).

Proteomic analysis of virions, sequencing of ovary cDNAs and sequencing of genomic BAC clones has also been conducted in two campoplegine (*Hyposoter didymator* and *Tranosema rostrale*) and one banchine ichneumonid wasp (*Glypta fumiferanae*) that produce HdIV, TrIV, and GfIV, respectively (Volkoff *et al.*, 2010; Béliveau *et al.*, 2015) (Fig. 8.1). Volkoff *et al.* (2010) identified several structural genes in HdIV and TrIV. None shared homology with structural genes in any known virus, but in *H. didymater* these genes reside in clusters of 9–16 genes in three regions of genomic DNA, ichnovirus structural protein encoding regions (IVSPERs), that suggest a viral origin (Volkoff *et al.*, 2010). Comparisons to GfIV further indicated the inventory of IVSPER-associated structural genes were very similar to HdIV and TrIV, which supports the presumption that campoplegine and banchine IVs have similar ancestors (Béliveau *et al.*, 2015). Altogether, these findings strongly support the conclusion that IVs are not related to BVs, but whether the IV ancestor is an extinct or undiscovered virus remains unknown.

The genes and DNA segments required to produce BVs are dispersed in the genomes of wasps

The third step in genome analysis of PDVs was determining how the DNA segments are packaged into virions and how the core genes required for virion formation are organized. Currently, most insights derive from BV-carrying braconids. Sequencing of bacterial artificial chromosome (BAC) libraries against four species of braconids (*Glyptapanteles indiensis*, *G. flavicoxis*, *C. congregata*, *C. sesamiae*), plus whole genome sequencing of *M. demolitor*, show that the DNA segments in corresponding virions (GiBV, GfBV, CcBV, CsBV, and MdBV) reside in 6–8 'proviral' loci as integrated, linear DNAs (Desjardens *et al.*, 2007, 2008; Bézier *et al.*, 2013; Burke *et al.*, 2014, 2018b). In each wasp species, two of these loci contain multiple segments arranged in tandem, while the others contain only one or two segments. Within species comparisons indicate adjoining segments in a given locus are more similar to one another than to segments in other loci, which suggests the DNAs packaged into virions primarily evolve through tandem duplication events followed by sequence divergence (Desjardins *et al.*, 2008; Bézier *et al.*, 2013; Burke *et al.*, 2014).

Early studies suggested that replication of proviral segments begins with excision from the wasp genome in calyx cells followed by circularization and amplification by a rolling-circle mechanism (Gruber *et al.*, 1996; Savary *et al.*, 1997; Pasquier-Barre *et al.*, 2002). However, subsequent studies showed that unknown factors amplify proviral loci at discrete boundaries, which results in the formation of large, concatenated DNA intermediates that are then processed and circularized before packaging into nucleocapsids (Louis *et al.*, 2013; Burke *et al.*, 2015). Specific nucleotides in repeat domains identify the boundaries of amplification for proviral loci, while *int*/*vlf-1*-dependent processing occurs in direct repeats containing the tetramer AGCT, which flanks each segment. This results in formation of circularized DNAs that are packaged into virions (Annaheim and Lanzrein, 2007; Louis *et al.*, 2013; Burke *et al.*, 2013, 2015). Wasp integration motifs (WIMs) or direct repeat junctions (DRJs) are the sites of processing and circularization that flank each segment (Beck *et al.*, 2011; Louis *et al.*, 2013). Most circularized DNA segments in BV virions also contain a single conserved inverted repeat domain, the host integration motif (HIM), where circularized segments integrate into the genome of infected host cells (Beck *et al.*, 2011; Chevignon *et al.*, 2018). This enables segments to persist and express associated genes in infected hosts for long periods in the absence of replication. The integration of BV DNAs into hosts has also potentially contributed to horizontal transfer of BV genes to Lepidoptera through presumptive infection of germ cells (Beck *et al.*, 2011; Gasmi *et al.*, 2015; Drezen *et al.*, 2017).

Sequencing BAC genomic library clones from *C. congregata* showed that some nudivirus-like genes reside in an 18 kb domain, with one gene (*odv-e66-like*) located between proviral loci 1 and 2 (Bézier *et al.*, 2009). However, most insights about the locations of BV proviral loci and nudivirus-like replication genes in the genomes of wasps derive from sequencing of *M. demolitor* (Burke *et al.*, 2014, 2018b). The most current assembly (Mdem2) indicates the *M. demolitor* genome is 241 Mb and has an N50 scaffold size of 1.1 Mb (Burke *et al.*, 2018b). The 8 MdBV proviral loci reside on 5 scaffolds, while all of the nudivirus-like replication genes reside on 24 scaffolds (Fig. 8.3A). Proviral locus 2, located on a 2.4 Mb scaffold, is ≈75 kb away from one nudivirus-like gene (*p74*) and 323 kb away from several others (*35a-8* to *35a-14*; *odv-e66–9* to *odv-e66–20*; *35a-6* and *35a-7*; and *helicase*) (Fig. 8.3A). Another nudivirus-like gene, *HZNVorf93-like*, is located close to DNA segment T (Fig. 8.3A). An additional 20 nudivirus-like replication genes reside in a 75 kb region of another scaffold with a subset of these genes present in the same order and orientation in the *C. congregata* genome (Fig. 8.3B). These data also indicate this nudivirus-like gene cluster has remained stable since divergence of *M. demolitor* and *C. congregata* approximately 53 million years ago, which suggests dispersal of the other nudivirus-like genes occurred relatively early in BV evolution (Burke *et al.*, 2014). However, this cluster likely resides more than 1 Mb away from any proviral locus. All remaining nudivirus-like genes, which include the 4 essential RNA polymerase subunits and several virion components, are flanked by insect genes on different scaffolds and thus are also distantly located from: (1) all proviral loci, (2) the scaffold on which the cluster of 20 nudivirus-like genes are located, and (3) one another except for local gene family expansion events (Burke *et al.*, 2018b) (Fig. 8.3A). No nudivirus-like genes are flanked by the conserved repeats that identify the amplification boundaries for proviral loci and processing boundaries for segment circularization. This architectural feature is why no DNA regions containing nudivirus-like genes are amplified and packaged into virions. Lastly, while all nudivirus and baculovirus genes are intronless, four nudivirus-like genes in *M. demolitor* (*lef-8*, *lef-9*, *HzNVorf128-like*, and *K425_12*) contain small introns, which, as previously noted, is also the case for some but not all virulence genes in proviral loci. Thus, acquisition of introns by some MdBV genes is likely a feature linked to transcription by wasp or host machinery. Overall, results from *M. demolitor* suggest proviral loci and nudivirus-like replication genes are likely dispersed in all microgastrine braconids. However, confirming this and assessing whether MdBV proviral loci and replication genes are fixed in the same or different locations across microgastroid genomes will require data from other species.

BVs are endogenous virus elements (EVEs) that retain many ancestral functions

Endogenous virus elements (EVEs) are DNA sequences from viruses that have integrated and become fixed in the germline of a host resulting in strict vertical inheritance (Katzourkakis and Gifford, 2010; see also Chapter 2). Most known EVEs derive from retroviruses (eukaryotes) and phages (prokaryotes), which is consistent with the former obligately integrating into hosts as part of replication and the latter either obligately or facultatively integrating into a broad range of bacteria and archaea (Calendar, 2005; Greenwood *et al.*, 2018). Other types of viruses, including several that are not known to integrate into the genomes of hosts, have also endogenized in different organisms through unknown mechanisms (Katzourkakis and Gifford, 2010; Holmes, 2011b; Aswad and Katzourkakis, 2012). Most EVEs are ancient and persist as non-functional fragments that have decayed through accumulation of neutral mutations. However, a few individual genes or regulatory elements of viral origin have been coopted by hosts for novel functions that provide fitness benefits in development and control of particular genes (Feschotte and Gilbert, 2012; Stoye, 2012).

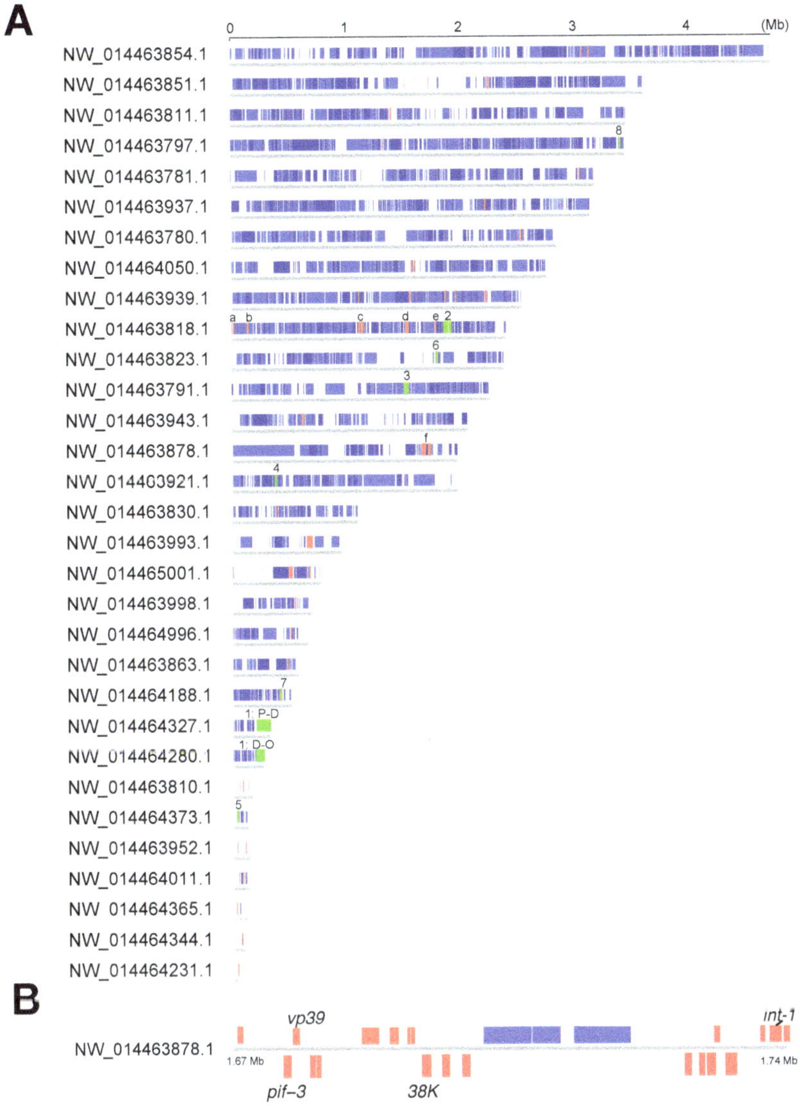

Figure 8.3 Dispersal of MdBV in the genome of *Microplitis demolitor.* (A) Scaffolds in the *M. demolitor* genome that contain proviral loci and/or nudivirus-like genes. Wasp genes are indicated using light (forward strand) or dark (reverse strand) blue bars; proviral loci are indicated using green bars, and nudivirus-like genes are indicated using pink bars. Each proviral locus is identified numerically as established in prior papers (Burke *et al.*, 2014). Proviral locus 1 is split across two scaffolds with the proviral segments that identify boundaries for the locus (segment P, D, O) indicated. Proviral segments in loci 2–8 are not shown. Select nudivirus-like genes on different scaffolds are identified by small case letters: (a) *helicase*; (b) *35a-6* and *-7*; (c) 3, *odv-e66-1*, *-2*, *-9*, *-10*, *-17*, *-19*, *-20*; (d) *35a-8* through *-14*; (e) *p74*; (f) *HzNVorf93-like*; and (g) the cluster of nudivirus-like replication genes on scaffold NW_014463878.1. (B) Higher resolution schematic of the nudivirus-like gene cluster on scaffold NW_014463878.1 with select nudivirus-like (pink) and wasp genes (blue) on the forward (upper) or reverse (lower) strand identified.

Current results strongly support that BVs evolved from a betanudivirus that infected the common ancestor of microgastroid braconids approximately 100 million years ago; IVs also likely evolved from a virus ancestor (Volkoff *et al.*, 2010; Bézier *et al.*, 2015). Fixation of all DNA components required to produce BVs in the genomes of wasps indicate these entities are now EVEs that have been coopted by wasps for beneficial functions in parasitism (Strand and Burke, 2013; Herniou *et al.*, 2013; Burke *et al.*,

2014). In turn, BVs and IVs can no longer be viewed as beneficial symbionts of wasps because no BV traits exist outside of the wasp genome.

Yet, BVs and IVs also differ from most known EVEs because they retain a remarkably large number of ancestral features. This begins with overall genome composition where nudiviruses encode a small proportion of conserved core genes and a large proportion of non-core genes from diverse sources with functions in infecting hosts. Most of the same core genes in nudiviruses and baculoviruses remain in BV-producing wasps while experimental studies indicate that BV core genes also have similar functions as determined primarily in experimental studies of baculoviruses. BV proviral loci contain virulence genes, which have functions in parasitism of hosts by wasps, that also potentially derive from the nudivirus ancestor. A crucial difference between nudiviruses and BVs is that all components of nudivirus genomes reside on a single, large dsDNA that can be amplified, packaged into virions, and horizontally transferred to another host, while BV components are dispersed in the genomes of wasps. Given results showing that the titres of MdBV virions produced in *M. demolitor* calyx cells greatly exceed baculovirus titres produced in permissive cell lines, dispersal of the ancestral genome is not a barrier for function (Burke *et al.*, 2012a, 2014). This is because the MdBV RNA polymerase subunits, once expressed and assembled into a holoenzyme, specifically transcribe other core genes required for virion formation through promoter recognition regardless of their location in the wasp genome (Burke *et al.*, 2013, 2014; Strand and Burke, 2015). In turn, incompletely characterized DNA replication machinery plus two nudivirus-like integrases (*int-1*, *vlf-1*) use the repeat boundaries that flank proviral segments for proper amplification, processing and packaging into virions. That motifs like WIMs are required for proper processing further suggests these flanking motifs derive from the nudivirus ancestor. In contrast, genome dispersal is a key factor in explaining how BVs amplify and package a specific population of DNAs containing virulence genes into infectious virions, but package no DNAs containing core genes, which results in BVs being replication defective (Strand and Burke, 2015).

A challenge then is what to call and how to classify PDVs. Some have suggested PDVs are a type of organelle (Federici and Bigot, 2003; Whitfield and Asgari, 2003) or could be named using a novel term like 'suppresson', which refers to their generally inhibitory effects on parasitized hosts (Kroemer and Webb, 2004). Evolution of BVs clearly involved gene transfer from an ancestral nudivirus to the germ line of a wasp, as occurred in the symbiogenesis of mitochondria from bacteria (Margulis, 1981). However, the word organelle does not accurately reflect what PDVs are because mitochondria still retain small autonomously replicating genomes that provide the potential for evolving separately from the rest of the host genome. Kroemer and Webb (2004) also noted that 'a derivative terminology', like suppresson, 'whether based on viruses or organelles, is likely to prove insufficient to describe PDVs, which clearly have a unique set of genetic, evolutionary, and physiological attributes.'

The closest parallel to PDVs outside of findings discussed below are gene transfer agents (GTAs), which are phage-like entities in certain proteobacteria, spirochaetes and archaea (Lang *et al.*, 2012). Like PDVs, GTAs produce particles from ancestral phage genes that are fixed in the bacterial genome and package DNA into particles that the GTA can transfer to another cell. However, GTAs differ from PDVs in that the DNAs they package are random, small ($\approx$ 5 kb or less) pieces of the host genome that sometimes contain GTA genes (Lang *et al.*, 2012). Whether GTAs have been coopted by their hosts for beneficial functions is also unclear.

Taken together, the PDV community continues to use the terms polydnavirus, bracovirus and ichnovirus but has largely coalesced in recognizing that BVs and IVs are complex EVEs that retain many functions of their virus ancestors (Strand and Burke, 2013, 2014; Bézier *et al.*, 2013; Drezen *et al.*, 2017). This is perhaps most analogous to endogenous retroviruses (ERVs), which share many traits with exogenous retroviruses (XRVs) (Greenwood *et al.*, 2018). ERVs and XRVs are both used in phylogenies for the *Retroviridae* (Greenwood *et al.*, 2018). Recent phylogenies for the *Nudiviridae* do so as well (Bézier *et al.*, 2015; Pichon *et al.*, 2015; Burke *et al.*, 2018a) (see Fig. 8.2 and below). Thus, endogenous nudivirus (ENV) is likely the most parsimonious name for BVs. If adopted, *M. demolitor bracovirus* (MdBV) would become MdENV or *M. demolitor* endogenous betanudivirus (MdEBENV) while other endogenized nudiviruses would be named by the host species and potential genus in which they reside. This approach is not possible for IVs because their exogenous virus ancestor remains unknown.

Other nudivirus-derived EVEs exist

Nudiviruses and baculoviruses infect a variety of insects in several orders but, similar to parasitoids, most isolates have narrow ranges (Rohrmann, 2013). Most isolates also cause systemic fatal infections. Baculovirus genomes remain episomal in infected host cells and productively replicate through the coordinated expression of early genes like *dnapol* that amplify the genome and RNA polymerase subunits that transcribe the structural genes needed to produce infectious virions. Nudiviruses replicate similarly, but two very similar betanudiviruses, HzNV-1 and HzNV-2, preferentially infect the reproductive system of hosts rather than establish systemic infections (Lu and Burand, 2001; Burand *et al.*, 2012). These isolates can also establish persistent infections *in vitro* by integrating into the genome of infected cells (Wu *et al.*, 2010, 2011). This suggests the betanudivirus ancestor of BVs initially established a persistent infection in the progenitor of the microgastroid complex by infecting its reproductive system and integrating into the germ line (Herniou *et al.*, 2013; Strand and Burke, 2013). Initial integration of multiple copies of the ancestral genome or duplication of the ancestral genome was then followed by rearrangements and mutation that resulted in fixation and dispersal in the genomes of wasps. Other alterations presumably restricted nudivirus-like replication gene expression to calyx cells in pupal and adult stage female parasitic wasps, which enabled the formation of replication defective virions that could be stored in the lumen of the reproductive tract and deliver virulence genes to the wasps' hosts.

It is impossible to know the precise series of events that led to the evolution of BVs given the ancestral infection event occurred approximately 100 million years ago (see above). Because all microgastrine braconids studied to date exhibit these traits, it is likely that genome dispersal, the architecture of proviral loci, and the restriction of replication to calyx cells occurred relatively early in the association. If nudiviruses more generally exhibit tropisms for infecting reproductive organs and/or the capacity to establish persistent infections, it is also possible other endogenization events have occurred that could provide insights into whether nudiviruses exhibit similar or different fates as BVs.

Notably, recent studies identify three examples of endogenized nudiviruses evolving independently of BVs. The first was discovered by sequencing the brown planthopper *Nilaparvata lugens* (Hemiptera), which identified 32 of the 33 known nudivirus core genes in the genome (Cheng *et al.*, 2014). Analysis of concatenated amino acid alignments for these genes indicate the ancestor of *N. lugens* endogenous nudivirus (NlENV) was an alphanudivirus (Bézier *et al.*, 2015) (Fig. 8.2). qPCR data show that some of the nudivirus core genes are transcribed in the gut and fat body of *N. lugens*, but EM studies detected no virions in any tissue including the reproductive tract (Cheng *et al.*, 2014).

The second example was identified in the parasitoid wasp *Venturia canescens* (Pichon *et al.*, 2015). This example is notable because this species is a campoplegine ichneumonid, which were all thought to carry IVs. *V. canescens* was one of the wasp species where virus-like particles were first observed (Rotheram, 1967). Like other campoplegine ichneumonids, *V. canescens* particles assemble in calyx cell nuclei and accumulate in the lumen of the reproductive tract. Particles also bind to the surface of wasp eggs and impede encapsulation by host immune cells after parasitism by wasps (Rotheram, 1967; Salt, 1973). The particles contain no nucleic acid and differ from IVs morphologically by having an amorphous electron dense centre surrounded by a single envelope (Rotheram, 1967; Bedwin, 1979). Subsequent studies identified three proteins (VcVLP1–3) in particles encoded by putative intron-containing wasp genes, suggesting they may not derive from any virus (Hellers *et al.*, 1996; Reineke *et al.*, 2002; Asgari *et al.*, 2003). Recent sequencing of the *V. canescens* genome, in contrast, identified 16 nudivirus-like genes in six dispersed clusters most closely related to alphanudiviruses (Figs. 8.2 and 8.4). Among these are nudivirus core genes including the four RNA polymerase subunit genes and structural genes, *per os* infectivity factors (*pifs*), that are envelope components in baculoviruses (Boogaard *et al.*, 2018) (Fig. 8.4). Immunolabeling studies confirmed the presence

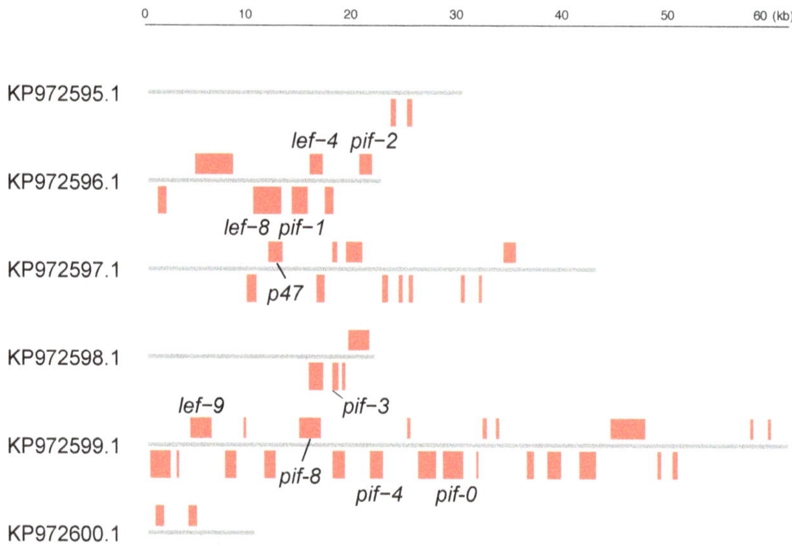

Figure 8.4 Dispersal of VcENV in the genome of *Venturia canescens*. Six scaffolds containing VcENV genes are shown with RNA polymerase subunit and *pif* genes on the forward (upper) or reverse (lower) strand identified.

of some *pif* gene encoded proteins in the envelope and VLP2 in the electron dense centre of *V. canescens* particles (Pichon *et al.*, 2015). In contrast, the *V. canescens* genome did not yield nudivirus-like genes with predicted functions in capsid formation (Pichon *et al.*, 2015). Sequence data also indicated that the IVSPER genes present in other campoplegine ichneumonids have been either pseudogenized or lost in *V. canescens* (Pichon *et al.*, 2015). These results collectively suggest alphanudivirus endogenization in *V. canescens* occurred subsequently to endogenization of the virus that gave rise to IVs (Pichon *et al.*, 2015). Whether VcENV is directly or indirectly responsible for the loss of IVSPERs is unclear.

The third example also derives from a parasitoid wasp, *Fopius arisanus*, which resides in the Braconidae in a subfamily (Opiniinae) outside the microgastroid complex (Burke *et al.*, 2018a) (Fig. 8.1). Sequencing of *F. arisanus* identified 55 nudivirus-like genes in 9 clusters and several other dispersed locations also derived from an alphanudivirus ancestor (Geib *et al.*, 2017; Burke *et al.*, 2018b) (Figs. 8.2 and 8.5). EM studies showed that particles assemble specifically in calyx cell nuclei of female wasps, which accumulate in the oviducts of ovaries (Burke *et al.*, 2018a). These particles consisting of long open-ended capsids that are individually surrounded by an envelope are strikingly different from BVs and the particles in *V. canescens*. The nudivirus genes present in *F. arisanus* include all of the nudivirus core genes except *dnapol*, *pif-5*, *p6.9*, and *vlf-1* (Fig. 8.5). Proteomic analysis confirmed that nudivirus envelope and capsid gene products are present in particles, but they do not contain nucleic acid (Burke *et al.*, 2018b).

Since *lef-8* is the most conserved gene among sequenced alphanudiviruses, Burke *et al.* (2018b) screened five other species of *Fopius* by PCR to determine if potentially other species in this genus contain an endogenized nudiviruses. Results identified *lef-8* orthologues in three species whose identity were confirmed by sequencing while no *lef-8* orthologues were identified in two others (Burke *et al.*, 2018a). Thus, detection of *lef-8* in only some *Fopius* species suggests endogenization likely occurred relatively recently in this genus of parasitoids (Burke *et al.*, 2018b). Whether the particles produced by *F. arisanus* have functions in parasitism of hosts in contrast is unknown.

These findings together indicate that alpha- and betanudiviruses independently endogenized in different insects. Endogenization can also result in loss of virion formation and infectivity (NlENV), formation of virions containing core gene products but no nucleic acid (VcENV and FaENV), or formation of infectious, replication-defective virions that package large DNAs containing virulence genes

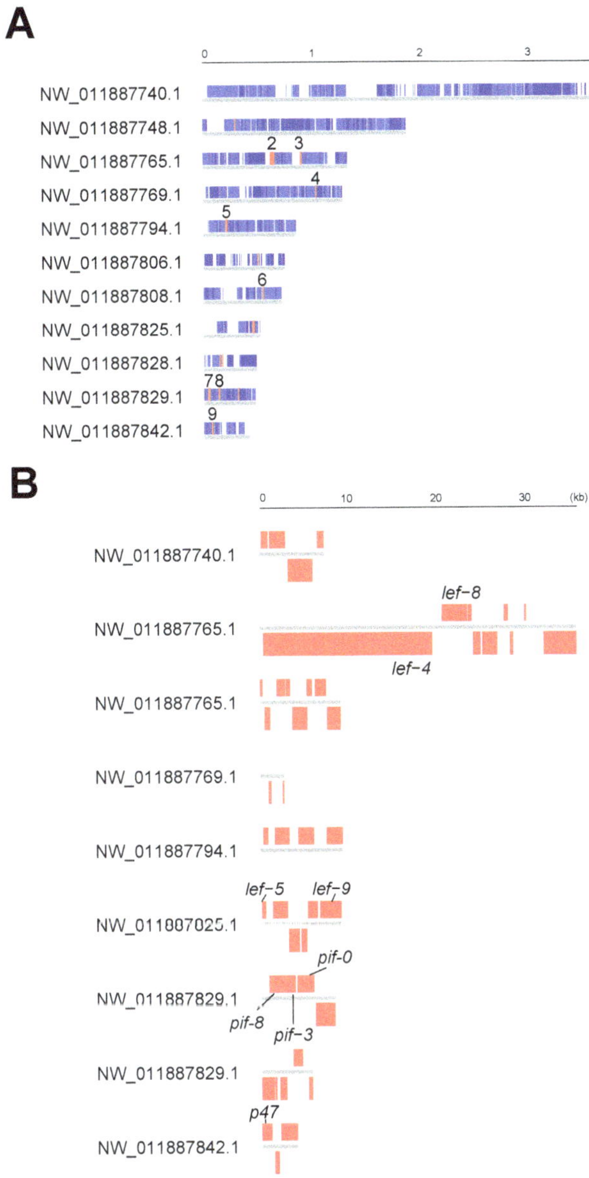

Figure 8.5 Dispersal of FaENV in the genome of *Fopius arisanus*. (A) The 11 scaffolds in the *F. arisanus* genome that contain FaENV genes. Wasp genes are indicated in light (forward strand) or dark (reverse strand) blue, while FaENV genes are indicated in pink. Numbers identify domains were multiple FaENV genes reside in close proximity to one another and form a cluster. (B) Higher resolution schematic of the nine FaENV gene clusters shown in (A). The location of the RNA polymerase subunit and *pif* genes on the forward (upper) or reverse (lower) strand of a given cluster are identified.

(BVs). A wide range in the age of these associations is suggested, with VcENV and FaENV showing features suggesting more recent endogenization than in microgastrine braconids. Despite varying fates and age, these endogenized nudiviruses share three similar architectural features: genome fragmentation, retention of multiple core genes, and loss of *dnapol*. NlENV, VcENV, and FaENV have only a single copy of each retained core gene, whereas some core genes in the more ancient association between BVs and microgastrine braconids form multimember families. As evidenced by insect genes flanking clusters of nudivirus genes, the most parsimonious explanation for the patterns observed in NIENV, VcENV and FaENV is that a single nudivirus genome integrated into the germline of each host species, followed by rearrangement. Dispersal of FaENV or VcENV is also less pronounced than for BVs, potentially reflecting an older association between BVs and microgastroids. The fact that all ENVs identified to date are dispersed suggests that genome fragmentation occurs relatively soon after endogenization.

Burke (2018) compared core gene inventories between parasitoid-associated ENVs and noted that the four subunits of the viral RNA polymerase (*lef-4*, *lef-8*, *lef-9*, *p47*) and select *pifs* are the only genes shared by all (see Figs. 8.3–8.5). However, no syntenic gene location exists between these ENVs (Figs. 8.3–8.5). RNA polymerase is essential for transcription of structural core genes in both baculoviruses and BVs (Rohrmann, 2013; Burke *et al.*, 2013). Selection also likely maintains functional RNA polymerase subunit genes in FaENV and VcENV because they are required for holoenzyme formation and expression of capsid and/or envelope genes. In baculoviruses, seven known *pif* genes produce envelope proteins that function together in a complex (*pif0–4*, *pif6*, *pif8*) while one (*pif-5*) encodes an envelope protein that does not interact with the others (Boogaard *et al.*, 2018). Retention of a complete set of PIFs in MdBV, VcENV and FaENV is consistent with each producing enveloped particles. In contrast, functional parallels to baculoviruses are less clear because BV-carrying wasps, *V. canescens*, and *F. arisanus* inject particles plus eggs into the hemocoel of hosts, which precludes interaction with midgut cells as *per os* infectivity factors. Moreover, RNAi knockdown of *pif-1* from MdBV has no effect on the number of DNA segments detected in host cells like haemocytes but does affect transcript abundance of virulence genes expressed in haemocytes (Burke *et al.*, 2013). This suggests *pif-1* may not be essential for entry of MdBV into haemocytes but instead may affect capsid migration and/or delivery of encapsidated DNAs to the nucleus where associated virulence genes are transcribed. The universal loss of *dnapol* among currently known ENVs may reflect a selective advantage to insects by preventing viral genomes from re-mobilizing, as previously hypothesized for BVs (Strand and Burke, 2013, 2014).

Future directions

When we wrote a chapter on PDVs for the first edition of *Insect Virology*, the first evidence that BVs evolved from a nudivirus had entered the literature (Bézier *et al.*, 2009). In contrast, the architecture of BV and IV genomes in wasps were largely unknown, no evidence existed that the DNAs in BV virions integrate into the genomes of hosts that wasps parasitize, and no other ENVs had been identified. Results now strongly support that all BVs derive from a betanudivirus and that IVs likely evolved from a virus whose identity remains unknown. Several additional ENVs have also been identified. Future revision of the Polydnaviridae is obviously needed but has been delayed because of issues related to revision of the Baculoviridae and Nudiviridae. Whether to include endogenous nudiviruses like BVs in classification of the Nudiviridae, as the retrovirus community does, must be decided as well as how to classify IVs in the absence of a known virus ancestor.

Mechanistically, many unanswered questions exist regarding how nudiviruses integrate to form persistent infections and the factors that regulate preferential infection of reproductive tissues by isolates like HzNV1 and 2. It is unclear whether the mechanism of integration by nudiviruses is in any way related to the features in BVs that regulate amplification and processing in wasps, or integration of circularized DNAs into the genomes of hosts that wasps parasitize. While both an early gene and a microRNA are implicated in maintenance and reactivation of HzNV1 persistent infections (Wu *et al.*, 2010, 2011), no orthologues of these products have been identified in BVs. The wasp genes required for amplification of BV proviral loci also remain unknown but are essential to understanding how these domains are mobilized in concert with particle formation in calyx cells. A final point of interest is whether the current pattern of most ENVs being associated with parasitic Hymenoptera reflects the biased efforts of individuals with interests in PDVs to identify ENVs, a legitimate bias in nudivirus infection of Hymenoptera, or a potential bias in parasitoids providing a selective environment that favours cooption of ENVs for roles in parasitism. The rapid increase in available genomes across different orders of insects should be able to address this question in the fairly near future.

Acknowledgements

Work discussed in this chapter includes research supported by grants from the National Institutes of Health (F32AI096552) (GRB), the US National Science Foundation (IOS-12611328) (MRS), (IOS-1748862) (GRB) and (DEB-1622986) (GRB), and the US Department of Agriculture (2009-35302-05250) (MRS)

References

Albrecht, U., Wyler, T., Pfister-Wilhelm, R., Gruber, A., Stettler, P., Heiniger, P., Schumperli, D., and Lanzrein, B. (1994). PDV of the parasitic wasp *Chelonus inanitus* (Braconidae): characterization, genome organization and time point of replication. J. Gen. Virol. *75*, 3353–3363. https://doi.org/10.1099/0022-1317-75-12-3353

Annaheim, M., and Lanzrein, B. (2007). Genome organization of the *Chelonus inanitus* polydnavirus: excision sites, spacers, and abundance of proviral and excised segments. J. Gen. Virol. *8*, 450–457. https://doi.org/10.1099/0022-1317-75-12-3353

Asgari, S., Reineke, A., Beck, M., and Schmidt, O. (2002). Isolation and characterization of a neprilysin-like protein from *Venturia canescens* virus-like particles. Insect Mol. Biol. *11*, 477–485.

Aswad, A., and Katzourakis, A. (2012). Paleovirology and virally derived immunity. Trends Ecol. Evol. *27*, 627–636. https://doi.org/10.1016/j.tree.2012.07.007

Beck, M., and Strand, M.R. (2005). Glc1.8 from *Microplitis demolitor* bracovirus induces a loss of adhesion and phagocytosis in insect high five and S2 cells. J. Virol. *79*, 1861–1870. https://doi.org/10.1128/JVI.79.3.1861-1870.2005

Beck, M.H., Inman, R.B., and Strand, M.R. (2007). *Microplitis demolitor* bracovirus genome segments vary in abundance and are individually packaged in virions. Virology *359*, 179–189. https://doi.org/10.1016/j.virol.2006.09.002

Beck, M.H., Zhang, S., Bitra, K., Burke, G.R., and Strand, M.R. (2011). The encapsidated genome of *Microplitis demolitor* bracovirus integrates into the host *Pseudoplusia includens*. J. Virol. *85*, 11685–11696. https://doi.org/10.1128/JVI.05726-11

Beckage, N.E., and Drezen, J.-M. (2012). Parasitoid Viruses. (Academic Press, San Diego, CA).

Bedwin, O. (1979). An insect glycoprotein: a study of the particles responsible for the resistance of a parasitoid's egg to the defence reactions of its insect host. Proc. R. Soc. Lond., B, Biol. Sci. *205*, 271–286. https://doi.org/10.1098/rspb.1979.0065

Béliveau, C., Cohen, A., Stewart, D., Periquet, G., Djoumad, A., Kuhn, L., Stoltz, D., Boyle, B., Volkoff, A.N., Herniou, E.A., *et al.* (2015). Genomic and proteomic analyses indicate that banchine and campoplegine polydnaviruses have similar, if not identical, viral ancestors. J. Virol. *89*, 8909–8921. https://doi.org/10.1128/JVI.01001-15

Bézier, A., Annaheim, M., Herbinière, J., Wetterwald, C., Gyapay, G., Bernard-Samain, S., Wincker, P., Roditi, I., Heller, M., Belghazi, M., *et al.* (2009). Polydnaviruses of braconid wasps derive from an ancestral nudivirus. Science *323*, 926–930. https://doi.org/10.1126/science.1166788

Bézier, A., Louis, F., Jancek, S., Periquet, G., Thézé, J., Gyapay, G., Musset, K., Lesobre, J., Lenoble, P., Dupuy, C., *et al.* (2013). Functional endogenous viral elements in the genome of the parasitoid wasp *Cotesia congregata*: insights into the evolutionary dynamics of bracoviruses. Philos. Trans. R. Soc. Lond., B, Biol. Sci. *368*, 20130047. https://doi.org/10.1098/rstb.2013.0047

Bézier, A., Thézé, J., Gavory, F., Gaillard, J., Poulain, J., Drezen, J.M., and Herniou, E.A. (2015). The genome of the nucleopolyhedrosis-causing virus from *Tipula oleracea* sheds new light on the Nudiviridae family. J. Virol. *89*, 3008–3025. https://doi.org/10.1128/JVI.02884-14

Bitra, K., Zhang, S., and Strand, M.R. (2011). Transcriptomic profiling of *Microplitis demolitor* bracovirus reveals host, tissue, and stage-specific patterns of activity. J. Gen. Virol. *92*, 2060–2071. https://doi.org/10.1099/vir.0.032680-0

Bitra, K., Suderman, R.J., and Strand, M.R. (2012). Polydnavirus Ank proteins function as IkB mimics that subvert the insect Imd signaling pathway. PLOS Pathog. *8*, e1002722. https://doi.org/10.1371/journal.ppat.1002722

Bitra, K., Burke, G.R., and Strand, M.R. (2016). Permissiveness of lepidopteran hosts is linked to differential expression of bracovirus genes. Virology *492*, 259–272. https://doi.org/10.1016/j.virol.2016.02.023

Burand, J.P., Kim, W., Afonso, C.L., Tulman, E.R., Kutish, G.F., Lu, Z., and Rock, D.L. (2012). Analysis of the genome of the sexually transmitted insect virus *Helicoverpa zea* nudivirus 2. Viruses *4*, 28–61. https://doi.org/10.3390/v4010028

Burke, G.R. (2018). Common themes in three independently derived endogenous nudivirus elements in parasitoid wasps. Curr. Opin. Insect Sci. *32*, 28–35. https://doi.org/10.1016/j.cois.2018.10.005

Burke, G.R., and Strand, M.R. (2012a). Polydnaviruses of parasitic wasps: domestication of viruses to act as gene delivery vectors. Insects *3*, 91–119. https://doi.org/10.3390/insects3010091

Burke, G.R., and Strand, M.R. (2012b). Deep sequencing identifies viral and wasp genes with potential roles in replication of *Microplitis demolitor* Bracovirus. J. Virol. *86*, 3293–3306. https://doi.org/10.1128/JVI.06434-11

Burke, G.R., and Strand, M.R. (2014). Systematic analysis of a wasp parasitism arsenal. Mol. Ecol. *23*, 890–901. https://doi.org/10.1111/mec.12648

Burke, G.R., Thomas, S.A., Eum, J.H., and Strand, M.R. (2013). Mutualistic polydnaviruses share essential replication gene functions with pathogenic ancestors. PLOS Pathog. *9*, e1003348. https://doi.org/10.1371/journal.ppat.1003348

Burke, G.R., Walton, K., Robertson, H., Whitfield, J.B., and Strand, M.R. (2014). Widespread genomic organization of an ancient viral integration event. PLOS Genetics 10, e1004660. https://doi.org/10.1371/journal.pgen.1004660

Burke, G.R., Simmonds, T.J., Thomas, S.A., and Strand, M.R. (2015). *Microplitis demolitor* Bracovirus proviral loci and clustered replication genes exhibit distinct dna amplification patterns during replication. J. Virol. *89*, 9511–9523. https://doi.org/10.1128/JVI.01388-15

Burke, G.R., Simmonds, T.J., Sharanowski, B.J., and Geib, S.M. (2018a). Rapid Viral Symbiogenesis via Changes in Parasitoid Wasp Genome Architecture. Mol. Biol. Evol. *35*, 2463–2474. https://doi.org/10.1093/molbev/msy148

Burke, G.R., Walden, K.K.O., Whitfield, J.B., Robertson, H.M., and Strand, M.R. (2018b). Whole genome sequence of the parasitoid wasp *Microplitis demolitor* that harbors an endogenous virus mutualist. G3 *8*, 2875–2880. https://doi.org/10.1534/g3.118.200308

Calender, R.L. (2005). The Bacteriophages (Oxford University Press, Oxford).

Chen, Y.F., Gao, F., Ye, X.Q., Wei, S.J., Shi, M., Zheng, H.J., and Chen, X.X. (2011). Deep sequencing of *Cotesia vestalis* bracovirus reveals the complexity of a polydnavirus genome. Virology *414*, 42–50. https://doi.org/10.1016/j.virol.2011.03.009

Cheng, R.L., Xi, Y., Lou, Y.H., Wang, Z., Xu, J.Y., Xu, H.J., and Zhang, C.X. (2014). Brown planthopper nudivirus DNA integrated in its host genome. J. Virol. *88*, 5310–5318. https://doi.org/10.1128/JVI.03166-13

Chevignon, G., Thézé, J., Cambier, S., Poulain, J., Da Silva, C., Bézier, A., Musset, K., Moreau, S.J., Drezen, J.M., and Huguet, E. (2014). Functional annotation of *Cotesia congregata* bracovirus: identification of viral genes expressed in parasitized host immune tissues. J. Virol. *88*, 8795–8812. https://doi.org/10.1128/JVI.00209-14

Chevignon, G., Periquet, G., Gyapay, G., Vega-Czarny, N., Musset, K., Drezen, J.M., and Huguet, E. (2018). *Cotesia congregata* bracovirus circles encoding PTP and Ankyrin genes integrate into the DNA of parasitized *Manduca sexta* hemocytes. J. Virol. *92*. https://doi.org/10.1128/JVI.00438-18

Cooper, T.H., Bailey-Hill, K., Leifert, W.R., McMurchie, E.J., Asgari, S., and Glatz, R.V. (2011). Identification of an in vitro interaction between an insect immune suppressor protein (CrV2) and G alpha proteins. J. Biol. Chem. *286*, 10466–10475. https://doi.org/10.1074/jbc.M110.214726

Davies, D.H., Strand, M.R., and Vinson, S.B. (1987). Changes in differential haemocyte count and *in vitro* behaviour of plasmatocytes from host *Heliothis virescens* caused by *Campoletis sonorensis* PDV. J. Insect Physiol. *33*, 143–153.

Deng, L., Stoltz, D.B., and Webb, B.A. (2000). A gene encoding a polydnavirus structural polypeptide is not encapsidated. Virology *269*, 440–450. https://doi.org/10.1006/viro.2000.0248

Desjardins, C.A., Gundersen-Rindal, D.E., Hostetler, J.B., Tallon, L.J., Fuester, R.W., Schatz, M.C., Pedroni, M.J., Fadrosh, D.W., Haas, B.J., Toms, B.S., *et al.* (2007). Structure and evolution of a proviral locus of *Glyptapanteles indiensis* bracovirus. BMC Microbiol. *7*, 61.

Desjardins, C.A., Gundersen-Rindal, D.E., Hostetler, J.B., Tallon, L.J., Fadrosh, D.W., Fuester, R.W., Pedroni, M.J., Haas, B.J., Schatz, M.C., Jones, K.M., *et al.* (2008). Comparative genomics of mutualistic viruses of *Glyptapanteles* parasitic wasps. Genome Biol. *9*, R183. https://doi.org/10.1186/gb-2008-9-12-r183

Dorémus, T., Cousserans, F., Gyapay, G., Jouan, V., Milano, P., Wajnberg, E., Darboux, I., Cônsoli, F.L., and Volkoff, A.N. (2014). Extensive transcription analysis of the *Hyposoter didymator* Ichnovirus genome in permissive and non-permissive lepidopteran host species. PLOS ONE *9*, e104072. https://doi.org/10.1371/journal.pone.0104072

Dover, B.A., Davies, D.H., Strand, M.R., Gray, R.S., Keeley, L.L., and Vinson, S.B. (1987). Ecdysteroid titer reduction and developmental arrest of last-instar *Heliothis virescens* larvae by calyx fluid from the parasitoid *Campoletis sonorensis*. J. Insect Physiol. *33*, 333–338.

Drezen, J.M., Leobold, M., Bézier, A., Huguet, E., Volkoff, A.N., and Herniou, E.A. (2017). Endogenous viruses of parasitic wasps: variations on a common theme. Curr. Opin. Virol. *25*, 41–48. https://doi.org/10.1016/j.coviro.2017.07.002

Edson, K.M., Vinson, S.B., Stoltz, D.B., and Summers, M.D. (1981). Virus in a parasitoid wasp: suppression of the cellular immune response in the parasitoid's host. Science *211*, 582–583.

Espagne, E., Dupuy, C., Huguet, E., Cattolico, L., Provost, B., Martins, N., Poirié, M., Periquet, G., and Drezen, J.M. (2004). Genome sequence of a polydnavirus: insights into symbiotic virus evolution. Science *306*, 286–289. https://doi.org/10.1126/science.1103066

Falabella, P., Caccialupi, P., Varricchio, P., Malva, C., and Pennacchio, F. (2006). Protein tyrosine phosphatases of *Toxoneuron nigriceps* bracovirus as potential disrupters of host prothoracic gland function. Arch. Insect Biochem. Physiol. *61*, 157–169. https://doi.org/10.1002/arch.20120

Feschotte, C., and Gilbert, C. (2012). Endogenous viruses: insights into viral evolution and impact on host biology. Nat. Rev. Genet. *13*, 283–296. https://doi.org/10.1038/nrg3199

Fleming, J.A. (1992). Polydnaviruses: mutualists and pathogens. Annu. Rev. Entomol. *37*, 401–425. https://doi.org/10.1146/annurev.en.37.010192.002153

Fleming, J.A., Blissard, G.W., Summers, M.D., and Vinson, S.B. (1983). Expression of *Campoletis sonorensis* virus in the parasitized host, *Heliothis virescens*. J. Virol. *48*, 74–78.

Fleming, J.G., and Summers, M.D. (1991). Polydnavirus DNA is integrated in the DNA of its parasitoid wasp host. Proc. Natl. Acad. Sci. U.S.A. *88*, 9770–9774.

Forbes, A.A., Bagley, R.K., Beer, M.A., Hippee, A.C., and Widmayer, H.A. (2018). Quantifying the unquantifiable: why Hymenoptera, not Coleoptera, is the most speciose animal order. BMC Ecol. *18*, 21. https://doi.org/10.1186/s12898-018-0176-x

Forterre, P. (2006). The origin of viruses and their possible roels in major evolutionary transitions. Virus Res. *117*, 5–16. https://doi.org/10.1016/j.virusres.2006.01.010

Garavaglia, M.J., Miele, S.A., Iserte, J.A., Belaich, M.N., and Ghiringhelli, P.D. (2012). The ac53, ac78, ac101, and ac103 genes are newly discovered core genes in the family Baculoviridae. J. Virol. *86*, 12069–12079. https://doi.org/10.1128/JVI.01873-12

Gasmi, L., Boulain, H., Gauthier, J., Hua-Van, A., Musset, K., Jakubowska, A.K., Aury, J.M., Volkoff, A.N., Huguet, E., Herrero, S., *et al.* (2015). Recurrent domestication by lepidoptera of genes from their parasites mediated by bracoviruses. PLOS Genet. *11*, e1005470. https://doi.org/10.1371/journal.pgen.1005470

Gauthier, J., Drezen, J.M., and Herniou, E.A. (2018). The recurrent domestication of viruses: major evolutionary transitions in parasitic wasps. Parasitology *145*, 713–723. https://doi.org/10.1017/S0031182017000725

Geib, S.M., Liang, G.H., Murphy, T.D., and Sim, S.B. (2017). Whole genome sequencing of the braconid parasitoid wasp *Fopius arisanus*, an important biocontrol agent of pest tepritid fruit flies. G3 *7*, 2407–2411. https://doi.org/10.1534/g3.117.040741

Godfray, H.C.J. (1994). Insect Parasitoids (Princeton University Press, Princeton, NJ).

Greenwood, A.D., Ishida, Y., O'Brien, S.P., Roca, A.L., and Eiden, M.V. (2018). Transmission, evolution, and endogenization: lessons learned from recent retroviral invasions. Microbiol. Mol. Biol. Rev. *82*, e00044-17.

Gruber, A., Stettler, P., Heiniger, P., Schümperli, D., and Lanzrein, B. (1996). Polydnavirus DNA of the braconid wasp *Chelonus inanitus* is integrated in the wasp's genome and excised only in later pupal and adult stages of the female. J. Gen. Virol. *77*, 2873–2879. https://doi.org/10.1099/0022-1317-77-11-2873

Gueguen, G., Kalamarz, M.E., Ramroop, J., Uribe, J., and Govind, S. (2013). Polydnaviral ankyrin proteins aid parasitic wasp survival by coordinate and selective inhibition of hematopoietic and immune NF-kappa B signaling in insect hosts. PLOS Pathog. *9*, e1003580. https://doi.org/10.1371/journal.ppat.1003580

Gundersen-Rindal D., Dupuy, C., Huguet, E., and Drezen, J.-M. (2013). Parasitoid polydnaviruses: evolution, pathology and applications. Biocontrol Sci. Tech. *23*, 1–61. https://doi.org/10.1080/09583157.2012.731497

Hellers, M., Beck, M., Theopold, U., Kamei, M., and Schmidt, O. (1996). Multiple alleles encoding a virus-like particle protein in the ichneumonid endoparasitoid *Venturia canescens*. Insect Mol. Biol. *5*, 239–249.

Heraty, J., Ronquist, F., Carpenter, J.M., Hawks, D., Schulmeister, S., Dowling, A.P., Murray, D., Munro, J., Wheeler, W.C., Schiff, N., *et al.* (2011). Evolution of the hymenopteran megaradiation. Mol. Phylogenet. Evol. *60*, 73–88. https://doi.org/10.1016/j.ympev.2011.04.003

Herniou, E.A., Huguet, E., Thézé, J., Bézier, A., Periquet, G., and Drezen, J.M. (2013). When parasitic wasps hijacked viruses: genomic and functional evolution of polydnaviruses. Philos. Trans. R. Soc. Lond., B, Biol. Sci. *368*, 20130051. https://doi.org/10.1098/rstb.2013.0051

Holmes, E.C. (2011a). What does virus evolution tell us about virus origins? J. Virol. *85*, 5247–5251. https://doi.org/10.1128/JVI.02203-10

Holmes, E.C. (2011b). The evolution of endogenous viral elements. Cell Host Microbe *10*, 368–377. https://doi.org/10.1016/j.chom.2011.09.002

Huguet, E., Serbielle, C., and Moreau, J.M. (2012). Evolution and origin of polydnavirus virulence genes. In Parasitoid Viruses Symbionts and Pathogens, Beckage, N.E., and Drezen, J.-M., eds. (Academic Press, San Diego, CA), pp. 63–78.

Ibrahim, A.M., and Kim, Y. (2008). Transient expression of protein tyrosine phosphatases encoded in *Cotesia plutellae* bracovirus inhibits insect cellular immune responses. Naturwissenschaften *95*, 25–32. https://doi.org/10.1007/s00114-007-0290-7

Ignesti, M., Ferrara, R., Romani, P., Valzania, L., Serafini, G., Pennacchio, F., Cavaliere, V., and Gargiulo, G. (2018). A polydnavirus-encoded ANK protein has a negative impact on steroidogenesis and development. Insect Biochem. Mol. Biol. 95, 26–32.

Iyer, L.M., Balaji, S., Koonin, E.V., and Aravind, L. (2006). Evolutionary genomics of nucleo-cytoplasmic large DNA viruses. Virus Res. *117*, 156–184. https://doi.org/10.1016/j.virusres.2006.01.009

Jehle, J.A., Herniou, E.A., Harrison, R., Arif, B., Thielmann, D., van Oers, M., and Becnel, J. (2013). Creation of a new family Nudiviridae including two new genera and three species. Taxonomy Proposals. Available online: http://talk.ictvonline.org/files/proposals/taxonomy_proposals_invertebrate1/m/inv04/4770.aspx. Accessed March 2019.

Katzourakis, A., and Gifford, R.J. (2010). Endogenous viral elements in animal genomes. PLOS Genet. *6*, e1001191. https://doi.org/10.1371/journal.pgen.1001191

Kim, J., Hepat, R., Lee, D., and Kim, Y. (2013). Protein tyrosine phosphatase encoded in *Cotesia plutellae* bracovirus suppresses a larva-to-pupa metamorphosis of the diamondback moth, *Plutella xylostella*. Comp. Biochem. Physiol., Part A Mol. Integr. Physiol. *166*, 60–69. https://doi.org/10.1016/j.cbpa.2013.04.025

Krell, P.J., and Stoltz, D.B. (1979). Unusual baculovirus of the parasitoid wasp *Apanteles melanoscelus*: Isolation and preliminary characterization. J. Virol. *29*, 1118–1130.

Krell, P.J., and Stoltz, D.B. (1980). Virus-like particles in the ovary of an ichneumonid wasp: purification and preliminary characterization. Virology *101*, 408–418.

Krell, P.J., Summers, M.D., and Vinson, S.B. (1982). Virus with a multipartite superhelical DNA genome from the ichneumonid parasitoid *Campoletis sonorensis*. J. Virol. *43*, 859–870.

Kroemer, J.A., and Webb, B.A. (2004). Polydnavirus genes and genomes: emerging gene families and new insights into polydnavirus replication. Annu. Rev. Entomol. *49*, 431–456. https://doi.org/10.1146/annurev.ento.49.072103.120132

Kwon, B., and Kim, Y. (2008). Transient expression of an EP1-like gene encoded in *Cotesia plutellae* bracovirus suppresses the hemocyte population in the diamondback moth, *Plutella xylostella*. Dev. Comp. Immunol. *32*, 932–942. https://doi.org/10.1016/j.dci.2008.01.005

Labropoulou, V., Douris, V., Stefanou, D., Magrioti, C., Swevers, L., and Iatrou, K. (2008). Endoparasitoid wasp bracovirus-mediated inhibition of hemolin function and lepidopteran host immunosuppression. Cell. Microbiol. *10*, 2118–2128. https://doi.org/10.1111/j.1462-5822.2008.01195.x

Lang, A.S., Zhaxybayeva, O., and Beatty, J.T. (2012). Gene transfer agents: phage-like elements of genetic exchange. Nat. Rev. Microbiol. *10*, 472–482. https://doi.org/10.1038/nrmicro2802

Lapointe, R., Tanaka, K., Barney, W.E., Whitfield, J.B., Banks, J.C., Béliveau, C., Stoltz, D., Webb, B.A., and Cusson, M. (2007). Genomic and morphological features of a banchine polydnavirus: comparison with bracoviruses and ichnoviruses. J. Virol. *81*, 6491–6501. https://doi.org/10.1128/JVI.02702-06

Li, X., and Webb, B.A. (1994). Apparent functional role for a cysteine-rich polydnavirus protein in suppression of the insect cellular immune response. J. Virol. *68*, 7482–7489.

Louis, F., Bézier, A., Periquet, G., Ferras, C., Drezen, J.-M., and Dupuy, C. (2013). The bracovirus genome of the parasitoid wasp *Cotesia congregata* is amplified within 13 replication units, including sequences not packaged into particles. J. Virol. *87*, 9649–9660. https://doi.org/10.1128/JVI.00886-13

Lu, H., and Burand, J.P. (2001). Replication of the gonad-specific virus Hz-2V in Ld652Y cells mimics replication in vivo. J. Invertebr. Pathol. *77*, 44–50. https://doi.org/10.1006/jipa.2000.4990

Magkrioti, C., Iatrou, K., and Labropoulou, V. (2011). Differential inhibition of BmRelish1-dependent transcription in lepidopteran cells by bracovirus ankyrin-repeat proteins. Insect Biochem. Mol. Biol. *41*, 993–1002. https://doi.org/10.1016/j.ibmb.2011.09.008

Margulis, L. (1981). Symbiosis in Cell Evolution. (W.H. Freman, New York, NY).

McGeoch, D.J., and Davison, A.J. (1999). The descent of human herpesvirus 8. Semin. Cancer Biol. *9*, 201–209. https://doi.org/10.1006/scbi.1999.0093

Murphy, N., Banks, J.C., Whitfield, J.B., and Austin, A.D. (2008). Phylogeny of the parasitic microgastroid subfamilies (Hymenoptera: Braconidae) based on sequence data from seven genes, with an improved time estimate of the origin of the lineage. Mol. Phylogenet. Evol. *47*, 378–395. https://doi.org/10.1016/j.ympev.2008.01.022

Norton, W.N., and Vinson, S.B. (1983). Correlating the initiation of virus replication with a specific pupal developmental phase of an ichneumonid parasitoid. Cell Tissue Res. *231*, 387–398.

Pasquier-Barre, F., Dupuy, C., Huguet, E., Moneiro, F., Moreau, A., Poire, M., and Drezen, J.-M. (2002). Polydnavirus replication: The EP1 segment of the parasitoid wasp *Cotesia congregata* is amplified within a larger precursor molecule. J. Gen. Virol. *83*, 2035–2045. https://doi.org/10.1099/0022-1317-83-8-2035

Payne, S. (2017). Viruses. (Academic Press, San Diego, CA).

Pennacchio, F., and Strand, M.R. (2006). Evolution of developmental strategies in parasitic hymenoptera. Annu. Rev. Entomol. *51*, 233–258. https://doi.org/10.1146/annurev.ento.51.110104.151029

Pichon, A., Bézier, A., Urbach, S., Aury, J.M., Jouan, V., Ravallec, M., Guy, J., Cousserans, F., Thézé, J., Gauthier, J., *et al.* (2015). Recurrent DNA virus domestication leading to different parasite virulence strategies. Sci. Adv. *1*, e1501150. https://doi.org/10.1126/sciadv.1501150

Prasad, S.V., Hepat, R., and Kim, Y. (2014). Selectivity of a translation-inhibitory factor, CpBV15β, in host mRNAs and subsequent alterations in host development and immunity. Dev. Comp. Immunol. *44*, 152–162. https://doi.org/10.1016/j.dci.2013.12.001

Provost, B., Varricchio, P., Arana, E., Espagne, E., Falabella, P., Huguet, E., La Scaleia, R., Cattolico, L., Poirié, M., Malva, C., *et al.* (2004). Bracoviruses contain a large multigene family coding for protein tyrosine phosphatases. J. Virol. *78*, 13090–13103. https://doi.org/10.1128/JVI.78.23.13090-13103.2004

Pruijssers, A.J., and Strand, M.R. (2007). PTP-H2 and PTP-H3 from *Microplitis demolitor* Bracovirus localize to focal adhesions and are antiphagocytic in insect immune cells. J. Virol. *81*, 1209–1219. https://doi.org/10.1128/JVI.02189-06

Pruijssers, A.J., Falabella, P., Eum, J.H., Pennacchio, F., Brown, M.R., and Strand, M.R. (2009). Infection by a symbiotic polydnavirus induces wasting and inhibits metamorphosis of the moth *Pseudoplusia includens*. J. Exp. Biol. *212*, 2998–3006. https://doi.org/10.1242/jeb.030635

Quicke, D.L.J., Laurenne, N.M., Fitton, M.G., and Broad, G.R., (2009). A thousand and one wasps: A 28S rDNA and morphological phylogeny of the Ichneumonidae (Insecta: Hymenoptera) with an investigation inot alignment parameter space and elision. J. Nat. Hist. *43*, 1305–1321. https://doi.org/10.1080/00222930902807783

Rasoolizadeh, A., Béliveau, C., Stewart, D., Cloutier, C., and Couson, M. (2009). *Tranosema rostrale* ichnovirus repeat element genes display distinct patterns in caterpillar and wasp hosts. J. Gen. Virol. *90*, 1505–1514. https://doi.org/10.1099/vir.0.008664-0

Reineke, A., Asgari, S., Ma, G., Beck, M., and Schmidt, O. (2002). Sequence analysis and expression of a virus-like particle protein, VLP2, from the parasitic wasp *Venturia canescens*. Insect Mol. Biol. *11*, 233–239.

Rohrmann, G.F. (2013). Baculovirus Molecular Biology. (National Library of Medicine, National Center for Biotechnology Information, Bethesda, MD).

Rotheram, S. (1967). Immune surface of eggs of a parasitic insect. Nature *214*, 700.

Salt, G. (1973). Experimental studies in insect parasitism. XVI. The mechanism of the resistance of *Nemeritis* to defence reactions. Proc. R. Soc. Lond. B. *183*, 337–350.

Savary, S., Beckage, N., Tan, F., Periquet, G., and Drezen, J.M. (1997). Excision of the polydnavirus chromosomal integrated EP1 sequence of the parasitoid wasp *Cotesia congregata* (Braconidae, Microgastinae) at potential recombinase binding sites. J. Gen. Virol. *78*, 3125–3134. https://doi.org/10.1099/0022-1317-78-12-3125

Serbielle, C., Dupas, S., Perdereau, E., Héricourt, F., Dupuy, C., Huguet, E., and Drezen, J.M. (2012). Evolutionary mechanisms driving the evolution of a large polydnavirus gene family coding for protein tyrosine phosphatases. BMC Evol. Biol. *12*, 253. https://doi.org/10.1186/1471-2148-12-253

Shackelton, L.A., and Holmes, E.C. (2004). The evolution of large DNA viruses: combining genomic information of viruses and their hosts. Trends Microbiol. *12*, 458–465. https://doi.org/10.1016/j.tim.2004.08.005

Stoltz, D.B. (1990). Evidence for chromosomal transmission of polydnavirus DNA. J. Gen. Virol. *71*, 1051–1056. https://doi.org/10.1099/0022-1317-71-5-1051

Stoltz, D.B., and Cook, D.I. (1983). Inhibition of host phenoloxidase activity by parasitoid Hymenoptera. Experientia *39*, 1022–1024.

Stoltz, D.B., and Vinson, S.B. (1979). Viruses and parasitism in insects. Adv. Virus Res. *24*, 125–171.

Stoltz, D.B., Vinson, S.B., and MacKinnon, E.A. (1976). Baculovirus-like particles in the reproductive tracts of female parasitoid wasps. Can. J. Microbiol. *22*, 1013–1023.

Stoltz, D.B., Krell, P., Summers, M.D., and Vinson, S.B. (1984). Polydnaviridae – a proposed family of insect viruses with segmented, double-stranded, circular DNA genomes. Intervirology *21*, 1–4. https://doi.org/10.1159/000149497

Stoye, J.P. (2012). Studies of endogenous retroviruses reveal a continuing evolutionary saga. Nat. Rev. Microbiol. *10*, 395–406. https://doi.org/10.1038/nrmicro2783

Strand, M.R. (1994). *Microplitis demolitor* polydnavirus infects and expresses in specific morphotypes of *Pseudoplusia includens* haemocytes. J. Gen. Virol. *75*, 3007–3020. https://doi.org/10.1099/0022-1317-75-11-3007

Strand, M.R. (2010). Polydnaviruses. In Insect Virology, Asgari, S., Johnson, K.N., eds. (Caister Academic Press, Norfolk), pp. 171–197.

Strand, M.R., and Burke, G.R. (2012). Polydnaviruses as symbionts and gene delivery systems. PLOS Pathog. *8*, e1002757. https://doi.org/10.1371/journal.ppat.1002757

Strand, M.R., and Burke, G.R. (2013). Polydnavirus-wasp associations: evolution, genome organization, and function. Curr. Opin. Virol. *3*, 587–594. https://doi.org/10.1016/j.coviro.2013.06.004

Strand, M.R., and Burke, G.R. (2014). Polydnaviruses: nature's genetic engineers. Annu. Rev. Virol. *1*, 333–354. https://doi.org/10.1146/annurev-virology-031413-085451

Strand, M.R., and Burke, G.R. (2015). Polydnaviruses: From discovery to current insights. Virology 479-480, 393–402. https://doi.org/10.1016/j.virol.2015.01.018

Strand, M.R., McKenzie, D.I., Grassl, V., Dover, B.A., and Aiken, J.M. (1992). Persistence and expression of *Microplitis demolitor* polydnavirus in *Pseudoplusia includens*. J. Gen. Virol. *73*, 1627–1635. https://doi.org/10.1099/0022-1317-73-7-1627

Tan, C.W., Peiffer, M., Hoover, K., Rosa, C., Acevedo, F.E., and Felton, G.W. (2018). Symbiotic polydnavirus of a parasite manipulates caterpillar and plant immunity. Proc. Natl. Acad. Sci. U.S.A. *115*, 5199–5204. https://doi.org/10.1073/pnas.1717934115

Tanaka, K., Lapointe, R., Barney, W.E., Makkay, A.M., Stoltz, D., Cusson, M., and Webb, B.A. (2007). Shared and species-specific features among ichnovirus genomes. Virology *363*, 26–35.

Theilmann, D.A., and Summers, M.D. (1986). Molecular analysis of *Campoletis sonorensis* virus DNA in the lepidopteran host *Heliothis virescens*. J. Gen. Virol. *67*, 1961–1969. https://doi.org/10.1099/0022-1317-67-9-1961

Thézé, J., Bézier, A., Periquet, G., Drezen, J.M., and Herniou, E.A. (2011). Paleozoic origin of insect large dsDNA viruses. Proc. Natl. Acad. Sci. U.S.A. *108*, 15931–15935. https://doi.org/10.1073/pnas.1105580108

Thoetkiattikul, H., Beck, M.H., and Strand, M.R. (2005). Inhibitor kappaB-like proteins from a polydnavirus inhibit NF-kappaB activation and suppress the insect immune response. Proc. Natl. Acad. Sci. U.S.A. *102*, 11426–11431.

Vinson, S.B., and Scott, J.R. (1975). Particles containing DNA associated with the oocyte of an insect parasitoid. J. Invert. Pathol. *25*, 375–378.

Volkoff, A.N., Jouan, V., Urbach, S., Samain, S., Bergoin, M., Wincker, P., Demettre, E., Cousserans, F., Provost, B., Coulibaly, F., *et al.* (2010). Analysis of virion structural components reveals vestiges of the ancestral ichnovirus genome. PLOS Pathog. *6*, e1000923. https://doi.org/10.1371/journal.ppat.1000923

Wang, Y., and Jehle, J.A. (2009). Nudiviruses and other large, double-stranded circular DNA viruses of invertebrates: new insights on an old topic. J. Invertebr. Pathol. *101*, 187–193. https://doi.org/10.1016/j.jip.2009.03.013

Wang, Z.Z., Ye, X.Q., Shi, M., Li, F., Wang, Z.H., Zhou, Y.N., Gu, Q.J., Wu, X.T., Yin, C.L., Guo, D.H., *et al.* (2018). Parasitic insect-derived miRNAs modulate host development. Nat. Commun. *9*, 2205. https://doi.org/10.1038/s41467-018-04504-1

Webb, B.A., Strand, M.R., Dickey, S.E., Beck, M.H., Hilgarth, R.S., Barney, W.E., Kadash, K., Kroemer, J.A., Lindstrom, K.G., Rattanadechakul, W., *et al.* (2006). Polydnavirus genomes reflect their dual roles as mutualists and pathogens. Virology *347*, 160–174.

Wetterwald, C., Roth, T., Kaeslin, M., Annaheim, M., Wespi, G., Heller, M., Maser, P., Roditi, I., Pfister-Wilhelm, R., Bézier, A., Gyapay, G., Drezen, J.M., and Lanzrein, B. (2010). Identification of bracovirus particle proteins and analysis of their transcript levels at the stage of virion formation. J. Gen. Virol. *91*, 2610–2619. https://doi.org/10.1099/vir.0.022699-0

Whitfield, J.B. (2002). Estimating the age of the polydnavirus/braconid wasp symbiosis. Proc. Natl. Acad. Sci. U.S.A. *99*, 7508–7513. https://doi.org/10.1073/pnas.112067199

Whitfield, J.B., and Asgari, S. (2003). Virus or not? Phylogenetics of polydnaviruses and their wasp carriers. J. Insect Physiol. *49*, 397–405.

Wu, Y.L., Wu, C.P., Lee, S.T., Tang, H., Chang, C.H., Chen, H.H., and Chao, Y.C. (2010). The early gene hhi1 reactivates *Heliothis zea* nudivirus 1 in latently infected cells. J. Virol. *84*, 1057–1065. https://doi.org/10.1128/JVI.01548-09

Wu, Y.L., Wu, C.P., Liu, C.Y., Hsu, P.W., Wu, E.C., and Chao, Y.C. (2011). A non-coding RNA of insect HzNV-1 virus establishes latent viral infection through microRNA. Sci. Rep. *1*, 60. https://doi.org/10.1038/srep00060

Wyler, T., and Lanzrein, B. (2003). Ovary development and polydnavirus morphogenesis in the parasitic wasp *Chelonus inanitus*. II. Ultrastructural analysis of calyx cell development, virion formation and release. J. Gen. Virol. *84*, 1151–1163. https://doi.org/10.1099/vir.0.18830-0

Yu, D.S., Chen, Y.B., Li, M., Yang, M.J., Yang, Y., Hu, J.S., and Luo, K.J. (2016). A polydnaviral genome of *Microplitis bicoloratus* bracovirus and molecular interactions between the host and virus involved in NF-κB signaling. Arch. Virol. *161*, 3095–3124. https://doi.org/10.1007/s00705-016-2988-3

Zhu, F., Cusumano, A., Bloem, J., Weldegergis, B.T., Villela, A., Fatouros, N.E., van Loon, J.J.A., Dicke, M., Harvey, J.A., Vogel, H., *et al.* (2018). Symbiotic polydnavirus and venom reveal parasitoid to its hyperparasitoids. Proc. Natl. Acad. Sci. U.S.A. *115*, 5205–5210. https://doi.org/10.1073/pnas.1717904115

Advances in the Molecular Biology of Baculoviruses

9

Manli Wang and Zhihong Hu*

State Key Laboratory of Virology, Wuhan Institute of Virology, Chinese Academy of Sciences, Wuhan, China.

*Correspondence: huzh@wh.iov.cn

https://doi.org/10.21775/9781912530083.09

Abstract

Baculoviridae constitutes a family of insect-specific, large DNA viruses with a unique life cycle characterized by the production of two morphologically distinct virions, the budded virus (BV) and the occlusion-derived virus (ODV). ODV and BV, with different tissue tropisms, have been widely applied in the areas of biological control and biotechnology, respectively. In nature, baculovirus infection of susceptible host larvae is initiated by ODV-mediated primary infection, followed by the production of BV for spreading infection within larval body. Across millions of years of co-evolution with their hosts, baculoviruses have developed dedicated mechanisms for efficient entry/egress, genome replication/transcription, and virion assembly by employing either their own proteins or host machineries. They have also adopted versatile strategies to precisely regulate the immunity, behaviours and physiology of hosts to facilitate their own replication and dispersal. In this chapter, research advances relating to key aspects of the baculovirus life cycle are reviewed, and the application of a newly-developed baculovirus synthetic biology technology is introduced. Finally, future avenues for baculovirus research are discussed.

Introduction

Baculoviridae constitutes a family of insect-specific large DNA viruses with rod-shaped, enveloped virions. The family is divided into four genera, *Alphabaculovirus* [lepidopteran-specific nucleopolyhedroviruses (NPVs)], *Betabaculovirus* [lepidopteran granuloviruses (GVs)], *Gammabaculovirus* (hymenopteran NPVs), and *Deltabaculovirus* (dipteran NPVs).

A typical baculovirus life cycle produces two morphologically distinct virion phenotypes: the budded virus (BV) and the occlusion-derived virus (ODV). The ODV is embedded within occlusion bodies (OBs) and is specifically responsible for mediating infection of midgut columnar epithelial cells (known as 'primary infection', 'oral infection' or '*per os* infection'). The BV is produced from midgut epithelial cells and infects other tissues within the larval body (termed as 'secondary infection' or 'systemic infection') (Slack and Arif, 2007). BVs and ODVs share identical genomes but differ with regard to their protein composition, especially the envelope proteins, which may determine the different tissue-tropisms of these two virion phenotypes (Hou *et al.*, 2013). A comparative model presenting protein compositions and localizations of BV and ODV based on experimental data for *Helicoverpa armigera nucleopolyhedrovirus* (HearNPV) is shown in Fig. 9.1. OBs containing ODVs have been widely used as bio-insecticides to control insect pests, whereas BVs have important applications as vectors for protein expression and gene delivery. A successful baculovirus infection depends on delicate interactions between virus and host at different stages of infection including

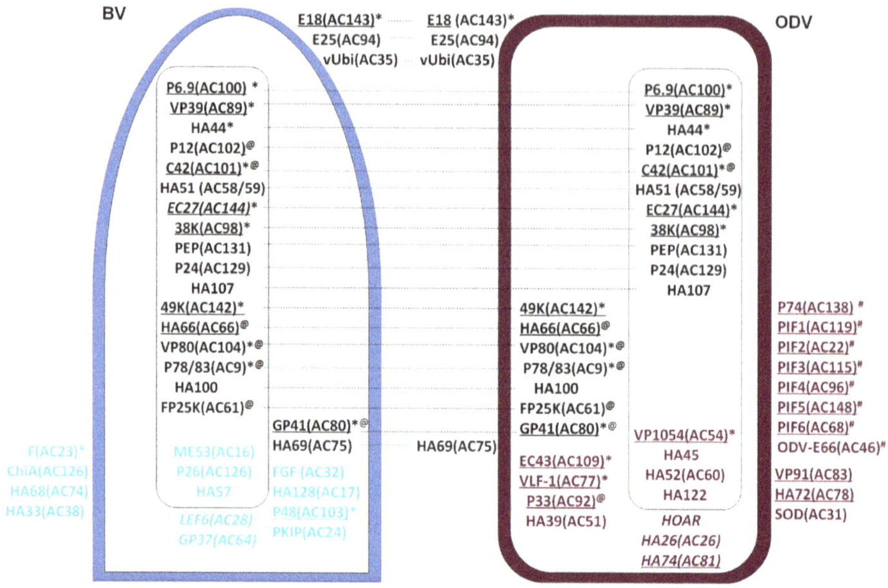

Figure 9.1 Comparative schematics showing protein composition and localization of HearNPV BV and ODV based on proteomics data. Envelope proteins are shown on the surfaces of BV and ODV, nucleocapsid proteins are shown inside each rectangle, and EC proteins are located in the space between the envelope and the nucleocapsid. Proteins shared by BV and ODV are in black; those proteins located in both BV and ODV are linked by dashed lines; BV-specific proteins are shown in blue; and the ODV-specific proteins are shown in maroon. The underlined proteins are conserved baculovirus core genes, * denotes proteins essential for BV production, @ denotes proteins involved in nucleocapsid and/or protein traffic, and # denotes proteins involved in oral infection. BV and ODV proteins with unknown locations are shown in italics. The ORF numbers of the AcMNPV homologues for all the HearNPV ORFs are indicated in brackets. (Reprinted from Hou *et al.*, 2013, J Virol. *87*, 829–839.)

entry, intracellular transport, genome replication/transcription, virion assembly and egress. In addition, baculoviruses have developed strategies to regulate the immunity, behaviours and physiology of hosts to facilitate the production of virus progeny and dispersal. Research advances in these areas are summarized below.

Baculovirus genome

To date, baculoviruses from more than 700 host species have been described. Among these, more than 172 sequenced complete baculovirus genomes have been reported, including 127 alphabaculoviruses, 41 betabaculoviruses, 3 gammabaculoviruses, and 1 deltabaculovirus (Wennmann *et al.*, 2018). The baculovirus genome consists of a double-stranded (ds), covalently closed, circular DNA molecule ranging from 80–180 kbp, with 90–180 predicted open reading frames (ORFs) tightly organized in either orientation. Among these, *Xestia c-nigrum granulovirus* (XecnGV) has the largest genome, with approximately 179 kbp encoding 181 ORFs; gammabaculovirus genomes are the smallest, with approximately 80 kbp and around 90 predicted ORFs. A total of 38 conserved core genes are present in all the baculoviruses sequenced to date (Javed *et al.*, 2017). Most baculovirus genomes also contain short homologous repeats (*hrs*). Fig. 9.2 shows the genome organization of the baculovirus type species, *Autographa californica multiple nucleopolyhedrovirus* (AcMNPV). Notably, a collinear region containing approximately 20 core genes is conserved across all lepidopteran baculoviruses (Zhu *et al.*, 2018) (Fig. 9.2).

Baculovirus core genes

Although different baculovirus genomes exhibit marked diversity with regard to gene content and order, the 38 core genes present in all sequenced baculovirus genomes can be classified into four categories: DNA replication, transcription, oral

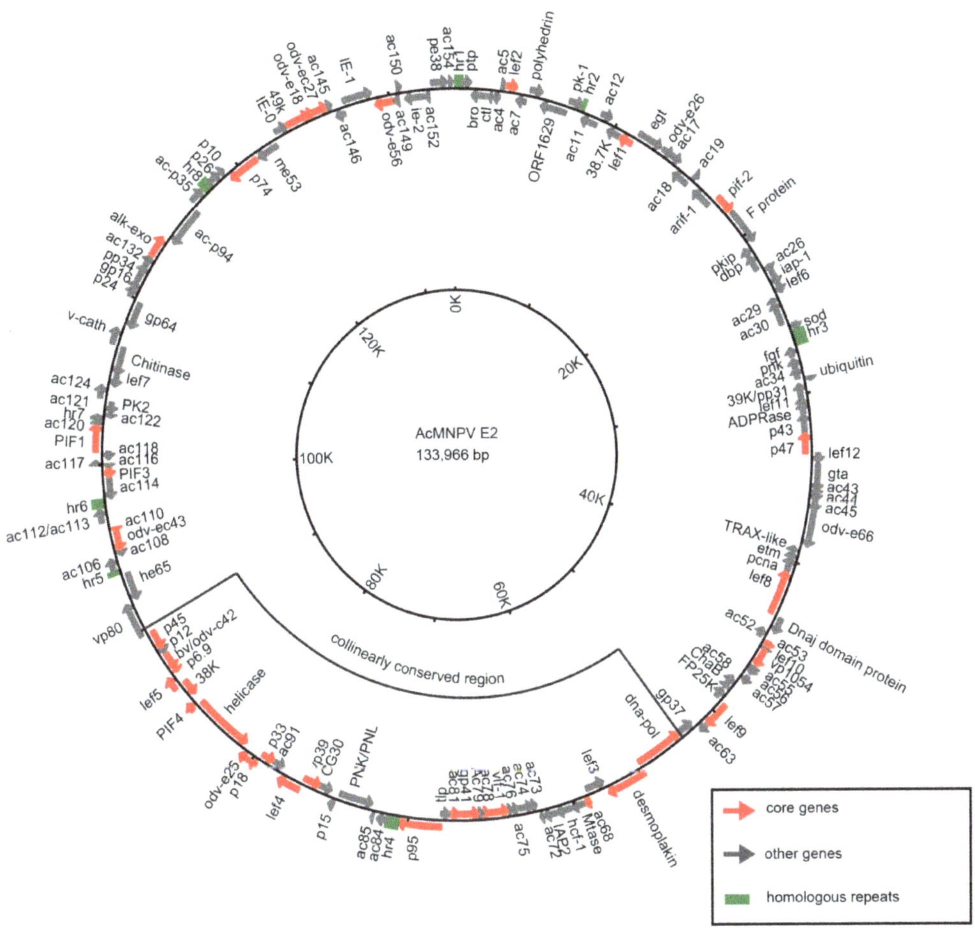

Figure 9.2 Circular map of AcMNPV genome. The complete circular dsDNA genome of AcMNPV E2 strain is 133,966 bp (GenBank accession number: KM667940) with a G+C content of 40.7%. The genome contains 154 ORFs including 38 core genes (red), and 8 *hrs* (green). The lepidopteran linearly conserved region is indicated.

infection, and structure/assembly (Table 9.1). Among the core genes, five are related to DNA replication, including *lef1*, *lef2*, DNA polymerase (*dnapol*), helicase (*p143*) and alkaline nuclease (*an*). Six core genes are involved in transcription, including the four subunits of viral RNA polymerase. Nine core genes are *per os* infectivity factors (*pifs*), including *p74* (*pif0*), *pif1–7* and *pif8* (*vp91*). The remaining 18 core genes are related to nucleocapsid assembly, BV formation or ODV morphogenesis. Notably, some of the core genes are also found in other invertebrate DNA viruses, including *White spot syndrome virus* (WSSV), ascovirus, nudivirus, and hytrosavirus, suggesting a common origin of these large DNA viruses and the occurrence of gene transfer between different viruses and their insect hosts.

Homologous repeats

Hrs are characterized by AT-rich and direct repeats along with imperfect palindromes that are interspersed in baculovirus genomes (Kool *et al.*, 1993; Hilton and Winstanley, 2008). The sequences and numbers of *hrs* vary considerably among different baculoviruses. Certain baculoviruses, e.g. *Trichoplusia ni single nucleopolyhedrovirus* (TnSNPV) and *Agrotis segetum granulovirus* (AgseGV), do not appear to contain *hrs*, whereas others contain many, such as *Cydia pomonella granulovirus* (CpGV), which has 13. *Hrs* function as transcriptional enhancers for early gene expression and as origins of DNA replication (Guarino and Summers, 1986; Pearson *et al.*, 1992, 1993). The transcriptional activator IE-1 can bind to *hrs* and enhance the activity of *hr*-adjacent viral promoters (Olson *et al.*, 2001, 2003). Although

Table 9.1 The 38 core genes of baculovirus

Catalogue (Total no.)	Gene (AcMNPV homolog)	Protein function
DNA replication (5)	*lef2* (*ac6*)	A primase-associated factor
	lef1 (*ac14*)	primase
	dnapol (*ac65*)	DNA polymerase
	helicase (*ac95*)	Helicase
	alkaline nucleus (*ac133*)	DNA recombination
Transcription (6)	*p47* (*ac40*)	RNA polymerase subunit
	lef8 (*ac50*)	RNA polymerase subunit
	lef9 (*ac62*)	RNA polymerase subunit
	vlf-1 (*ac77*)	Very late gene expression
	lef4 (*ac90*)	RNA polymerase subunit
	lef5 (*ac99*)	Late gene transcription
Oral infection (9)	*p74/pif0* (*ac138*)	PIF complex, ODV binding?
	pif1 (*ac119*)	PIF complex, ODV binding?
	pif2 (*ac22*)	PIF complex, ODV binding?
	pif3 (*ac115*)	PIF complex, unknown function
	pif4 (*ac96*)	PIF complex, unknown function
	pif5 (*ac148*)	unknown function
	pif6 (*ac68*)	PIF complex, unknown function
	pif7 (*ac110*)	PIF complex, unknown function
	vp91/pif8 (*ac83*)*	PIF complex, unknown function
Structure-related (18)	*ac53*	Nucleocapsid assembly
	vp1054 (*ac54*)	Nucleocapsid assembly
	desmoplakin (*ac66*)	Nucleocapsid egress
	ac78	BV production and ODV morphogenesis
	gp41 (*ac80*)	Tegument protein, BV production and ODV morphogenesis
	ac81	BV production and ODV envelopment
	vp39 (*ac89*)	Major capsid protein, nucleocapsid assembly
	p33 (*ac92*)	sulfhydryl oxidase, BV production and ODV morphogenesis
	p18 (*ac93*)	BV production and ODV envelopment
	odv-e25 (*ac94*)	BV infectivity and ODV formation
	38k (*ac98*)	Phosphatase, nucleocapsid assembly
	p6.9 (*ac100*)	DNA condensation
	bv/odv-c42 (*ac101*)	Nucleocapsid assembly and actin polymerization
	p48/p45 (*ac103*)	BV production and ODV envelopment
	odv-ec43 (*ac109*)	BV production and ODV envelopment
	49k (*ac142*)	BV production and ODV envelopment
	odv-e18 (*ac143*)	BV production
	odv-ec27 (*ac144*)	Nucleocapsid assembly and cell cycle arrest

**vp91* is also involved in nucleocapsid assembly.

deletion of each of the 8 *hrs* or a double *hr* deletion (*hr*2 and *hr*3) in AcMNPV did not impair BV infectivity (Carstens and Wu, 2007), simultaneous elimination of all eight *hrs* of AcMNPV using a sequential deletion strategy resulted in a 100-fold reduction of infectious BV production, as well as severe impairment of OB production (Bossert and Carstens, 2018). It was therefore proposed that three *hrs* may be sufficient to mediate optimal virus infection (Bossert and Carstens, 2018). *Hrs* are also present in the genomes of other invertebrate DNA viruses (e.g. nimavirus, ascovirus, nudivirus, and iridovirus), suggesting a common function of *hrs* in these invertebrate large dsDNA viruses.

PIF-mediated ODV entry into midgut cells (primary infection)

In nature, ODVs initiate primary infection when they are released from ingested OBs in the alkaline conditions within insect midguts. Unlike BVs, which use only one envelope fusion protein (EFP) (GP64 or F protein) for infection, ODVs employ a group of specific envelope proteins called *per os* infectivity factors (PIFs) to mediate primary infection. Deletion of any one of the individual PIFs leads to the complete loss of oral infectivity. PIFs form a stable protein complex on the surface of ODV. Interestingly, certain PIFs are also present in other invertebrate large DNA viruses, suggesting a common origin and conserved functions of these PIFs. The current knowledge of the *per os* infection process and PIFs has been summarized in detail in recent reviews (Boogaard *et al.*, 2018; Wang, X. *et al.*, 2017; Wei *et al.*, 2014).

PIFs

Nine reported PIFs (P74 or PIF0, and PIF1–8) are conserved in all baculoviruses sequenced to date (Table 9.1). They were named in the order in which they were identified as being essential for oral infectivity. These PIFs do not show obvious sequence similarity and vary considerably in their molecular mass (ranging from ≈ 7 kDa to ≈ 95 kDa) (Wang, X. *et al.*, 2017). However, they do share some common features. All PIFs contain one or more hydrophobic transmembrane (TM) domains. Some of these TMs serve as an inner nuclear membrane sorting motif (INM-SM) and are responsible for the trafficking of PIFs from the cytoplasm to the inner nuclear membrane and ODV envelope (Braunagel and Summers, 2007). Additionally, most PIFs contain conserved cysteines, which may be involved in disulfide bond formation, thus stabilizing the interaction between PIFs or facilitating the proper folding of individual PIFs (Wang, X. *et al.*, 2017).

Direct membrane fusion between the ODV envelope and the midgut cell plasma membrane is likely involved in ODV entry into cells. However, the precise role of each PIF during *per os* infection remains largely enigmatic. Among PIFs, P74, PIF1, and PIF2 have been suggested to play a role in the specific binding of ODV to midgut cells (Ohkawa *et al.*, 2005). However, another study by using EGFP-labelled ODVs showed that the deletion of either *pif1* or *pif2* did not prevent virus binding to midgut cells (Mu *et al.*, 2014). P74 shares a common feature with class I viral fusion proteins in that it undergoes proteolysis for activation (Slack *et al.*, 2008; Peng *et al.*, 2011). However, whether P74 serves as an actual fusion protein for ODV entry is unclear.

PIF complex

PIFs are likely to orchestrate the entry of ODVs into midgut epithelial cells by formation of a complex, a similar mechanism to that utilized by other large DNA viruses such as poxviruses and herpesviruses. Peng *et al.* (2010a) provided the first evidence for the presence of a stable PIF complex (> 170 kDa) under partially denaturing conditions. In 2012, the same group determined that the PIF complex is as large as ≈ 480 kDa under non-denaturing conditions (Peng *et al.*, 2012). The components of the PIF complex have been discovered successively, including 8 PIFs (PIFs 0 to 4 and PIFs 6 to 8), i.e. all the PIFs except PIF5 (Peng *et al.*, 2010a, 2012; Javed *et al.*, 2017; Boogaard *et al.*, 2018). Among these, PIFs 1–4 are proposed to form a stable core (Boogaard *et al.*, 2018). PIFs interact extensively with each other, as well as with themselves (Peng *et al.*, 2010b; Dong *et al.*, 2014; Zheng *et al.*, 2017). HearNPV PIFs 0 through 3 could not be functionally substituted with homologues from other baculoviruses, suggesting that the interactions between PIFs are highly specific (Song *et al.*, 2016; Makalliwa *et al.*, 2018). It has been suggested that the formation of an intact PIF complex may prevent degradation of individual PIF proteins in the proteolytic environment of the insect midgut lumen (Boogaard *et al.*, 2017). Many details of the formation and structure of the PIF complex remain unknown.

Conservation of PIFs

Interestingly, homologues of certain PIFs are present in other invertebrate large DNA viruses, including WSSV (P74 and PIF1–3), salivary gland hypertrophy viruses (P74, PIF1–4), Apis mellifera filamentous virus (P74, PIF1–4 and PIF8), bracovirus and nudivirus (P74, PIF1–6 and PIF8) (Wang, X. *et al.*, 2017). These viruses have co-evolved with their invertebrate hosts over 300 million years ago, implying that PIF-mediated virus entry is an ancient and evolutionarily conserved mechanism shared by these invertebrate DNA viruses. Whether a similar PIF-complex also exists in other invertebrate large DNA viruses and whether such PIFs also play pivotal roles during virus infection would be of great interest.

Envelope fusion protein (EFP)-mediated BV entry (systemic infection)

Secondary infection of baculoviruses from midgut ephithelia into other tissues within the susceptible larval body is mediated by the BV phenotype. BVs acquire envelopes containing specific viral EFPs during nucleocapsid budding from the host plasma membrane. EFPs have been found in all baculoviruses, except for members of the *Gammabaculovirus* genus. Baculoviruses encode two types of EFPs: GP64 and F. These two proteins are quite distinct in their distribution, structure, and mode of activation, although they perform similar essential roles in mediating BV entry.

GP64

GP64 homologues only occur in Group I alphabaculoviruses and a betabaculovirus and they are closely related in amino acid (aa) sequence (> 74% aa identity). The GP64 protein was first identified via a neutralizing monoclonal antibody (AcV1) that could block BV entry of AcMNPV (Hohmann and Faulkner, 1983). GP64 is synthesized as an N-glycosylated protein with molecular mass of approximately 64 kDa and is sufficient to mediate low-pH-dependent membrane fusion (Blissard and Wenz, 1992; Jarvis and Garcia, 1994). The crystal structure revealed that the post-fusion conformation of AcMNPV GP64 is comprised of a trimer with a central helix-core and elongated β-sheets, which resembles the features of class III viral fusion proteins such as herpesvirus gB protein and vesicular stomatitis virus G protein (Kadlec *et al.*, 2008). The N-terminal region (21 to 159 aa) of GP64 is implicated as a putative receptor binding domain (RBD) based on neutralization and truncation analyses (Zhou and Blissard, 2008). The C-terminal-most regions of AcMNPV GP64 include a 23-aa pre-transmembrane (PTM) domain, a 23-aa transmembrane (TM), and a very short (7 aa) cytoplasmic tail domain (CTD). Functional analyses show that the PTM and TM domains are crucial for the membrane fusion activity of GP64 and BV infectivity. The 7-aa CTD is not essential for infectious BV production but plays a role in virus budding (Oomens and Blissard, 1999; Li and Blissard, 2008, 2009a,b). Two fusion loops (Y75–T86 and N149–H156) located at the tip of the β-sheet domains of the GP64 trimer are critical for GP64-induced membrane fusion events (hemifusion and fusion pore expansion) (Kadlec *et al.*, 2008; Dong and Blissard, 2012). Notably, the fusion loops overlap with the RBD at least at some key residues and certain mutations abolish GP64–liposome interaction at neutral pH. This finding suggests dual functions of the GP64 fusion loops in both receptor binding and membrane fusion (Dong and Blissard, 2012).

GP64 also undergoes various post-translational modifications. Four N-glycosylation sites (N198, 355, 385, and 426) are distributed in AcMNPV GP64, some of which are involved in infectious BV production (Jarvis and Garcia, 1994; Jarvis *et al.*, 1998). A palmitoylation site (C503) is located in the TM of AcMNPV GP64, although it does not appear to be involved in BV entry or budding (Zhang *et al.*, 2003). Of the 15 total conserved cysteines of GP64, 14 are involved in disulfide bond formation. Among these, the trimeric structure is stabilized by an intermolecular disulfide bond (C24–C372) that is not essential for membrane fusion but is required for BV infectivity. The six conserved intramolecular disulfide bonds are all indispensable for GP64 fusogenicity (Li and Blissard, 2010). Some of these key domains/residues of GP64 are highlighted in the post-fusion structure of AcMNPV GP64 (PDB No. 3DUZ) (Fig. 9.3A).

F protein

Unlike GP64 Group I alphabaculoviruses, which use GP64 as an EFP, Group II alpha-, beta-, and delta-baculoviruses use F proteins for entry. Compared to those of GP64, F protein homologues are much less conserved (20–40% aa identity). Similar to GP64, the F protein homologues are

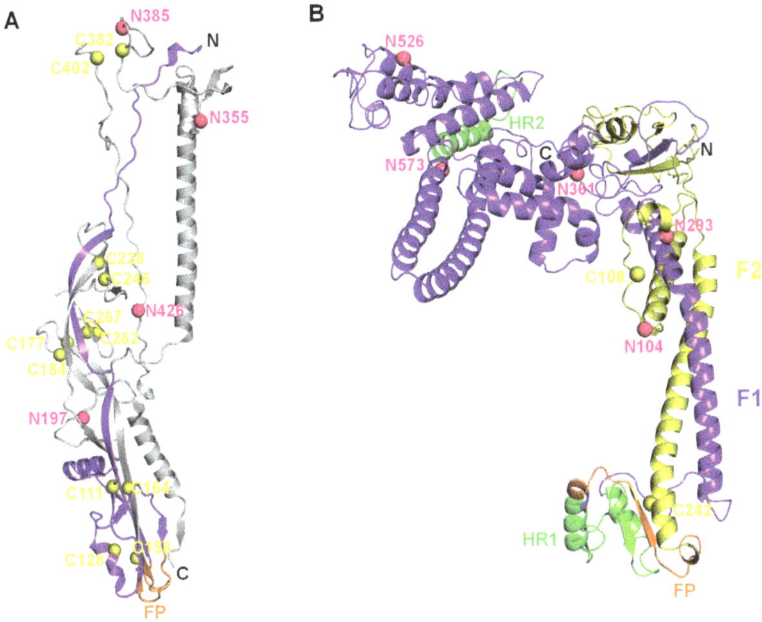

Figure 9.3 The key structural features of GP64 and F protein. (A) Ribbon diagram of the post-fusion structure of AcMNPV GP64 monomer (PDB entry: 3DUZ). Two fusion loops (residues 75–86 and residues 149–156) and the receptor binding domain (residues 21–159) are coloured orange and purple, respectively. The cysteines forming six intramolecular disulfide bonds and four N-glycosylation sites are shown as spheres and coloured in yellow and crimson respectively. (B) The structure of HearNPV F protein predicted by online I-TASSER server (https://zhanglab.ccmb.med.umich.edu/I-TASSER/). The cysteines (C108, C242) forming inter-subunit disulfide bonds and five N-glycosylation sites are shown as spheres. Colouring scheme: F_2 (residues 1–173) in yellow, F_1 (residues 174–677) in purple, fusion loop (residues 174–192) in orange, heptad repeat 1 (HR1) (residues 194–236) and heptad repeat 2 (HR2) (residues 541–568) in blue, cysteines in yellow, N-glycosylation sites in carmine.

N-glycosylated proteins that accumulate at the plasma membranes and possess low-pH induced membrane fusion activity (Pearson *et al.*, 2000; IJkel *et al.*, 2000). The F proteins are expressed as a precursor (F_0, approximately 80 kDa), which then undergo proteolysis by host furin-like pro-protein convertase to generate a disulfide-linked N-terminus (F_2, approximately 20 kDa) and C-terminus (F_1, approximately 60 kDa) that are required for activation of F fusogenicity and BV infectivity (Westenberg *et al.*, 2002). A fusion peptide (FP) domain (approximately 20 aa) directly downstream of the furin cleavage site is essential for insertion of the F protein into the cellular membrane, thus facilitating the subsequent fusion between viral envelope and cell membrane (Westenberg *et al.*, 2004; Tan *et al.*, 2008).

Nuclear magnetic resonance (NMR) analysis determined that the FP of HearNPV F (HaF) protein is an amphiphilic structure composed of a 'helix–turn–helix' motif (Tan *et al.*, 2008). At least two heptad repeats (HRs), one located downstream of the fusion peptide and the other located upstream of the transmembrane domain, are present in F proteins, which are presumed to form a six-helix bundle structure for mediating membrane fusion (Long *et al.*, 2008). An inter-subunit disulfide bond (C108–C242) and five N-glycosylation sites (one in the F_2 subunit and four in the F_1 subunit) were identified in HaF by mutagenesis analysis as being essential for proper function of the F protein and BV infectivity (Yin *et al.*, 2014; Shen *et al.*, 2016). These features suggest the baculovirus F protein is a class I viral fusion protein. However, the 3D structure of F protein, which is important for further in-depth elucidation of its function and fusion mechanism, has yet to be resolved. A post-fusion conformation of the HaF protein has been predicted using the I-TASSER method (Shen *et al.*, 2016) (Fig. 9.3B).

F-like protein and the evolutionary hypothesis

Notably, Group I alphabaculoviruses also encode an F protein homologue in their genome: F-like protein. F-like proteins share 20–30% aa identity with the F proteins of Group II alpha- and betabaculoviruses. Unlike the functional F protein and GP64, F-like proteins are present in the envelope of both BV and ODV phenotypes (Pearson *et al.*, 2001). Strikingly, F-like proteins accumulate mutations in their furin cleavage sites and are thus unable to induce membrane fusion. These remnant F-like proteins appear to have an auxiliary function in ODV morphogenesis and viral infectivity both *in vivo* and *in vitro* (Lung *et al.*, 2003; Wang, M. *et al.*, 2008; Yu *et al.*, 2009).

F proteins are hypothesized to be the ancestral baculovirus fusion protein and GP64 is likely to have been more recently acquired by Group I alphabaculoviruses. This may have led to the inactivation of F proteins and change into non-fusogenic F-like proteins (Jiang *et al.*, 2009). The first experimental data for this hypothesis was provided by construction of a series of recombinant viruses to mimic the important evolutionary event of acquisition of *gp64* by an ancestral baculovirus (Wang *et al.*, 2014). This hypothesis is further supported by the findings that an ancient deltabaculovirus from mosquito encodes a functional F protein and a newly discovered betabaculovirus possesses not only F protein, but also a functional GP64 homologue (Ardisson-Araujo *et al.*, 2016; Wang, M. *et al.*, 2017). The F protein appears to harbour additional functions over those of GP64 in mediating Group II alphabaculovirus infection, as evidenced by the ability of F proteins to readily replace the function of GP64, but not vice versa (Lung *et al.*, 2002; Long *et al.*, 2006; Westenberg and Vlak, 2008; Wang *et al.*, 2010b). In particular, GP64 was suggested to have been captured from *Thogotovirus*, an arbovirus that also utilizes a GP64 homologue (GP75) as EFP (Pearson *et al.*, 2002). In contrast, F proteins may be derived from their hosts or insect retroviruses, although their functions have since been substantially adapted to baculovirus infection (Lung and Blissard, 2005; Pearson and Rohrmann, 2006).

Role of EFPs in BV entry

BV entry is initiated by the specific binding of EFPs to susceptible cells (Hefferon *et al.*, 1999). In the case of GP64, abundant cell surface molecules, including phospholipids and heparin, are involved in AcMNPV entry into insect or mammalian cells, respectively (Duisit *et al.*, 1999; Tani *et al.*, 2001; Wu and Wang, 2012). In addition, the host cellular membrane protein BmREEPa facilitates *Bombyx mori nucleopolyhedrovirus* (BmNPV) entry into insect cells through interaction with GP64 (Dong, X.L. *et al.*, 2015). Although it is recognized that F proteins may exploit different cellular receptors from those of GP64 to infect insect cells (Westenberg *et al.*, 2007), the specific receptors for both F protein and GP64 have yet to be identified.

During AcMNPV infection, at 10–20 minutes post-binding, BV particles are internalized into cells by clathrin-mediated endocytosis. Viral nucleocapsids are released from endosomes into the cytoplasm at 15–30 minutes post-infection (min p.i.) (Volkman and Goldsmith, 1985; Hefferon *et al.*, 1999). The low-pH condition inside endosomes is critical for triggering the conformational change of EFPs, which results in fusion between the viral envelope and endosomal membranes. The protonation of histidine residues (H245, 304, and 430) may comprise an important trigger for the low-pH-induced conformational change of GP64 (Li and Blissard, 2011); in turn, five to eight GP64 trimers may act in concert to mediate the initial fusion pore formation (Markovic *et al.*, 1998). The optimal fusion pH for GP64 is around 5.5, implying that virus fusion may occur in late endosomes (Blissard and Wenz, 1992). Furthermore, low-pH-triggered direct fusion with the plasma membrane appears to serve as an alternative pathway for AcMNPV entry into insect and mammalian cells (Dong *et al.*, 2010).

Essential functions of actin polymerization in baculovirus infection

Baculovirus infection induces the sequential rearrangement of host actin in both the cytoplasm and nucleus (Charlton and Volkman, 1991). Unlike most other viruses, which use microtubules for inward transport and microtubules/actin for egress, both the retrograde (transport towards nucleus) and anterograde transport (egress of progeny virions) of baculovirus are dependent on the actin cytoskeleton for multiple steps (Fig. 9.4A). Thus,

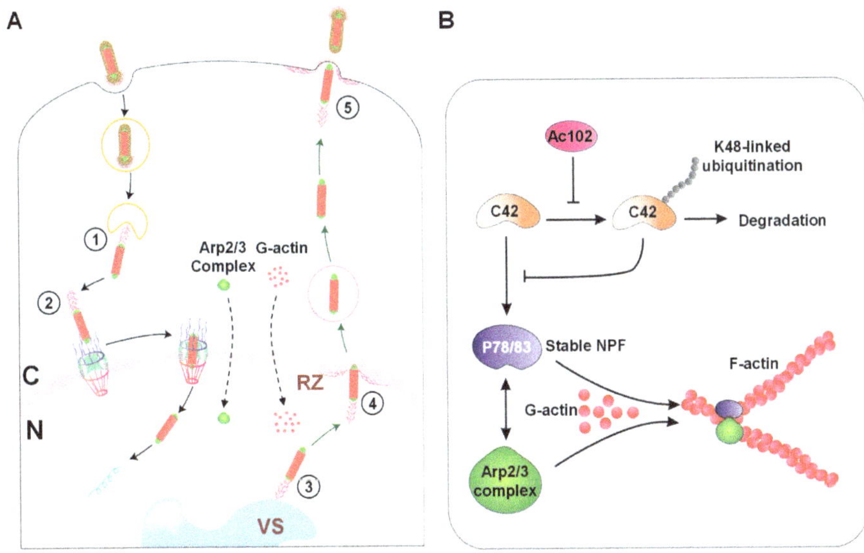

Figure 9.4 Baculovirus-induced actin polymerization. (A) The multiple functions of actin polymerization during baculovirus infection. After penetration of the nucleocapsids into the cytoplasm, actin polymerization is immediately induced for transport of nucleocapsids towards the nucleus (step ①) and passage through NPC (step ②). Within the infected nucleus, the nuclear F-actin network transports progeny nucleocapsids from the virogenic stroma (VS) to the nuclear periphery (ring zone, RZ) (step ③) and further provides the driving force for nucleocapsid egress from the nucleus (step ④). In the cytoplasm, F-actin is again exploited for the trafficking of nucleocapsids to the plasma membrane for further BV budding (step ⑤). (B) The regulatory mechanism of baculovirus-induced actin polymerization. Ac102 suppresses K48-linked ubiquitination and proteasomal degradation of C42, thus allowing P78/83 to function as a stable nucleation promoting factor (NPF) to interact with host Arp2/3 complex for actin nucleation and polymerization. C, cytoplasm; N, nucleus.

baculoviruses appear to have developed a unique and dedicated regulatory mechanism for usurping the cellular actin cytoskeleton for virus morphogenesis and intracellular transport (Fig. 9.4B).

Retrograde cytoplasmic transport of nucleocapsid

The dynamic transition between globular actin (G-actin) monomers and filamentous actin (F-actin) forms for actin cytoskeleton remodelling is critical for diverse cellular activities. Actin polymerization is required for transport of nucleocapsids to the nucleus where virus replication takes place (Fig. 9.4A, step ①). Immediately following penetration of the nucleocapsids into the cytoplasm (approximately 30 min p.i.), thick actin cables are induced (Charlton and Volkman, 1993). Conversely, actin cable formation was not observed when virus–endosome fusion was inhibited by drugs, suggesting that nucleocapsid proteins induce the polymerization of actin (Charlton and Volkman, 1993). Ohkawa *et al.* (2011) demonstrated that AcMNPV undergoes actin-based motility in the cytoplasm, colliding with and sticking to the nuclear envelope, which could be blocked by an actin-stabilizing drug (Ohkawa *et al.*, 2010). In eukaryotic cells, the actin polymerization process normally requires an N-Wiskott-Aldrich syndrome protein (N-WASP) acting as the nucleation promoting factor (NPF). N-WASP activates the actin-related protein 2/3 (Arp2/3) complex to promote G-actin nucleation and form F-actin. The baculovirus genome encodes an N-WASP-like protein, P78/83, which localizes to the blunt end of the nucleocapsid. P78/83 activates the host Arp2/3 complex for actin polymerization (Goley *et al.*, 2006). Electron tomography showed that the AcMNPV nucleocapsid was propelled by continuously assembled actin comet tails, with an average of four fishbone-like actin filaments per nucleocapsid (Mueller *et al.*, 2014).

Virus passage through the nuclear pore complex

It is now generally accepted that baculoviruses enter the nucleus via the nuclear pore complex (NPC), although the entry modes exploited may vary among different baculoviruses (Fig. 9.4A, step ②). For betabaculoviruses, early electron microscopy studies observed the attachment of empty nucleocapsids to the cytoplasmic side of the NPC and the absence of intact nucleocapsids in the nucleus, suggesting that virus uncoating took place at the NPC (Summers, 1969, 1971). In contrast, alphabaculoviruses appear to traverse the NPC with intact nucleocapsids, both in insect and mammalian cells (Granados and Lawler, 1981). AcMNPV nuclear entry was suggested to be dependent on NPC based on the use of a truncated form of importin-β or wheat germ agglutinin (a nuclear transport inhibitor) to interfere with the function of NPC (Ohkawa *et al.*, 2010). Using microinjection of *Xenopus* oocytes and electron tomography, Au *et al.* (2013) clearly visualized an AcMNPV capsid vertically docking at the cytoplasmic side of NPCs with its apical cap end, crossing the NPC, and finally reaching the nucleus in intact form (Au and Panté, 2012). During this process, the NPC central channel switched from a 'closed' to an 'open' state to allow the passage of the nucleocapsids (Au and Panté, 2012; Au *et al.*, 2013).

The nuclear import of large cargoes (proteins larger than 40 kDa or molecules larger than 9 nm in diameter) requires the recognition of special signals within the cargoes by cellular nuclear import receptors (importins or karyopherins), and further translocation of the cargoes through the NPC (Cohen *et al.*, 2011). To date, the nuclear import mechanism of baculovirus nucleocapsids remains unclear. Two viral nucleocapsid proteins, Ac109 and Ac132, may be involved in the entry of viral nucleocapsids into the nucleus (Alfonso *et al.*, 2012; Fang *et al.*, 2016). Deletion of either gene produces non-infectious BVs that are able to enter the cytoplasm but fail to enter the nucleus. However, both of the proteins lack a predicted nuclear localization signal (NLS), therefore being unlikely to directly interact with classical cellular nuclear transport receptors. It is also not known whether Ac109 and Ac132 are involved in the interaction between nucleocapsids with NPC or have a role in a step following NPC docking. Another viral nucleocapsid protein, VP80, is hypothesized to play a role in mediating the nuclear import of nucleocapsids because it is the only viral nucleocapsid protein containing a predicted NLS, although this requires further investigation (Au *et al.*, 2013).

An alternative unconventional nuclear import mechanism was recently proposed for AcMNPV. Using a well-established nuclear import assay, the nuclear entry of AcMNPV was found to be independent of importin β and Ran-GTPase used in the classical nuclear import pathway. The nuclear import of nucleocapsids was recreated instead under conditions supporting actin polymerization, suggesting that F-actin not only directs the motility of nucleocapsids towards the nuclear periphery within the cytoplasm, but also provides the driving force for the translocation of nucleocapsids through the NPC (Au *et al.*, 2016).

Nuclear localization of G-actin and Arp2/3 complex

The nuclear localization of G-actin (NLA) is mediated by six AcMNPV early genes (*ie-1*, *pe38*, *he65*, *ac004*, *ac102*, and *ac152*) as shown via transient transfection assay (Ohkawa *et al.*, 2002). Among these genes, only *ac102* is essential, albeit not sufficient for mediating NLA. Co-transfection of an *ac102*-deleted bacmid with a plasmid expressing NLS-GFP-actin did not rescue infectious progeny BV production, suggesting that Ac102 has other essential roles beyond mediating NLA (Gandhi *et al.*, 2012). Based on evidence that an NLA-blocking drug inhibits late gene expression and the deletion of *ac102* does not have an obvious impact on actin nuclear accumulation, an AcMNPV late gene other than *ac102* could play a key role in mediating NLA (Zhang, Y. *et al.*, 2018). In turn, the nuclear accumulation of the Arp2/3 complex is facilitated by a viral encoded late protein, Ac34. This protein inhibits the CRM1-dependent nuclear exportin pathway, leading to the retention of Arp2/3 in the nucleus to promote actin polymerization (Mu *et al.*, 2016a,b).

Anterograde transport of nucleocapsid

A unique feature of baculovirus infection is that substantial nuclear actin polymerization is induced at the late stage of infection and plays a critical role in progeny virus morphogenesis (Ohkawa and

Volkman, 1999; Goley *et al.*, 2006). The nuclear F-actin cytoskeleton interacts with the virus-encoded motor molecule VP80 to form a network for transport of progeny nucleocapsids from the virogenic stroma (VS) to the nuclear periphery (ring zone, RZ) (Marek *et al.*, 2011) (Fig. 9.4A, step ③). Further nuclear egress is also dependent on actin polymerization, including the direction of nucleocapsids into nuclear envelope protrusions and local disruption of nuclear envelope integrity (Ohkawa and Welch, 2018) (Fig. 9.4A, step ④). Notably, AcMNPV egress from the nucleus to the cytoplasm via NPC and by the classical nuclear export pathway was excluded by using specific inhibitors (Ohkawa and Welch, 2018). AcMNPV again exploits F-actin in the cytoplasm, likely together with microtubules, for the trafficking of nucleocapsids to the plasma membrane for further BV budding (Ohkawa *et al.*, 2010; Ohkawa and Welch, 2018) (Fig. 9.4A, step ⑤).

Regulatory mechanism of baculovirus-induced actin polymerization

Baculovirus-induced actin polymerization is orchestrated by G-actin, the nucleator Arp2/3 complex, and the virus NPF P78/83 (Goley *et al.*, 2006). AcMNPV has developed a sophisticated regulatory cascade to control the activity of viral NPF P78/83 for nuclear actin polymerization. BV/ODV-C42 (C42) is responsible for recruiting P78/83 to the nucleus (Wang, Y. *et al.*, 2008). C42 also protects P78/83 from ubiquitin-independent proteasomal degradation by masking a degron in P78/83, thus ensuring its NPF activity in mediating actin polymerization (Li *et al.*, 2010; Wang *et al.*, 2015). Ac102 suppresses ubiquitination and proteasomal degradation of C42, thus allowing P78/83 to function as a stable NPF to induce actin polymerization (Hepp *et al.*, 2018; Zhang, Y. *et al.*, 2018). A dynamic regulatory mechanism for P78/83-induced actin polymerization and nucleocapsid decapsidation has been proposed. This includes the degradation of P78/83 by exposing the degron for detachment of F-actin from the nucleocapsids and further release of the viral genome in the nucleus, as well as the stabilization of P78/83 by C42 to facilitate nuclear actin polymerization allowing progeny virus morphogenesis and transport (Wang *et al.*, 2015) (Fig. 9.4B).

Genome replication

Virus genome replication begins on entry into the nucleus. As a large DNA virus, the replication of baculovirus presents a complex picture. Viral genomes usually contain multiple separate origins of replication for initiation of DNA synthesis, which may ensure replication efficiency. Baculovirus genomes also encode a complicated and dedicated replication machinery that creates a factory within the infected nucleus dedicated to viral genome replication.

Baculovirus origin of replication

Two different approaches, generation of defective genomes and replication assays, have been employed to identify viral origins of DNA replication. In the former, BV is continuously passaged at high multiplicity of infection, resulting in the enrichment of defective interfering (DI) particles with much smaller genome sizes. The generation of defective genomes with enriched *hr* sequences, implicates them as origins of replication (Pearson *et al.*, 1992; Kool *et al.*, 1993). In replication assays, plasmids containing various *hrs* are transfected into virus-infected cells, and newly synthesized DNA is distinguished from input DNA by *Dpn* I digestion. Using this approach, all eight *hrs* of AcMNPV have shown the capacity to act as origins of replication (Kool *et al.*, 1993; Leisy and Rohrmann, 1993). Similar methods have also identified non-*hr* sequences and a variety of early gene promoters as origins of replication (Lee and Krell, 1992; Pearson *et al.*, 1993; Wu *et al.*, 1999). Recently, a novel replication origin within the *p143* (helicase) coding sequence was identified to replicate more efficiently in insect cells than the known *hr*- or non-*hr* origins and could function even in mammalian cells (Wu *et al.*, 2014). It has been suggested that DNA replication could initiate at any unwound sequences that allow for entry of the replication complex. Deletion of individual *hr* sites did not have an obvious impact on AcMNPV replication and simultaneous elimination of all eight *hrs* did not prevent the production of infectious progeny virus, although virus titre was reduced (Carstens and Wu, 2007; Bossert and Carstens, 2018). However, some degree of viral-host specificity for *hr* origins does appear to exist, as evidenced by the origin of one virus replicating poorly in cells infected by another virus (Pearson *et al.*, 1993). Thus, the presence of redundant origins of replication in the baculovirus genome may provide functional complementarity and increase the speed of the virus replication cycle.

Viral genes involved in DNA replication

Six genes essential for DNA replication have been identified using a transient replication assay with overlapping cosmid clones encompassing the complete genome of AcMNPV (Kool *et al.*, 1994). These include genes for DNA polymerase (*dnapol*), helicase (*p143*), *lef-1*, *lef-2*, *lef-3*, and *ie-1*. The first four are core genes of baculoviruses (Table 9.1). *In vivo* study confirmed that *dnapol* (Vanarsdall *et al.*, 2005), *p143* (Lu and Carstens, 1991), *lef-3* (Yu and Carstens, 2010), as well as a new member, *lef-11* (Lin and Blissard, 2002), are essential for DNA replication.

DNA polymerase

The baculovirus DNA polymerase is in the DNA polymerase B family (Tomalski *et al.*, 1988; Braithwaite and Ito, 1993). It also exhibits a 3′ to 5′ exonuclease activity for proofreading of mispaired nucleotides (Hang and Guarino, 1999). In AcMNPV infected cells, the transcription of *dnapol* was detected in a narrow period between 2–8 h.p.i. with peak at 4–6 h.p.i (Tomalski *et al.*, 1988). The coding sequence of AcMNPV DNA polymerase contains 948 aa with an exonuclease domain (aa 185–401), a polymerase domain (aa 510–750), and two C-terminal NLSs (Feng and Krell, 2014). The N-terminal region (aa 1–185) of AcMNPV DNA polymerase is also conserved in other baculovirus DNA polymerase sequences and is essential for DNA replication (Chen *et al.*, 2018).

Helicase

Helicase (P143) of baculovirus constitutes a motor protein that unwinds DNA by disrupting the hydrogen bonds between bases in the DNA double helix. It contains both ATPase and helicase activity and can bind non-specifically to single-stranded DNA (ssDNA) and dsDNA (Lu and Carstens, 1991; McDougal and Guarino, 2000). An inchworm DNA unwinding mechanism has been proposed for baculovirus helicase wherein the protein binds to dsDNA with high affinity, then the hydrolysis of ATP causes a conformational change of the protein leading to its disassociation from DNA. Through cycles of ATP hydrolysis and helicase translocation, dsDNA is unwound (McDougal and Guarino, 2001). Helicase interacts with LEF-3 and depends on the latter to be transported into the nucleus (Wu and Carstens, 1998). Notably, mutation of two key residues in the helicase sequence expands the host range of an AcMNPV isolate to *Bombyx mori*, an original non-permissive host for that strain of AcMNPV (Argaud *et al.*, 1998).

LEF-1

LEF-1 contains a conserved primase domain (WVVDAD), and purified LEF-1 exhibits primase activity (Mikhailov and Rohrmann, 2002). Although LEF-1 shows low affinity for ssDNA, it interacts with LEF-2 (Evans *et al.*, 1997) and co-purified LEF-1 and LEF-2 exhibit high affinity for ssDNA (Mikhailov and Rohrmann, 2002). *Lef-1* appears to be an essential gene in BmNPV (Ono *et al.*, 2012), but its requirement for viral DNA replication *in vivo* has not been confirmed.

LEF-2

LEF-2, a primase accessory factor, interacts with both LEF-1 and DNA and may stabilize LEF-1 binding to DNA (Mikhailov and Rohrmann, 2002). LEF-2 serves as a capsid protein of both BV and ODV, suggesting that the presence of LEF-2 immediately after entry is important for DNA replication (Wu *et al.*, 2010). Deletion of *lef-2* showed that it is essential for viral DNA amplification but not for initiating DNA replication (Wu *et al.*, 2010). LEF-2 is also involved in late gene expression (Passarelli and Miller, 1993; Merrington *et al.*, 1996).

LEF-3

LEF-3 constitutes an ssDNA binding protein (SSB) (Hang *et al.*, 1995) that is conserved in lepidopteran baculoviruses but is not found in gamma- and delta baculoviruses. LEF-3 interacts with and mediates nuclear import of the helicase (Wu and Carstens, 1998); consistent with this, a knock-out study showed that LEF-3 is essential for viral DNA replication *in vivo* (Yu and Carstens, 2010). Specifically, LEF-3 forms homo-oligomers that may facilitate its function (Mikhailov *et al.*, 2006; Downie *et al.*, 2013) and forms a complex with alkaline nuclease (AN) that exhibits 5′–3′ exonuclease activity (Mikhailov *et al.*, 2003).

IE-1

IE-1 is an immediate early gene that binds to *hrs* as a transcriptional activator (Choi and Guarino, 1995; Rodems and Friesen, 1995). The N-terminal 2–23 aa of AcMNPV IE-1 are critical for DNA replication. This domain contains a consensus cyclin-dependent kinase phosphorylation site, the mutation of which results in the loss of viral DNA replication. This suggests that phosphorylation of IE-1 may be involved in recruiting replicative factors to the *hr* origins of DNA replication (Taggart *et al.*, 2012).

LEF-11

LEF-11 is conserved at the gene level in all baculoviruses except *Culex nigripalpus nucleopolyhedrovirus* (CuniNPV, deltabaculovirus) and has been identified as essential for DNA replication via knock-out assay *in vivo* (Lin and Blissard, 2002). In BmNPV, LEF-11 co-localizes with IE-1 (Zhang *et al.*, 2014), with oligomerization apparently critical for DNA replication (Dong, Z.Q. *et al.*, 2015). LEF-11 also interacts with two host ATPase family members, ATAD3A and HSPD1 (HSP60), to facilitate virus manipulation (Dong *et al.*, 2017). Whether this is related to the function of LEF-11 in DNA replication requires further investigation.

Other baculoviral genes are also involved in baculovirus DNA replication, including a gene encoding for DNA binding protein (DBP) (Vanarsdall *et al.*, 2007a), *me53* (Xi *et al.*, 2007) and *lef-7* (Lu and Miller, 1995).

The virogenic stroma

Baculovirus infection induces a special subnuclear structure, termed the virogenic stroma (VS), that consists of a network of electron-dense cords with electron-translucent spaces sensitive to salt and RNase (Young *et al.*, 1993). The VS constitutes the factory for viral DNA replication (Kawasaki *et al.*, 2004). Viral proteins involved in DNA replication, including IE-1, DBP, LEF-3, and PP31 (39K), localize to the VS, with the sizes of the foci dramatically increasing during infection to occupy over half of the nucleoplasm (Guarino *et al.*, 1992; Okano *et al.*, 1999). LEF-2 also localizes to the VS (Wu *et al.*, 2010). Inhibition of viral DNA replication by aphidicolin or deletion of *dbp* disturbs the proper formation of the VS (Vanarsdall *et al.*, 2007a). Co-expression of four viral proteins including IE-1, LEF-3, P143, and PP31 is sufficient to induce the formation of a cellular chromatin-containing VS-like structure in the nuclei of uninfected cells (Nagamine *et al.*, 2011). The VS also serves as the factory for transcription and nucleocapsid assembly.

Viral gene transcription

Baculovirus gene transcription is divided into early and late (including late and very late) stages separated by the onset of DNA replication. Two distinct RNA polymerase machineries encoded by the host or the virus itself synthesize early- and late/very late-mRNAs, respectively. These two RNA polymerase complexes recognize different promoter motifs and directly program viral gene expression.

Early gene transcription

Early genes are transcribed by host RNA polymerase II, which is sensitive to alpha amanitin. Early gene products are generally required for initiating DNA replication or executing other early events, including factors for viral replication, GP64 for establishing early systemic infection and P35 for blocking host apoptotic response. Both transcriptional enhancers (*hrs*) and activators play a major role in early transcription.

IE-1

IE-1, in addition to being essential for DNA replication, is the major activator required for early gene transcription. It contains an N-terminal transcription activation domain and a C-terminal DNA binding domain (Kovacs *et al.*, 1992). IE-1 forms dimers and binds to the palindromes of *hrs* in the absence of other proteins, leading to elevation of the transactivation activity of IE-1 as much as 200-fold. A helix–loop–helix domain at aa 543–568 of AcMNPV IE1 mediates dimerization and is required for *hr* binding (Olson *et al.*, 2001). A conserved basic domain at aa 152–161 is the enhancer binding domain for IE1 transcription activation (Olson *et al.*, 2003). IE-1 can also activate transcription independent of *hrs*, most likely by directly interacting with the transcription complex.

IE-0

IE-0 is structurally identical to IE-1, except it contains an additional 54 aa at its N-terminus. IE-0 is produced by splicing a segment of the exon0 transcript encoding 38 aa to the 5′ untranslated region of the *ie-1* mRNA (encoding 16 aa and the entire IE-1). Both IE-0 and IE-1 can regulate transcription of early- and late-genes. IE-1 can also down-regulate transcription under the control of the *ie-0* promoter. IE-0 expression peaks at the very early stage of infection and declines thereafter, while IE-1 becomes more abundant than IE-0 after the onset of DNA replication. Some viral promoters are preferably transactivated by IE-0 to IE-1, suggesting that at the very early stage of infection certain genes are specifically regulated by IE-0 (Sokal *et al.*, 2014).

IE-2

IE-2, an additional transcriptional activator that occurs only in some Group I alphabaculoviruses, contains a predicted RING finger domain that can augment IE-1-mediated transactivation when co-transfected with IE-1 (Carson *et al.*, 1988; Yoo and Guarino, 1994). The transactivation effect of IE-2 appears to be cell-line specific or host-specific (Lu and Miller, 1995; Prikhod'ko *et al.*, 1999). IE2 appears to have ubiquitin-ligase (E3) activity (Imai *et al.*, 2003), whether this is related to its transactivation function still awaits further investigation.

Late gene transcription

In contrast to early gene transcription, late/very late genes are transcribed by virus-encoded transcription factors and thus are insensitive to amanitin. Late transcription likely starts after DNA replication, based on the fact that inhibitors of DNA replication block late gene transcription. The results of transient expression assays identified 19 genes involved in late gene transcription, known as late expression factors (*lefs*) (Rapp *et al.*, 1998). Ten of these are required for optimal DNA replication and are known as replication *lef* genes (Kool *et al.*, 1994; Lu and Miller, 1995). The remaining nine genes, including *lef4–6*, *lef8–10*, *lef-12*, *p47* and *39k*, are termed transcription *lef* genes.

In 1998, a virus-encoded RNA polymerase complex that can support transcription from late/very late promoters was purified from AcMNPV-infected cells (Guarino *et al.*, 1998). This approximately 560 kDa protein complex consists of four subunits: LEF-4 (54 kDa), LEF-8 (102 kDa), LEF-9 (55 kDa), and P47 (47 kDa), markedly simpler than bacterial and eukaryotic RNA polymerases. The four subunits are present in almost equimolar amounts, so this complex may contain two copies of each protein. Overexpression of the four subunits significantly increases transcriptional activity at the late stage of infection (Guarino *et al.*, 1998).

LEF-4

LEF-4 constitutes a metal-dependent mRNA capping enzyme. In eukaryotes, three enzymatic activities are needed for cap formation: an RNA triphosphatase to hydrolyse the 5′-triphosphate end of mRNA to a diphosphate, an RNA cap guanylyltransferase to cap the diphosphate end with GMP, and an RNA cap methyltransferase to catalyse N-7 methylation of the guanine cap. LEF-4 contains both triphosphatase and guanylyltransferase activities, located in the N-terminal and C-terminal region of the protein, respectively (Gross and Shuman, 1998; Jin *et al.*, 1998; Martins and Shuman, 2003). The guanylyltransferase function but not the RNA triphosphatase function of LEF-4 is essential for productive infection (Knebel-Mörsdorf *et al.*, 2006; Li and Guarino, 2008).

LEF-8

LEF-8 is the largest subunit of the baculovirus RNA polymerase complex and contains a conserved motif, $GXKX_4HGQ$/NKG, which is found in the β or β' subunit of DNA-directed RNA polymerases from a wide range of organisms (Passarelli *et al.*, 1994). This motif is generally located at the C-terminus of the enzyme and comprises part of the catalytic site. Mutagenesis analysis shows that the conserved C-terminal motif is critical for late gene expression (Titterington *et al.*, 2003). *lef-8* is considered an essential gene for BV production/late gene transcription (Gauthier *et al.*, 2012; Ioannidis *et al.*, 2016).

LEF-9

LEF-9 contains a 7-aa motif (NTDCDGD) similar to the Mg^{2+} binding sequence (NADFDGD) that comprises the catalytic domain present in the largest subunits of some RNA polymerases. This motif is essential for LEF-9 mediated late gene expression (Crouch *et al.*, 2007). *lef-9* is considered an essential gene, as its deletion in BmNPV fails to produce infectious progeny virus (Gomi *et al.*, 1997; Ono *et al.*, 2012).

P47

P47 was originally identified via a *ts* mutant that caused a severe reduction in late gene expression (Partington *et al.*, 1990; Carstens *et al.*, 1993). This protein shows low sequence similarity to the α subunit of bacterial RNA polymerase, which is involved in the initiation of RNA polymerase assembly (Rohrmann, 2013). The exact role of P47 in the RNA polymerase complex has yet to be determined.

LEF-5

LEF-5 is conserved in all sequenced baculoviruses. LEF-5 strongly stimulates transcription activity of purified baculovirus RNA polymerase complex (Guarino *et al.*, 2002). LEF-5 interacts with itself (Harwood *et al.*, 1998) and is essential for productive infection (Su *et al.*, 2011). It contains a motif similar to the zinc ribbon domain of eukaryotic RNA polymerase II elongation factor SII, deletion of which resulted in a 70% reduction in LEF-5 activity (Harwood *et al.*, 1998). An *in vitro* activity assay indicates that LEF-5 may not act as an elongation factor as SII, but rather play a role in the transcription initiation (Guarino *et al.*, 2002).

VLF-1

VLF-1 (very late expression factor 1), a member of the lambda integrase family encoded by the core gene *vlf-1*, is required for high-level expression of very late genes (polyhedrin (*polh*) and *p10*) (McLachlin and Miller, 1994). The promoters of the very late genes differ from those of the late genes; very late genes possess an A+T rich burst sequence downstream of the transcription initiation site. *In vitro* transcription assays show that VLF-1 can stimulate transcription by binding to these 'burst' sequences in very late promoters (Yang and Miller, 1998; Mistretta and Guarino, 2005). A mutation from cysteine to a tyrosine at residue 202 appears to abolish the activity of AcMNPV VLF-1 (McLachlin and Miller, 1994). VLF-1 is also involved in processing of the nucleocapsid assembly and DNA packaging (Vanarsdall *et al.*, 2006).

Global patterns of gene transcription

A comprehensive single nucleotide resolution analysis to globally map transcription start sites (TSS), polyadenylation sites (PAS), and mRNA abundance of the AcMNPV transcription has been performed in cultured *Trichoplusia ni* cells (Chen *et al.*, 2013). A total of 218 TSS for 156 AcMNPV ORFs were identified, with 46 ORFs containing multiple TSS. Among these, 126 TSS for 101 genes (65%) were mapped at the consensus later promoter motifs (TAAG); the remaining 92 TSS associated with 77 genes (49%) were non-TAAG sites, thus being likely to generate early transcripts. Twenty-one genes (13%) appear to contain both early and late promoters. For the 126 identified TAAG motifs, transcription exclusively initiated from the second position (TAAG, underlined). The sequences of non-TAAG motif-containing TSS were much more diverse. Only 29 of the 92 TSS contained upstream TATA boxes and 14 initiated within or near the previously described CAGT start site, although 84% initiated at an A residue. The average 5′-UTR lengths of early (non-TAAG motif) and late (TAAG motif) genes were 155 and 169 nt, respectively.

Analysis of 3′PAS identified a total of 120 PAS corresponding to 135 ORFs. Moreover, 16 genome regions were likely to terminate at a common PAS. The majority (approximately 77%) of the identified PAS were located within 30 nt downstream of a consensus termination signal (AUUAAA/AUUAAA). The average length from the stop codon to the downstream of PAS was 338 nt. In 58 transcripts, a U-rich region was located 2–10 nt downstream of the PAS.

Temporal gene expression was analysed by strand-specific RNA sequencing (RNA-Seq). At 6 h.p.i. the majority of transcripts were early genes, including those involved in DNA replication (*lef-2*, *lef-3*, *dbp*, and *pp31*) and anti-apoptosis (*p35*), as well as the major EFP *gp64*. From 6–12 h.p.i. virus transcription was highly activated, with virus mRNA increasing from 3% (6 h.p.i.) to 38% (12 h.p.i.) of the total cellular mRNA. This may be associated with the onset of virus DNA replication. At 12 h.p.i. 109 genes were transcriptionally active; the most abundant of which encode virus structural proteins (*p6.9*, *odv-ec27*, and *odv-e18*). From 24 h.p.i., *polh* and *p10* were the most abundant mRNAs, accounting for 24 and 7.5% of the total cellular mRNA. At the late stage of infection, viral transcripts comprised approximately 80% of the total mRNA of infected host cells. Overall, genes with early promoters peaked at 6–12 h.p.i., whereas those containing late promoters were expressed at the highest levels at 12–18 h.p.i. or later. Interestingly, low levels of mRNA of 40 genes were already present at 0 h.p.i. As most of these show highly abundant late transcripts, these late mRNAs may be incorporated into BVs during the nucleocapsid assembly and budding process. Twelve spliced transcripts (including the previously characterized *ie-0*) and antisense RNAs to 50 genes were also identified, suggesting possibly significant roles in regulating AcMNPV gene expression.

A transcriptomic analysis of AcMNPV gene expression in the midgut of *T. ni* larvae was published by the same research group (Shrestha *et al.*, 2018). Viral gene expression profiles in ODV-infected midgut share marked similarities with those in BV-infected cell lines. However, several genes associated with enhanced F-actin mediated viral motility (*arif-1* and *ac102*), accelerated systemic infection (*v-fgf*), DNA replication (*lef-3* and *an*) and nucleocapsid assembly (*ac54*) were expressed at much higher levels in the midgut than in cell lines. In contrast, *fp-25k*, a negative factor for BV production, was expressed at a lower level in the midgut. These differential expression patterns represent specific viral adaptations for rapid egress from the midgut to disseminate systemic infection.

Nucleocapsid assembly

The next stage of virus infection is the assembly of progeny nucleocapsids, followed by the formation of BVs. Baculovirus nucleocapsids (30–60 nm × 250–300 nm) are characterized by cylindrical capsids with distinct apical cap and basal structures, with an electron-dense internal nucleoprotein core. In AcMNPV, 24 proteins are associated with the nucleocapsids of both BV and ODV phenotypes, including 11 core proteins (Table 9.1) (Blissard and Theilmann, 2018). Nucleocapsid assembly is a highly organized (both in time and space) and complex process. A brief model for baculovirus nucleocapsid assembly has been proposed where viral genomic DNA is condensed by viral proteins and the resulting nucleoprotein core is packaged into pre-assembled empty capsids (Fig. 9.5).

Nucleocapsid morphogenesis

The progeny nucleocapsids dispersed within the VS were first observed at 8 h.p.i. in the midgut epithelial cells of *T. ni* larvae infected with AcMNPV (Granados and Lawler, 1981). The kinetics of virus replication and morphogenesis appears to be slower in AcMNPV-infected *Mamestra brassicae* cells, with formation of VS observed at 7 h.p.i., empty capsids appearing at 10 h.p.i. and partially or fully filled nucleocapsids observed at 10–11 h.p.i. (Bassemir *et al.*, 1983) A low-resolution structural model of the nucleocapsid of *Spodoptera litura granulovirus* (SpliGV) was created based on electron microscopy and x-ray scattering in 1982. The cylindrical body of the viral capsid is composed of a 12-part helix of monomers. Structural proteins, which form caps at both ends of the nucleocapsid, appear to be distinct from those constituting the cylindrical body. Within the nucleocapsid, the dsDNA associates with a highly basic protein (P6.9) to form a cylindrical core (Burley *et al.*, 1982). An investigation of AcMNPV genome packaging and virion morphogenesis in Sf9 cells showed the formation of empty capsids of nearly full length, most likely initiating with the basal structures; the association of these empty capsids with their apical cap ends to the matrix of VS, composed of newly synthesized DNA and nucleoproteins; and the packaging of the nucleoprotein core from the base to the apex of the capsid sheath to form the filled nucleocapsids, apparently directed by the apical cap structure (Fraser, 1986). Recent cryo-EM analysis of AcMNPV BV confirms earlier observations that the helical structure of nucleocapsids contains a high electron density core with an apical nipple-like structure and a basal claw-like structure at different ends. At least three viral proteins, including P78/83, VLF-1 and VP80, are associated with the basal end region (Rohrmann, 2013). Empty capsids possess a rigid tubular structure and share identical shape and size with filled nucleocapsids, suggesting that viral capsid assembly is independent of DNA packaging (Wang *et al.*, 2016). This independence is further supported by the discovery that long, branched tubules were common in *Trichoplusia ni granulovirus* (TnGV)-infected *T. ni* larvae; these may be self-assembled capsid proteins (Summers, 1971). The elimination of genes such as *p6.9* (Wang *et al.*, 2010a), *38k* (Wu *et al.*, 2006), *vlf-1* (Vanarsdall *et al.*, 2006), *vp1054* (Guan *et al.*, 2016), *p78/83* (Li *et al.*, 2010), *c42* (Li

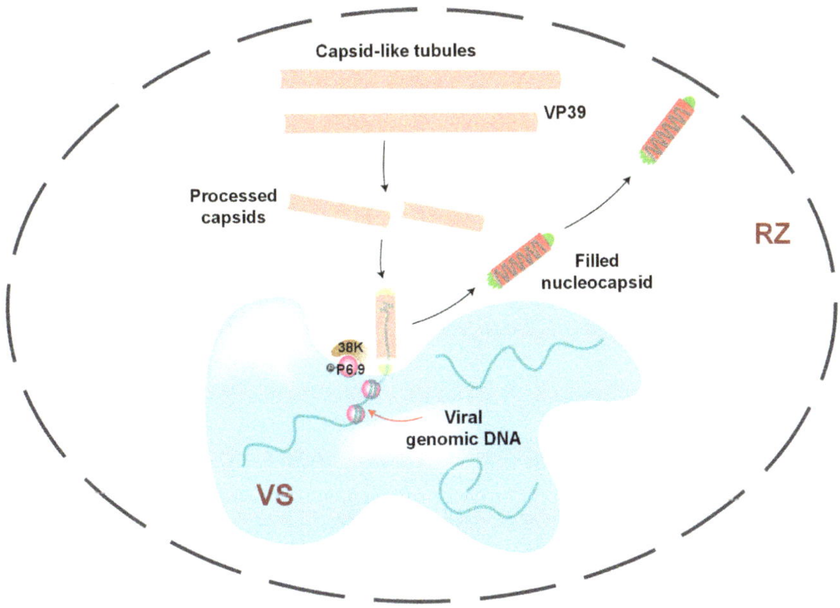

Figure 9.5 Simplified model for baculovirus nucleocapsid assembly within infected nucleus. Capsid-like tubules formed by VP39 are processed into empty capsids. The viral DNA binding protein P6.9 is dephosphorylated by viral phosphatase 38K, allowing P6.9 to bind and condense newly synthesized viral genomic DNA. Subsequently, the nucleoprotein core is packaged from the base to the apex of the pre-assembled capsid sheath to form filled nucleocapsids. Nucleocapsids are then assembled in the virogenic stroma (VS) and filled nucleocapsids are transported to the ring zone (RZ) to form BV or ODV.

et al., 2010; Vanarsdall *et al.*, 2007b), *ac53* (Liu *et al.*, 2008), *ac142* (Vanarsdall *et al.*, 2007b) and *ac144* (Vanarsdall *et al.*, 2007b) resulted in accumulation of these elongated, empty capsid-like tubules. Therefore, the intact capsid sheath may represent the processed product of these long tubules, with capsid assembly occurring prior to DNA packaging, rather than simultaneously. The insertion of an ≈15 kb DNA fragment into the genome of AcMNPV led to the increase of capsid length by about 13 nm (from 260 to 283 nm) (Fraser, 1986). Furthermore, the capsid length was shorter with defective genomes (Kool *et al.*, 1991). In combination, this suggests that the length of the baculovirus capsid is flexible in response to genome size.

Key proteins involved in nucleocapsid assembly

During the process of nucleocapsid assembly, a virus-encoded protamine-like DNA binding protein (P6.9) condenses newly synthesized viral DNA ≈100-fold, followed by packaging into the pre-assembled capsids made of the major capsid protein VP39 and other minor capsid proteins.

VP39

VP39 is the most abundant viral capsid protein with homologues in all sequenced baculovirus genomes. Immunoelectron microscopy analysis showed that the electron-lucent capsid-like tubular structures reacted specifically with a VP39 antibody (Vanarsdall *et al.*, 2006; Wu *et al.*, 2006). Consistent with this, when VP39 of HearNPV was expressed using a prokaryotic system, it was able to self-assemble into viral capsid-like helical tubes *in vitro* (Rao *et al.*, 2018), indicating VP39 has a major role in forming viral capsid structures. Specifically, a conserved glycine residue (Gly-276) of VP39 appeared to be essential for nucleocapsid assembly and very late gene expression (Katsuma and Kokusho, 2017).

Recently, Cryo-EM tomography was used to construct a higher resolution 3D structure of the HearNPV VP39 helical tubes. Two slightly different structural forms were observed: a narrow (N)-tube type (the major type, comprising approximately 90%) and a wide (W)-tube type (the minor type, approximately 10%). The diameters of the two types (25 nm for N-type and 28 nm for W-type) of helix tubes are close to the size of the native capsids. N-tubes are organized as a continuous helix composed of 11⅓ subunits per turn, whereas W-tubes are arranged as ring-shape building blocks with 12 subunits per turn (Fig. 9.6). The structure of the W-tube type is consistent with the previously proposed nucleocapsid model of SpliGV (Burley *et al.*, 1982). These data provide an important basis for understanding the mechanism of baculovirus nucleocapsid assembly and supports the application of VP39-based novel nanomaterials for loading functional motifs (Rao *et al.*, 2018).

P6.9

P6.9 is a virus encoded DNA binding protein found in all sequenced baculoviruses. P6.9 is a small basic protein rich in arginine, serine, and threonine responsible for binding and condensing viral DNA for genome encapsidation. P6.9 is highly abundant in the nuclear matrix and is believed to replace host cellular histones from newly synthesized DNA prior to genome packaging, as it is considerably more basic than host histones (Wilson and Miller, 1986). P6.9 is a hyper-phosphorylated protein, with 22 phosphorylation sites identified by mass spectrometry (Li *et al.*, 2015). The C-terminal basic domain plays a key role in nucleocapsid assembly and specificity (Wang *et al.*, 2010a). Notably, the phosphorylation state of P6.9 is under dynamic regulation in different stages of the virus life cycle. Distinct forms of P6.9 phosphorylated at different sites were identified in infected cells, whereas only dephosphorylated P6.9 was found in BVs (Liu *et al.*, 2012). During nucleocapsid uncoating, P6.9 is phosphorylated by the viral protein kinase PK1 and potentially other host kinases to dissociate from viral DNA, thus facilitating the release of the viral genome (Li *et al.*, 2015). In contrast, upon nucleocapsid assembly at VS, P6.9 is dephosphorylated by 38K to promote DNA binding (Lai *et al.*, 2018).

38K

The protein 38K is a nucleocapsid protein distributed over the cylindrical capsid sheath of both BV and ODV (Wu *et al.*, 2008). 38K is conserved in baculovirus, nudivirus, and polydnavirus genomes. Deletion of the AcMNPV *38K* gene led to the formation of aberrant, elongated, empty capsid-like tubules (Wu *et al.*, 2006). Lai *et al.* (2018) suggested that 38K was a member of the viral phosphatase superfamily and identified P6.9 as one of its substrates. The aa 135–288 of AcMNPV 38K contains haloacid dehalogenase (HAD) motifs that are involved in phosphoryl transfer. Mutating the key amino acids of these motifs resulted in empty tube-like structures in the VS and absence of nucleocapsids. A functional study suggested that 38K was involved in genome encapsidation by mediating the dephosphorylation of five phosphorylated sites at the C terminus of P6.9 (Lai *et al.*, 2018). 38K may have other functions based on its interaction with multiple viral capsid proteins, including VP39, VP80, and VP1054 (Wu *et al.*, 2008).

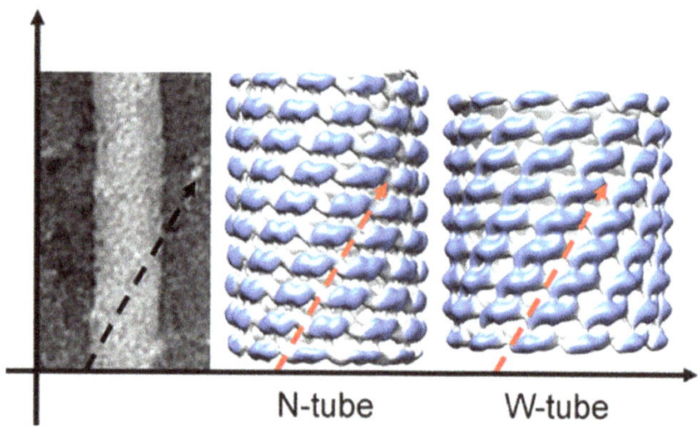

Figure 9.6 Cryo-EM reconstructions of HearNPV VP39 N-tubes and W-tubes. Cryo-EM tomography was used to reconstruct 3D structures of the HearNPV VP39 helical tubes (left). Two slightly different structural forms, the narrow (N)-tube (25 nm in diameter, the major type) (middle) and wide (W)-tube (28 nm in diameter, the minor type) (right) were detected. N-tubes are organized as a continuous helical line wraps around with 11⅓ subunits per turn, whereas W-tubes are arranged as ring-shape building blocks by stable lateral interactions between 12 subunits. (Reprint from Rao *et al.*, 2018, ACS Appl. Mater. Interfaces *10*, 25135–25145.)

VP1054

VP1054 (Ac54) is a capsid protein encoded by a baculovirus core gene that functions in nucleocapsid assembly. A ts mutant of this protein produced spherical electron-dense structures in the VS in place of nucleocapsids at the non-permissive temperature (Olszewski and Miller, 1997). When *ac54* was knocked out, abnormal elongated empty capsids formed by VP39 were found at the RZ region (although no capsid or VP39 was found at VS), suggesting that VP1054 is involved in the localization of VP39 to VS (Guan *et al.*, 2016). Immunoelectron microscopy showed that VP1054 is critical for the transport of capsid proteins, including VP39, 38K, C42, and P78/83 to the VS for nucleocapsid assembly (Guan *et al.*, 2016).

Ac53

Ac53 is conserved in all sequenced baculoviruses. Knockout of *ac53* resulted in abnormal electron-lucent tubular structures in both the VS and RZ regions, and BV and ODV production were abolished (Liu *et al.*, 2008). Ac53 was also found to be associated with BVs of AcMNPV (Liu *et al.*, 2008). Although Ac53 is predicted to contain a conserved U-box/RING-like domain of E3 ubiquitin ligases (Garavaglia *et al.*, 2012), the exact role of Ac53 in nucleocapsid assembly requires further investigation.

BV budding and ODV morphogenesis

Assembled nucleocapsids either egress from nuclei and bud from plasma membrane to form BVs in the early stages of infection or remain within infected nuclei to form ODVs, which are further embedded into OBs in the late stages of infection. It has been suggested that the majority of synthesized viral DNA remains in the nucleus to form ODVs, with only a small portion used to generate BVs. The mechanism of how baculovirus regulates the production of BV and ODV is still unknown. Viral encoded FP25K plays an important role in the switch from BV to ODV production through its effect on the nuclear trafficking of occlusion-associated proteins (Rosas-Acosta *et al.*, 2001; Garretson *et al.*, 2016). Ubiquitination of a nucleocapsid protein Ac66 by virus encoded ubiquitin (v-Ubi) and potential E3 ligase (Ac141, Exon 0) may serve as a signal to determine if a nucleocapsid is destined to form BV or ODV (Biswas *et al.*, 2018; Blissard and Theilmann, 2018).

BV budding

BV budding is relatively poorly understood. During BV budding, the pre-existing GP64 proteins on the plasma membrane drive nucleocapsid budding. Deletion of *gp64* resulted in an approximately 98% reduction in progeny BV production (Oomens and Blissard, 1999). Host cellular vesicular biogenesis, fusion, and trafficking machineries, including the endosomal sorting complex required for transport (ESCRT) and the soluble N-ethylmaleimide-sensitive factor (NSF) attachment protein receptor (SNARE), are involved in the entry and nuclear egress of BVs (Li and Blissard, 2012; Guo *et al.*, 2017; Yue *et al.*, 2018). Additional work is needed to elucidate the details of the intracellular vesicle transport process of baculoviruses. Data from fluorescence resonance energy transfer-fluorescence lifetime imaging microscopy (FRET-FLIM) showed that VP39 interacts directly with kinesin-1, suggesting that microtubules are involved in anterograde trafficking (towards the plasma membrane) of baculoviruses (Danquah *et al.*, 2012). As previously mentioned, F-actin is a key driver of the nuclear egress of newly assembled nucleocapsids and further BV budding at the plasma membrane (see 'Anterograde transport of nucleocapsid').

ODV morphogenesis

In the late stage of infection, nucleocapsids are transported to the RZ, a region between the margins of nuclei and the VS, where they acquire envelopes from virus-induced intranuclear membrane vesicles (MVs) to form ODVs. Early electron microscopy studies showed the formation of MVs within the RZ and the association of nucleocapsids with these vesicles via their apical cap ends (Kawamoto *et al.*, 1977; Fraser, 1986). The inner nuclear membranes (INMs) of infected cells bleb, fold in and vesiculate; MVs are derived from the budding of INMs into the nucleoplasm (Tanada and Hess, 1976; Shi *et al.*, 2015). A model for the formation of MVs has been proposed, which includes: (1) disruption of the nuclear lamina underlying the nucleoplasmic face; (2) membrane curvature at the budding sites; and (3) membrane scission at the bud neck to form MVs (Shi *et al.*, 2018). Several viral genes have been reported to be required for ODV envelopment; however, only *ac75*, *ac76*, and *ac93* have been shown to be required for intranuclear microvesicle formation (Hu *et al.*, 2010; Yuan *et al.*, 2011; Shi *et al.*, 2018). These proteins may participate in the regional disruption of the nuclear lamina or modify the nuclear membrane to promote bending or membrane scission. The morphogenesis of these intranuclear MVs has yet to be demonstrated.

Recently, electron tomography (ET) revealed the dynamic process of microvesicle formation and ODV envelopment in AcMNPV (Shi *et al.*, 2015). The ODV envelopment process can be divided into five steps: (1) the attachment of nucleocapsids to large globular intranuclear vesicles in a vertical fashion; (2) the deformation and rupture of nucleocapsid-associated vesicles; (3) formation of arch-shaped membranes, which grow in size by fusion with small microvesicles; (4) gradual envelopment of the attached nucleocapsids to form immature ODVs containing multiple bundles of nucleocapsids; and (5) invagination of the immature ODV membrane and separation of the bundles of nucleocapsids into individual ODVs, each with a single bundle of parallel nucleocapsids. Whereas it had been generally believed that MVs were derived from the INMs, in this study, the authors found that both the inner and the outer nuclear membranes directly served as the source of the ODV envelopes.

In the final stage, mature ODVs are embedded into OBs. Polymerization of polyhedrin protein appears to follow formation of ODVs in a fairly rapid process (Fraser, 1986; Sajjan and Hinchigeri, 2016). The crystal structure of baculovirus OBs revealed polyhedrin proteins folded into a trimer, with the trimeric interface forming predominantly hydrophobic surfaces. Four identical trimers assemble into a tetrahedral cluster, stabilized mainly by an intermolecular disulfide bond and N-terminal domains. The tetrahedral clusters are further tightly packed in OBs and form unit cells interlocked by the C-terminal arms (Coulibaly *et al.*, 2009). The envelope of mature OBs is composed of polyhedron envelope protein (PEP) and carbohydrate. Deletion of *pep* resulted in OBs with rough pitted surfaces, with many cavities owing to the dislodgment of ODVs (Gross *et al.*, 1994). A recent study employing AFM revealed a multilayered structure for PEP, suggesting that it confers OB stability from environmental adversities (Sajjan and Hinchigeri, 2016).

Regulation of host immunity, behaviour and physiology

As a result of millions of years of co-evolution with their hosts, baculoviruses have developed different strategies to suppress host defence systems, manipulate host behaviour (induce 'tree-top disease' or 'Wipfelkrankheit') and modulate host physiology (prevent host moulting and induce liquefaction of larval body) for optimal virus replication and dissemination.

Suppression of host anti-viral immune responses

Insects possess powerful innate immune systems to combat virus infection (Sabin *et al.*, 2010). To date, three innate immune responses, RNA interference (RNAi), apoptosis and melanization, have been implicated as host defences against baculoviruses. To successfully infect their hosts, baculoviruses have evolved multiple strategies to overcome these defence systems.

Apoptosis

Apoptosis is programmed cell death executed by a cascade of caspase activation. This process is known to be negatively regulated by cellular inhibitors of apoptosis proteins (IAPs). During baculovirus infection, viral DNA replication triggers host DNA damage responses, which leads to the depletion of host cellular IAPs and the activation of apoptosis (Vandergaast *et al.*, 2011; Mitchell and Friesen, 2012). Baculoviruses encode several types of apoptosis suppressors. P35 is the first identified apoptosis inhibitor (Clem *et al.*, 1991), and was found to abolish a wide range of effector caspases. A divergent homologue of P35, termed P49, can inhibit both initiator and effector caspases. Another baculovirus apoptosis suppressor, Apsup, inhibits initiator caspase (Yamada *et al.*, 2013). Individual baculovirus genomes generally contain only one of these three classes of apoptosis suppressors. Apart from these, most baculoviruses also encode cellular IAP homologues. In fact, IAPs were initially identified in baculoviruses and have since been found in cells (Crook *et al.*,

1993). They are phylogenetically classified into five clades (*iap*1–5). Among these, IAP3 exhibits the broadest anti-apoptotic activity (Ikeda *et al.*, 2013) and likely acts by stabilizing cellular IAPs for their proper functions as evidence by the study in *Orgyia pseudotsugata multiple NPV* (Byers *et al.*, 2016). Not all IAPs block apoptosis; some of them can even have the opposite effect under certain conditions (Ikeda *et al.*, 2011).

RNAi

RNAi is a critical anti-viral immune response for insects. High throughput sequencing has identified a large number of viral short interfering RNAs (vsRNA) in baculovirus infected cells. Knock-down of *Dicer-2*, an enzyme required for siRNA biogenesis, significantly enhanced viral DNA replication, suggesting that RNAi acts as an important host defence mechanism against baculovirus infection (Jayachandran *et al.*, 2012). To inhibit the defence mechanism, AcMNPV exploits viral protein P35 as a broadly active viral suppressor of RNAi (VSR). The function of P35 in suppression of RNAi is downstream in the RNAi pathway and independent of the P35 anti-apoptotic activity, although the details of its mechanism remains largely unknown (Mehrabadi *et al.*, 2015).

Melanization

Melanization is a unique and conserved innate defence mechanism in invertebrates. It consists of a cascade of serine proteases that convert prophenoloxidase (PPO) zymogens into the active phenoloxidase (PO) form, which in turn catalyses the formation of melanin to encapsulate and kill invading pathogens. Both *in vitro* and *in vivo* evidence suggest that melanization plays a crucial role in the protection of insects against baculovirus infection (Washburn *et al.*, 1996; Yuan *et al.*, 2017). To counteract this host response, HearNPV appears to have developed two mechanisms: (1) global shutdown of the expression of serine proteases in the melanization cascade; (2) up-regulation of two specific negative regulators (serine protease inhibitors, serpins) (Yuan *et al.*, 2017). Interestingly, certain baculoviruses also encode a *serpin* orthologue and a *conotoxin*-like gene to antagonize host melanization (Cao *et al.*, 2012; Ardisson-Araujo *et al.*, 2015).

Manipulation of host behaviour

Baculovirus infection induces enhanced locomotor activity (ELA) of host insects, including 'hyperactivity' and 'tree-top disease'. These behaviour changes are thought to increase the dispersal of OBs, since insects climbing to the top of the tree enhance visibility to predators. Two viral proteins have been implicated in inducing ELA in infected hosts.

PTP

PTP, a virus-encoded protein tyrosine phosphatase, was identified as responsible for 'wandering-like' ELA via screening of a gene-knockout library of BmNPV (Kamita *et al.*, 2005). PTP-induced ELA is likely to be conserved in a subset of baculoviruses (van Houte *et al.*, 2012; Katsuma *et al.*, 2012). However, the phosphatase activity of PTP may not be conservatively required for ELA induction, since there are contradictory reports in different virus–host systems. Homologues of *ptp* only occur in Group I alphabaculoviruses; this gene could also be captured from insect hosts. In fact, a host PTP homologue was found to partially replace the function of BmNPV PTP (Kamita *et al.*, 2005).

EGT

EGT, a virus encoded ecdysteroid (UDP)-glucosyltransferase, may be involved in tree-top disease. The deletion of the *egt* gene in *Lymantria dispar multiple nucleopolyhedrovirus* (LdMNPV) and *Spodoptera exigua multiple nucleopolyhedrovirus* (SeMNPV) led to the death of host larvae at lower positions compared to the wild-type viruses (Hoover *et al.*, 2011; Han *et al.*, 2015). However, this does not appear to be the case in *T. ni* and *S. exigua* larvae infected by an *egt*-deleted AcMNPV (Ros *et al.*, 2015). It has recently been suggested that EGT-induced tree-top disease could be an indirect effect of moulting (Zhang, S. *et al.*, 2018).

Regulation of host physiology

Larvae stop feeding and become quiescent during moulting, which reduces baculovirus proliferation. EGT can inactivate host moulting hormones by catalysing the conjugation of glucose and galactose to ecdysteroids, which interferes with normal insect moulting (O'Reilly and Miller, 1989). This leads to prolonged insect feeding times and increased yield of viral progeny. Accordingly, deletion of *egt* improves the efficacy of baculoviral pesticides, significantly reducing foliage consumption and increasing speed of action as compared to wild type viruses (O'Reilly and Miller, 1991). This gene likely has been captured by baculoviruses from their insect hosts for optimal virus proliferation.

The rigid exoskeleton of chitin poses a barrier for the release of virus progeny. At the late stage of infection, larval tissues undergo liquefaction, which is executed by two virus-encoded enzymes, chitinase and cathepsin (Ohkawa *et al.*, 1994; Hawtin *et al.*, 1997). Insects infected with recombinant viruses lacking either *chitinase* or *cathepsin* genes remain intact after death (Hawtin *et al.*, 1997). Chitinase and cathepsin are stored in the endoplasmic reticulum (ER) to prevent pre-mature degradation of host tissues and are liberated upon host death. Thus, OBs are released from the liquefied larval corpse to contaminate surrounding foliage or soils, for ingestion by new host insects to begin another infection cycle.

Synthetic biology in baculoviruses

Viruses are ideal models for synthetic biology research because of their small genome size. Viruses are likely the most readily reconstituted form of life. A small RNA virus (poliovirus) was the first organism to be synthesized, in 2002 (Cello *et al.*, 2002). The unique life cycle of baculovirus with its two phenotypes (BV and ODV) provides an excellent model system for synthetic biology. For example, some of the ODV-specific genes that are non-essential for BV production can be deleted from the virus genome for both fundamental and applied research.

Shang *et al.* (2017) reported the first synthesis of the prototypical baculovirus, AcMNPV, by using a combination of overlap extension PCR and transformation-associated recombination (TAR) in yeast. The synthetic genome, designated AcMNPV-WIV-Syn1, is approximately 145 kb, encompassing almost the complete genome of AcMNPV. A cassette of about 11.5 kb including bacterial and yeast replication elements and an *egfp* gene in the *hr4a* locus were added to the synthetic genome. Further transfection of AcMNPV-WIV-Syn1 DNA into insect cells produced infectious virus that retained similar biological properties (infectivity and virion morphology) to the native virus, a major step towards genome-wide editing of baculovirus.

This new synthetic platform offers the opportunity to manipulate any or multiple loci simultaneously, thus providing a powerful tool for further investigation of the basic biology as well as engineering of baculoviruses. Potential future research directions include (1) construction of a mini-genome of baculovirus to understand the elements required for a minimal functional baculovirus; (2) deletion of irrelevant genes for BV production from the baculovirus genome to engineer a vector with larger insertion capacities for gene expression and delivery; (3) adding entire mammalian protein modification pathways into the baculovirus genome to facilitate better production of foreign proteins and vaccines; and (4) substitution of a whole set of heterologous PIFs for studying the oral infection mechanism and expansion of the host ranges of baculoviral pesticides.

To date, approximately 45% of the AcMNPV genome has been classified as inessential for BV production. This might be verified by the 'Design-Build-Test Cycle' strategy used by the laboratory of Craig Venter for synthesis of a minimal bacterial genome (Hutchison *et al.*, 2016). Both 'top-down' and 'bottom-up' synthetic biology approaches might be used to synthesize such a mini-genome for AcMNPV (Wang and Maranas, 2018) (Fig. 9.7). The resulting minimal functional baculovirus would provide pivotal fundamental information for understanding baculovirus functional genomics and evolution, as well as being beneficial for designing novel gene expression and delivery vectors with enhanced stability and greater packaging capacities.

Conclusions and future perspectives

Over the past few decades, remarkable progress has been made towards understanding baculovirus genome organization, structures of both virion phenotypes, functional genomics, regulation of viral gene expression, and baculovirus–host interaction. However, numerous fundamental questions remain. The structure and function of individual PIFs and of the PIF complex have yet to be determined. The cellular receptors for ODV and BV entry are likewise still mysteries. Viral proteins and host factors for mediating the nuclear import of nucleocapsids

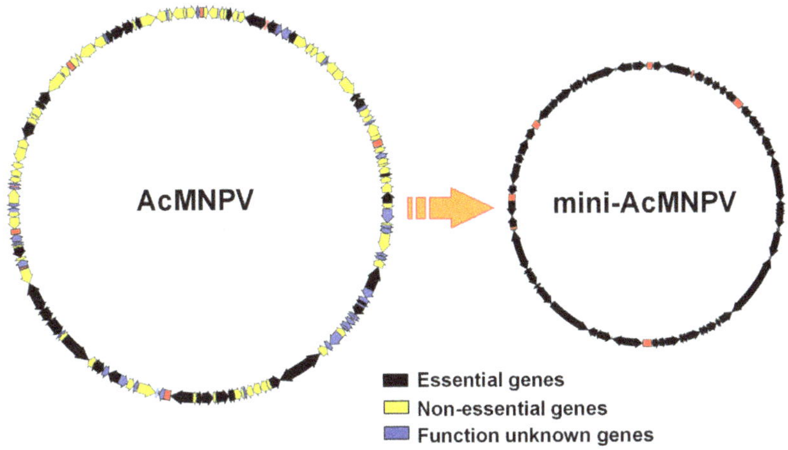

Figure 9.7 Schematic representation of an AcMNPV mini-genome. Approximately 45% of the AcMNPV genome is estimated to be non-essential (yellow) for BV production. After validating the impacts of unknown genes on BV production, a mini-genome of AcMNPV can be designed and constructed by removal of the non-essential genes using synthetic biology approaches.

are completely unknown. In addition, knowledge of how baculoviruses modulate the host immune system and manipulate host physiology/behaviour for better virus replication and dissemination is limited. With the expanded potential of technologies such as Cryo-EM, CRISPR/Cas9, and synthetic biology platforms, these questions will likely be addressed in the near future. The answers will shed light on the baculovirus infection mechanisms, providing useful information and important strategies for further engineering of baculoviruses into more efficient protein expression vectors and viral pesticides.

References

Alfonso, V., Maroniche, G.A., Reca, S.R., López, M.G., del Vas, M., and Taboga, O. (2012). AcMNPV core gene ac109 is required for budded virion transport to the nucleus and for occlusion of viral progeny. PLOS ONE *7*, e46146. https://doi.org/10.1371/journal.pone.0046146

Ardisson-Araujo, D.M., Rohrmann, G.F., Ribeiro, B.M., and Clem, R.J. (2015). Functional characterization of hesp018, a baculovirus-encoded serpin gene. J. Gen. Virol. *96*, 1150–1160. https://doi.org/10.1099/vir.0.000041

Ardisson-Araújo, D.M., Melo, F.L., Clem, R.J., Wolff, J.L., and Ribeiro, B.M. (2016). A Betabaculovirus-encoded gp64 homolog codes for a functional envelope fusion protein. J. Virol. *90*, 1668–1672. https://doi.org/10.1128/JVI.02491-15

Argaud, O., Croizier, L., López-Ferber, M., and Croizier, G. (1998). Two key mutations in the host-range specificity domain of the p143 gene of Autographa californica nucleopolyhedrovirus are required to kill Bombyx mori larvae. J. Gen. Virol. *79*, 931–935. https://doi.org/10.1099/0022-1317-79-4-931

Au, S., and Panté, N. (2012). Nuclear transport of baculovirus: revealing the nuclear pore complex passage. J. Struct. Biol. *177*, 90–98. https://doi.org/10.1016/j.jsb.2011.11.006

Au, S., Wu, W., and Panté, N. (2013). Baculovirus nuclear import: open, nuclear pore complex (NPC) sesame. Viruses *5*, 1885–1900. https://doi.org/10.3390/v5071885

Au, S., Wu, W., Zhou, L., Theilmann, D.A., and Panté, N. (2016). A new mechanism for nuclear import by actin-based propulsion used by a baculovirus nucleocapsid. J. Cell. Sci. *129*, 2905–2911. https://doi.org/10.1242/jcs.191668

Bassemir, U., Miltenburger, H.G., and David, P. (1983). Morphogenesis of nuclear polyhedrosis virus from Autographa californica in a cell line from Mamestra brassicae (cabbage moth). Further aspects on baculovirus assembly. Cell Tissue Res. *228*, 587–595.

Biswas, S., Willis, L.G., Fang, M., Nie, Y., and Theilmann, D.A. (2018). Autographa californica Nucleopolyhedrovirus AC141 (Exon0), a Potential E3 Ubiquitin Ligase, Interacts with Viral Ubiquitin and AC66 To Facilitate Nucleocapsid Egress. J. Virol. *92*, e01713–17. https://doi.org/10.1128/JVI.01713-17

Blissard, G.W., and Theilmann, D.A. (2018). Baculovirus Entry and egress from insect cells. Annu. Rev. Virol. *5*, 113–139. https://doi.org/10.1146/annurev-virology-092917-043356

Blissard, G.W., and Wenz, J.R. (1992). Baculovirus gp64 envelope glycoprotein is sufficient to mediate pH-dependent membrane fusion. J. Virol. *66*, 6829–6835.

Boogaard, B., van Lent, J.W.M., Theilmann, D.A., Erlandson, M.A., and van Oers, M.M. (2017). Baculoviruses require an intact ODV entry-complex to resist proteolytic degradation of per os infectivity factors by co-occluded proteases from the larval host. J. Gen. Virol. [Epub ahead of print]. https://doi.org/10.1099/jgv.0.000974

Boogaard, B., van Oers, M.M., and van Lent, J.W.M. (2018). An advanced view on baculovirus per os infectivity factors. Insects *9*, 84. https://doi.org/10.3390/insects9030084

Bossert, M., and Carstens, E.B. (2018). Sequential deletion of AcMNPV homologous regions leads to reductions in budded virus production and late protein expression. Virus Res. *256*, 125–133. https://doi.org/10.1016/j.virusres.2018.08.010

Braithwaite, D.K., and Ito, J. (1993). Compilation, alignment, and phylogenetic relationships of DNA polymerases. Nucleic Acids Res. *21*, 787–802.

Braunagel, S.C., and Summers, M.D. (2007). Molecular biology of the baculovirus occlusion-derived virus envelope. Curr. Drug Targets *8*, 1084–1095.

Burley, S.K., Miller, A., Harrap, K.A., and Kelly, D.C. (1982). Structure of the Baculovirus nucleocapsid. Virology *120*, 433–440.

Byers, N.M., Vandergaast, R.L., and Friesen, P.D. (2016). Baculovirus inhibitor-of-apoptosis Op-IAP3 blocks apoptosis by interaction with and stabilization of a host insect cellular IAP. J. Virol. *90*, 533–544. https://doi.org/10.1128/JVI.02320-15

Cao, Q., Zhu, S.Y., Wu, Y., Liu, Y., Zhu, J., and Wang, W. (2012). The Effect of a Small Conotoxin-Like ctx gene from autographa californica nuclear polyhedrosis virus (AcMNPV) on insect hemolymph melanization. Pol. J. Microbiol. *61*, 183–189.

Carson, D.D., Guarino, L.A., and Summers, M.D. (1988). Functional mapping of an AcNPV immediately early gene which augments expression of the IE-1 trans-activated 39K gene. Virology *162*, 444–451.

Carstens, E.B., and Wu, Y. (2007). No single homologous repeat region is essential for DNA replication of the baculovirus Autographa californica multiple nucleopolyhedrovirus. J. Gen. Virol. *88*, 114–122. https://doi.org/10.1099/vir.0.82384-0

Carstens, E.B., Lu, A.L., and Chan, H.L. (1993). Sequence, transcriptional mapping, and overexpression of p47, a baculovirus gene regulating late gene expression. J. Virol. *67*, 2513–2520.

Cello, J., Paul, A.V., and Wimmer, E. (2002). Chemical synthesis of poliovirus cDNA: generation of infectious virus in the absence of natural template. Science *297*, 1016–1018. https://doi.org/10.1126/science.1072266

Charlton, C.A., and Volkman, L.E. (1991). Sequential rearrangement and nuclear polymerization of actin in baculovirus-infected Spodoptera frugiperda cells. J. Virol. *65*, 1219–1227.

Charlton, C.A., and Volkman, L.E. (1993). Penetration of Autographa californica nuclear polyhedrosis virus nucleocapsids into IPLB Sf 21 cells induces actin cable formation. Virology *197*, 245–254. https://doi.org/10.1006/viro.1993.1585

Chen, G., Yan, Q., Fang, Y., Wu, L., Krell, P.J., and Feng, G. (2018). The N terminus of Autographa californica multiple Nucleopolyhedrovirus DNA Polymerase is required for efficient viral DNA replication and virus and occlusion body production. J. Virol. *92*, e00398-18. https://doi.org/10.1128/JVI.00398-18

Chen, Y.R., Zhong, S., Fei, Z., Hashimoto, Y., Xiang, J.Z., Zhang, S., and Blissard, G.W. (2013). The transcriptome of the baculovirus Autographa californica multiple nucleopolyhedrovirus in Trichoplusia ni cells. J. Virol. *87*, 6391–6405. https://doi.org/10.1128/JVI.00194-13

Choi, J., and Guarino, L.A. (1995). The baculovirus transactivator IE1 binds to viral enhancer elements in the absence of insect cell factors. J. Virol. *69*, 4548–4551.

Clem, R.J., Fechheimer, M., and Miller, L.K. (1991). Prevention of apoptosis by a baculovirus gene during infection of insect cells. Science *254*, 1388–1390.

Cohen, S., Au, S., and Panté, N. (2011). How viruses access the nucleus. Biochim. Biophys. Acta *1813*, 1634–1645. https://doi.org/10.1016/j.bbamcr.2010.12.009

Coulibaly, F., Chiu, E., Gutmann, S., Rajendran, C., Haebel, P.W., Ikeda, K., Mori, H., Ward, V.K., Schulze-Briese, C., and Metcalf, P. (2009). The atomic structure of baculovirus polyhedra reveals the independent emergence of infectious crystals in DNA and RNA viruses. Proc. Natl. Acad. Sci. U.S.A. *106*, 22205–22210. https://doi.org/10.1073/pnas.0910686106

Crook, N.E., Clem, R.J., and Miller, L.K. (1993). An apoptosis-inhibiting baculovirus gene with a zinc finger-like motif. J. Virol. *67*, 2168–2174.

Crouch, E.A., Cox, L.T., Morales, K.G., and Passarelli, A.L. (2007). Inter-subunit interactions of the Autographa californica M nucleopolyhedrovirus RNA polymerase. Virology *367*, 265–274. https://doi.org/10.1016/j.virol.2007.05.026

Danquah, J.O., Botchway, S., Jeshtadi, A., and King, L.A. (2012). Direct interaction of baculovirus capsid proteins VP39 and EXON0 with kinesin-1 in insect cells determined by fluorescence resonance energy transfer-fluorescence lifetime imaging microscopy. J. Virol. *86*, 844–853. https://doi.org/10.1128/JVI.06109-11

Dong, S., and Blissard, G.W. (2012). Functional analysis of the Autographa californica multiple nucleopolyhedrovirus GP64 terminal fusion loops and interactions with membranes. J. Virol. *86*, 9617–9628. https://doi.org/10.1128/JVI.00813-12

Dong, S., Wang, M., Qiu, Z., Deng, F., Vlak, J.M., Hu, Z., and Wang, H. (2010). Autographa californica multicapsid nucleopolyhedrovirus efficiently infects Sf9 cells and transduces mammalian cells via direct fusion with the plasma membrane at low pH. J. Virol. *84*, 5351–5359. https://doi.org/10.1128/JVI.02517-09

Dong, X.L., Liu, T.H., Wang, W., Pan, C.X., Wu, Y.F., Du, G.Y., Chen, P., Lu, C., and Pan, M.H. (2015). BmREEPa Is a Novel Gene that Facilitates BmNPV Entry into Silkworm Cells. PLOS ONE *10*, e0144575. https://doi.org/10.1371/journal.pone.0144575

Dong, Z.Q., Zhang, J., Chen, X.M., He, Q., Cao, M.Y., Wang, L., Li, H.Q., Xiao, W.F., Pan, C.X., Lu, C., *et al.* (2014). Bombyx mori nucleopolyhedrovirus ORF79 is a per os infectivity factor associated with the PIF complex. Virus Res. *184*, 62–70. https://doi.org/10.1016/j.virusres.2014.02.009

Dong, Z.Q., Hu, N., Zhang, J., Chen, T.T., Cao, M.Y., Li, H.Q., Lei, X.J., Chen, P., Lu, C., and Pan, M.H. (2015). Oligomerization of Baculovirus LEF-11 Is Involved in Viral DNA Replication. PLOS ONE *10*, e0144930. https://doi.org/10.1371/journal.pone.0144930

Dong, Z.Q., Hu, N., Dong, F.F., Chen, T.T., Jiang, Y.M., Chen, P., Lu, C., and Pan, M.H. (2017). Baculovirus LEF-11 Hijack Host ATPase ATAD3A to Promote Virus Multiplication in Bombyx mori cells. Sci. Rep. 7, 46187. https://doi.org/10.1038/srep46187

Downie, K., Adetola, G., and Carstens, E.B. (2013). Characterization of protein-protein interaction domains within the baculovirus Autographa californica multiple nucleopolyhedrovirus late expression factor LEF-3. J. Gen. Virol. *94*, 2530–2535. https://doi.org/10.1099/vir.0.056267-0

Duisit, G., Saleun, S., Douthe, S., Barsoum, J., Chadeuf, G., and Moullier, P. (1999). Baculovirus vector requires electrostatic interactions including heparan sulfate for efficient gene transfer in mammalian cells. J. Gene Med. *1*, 93–102. https://doi.org/10.1002/(SICI)1521-2254(199903/04)1:2<93::AID-JGM19>3.0.CO;2-1

Evans, J.T., Leisy, D.J., and Rohrmann, G.F. (1997). Characterization of the interaction between the baculovirus replication factors LEF-1 and LEF-2. J. Virol. *71*, 3114–3119.

Fang, Z., Li, C., Wu, W., Yuan, M., and Yang, K. (2016). The Autographa californica multiple nucleopolyhedrovirus Ac132 plays a role in nuclear entry. J. Gen. Virol. *97*, 3030–3038. https://doi.org/10.1099/jgv.0.000602

Feng, G., and Krell, P.J. (2014). Autographa californica multiple nucleopolyhedrovirus DNA polymerase C terminus is required for nuclear localization and viral DNA replication. J. Virol. *88*, 10918–10933. https://doi.org/10.1128/JVI.01167-14

Fraser, M.J. (1986). Ultrastructural observations of virion maturation in Autographa californica nuclear polyhedrosis virus infected Spodoptera frugiperda cell cultures. J. Ultrastruct. Mol. Struct. Res. *95*, 189–195.

Gandhi, K.M., Ohkawa, T., Welch, M.D., and Volkman, L.E. (2012). Nuclear localization of actin requires AC102 in Autographa californica multiple nucleopolyhedrovirus-infected cells. J. Gen. Virol. *93*, 1795–1803. https://doi.org/10.1099/vir.0.041848-0

Garavaglia, M.J., Miele, S.A., Iserte, J.A., Belaich, M.N., and Ghiringhelli, P.D. (2012). The ac53, ac78, ac101, and ac103 genes are newly discovered core genes in the family Baculoviridae. J. Virol. *86*, 12069–12079. https://doi.org/10.1128/JVI.01873-12

Garretson, T.A., McCoy, J.C., and Cheng, X.W. (2016). Baculovirus FP25K Localization: Role of the Coiled-Coil Domain. J. Virol. *90*, 9582–9597. https://doi.org/10.1128/JVI.01241-16

Gauthier, D., Thirunavukkarasu, K., Faris, B.L., Russell, D.L., and Weaver, R.F. (2012). Characterization of an Autographa californica multiple nucleopolyhedrovirus dual mutant: ORF82 is required for budded virus production, and a point mutation in LEF-8 alters late and abolishes very late transcription. J. Gen. Virol. 93, 364–373. https://doi.org/10.1099/vir.0.037028-0

Goley, E.D., Ohkawa, T., Mancuso, J., Woodruff, J.B., D'Alessio, J.A., Cande, W.Z., Volkman, L.E., and Welch, M.D. (2006). Dynamic nuclear actin assembly by Arp2/3 complex and a baculovirus WASP-like protein. Science *314*, 464–467. https://doi.org/10.1126/science.1133348

Gomi, S., Zhou, C.E., Yih, W., Majima, K., and Maeda, S. (1997). Deletion analysis of four of eighteen late gene expression factor gene homologues of the baculovirus, BmNPV. Virology *230*, 35–47. https://doi.org/10.1006/viro.1997.8457

Granados, R.R., and Lawler, K.A. (1981). In vivo pathway of Autographa californica baculovirus invasion and infection. Virology *108*, 297–308.

Gross, C.H., and Shuman, S. (1998). RNA 5'-triphosphatase, nucleoside triphosphatase, and guanylyltransferase activities of baculovirus LEF-4 protein. J. Virol. *72*, 10020–10028.

Gross, C.H., Russell, R.L., and Rohrmann, G.F. (1994). Orgyia pseudotsugata baculovirus p10 and polyhedron envelope protein genes: analysis of their relative expression levels and role in polyhedron structure. J. Gen. Virol. *75*, 1115–1123. https://doi.org/10.1099/0022-1317-75-5-1115

Guan, Z., Zhong, L., Li, C., Wu, W., Yuan, M., and Yang, K. (2016). The Autographa californica multiple nucleopolyhedrovirus ac54 gene is crucial for localization of the major capsid protein VP39 at the site of nucleocapsid assembly. J. Virol. *90*, 4115–4126. https://doi.org/10.1128/JVI.02885-15

Guarino, L.A., and Summers, M.D. (1986). Interspersed Homologous DNA of Autographa californica nuclear polyhedrosis virus enhances delayed-early gene expression. J. Virol. *60*, 215–223.

Guarino, L.A., Dong, W., Xu, B., Broussard, D.R., Davis, R.W., and Jarvis, D.L. (1992). Baculovirus phosphoprotein pp31 is associated with virogenic stroma. J. Virol. *66*, 7113–7120.

Guarino, L.A., Xu, B., Jin, J., and Dong, W. (1998). A virus-encoded RNA polymerase purified from baculovirus-infected cells. J. Virol. 72, 7985–7991.

Guarino, L.A., Dong, W., and Jin, J. (2002). In vitro activity of the baculovirus late expression factor LEF-5. J. Virol. *76*, 12663–12675.

Guo, Y., Yue, Q., Gao, J., Wang, Z., Chen, Y.R., Blissard, G.W., Liu, T.X., and Li, Z. (2017). Roles of Cellular NSF Protein in Entry and Nuclear Egress of Budded Virions of Autographa californica Multiple Nucleopolyhedrovirus. J. Virol. *9*, e01111–17. https://doi.org/10.1128/JVI.01111-17

Han, Y., van Houte, S., Drees, G.F., van Oers, M.M., and Ros, V.I. (2015). Parasitic manipulation of host behaviour: Baculovirus SeMNPV EGT facilitates tree-top disease in spodoptera exigua larvae by extending the time to death. Insects *6*, 716–731. https://doi.org/10.3390/insects6030716

Hang, X., and Guarino, L.A. (1999). Purification of Autographa californica nucleopolyhedrovirus DNA polymerase from infected insect cells. J. Gen. Virol. *80*, 2519–2526. https://doi.org/10.1099/0022-1317-80-9-2519

Hang, X., Dong, W., and Guarino, L.A. (1995). The lef-3 gene of Autographa californica nuclear polyhedrosis virus encodes a single-stranded DNA-binding protein. J. Virol. *69*, 3924–3928.

Harwood, S.H., Li, L., Ho, P.S., Preston, A.K., and Rohrmann, G.F. (1998). AcMNPV late expression factor-5 interacts with itself and contains a zinc ribbon domain that is required for maximal late transcription activity and is homologous to elongation factor TFIIS.

Virology *250*, 118–134. https://doi.org/10.1006/viro.1998.9334

Hawtin, R.E., Zarkowska, T., Arnold, K., Thomas, C.J., Gooday, G.W., King, L.A., Kuzio, J.A., and Possee, R.D. (1997). Liquefaction of Autographa californica nucleopolyhedrovirus-infected insects is dependent on the integrity of virus-encoded chitinase and cathepsin genes. Virology *238*, 243–253. https://doi.org/10.1006/viro.1997.8816

Hefferon, K.L., Oomens, A.G., Monsma, S.A., Finnerty, C.M., and Blissard, G.W. (1999). Host cell receptor binding by baculovirus GP64 and kinetics of virion entry. Virology *258*, 455–468. https://doi.org/10.1006/viro.1999.9758

Hepp, S.E., Borgo, G.M., Ticau, S., Ohkawa, T., and Welch, M.D. (2018). Baculovirus AC102 is a nucleocapsid protein that is crucial for nuclear actin polymerization and nucleocapsid morphogenesis. J. Virol. *92*, e00111-18. https://doi.org/10.1128/JVI.00111-18

Hilton, S., and Winstanley, D. (2008). The origins of replication of granuloviruses. Arch. Virol. *153*, 1527–1535. https://doi.org/10.1007/s00705-008-0148-0

Hohmann, A.W., and Faulkner, P. (1983). Monoclonal antibodies to baculovirus structural proteins: determination of specificities by Western blot analysis. Virology *125*, 432–444.

Hoover, K., Grove, M., Gardner, M., Hughes, D.P., McNeil, J., and Slavicek, J. (2011). A gene for an extended phenotype. Science *333*, 1401. https://doi.org/10.1126/science.1209199

Hou, D., Zhang, L., Deng, F., Fang, W., Wang, R., Liu, X., Guo, L., Rayner, S., Chen, X., Wang, H., *et al.* (2013). Comparative proteomics reveal fundamental structural and functional differences between the two progeny phenotypes of a baculovirus. J. Virol. *87*, 829–839. https://doi.org/10.1128/JVI.02329-12

Hu, Z., Yuan, M., Wu, W., Liu, C., Yang, K., and Pang, Y. (2010). Autographa californica multiple nucleopolyhedrovirus ac76 is involved in intranuclear microvesicle formation. J. Virol. *84*, 7437–7447. https://doi.org/10.1128/JVI.02103-09

Hutchison, C.A., Chuang, R.Y., Noskov, V.N., Assad-Garcia, N., Deerinck, T.J., Ellisman, M.H., Gill, J., Kannan, K., Karas, B.J., Ma, L., *et al.* (2016). Design and synthesis of a minimal bacterial genome. Science *351*, aad6253. https://doi.org/10.1126/science.aad6253

IJkel, W.F., Westenberg, M., Goldbach, R.W., Blissard, G.W., Vlak, J.M., and Zuidema, D. (2000). A novel baculovirus envelope fusion protein with a proprotein convertase cleavage site. Virology *275*, 30–41. https://doi.org/10.1006/viro.2000.0483

Ikeda, M., Yamada, H., Ito, H., and Kobayashi, M. (2011). Baculovirus IAP1 induces caspase-dependent apoptosis in insect cells. J. Gen. Virol. *92*, 2654–2663. https://doi.org/10.1099/vir.0.033332-0

Ikeda, M., Yamada, H., Hamajima, R., and Kobayashi, M. (2013). Baculovirus genes modulating intracellular innate antiviral immunity of lepidopteran insect cells. Virology *435*, 1–13. https://doi.org/10.1016/j.virol.2012.10.016

Imai, N., Matsuda, N., Tanaka, K., Nakano, A., Matsumoto, S., and Kang, W. (2003). Ubiquitin ligase activities of Bombyx mori nucleopolyhedrovirus RING finger proteins. J. Virol. *77*, 923–930.

Ioannidis, K., Swevers, L., and Iatrou, K. (2016). Bombyx mori nucleopolyhedrovirus lef8 gene: effects of deletion and implications for gene transduction applications. J. Gen. Virol. *97*, 786–796. https://doi.org/10.1099/jgv.0.000383

Jarvis, D.L., and Garcia, A. (1994). Biosynthesis and processing of the Autographa californica nuclear polyhedrosis virus gp64 protein. Virology *205*, 300–313. https://doi.org/10.1006/viro.1994.1646

Jarvis, D.L., Wills, L., Burow, G., and Bohlmeyer, D.A. (1998). Mutational analysis of the N-linked glycans on Autographa californica nucleopolyhedrovirus gp64. J. Virol. *72*, 9459–9469.

Javed, M.A., Biswas, S., Willis, L.G., Harris, S., Pritchard, C., van Oers, M.M., Donly, B.C., Erlandson, M.A., Hegedus, D.D., and Theilmann, D.A. (2017). Autographa californica Multiple Nucleopolyhedrovirus AC83 is a *Per Os* Infectivity Factor (PIF) Protein Required for Occlusion-Derived Virus (ODV) and Budded Virus Nucleocapsid Assembly as well as Assembly of the PIF Complex in ODV Envelopes. J. Virol. *91*, e02115-16.

Jayachandran, B., Hussain, M., and Asgari, S. (2012). RNA interference as a cellular defense mechanism against the DNA virus baculovirus. J. Virol. *86*, 13729–13734. https://doi.org/10.1128/JVI.02041-12

Jiang, Y., Deng, F., Rayner, S., Wang, H., and Hu, Z. (2009). Evidence of a major role of GP64 in group I alphabaculovirus evolution. Virus Res. *142*, 85–91. https://doi.org/10.1016/j.virusres.2009.01.015

Jin, J., Dong, W., and Guarino, L.A. (1998). The LEF-4 subunit of baculovirus RNA polymerase has RNA 5'-triphosphatase and ATPase activities. J. Virol. *72*, 10011–10019.

Kadlec, J., Loureiro, S., Abrescia, N.G., Stuart, D.I., and Jones, I.M. (2008). The postfusion structure of baculovirus gp64 supports a unified view of viral fusion machines. Nat. Struct. Mol. Biol. *15*, 1024–1030. https://doi.org/10.1038/nsmb.1484

Kamita, S.G., Nagasaka, K., Chua, J.W., Shimada, T., Mita, K., Kobayashi, M., Maeda, S., and Hammock, B.D. (2005). A baculovirus-encoded protein tyrosine phosphatase gene induces enhanced locomotory activity in a lepidopteran host. Proc. Natl. Acad. Sci. U.S.A. *102*, 2584–2589.

Katsuma, S., and Kokusho, R. (2017). A conserved glycine residue is required for proper functioning of a Baculovirus VP39 protein. J. Virol. *91*, e02253-16.

Katsuma, S., Koyano, Y., Kang, W., Kokusho, R., Kamita, S.G., and Shimada, T. (2012). The baculovirus uses a captured host phosphatase to induce enhanced locomotory activity in host caterpillars. PLOS Pathog. *8*, e1002644. https://doi.org/10.1371/journal.ppat.1002644

Kawamoto, F., Suto, C., Kumada, N., and Kobayashi, M. (1977). Cytoplasmic budding of a nuclear polyhedrosis virus and comparative ultrastructural studies of envelopes. Microbiol. Immunol. *21*, 255–265.

Kawasaki, Y., Matsumoto, S., and Nagamine, T. (2004). Analysis of baculovirus IE1 in living cells: dynamics and spatial relationships to viral structural proteins. J. Gen Virol. *85*, 3575–3583. https://doi.org/10.1099/vir.0.80418-0

Knebel-Mörsdorf, D., Quadt, I., Li, Y., Montier, L., and Guarino, L.A. (2006). Expression of baculovirus late and very late genes depends on LEF-4, a component of the viral RNA polymerase whose guanyltransferase function is essential. J. Virol. *80*, 4168–4173. https://doi.org/10.1128/JVI.80.8.4168-4173.2006

Kool, M., Voncken, J.W., van Lier, F.L., Tramper, J., and Vlak, J.M. (1991). Detection and analysis of Autographa californica nuclear polyhedrosis virus mutants with defective interfering properties. Virology *183*, 739–746.

Kool, M., Voeten, J.T., Goldbach, R.W., Tramper, J., and Vlak, J.M. (1993). Identification of seven putative origins of Autographa californica multiple nucleocapsid nuclear polyhedrosis virus DNA replication. J. Gen. Virol. *74*, 2661–2668. https://doi.org/10.1099/0022-1317-74-12-2661

Kool, M., Ahrens, C.H., Goldbach, R.W., Rohrmann, G.F., and Vlak, J.M. (1994). Identification of genes involved in DNA replication of the Autographa californica baculovirus. Proc. Natl. Acad. Sci. U.S.A. *91*, 11212–11216.

Kovacs, G.R., Choi, J., Guarino, L.A., and Summers, M.D. (1992). Functional dissection of the Autographa californica nuclear polyhedrosis virus immediate-early 1 transcriptional regulatory protein. J. Virol. *66*, 7429–7437.

Lai, Q., Wu, W., Li, A., Wang, W., Yuan, M., and Yang, K. (2018). The 38K-mediated specific dephosphorylation of the viral core protein P6.9 plays an important role in the nucleocapsid assembly of Autographa californica multiple nucleopolyhedrovirus. J. Virol. *92*, e01989-17. https://doi.org/10.1128/JVI.01989-17

Lee, H.Y., and Krell, P.J. (1992). Generation and analysis of defective genomes of Autographa californica nuclear polyhedrosis virus. J. Virol. *66*, 4339–4347.

Leisy, D.J., and Rohrmann, G.F. (1993). Characterization of the replication of plasmids containing hr sequences in baculovirus-infected Spodoptera frugiperda cells. Virology *196*, 722–730. https://doi.org/10.1006/viro.1993.1529

Li, A., Zhao, H., Lai, Q., Huang, Z., Yuan, M., and Yang, K. (2015). Posttranslational Modifications of Baculovirus Protamine-Like Protein P6.9 and the Significance of Its Hyperphosphorylation for Viral Very Late Gene Hyperexpression. J. Virol. *89*, 7646–7659. https://doi.org/10.1128/JVI.00333-15

Li, K., Wang, Y., Bai, H., Wang, Q., Song, J., Zhou, Y., Wu, C., and Chen, X. (2010). The putative pocket protein binding site of Autographa californica nucleopolyhedrovirus BV/ODV-C42 is required for virus-induced nuclear actin polymerization. J. Virol. *84*, 7857–7868. https://doi.org/10.1128/JVI.00174-10

Li, Y., and Guarino, L.A. (2008). Roles of LEF-4 and PTP/BVP RNA triphosphatases in processing of baculovirus late mRNAs. J. Virol. *82*, 5573–5583. https://doi.org/10.1128/JVI.00058-08

Li, Z., and Blissard, G.W. (2008). Functional analysis of the transmembrane (TM) domain of the Autographa californica multicapsid nucleopolyhedrovirus GP64 protein: substitution of heterologous TM domains. J. Virol. *82*, 3329–3341. https://doi.org/10.1128/JVI.02104-07

Li, Z., and Blissard, G.W. (2009a). The Autographa californica multicapsid nucleopolyhedrovirus GP64 protein: analysis of transmembrane domain length and sequence requirements. J. Virol. *83*, 4447–4461. https://doi.org/10.1128/JVI.02252-08

Li, Z., and Blissard, G.W. (2009b). The pre-transmembrane domain of the Autographa californica multicapsid nucleopolyhedrovirus GP64 protein is critical for membrane fusion and virus infectivity. J. Virol. *83*, 10993–11004. https://doi.org/10.1128/JVI.01085-09

Li, Z., and Blissard, G.W. (2010). Baculovirus GP64 disulfide bonds: the intermolecular disulfide bond of Autographa californica multicapsid nucleopolyhedrovirus GP64 is not essential for membrane fusion and virion budding. J. Virol. *84*, 8584–8595. https://doi.org/10.1128/JVI.00264-10

Li, Z., and Blissard, G.W. (2011). Autographa californica multiple nucleopolyhedrovirus GP64 protein: roles of histidine residues in triggering membrane fusion and fusion pore expansion. J. Virol. *85*, 12492–12504. https://doi.org/10.1128/JVI.05153-11

Li, Z., and Blissard, G.W. (2012). Cellular VPS4 is required for efficient entry and egress of budded virions of Autographa californica multiple nucleopolyhedrovirus. J. Virol. *86*, 459–472. https://doi.org/10.1128/JVI.06049-11

Lin, G., and Blissard, G.W. (2002). Analysis of an Autographa californica nucleopolyhedrovirus lef-11 knockout: LEF-11 is essential for viral DNA replication. J. Virol. *76*, 2770–2779.

Liu, C., Li, Z., Wu, W., Li, L., Yuan, M., Pan, L., Yang, K., and Pang, Y. (2008). Autographa californica multiple nucleopolyhedrovirus ac53 plays a role in nucleocapsid assembly. Virology *382*, 59–68. https://doi.org/10.1016/j.virol.2008.09.003

Liu, X., Zhao, H., Fang, Z., Yuan, M., Yang, K., and Pang, Y. (2012). Distribution and phosphorylation of the basic protein P6.9 of Autographa californica nucleopolyhedrovirus. J. Virol. *86*, 12217–12227. https://doi.org/10.1128/JVI.00438-12

Long, G., Westenberg, M., Wang, H., Vlak, J.M., and Hu, Z. (2006). Function, oligomerization and N-linked glycosylation of the Helicoverpa armigera single nucleopolyhedrovirus envelope fusion protein. J. Gen. Virol. *87*, 839–846.

Long, G., Pan, X., and Vlak, J.M. (2008). Conserved leucines in N-terminal heptad repeat HR1 of envelope fusion protein F of group II nucleopolyhedroviruses are important for correct processing and essential for fusogenicity. J. Virol. *82*, 2437–2447.

Lu, A., and Carstens, E.B. (1991). Nucleotide sequence of a gene essential for viral DNA replication in the baculovirus Autographa californica nuclear polyhedrosis virus. Virology *181*, 336–347.

Lu, A., and Miller, L.K. (1995). Differential requirements for baculovirus late expression factor genes in two cell lines. J. Virol. *69*, 6265–6272.

Lung, O., and Blissard, G.W. (2005). A cellular *Drosophila melanogaster* protein with similarity to baculovirus F envelope fusion proteins. J. Virol. *79*, 7979–7989.

Lung, O., Westenberg, M., Vlak, J.M., Zuidema, D., and Blissard, G.W. (2002). Pseudotyping Autographa californica multicapsid nucleopolyhedrovirus

(AcMNPV): F proteins from group II NPVs are functionally analogous to AcMNPV GP64. J. Virol. *76*, 5729–5736.
Lung, O.Y., Cruz-Alvarez, M., and Blissard, G.W. (2003). Ac23, an envelope fusion protein homolog in the baculovirus Autographa californica multicapsid nucleopolyhedrovirus, is a viral pathogenicity factor. J. Virol. *77*, 328–339.
Makalliwa, G.A., Wang, X., Zhang, H., Zhang, N., Chen, C., Li, J., Deng, F., Wang, H., Wang, M., and Hu, Z. (2018). HearNPV Pseudotyped with PIF1, 2, and 3 from MabrNPV: Infectivity and Complex Stability. Virol. Sin. *33*, 187–196. https://doi.org/10.1007/s12250-018-0014-5
Marek, M., Merten, O.W., Galibert, L., Vlak, J.M., and van Oers, M.M. (2011). Baculovirus VP80 protein and the F-actin cytoskeleton interact and connect the viral replication factory with the nuclear periphery. J. Virol. *85*, 5350–5362. https://doi.org/10.1128/JVI.00035-11
Markovic, I., Pulyaeva, H., Sokoloff, A., and Chernomordik, L.V. (1998). Membrane fusion mediated by baculovirus gp64 involves assembly of stable gp64 trimers into multiprotein aggregates. J. Cell Biol. *143*, 1155–1166.
Martins, A., and Shuman, S. (2003). Mapping the triphosphatase active site of baculovirus mRNA capping enzyme LEF4 and evidence for a two-metal mechanism. Nucleic Acids Res. *31*, 1455–1463.
McDougal, V.V., and Guarino, L.A. (2000). The Autographa californica nuclear polyhedrosis virus p143 gene encodes a DNA helicase. J. Virol. *74*, 5273–5279.
McDougal, V.V., and Guarino, L.A. (2001). DNA and ATP binding activities of the baculovirus DNA helicase P143. J. Virol. *75*, 7206–7209. https://doi.org/10.1128/JVI.75.15.7206-7209.2001
McLachlin, J.R., and Miller, L.K. (1994). Identification and characterization of vlf-1, a baculovirus gene involved in very late gene expression. J. Virol. *68*, 7746–7756.
Mehrabadi, M., Hussain, M., Matindoost, L., and Asgari, S. (2015). The Baculovirus Antiapoptotic p35 Protein Functions as an Inhibitor of the Host RNA Interference Antiviral Response. J. Virol. *89*, 8182–8192. https://doi.org/10.1128/JVI.00802-15
Merrington, C.L., Kitts, P.A., King, L.A., and Possee, R.D. (1996). An Autographa californica nucleopolyhedrovirus lef-2 mutant: consequences for DNA replication and very late gene expression. Virology *217*, 338–348. https://doi.org/10.1006/viro.1996.0121
Mikhailov, V.S., and Rohrmann, G.F. (2002). Baculovirus replication factor LEF-1 is a DNA primase. J. Virol. *76*, 2287–2297.
Mikhailov, V.S., Okano, K., and Rohrmann, G.F. (2003). Baculovirus alkaline nuclease possesses a 5'→3' exonuclease activity and associates with the DNA-binding protein LEF-3. J. Virol. *77*, 2436–2444.
Mikhailov, V.S., Okano, K., and Rohrmann, G.F. (2006). Structural and functional analysis of the baculovirus single-stranded DNA-binding protein LEF-3. Virology *346*, 469–478. https://doi.org/10.1016/j.virol.2005.11.027
Mistretta, T.A., and Guarino, L.A. (2005). Transcriptional activity of baculovirus very late factor 1. J. Virol. *79*, 1958–1960. https://doi.org/10.1128/JVI.79.3.1958-1960.2005
Mitchell, J.K., and Friesen, P.D. (2012). Baculoviruses modulate a proapoptotic DNA damage response to promote virus multiplication. J. Virol. *86*, 13542–13553. https://doi.org/10.1128/JVI.02246-12
Mu, J., van Lent, J.W., Smagghe, G., Wang, Y., Chen, X., Vlak, J.M., and van Oers, M.M. (2014). Live imaging of baculovirus infection of midgut epithelium cells: a functional assay of per os infectivity factors. J. Gen. Virol. *95*, 2531–2539. https://doi.org/10.1099/vir.0.068262-0
Mu, J., Zhang, Y., Hu, Y., Hu, X., Zhou, Y., Chen, X., and Wang, Y. (2016a). The role of viral protein Ac34 in nuclear relocation of subunits of the actin-related protein 2/3 complex. Virol. Sin. *31*, 480–489. https://doi.org/10.1007/s12250-016-3912-4
Mu, J., Zhang, Y., Hu, Y., Hu, X., Zhou, Y., Zhao, H., Pei, R., Wu, C., Chen, J., Zhao, H., *et al.* (2016b). Autographa californica Multiple Nucleopolyhedrovirus Ac34 Protein Retains Cellular Actin-Related Protein 2/3 Complex in the Nucleus by Subversion of CRM1-Dependent Nuclear Export. PLOS Pathog. *12*, e1005994. https://doi.org/10.1371/journal.ppat.1005994
Mueller, J., Pfanzelter, J., Winkler, C., Narita, A., Le Clainche, C., Nemethova, M., Carlier, M.F., Maeda, Y., Welch, M.D., Ohkawa, T., *et al.* (2014). Electron tomography and simulation of baculovirus actin comet tails support a tethered filament model of pathogen propulsion. PLOS Biol. *12*, e1001765. https://doi.org/10.1371/journal.pbio.1001765
Nagamine, T., Abe, A., Suzuki, T., Dohmae, N., and Matsumoto, S. (2011). Co-expression of four baculovirus proteins, IE1, LEF3, P143, and PP31, elicits a cellular chromatin-containing reticulate structure in the nuclei of uninfected cells. Virology *417*, 188–195. https://doi.org/10.1016/j.virol.2011.06.001
O'Reilly, D.R., and Miller, L.K. (1989). A baculovirus blocks insect molting by producing ecdysteroid UDP-glucosyl transferase. Science *245*, 1110–1112.
O'Reilly, D.R., and Miller, L.K. (1991). Improvement of a baculovirus pesticide by deletion of the EGT gene. Nat. Biotech. *9*, 1086–1089.
Ohkawa, T., and Volkman, L.E. (1999). Nuclear F-actin is required for AcMNPV nucleocapsid morphogenesis. Virology *264*, 1–4. https://doi.org/10.1006/viro.1999.0008
Ohkawa, T., and Welch, M.D. (2018). Baculovirus actin-based motility drives nuclear envelope disruption and nuclear egress. Curr. Biol. *28*, 2153–2159.e4. https://doi.org/10.1016/j.cub.2018.05.027
Ohkawa, T., Majima, K., and Maeda, S. (1994). A cysteine protease encoded by the baculovirus Bombyx mori nuclear polyhedrosis virus. J. Virol. *68*, 6619–6625.
Ohkawa, T., Rowe, A.R., and Volkman, L.E. (2002). Identification of six Autographa californica multicapsid nucleopolyhedrovirus early genes that mediate nuclear localization of G-actin. J. Virol. *76*, 12281–12289.
Ohkawa, T., Washburn, J.O., Sitapara, R., Sid, E., and Volkman, L.E. (2005). Specific binding of Autographa californica M nucleopolyhedrovirus occlusion-derived virus to midgut cells of Heliothis virescens larvae is mediated by products of pif genes Ac119 and Ac022 but not by Ac115. J. Virol. *79*, 15258–15264. https://doi.org/10.1128/JVI.79.24.15258-15264.2005

Ohkawa, T., Volkman, L.E., and Welch, M.D. (2010). Actin-based motility drives baculovirus transit to the nucleus and cell surface. J. Cell Biol. *190*, 187–195. https://doi.org/10.1083/jcb.201001162

Okano, K., Mikhailov, V.S., and Maeda, S. (1999). Colocalization of baculovirus IE-1 and two DNA-binding proteins, DBP and LEF-3, to viral replication factories. J. Virol. 73, 110–119.

Olson, V.A., Wetter, J.A., and Friesen, P.D. (2001). Oligomerization mediated by a helix-loop-helix-like domain of baculovirus IE1 is required for early promoter transactivation. J. Virol. *75*, 6042–6051. https://doi.org/10.1128/JVI.75.13.6042-6051.2001

Olson, V.A., Wetter, J.A., and Friesen, P.D. (2003). The highly conserved basic domain I of baculovirus IE1 is required for hr enhancer DNA binding and hr-dependent transactivation. J. Virol. *77*, 5668–5677.

Olszewski, J., and Miller, L.K. (1997). Identification and characterization of a baculovirus structural protein, VP1054, required for nucleocapsid formation. J. Virol. *71*, 5040–5050.

Ono, C., Kamagata, T., Taka, H., Sahara, K., Asano, S., and Bando, H. (2012). Phenotypic grouping of 141 BmNPVs lacking viral gene sequences. Virus Res. *165*, 197–206. https://doi.org/10.1016/j.virusres.2012.02.016

Oomens, A.G., and Blissard, G.W. (1999). Requirement for GP64 to drive efficient budding of Autographa californica multicapsid nucleopolyhedrovirus. Virology *254*, 297–314. https://doi.org/10.1006/viro.1998.9523

Partington, S., Yu, H., Lu, A., and Carstens, E.B. (1990). Isolation of temperature sensitive mutants of Autographa californica nuclear polyhedrosis virus: phenotype characterization of baculovirus mutants defective in very late gene expression. Virology *175*, 91–102.

Passarelli, A.L., and Miller, L.K. (1993). Identification and characterization of lef-1, a baculovirus gene involved in late and very late gene expression. J. Virol. *67*, 3481–3488.

Passarelli, A.L., Todd, J.W., and Miller, L.K. (1994). A baculovirus gene involved in late gene expression predicts a large polypeptide with a conserved motif of RNA polymerases. J. Virol. *68*, 4673–4678.

Pearson, M.N., and Rohrmann, G.F. (2006). Envelope gene capture and insect retrovirus evolution: the relationship between errantivirus and baculovirus envelope proteins. Virus Res. *118*, 7–15.

Pearson, M., Bjornson, R., Pearson, G., and Rohrmann, G. (1992). The Autographa californica baculovirus genome: evidence for multiple replication origins. Science *257*, 1382–1384.

Pearson, M.N., Bjornson, R.M., Ahrens, C., and Rohrmann, G.F. (1993). Identification and characterization of a putative origin of DNA replication in the genome of a baculovirus pathogenic for Orgyia pseudotsugata. Virology *197*, 715–725. https://doi.org/10.1006/viro.1993.1647

Pearson, M.N., Groten, C., and Rohrmann, G.F. (2000). Identification of the lymantria dispar nucleopolyhedrovirus envelope fusion protein provides evidence for a phylogenetic division of the Baculoviridae. J. Virol. 74, 6126–6131.

Pearson, M.N., Russell, R.L., and Rohrmann, G.F. (2001). Characterization of a baculovirus-encoded protein that is associated with infected-cell membranes and budded virions. Virology *291*, 22–31. https://doi.org/10.1006/viro.2001.1191

Pearson, M.N., Russell, R.L., and Rohrmann, G.F. (2002). Functional analysis of a conserved region of the baculovirus envelope fusion protein, LD130. Virology *304*, 81–88.

Peng, K., van Oers, M.M., Hu, Z., van Lent, J.W., and Vlak, J.M. (2010a). Baculovirus per os infectivity factors form a complex on the surface of occlusion-derived virus. J. Virol. *84*, 9497–9504. https://doi.org/10.1128/JVI.00812-10

Peng, K., Wu, M., Deng, F., Song, J., Dong, C., Wang, H., and Hu, Z. (2010b). Identification of protein-protein interactions of the occlusion-derived virus-associated proteins of Helicoverpa armigera nucleopolyhedrovirus. J. Gen. Virol. *91*, 659–670. https://doi.org/10.1099/vir.0.017103-0

Peng, K., van Lent, J.W., Vlak, J.M., Hu, Z., and van Oers, M.M. (2011). In situ cleavage of baculovirus occlusion-derived virus receptor binding protein P74 in the peroral infectivity complex. J. Virol. *85*, 10710–10718. https://doi.org/10.1128/JVI.05110-11

Peng, K., van Lent, J.W., Boeren, S., Fang, M., Theilmann, D.A., Erlandson, M.A., Vlak, J.M., and van Oers, M.M. (2012). Characterization of novel components of the baculovirus per os infectivity factor complex. J. Virol. *86*, 4981–4988. https://doi.org/10.1128/JVI.06801-11

Prikhod'ko, E.A., Lu, A., Wilson, J.A., and Miller, L.K. (1999). In vivo and in vitro analysis of baculovirus ie-2 mutants. J. Virol. 73, 2460–2468.

Rao, G., Fu, Y., Li, N., Yin, J., Zhang, J., Wang, M., Hu, Z., and Cao, S. (2018). Controllable assembly of flexible protein nanotubes for loading multifunctional modules. ACS Appl. Mater. Interfaces *10*, 25135–25145. https://doi.org/10.1021/acsami.8b07611

Rapp, J.C., Wilson, J.A., and Miller, L.K. (1998). Nineteen baculovirus open reading frames, including LEF-12, support late gene expression. J. Virol. 72, 10197–10206.

Rodems, S.M., and Friesen, P.D. (1995). Transcriptional enhancer activity of hr5 requires dual-palindrome half sites that mediate binding of a dimeric form of the baculovirus transregulator IE1. J. Virol. *69*, 5368–5375.

Rohrmann, G.F. (2013). Baculovirus Molecular Biology, 3rd edn (National Center for Biotechnology Information, Bethesda, MD).

Ros, V.I., van Houte, S., Hemerik, L., and van Oers, M.M. (2015). Baculovirus-induced tree-top disease: how extended is the role of egt as a gene for the extended phenotype? Mol. Ecol. *24*, 249–258. https://doi.org/10.1111/mec.13019

Rosas-Acosta, G., Braunagel, S.C., and Summers, M.D. (2001). Effects of deletion and overexpression of the Autographa californica nuclear polyhedrosis virus FP25K gene on synthesis of two occlusion-derived virus envelope proteins and their transport into virus-induced intranuclear membranes. J. Virol. *75*, 10829–10842. https://doi.org/10.1128/JVI.75.22.10829-10842.2001

Sabin, L.R., Hanna, S.L., and Cherry, S. (2010). Innate antiviral immunity in *Drosophila*. Curr. Opin. Immunol. *22*, 4–9. https://doi.org/10.1016/j.coi.2010.01.007

Sajjan, D.B., and Hinchigeri, S.B. (2016). Structural organization of baculovirus occlusion bodies and protective role of multilayered polyhedron envelope protein. food environ. Virol. *8*, 86–100. https://doi.org/10.1007/s12560-016-9227-7

Shang, Y., Wang, M., Xiao, G., Wang, X., Hou, D., Pan, K., Liu, S., Li, J., Wang, J., Arif, B.M., *et al.* (2017). Construction and Rescue of a Functional Synthetic Baculovirus. ACS Synth. Biol. *6*, 1393–1402. https://doi.org/10.1021/acssynbio.7b00028

Shen, S., Wang, M., Li, X., Li, S., van Oers, M.M., Vlak, J.M., Braakman, I., Hu, Z., Deng, F., and Wang, H. (2016). Mutational and functional analysis of N-linked glycosylation of envelope fusion protein F of Helicoverpa armigera nucleopolyhedrovirus. J. Gen. Virol. *97*, 988–999. https://doi.org/10.1099/jgv.0.000404

Shi, A., Hu, Z., Zuo, Y., Wang, Y., Wu, W., Yuan, M., and Yang, K. (2018). Autographa californica multiple nucleopolyhedrovirus ac75 is required for the nuclear egress of nucleocapsids and intranuclear microvesicle formation. J. Virol. *92*, 01509-17. https://doi.org/10.1128/JVI.01509-17

Shi, Y., Li, K., Tang, P., Li, Y., Zhou, Q., Yang, K., and Zhang, Q. (2015). Three-dimensional visualization of the Autographa californica multiple nucleopolyhedrovirus occlusion-derived virion envelopment process gives new clues as to its mechanism. Virology *476*, 298–303. https://doi.org/10.1016/j.virol.2014.11.030

Shrestha, A., Bao, K., Chen, Y.R., Chen, W., Wang, P., Fei, Z., and Blissard, G.W. (2018). Global analysis of baculovirus AcMNPV gene expression in the midgut of the lepidopteran host, Trichoplusia ni. J. Virol. *92*, 01277-18. https://doi.org/10.1128/JVI.01277-18

Slack, J., and Arif, B.M. (2007). The baculoviruses occlusion-derived virus: virion structure and function. Adv. Virus Res. *69*, 99–165. https://doi.org/10.1016/S0065-3527(06)69003-9

Slack, J.M., Lawrence, S.D., Krell, P.J., and Arif, B.M. (2008). Trypsin cleavage of the baculovirus occlusion-derived virus attachment protein P74 is prerequisite in per os infection. J. Gen. Virol. *89*, 2388–2397. https://doi.org/10.1099/vir.0.2008/002543-0

Sokal, N., Nie, Y., Willis, L.G., Yamagishi, J., Blissard, G.W., Rheault, M.R., and Theilmann, D.A. (2014). Defining the roles of the baculovirus regulatory proteins IE0 and IE1 in genome replication and early gene transactivation. Virology *468-470*, 160–171. https://doi.org/10.1016/j.virol.2014.07.044

Song, J., Wang, X., Hou, D., Huang, H., Liu, X., Deng, F., Wang, H., Arif, B.M., Hu, Z., and Wang, M. (2016). The Host Specificities of Baculovirus per os Infectivity Factors. PLOS ONE *11*, e0159862. https://doi.org/10.1371/journal.pone.0159862

Su, J., Lung, O., and Blissard, G.W. (2011). The Autographa californica multiple nucleopolyhedrovirus lef-5 gene is required for productive infection. Virology *416*, 54–64. https://doi.org/10.1016/j.virol.2011.04.019

Summers, M.D. (1969). Apparent in vivo pathway of granulosis virus invasion and infection. J. Virol. *4*, 188–190.

Summers, M.D. (1971). Electron microscopic observations on granulosis virus entry, uncoating and replication processes during infection of the midgut cells of Trichoplusia ni. J. Ultrastruct. Res. 35, 606–625.

Taggart, D.J., Mitchell, J.K., and Friesen, P.D. (2012). A conserved N-terminal domain mediates required DNA replication activities and phosphorylation of the transcriptional activator IE1 of Autographa californica multicapsid nucleopolyhedrovirus. J. Virol. *86*, 6575–6585. https://doi.org/10.1128/JVI.00373-12

Tan, Y., Jiang, L., Wang, M., Yin, F., Deng, F., Liu, M., Hu, Z., and Wang, H. (2008). Mutagenesis and nuclear magnetic resonance analyses of the fusion peptide of Helicoverpa armigera single nucleocapsid nucleopolyhedrovirus F protein. J. Virol. *82*, 8138–8148. https://doi.org/10.1128/JVI.00368-08

Tanada, Y., and Hess, R.T. (1976). Development of a nuclear polyhedrosis virus in midgut cells and penetration of the virus into the hemocoel of the armyworm, Pseudaletia unipuncta. J. Invertebr. Pathol. *28*, 67–76.

Tani, H., Nishijima, M., Ushijima, H., Miyamura, T., and Matsuura, Y. (2001). Characterization of cell-surface determinants important for baculovirus infection. Virology *279*, 343–353. https://doi.org/10.1006/viro.2000.0699

Titterington, J.S., Nun, T.K., and Passarelli, A.L. (2003). Functional dissection of the baculovirus late expression factor-8 gene: sequence requirements for late gene promoter activation. J. Gen. Virol. *84*, 1817–1826. https://doi.org/10.1099/vir.0.19083-0

Tomalski, M.D., Wu, J.G., and Miller, L.K. (1988). The location, sequence, transcription, and regulation of a baculovirus DNA polymerase gene. Virology *167*, 591–600.

Vanarsdall, A.L., Okano, K., and Rohrmann, G.F. (2005). Characterization of the replication of a baculovirus mutant lacking the DNA polymerase gene. Virology *331*, 175–180. https://doi.org/10.1016/j.virol.2004.10.024

Vanarsdall, A.L., Okano, K., and Rohrmann, G.F. (2006). Characterization of the role of very late expression factor 1 in baculovirus capsid structure and DNA processing. J. Virol. *80*, 1724–1733. https://doi.org/10.1128/JVI.80.4.1724-1733.2006

Vanarsdall, A.L., Mikhailov, V.S., and Rohrmann, G.F. (2007a). Characterization of a baculovirus lacking the DBP (DNA-binding protein) gene. Virology *364*, 475–485. https://doi.org/10.1016/j.virol.2007.03.024

Vanarsdall, A.L., Pearson, M.N., and Rohrmann, G.F. (2007b). Characterization of baculovirus constructs lacking either the Ac 101, Ac 142, or the Ac 144 open reading frame. Virology *367*, 187–195. https://doi.org/10.1016/j.virol.2007.05.003

Vandergaast, R., Schultz, K.L., Cerio, R.J., and Friesen, P.D. (2011). Active depletion of host cell inhibitor-of-apoptosis proteins triggers apoptosis upon baculovirus DNA replication. J. Virol. *85*, 8348–8358. https://doi.org/10.1128/JVI.00667-11

van Houte, S., Ros, V.I., Mastenbroek, T.G., Vendrig, N.J., Hoover, K., Spitzen, J., and van Oers, M.M. (2012). Protein tyrosine phosphatase-induced hyperactivity is a conserved strategy of a subset of baculoviruses to manipulate lepidopteran host behavior. PLOS ONE *7*, e46933. https://doi.org/10.1371/journal.pone.0046933

Volkman, L.E., and Goldsmith, P.A. (1985). Mechanism of neutralization of budded Autographa californica nuclear polyhedrosis virus by a monoclonal antibody:

Inhibition of entry by adsorptive endocytosis. Virology *143*, 185–195.

Wang, L., and Maranas, C.D. (2018). MinGenome: An In silico top-down approach for the synthesis of minimized genomes. ACS Synth. Biol. *7*, 462–473. https://doi.org/10.1021/acssynbio.7b00296

Wang, M., Tan, Y., Yin, F., Deng, F., Vlak, J.M., Hu, Z., and Wang, H. (2008). The F-like protein Ac23 enhances the infectivity of the budded virus of gp64-null Autographa californica multinucleocapsid nucleopolyhedrovirus pseudotyped with baculovirus envelope fusion protein F. J. Virol. *82*, 9800–9804. https://doi.org/10.1128/JVI.00759-08

Wang, M., Tuladhar, E., Shen, S., Wang, H., van Oers, M.M., Vlak, J.M., and Westenberg, M. (2010a). Specificity of baculovirus P6.9 basic DNA-binding proteins and critical role of the C terminus in virion formation. J. Virol. *84*, 8821–8828. https://doi.org/10.1128/JVI.00072-10

Wang, M., Yin, F., Shen, S., Tan, Y., Deng, F., Vlak, J.M., Hu, Z., and Wang, H. (2010b). Partial functional rescue of Helicoverpa armigera single nucleocapsid nucleopolyhedrovirus infectivity by replacement of F protein with GP64 from Autographa californica multicapsid nucleopolyhedrovirus. J. Virol. *84*, 11505–11514. https://doi.org/10.1128/JVI.00862-10

Wang, M., Wang, J., Yin, F., Tan, Y., Deng, F., Chen, X., Jehle, J.A., Vlak, J.M., Hu, Z., and Wang, H. (2014). Unraveling the entry mechanism of baculoviruses and its evolutionary implications. J. Virol. *88*, 2301–2311. https://doi.org/10.1128/JVI.03204-13

Wang, M., Shen, S., Wang, H., Hu, Z., Becnel, J., and Vlak, J.M. (2017). Deltabaculoviruses encode a functional type I budded virus envelope fusion protein. J. Gen. Virol. *98*, 847–852. https://doi.org/10.1099/jgv.0.000745

Wang, Q., Bosch, B.J., Vlak, J.M., van Oers, M.M., Rottier, P.J., and van Lent, J.W.M. (2016). Budded baculovirus particle structure revisited. J. Invertebr. Pathol. *134*, 15–22. https://doi.org/10.1016/j.jip.2015.12.001

Wang, X., Liu, X., Makalliwa, G.A., Li, J., Wang, H., Hu, Z., and Wang, M. (2017). Per os infectivity factors: a complicated and evolutionarily conserved entry machinery of baculovirus. Sci. China Life Sci. *60*, 806–815. https://doi.org/10.1007/s11427-017-9127-1

Wang, Y., Wang, Q., Liang, C., Song, J., Li, N., Shi, H., and Chen, X. (2008). Autographa californica multiple nucleopolyhedrovirus nucleocapsid protein BV/ODV-C42 mediates the nuclear entry of P78/83. J. Virol. *82*, 4554–4561. https://doi.org/10.1128/JVI.02510-07

Wang, Y., Zhang, Y., Han, S., Hu, X., Zhou, Y., Mu, J., Pei, R., Wu, C., and Chen, X. (2015). Identification of a novel regulatory sequence of actin nucleation promoting factor encoded by Autographa californica multiple nucleopolyhedrovirus. J. Biol. Chem. *290*, 9533–9541. https://doi.org/10.1074/jbc.M114.635441

Washburn, J.O., Kirkpatrick, B.A., and Volkman, L.E. (1996). Insect protection against viruses. Nature *383*, 767.

Wei, D., Wang, Y., Zhang, X., Hu, Z., Yuan, M., and Yang, K. (2014). Autographa californica Nucleopolyhedrovirus Ac76: a dimeric type II integral membrane protein that contains an inner nuclear membrane-sorting motif. J. Virol. *88*, 1090–1103. https://doi.org/10.1128/JVI.02392-13

Wennmann, J.T., Keilwagen, J., and Jehle, J.A. (2018). Baculovirus Kimura two-parameter species demarcation criterion is confirmed by the distances of 38 core gene nucleotide sequences. J. Gen. Virol. *99*, 1307–1320. https://doi.org/10.1099/jgv.0.001100

Westenberg, M., and Vlak, J.M. (2008). GP64 of group I nucleopolyhedroviruses cannot readily rescue infectivity of group II f-null nucleopolyhedroviruses. J. Gen. Virol. *89*, 424–431.

Westenberg, M., Wang, H., IJkel, W.F., Goldbach, R.W., Vlak, J.M., and Zuidema, D. (2002). Furin is involved in baculovirus envelope fusion protein activation. J. Virol. *76*, 178–184.

Westenberg, M., Veenman, F., Roode, E.C., Goldbach, R.W., Vlak, J.M., and Zuidema, D. (2004). Functional analysis of the putative fusion domain of the baculovirus envelope fusion protein F. J. Virol. *78*, 6946–6954. https://doi.org/10.1128/JVI.78.13.6946-6954.2004

Westenberg, M., Uijtdewilligen, P., and Vlak, J.M. (2007). Baculovirus envelope fusion proteins F and GP64 exploit distinct receptors to gain entry into cultured insect cells. J. Gen. Virol. *88*, 3302–3306. https://doi.org/10.1099/vir.0.83240-0

Wilson, M.E., and Miller, L.K. (1986). Changes in the nucleoprotein complexes of a baculovirus DNA during infection. Virology *151*, 315–328.

Wu, C., and Wang, S. (2012). A pH-sensitive heparin-binding sequence from Baculovirus gp64 protein is important for binding to mammalian cells but not to Sf9 insect cells. J. Virol. *86*, 484–491. https://doi.org/10.1128/JVI.06357-11

Wu, C.P., Huang, Y.J., Wang, J.Y., Wu, Y.L., Lo, H.R., Wang, J.C., and Chao, Y.C. (2010). Autographa californica multiple nucleopolyhedrovirus LEF-2 is a capsid protein required for amplification but not initiation of viral DNA replication. J. Virol. *84*, 5015–5024. https://doi.org/10.1128/JVI.02423-09

Wu, W., Lin, T., Pan, L., Yu, M., Li, Z., Pang, Y., and Yang, K. (2006). Autographa californica multiple nucleopolyhedrovirus nucleocapsid assembly is interrupted upon deletion of the 38K gene. J. Virol. *80*, 11475–11485. https://doi.org/10.1128/JVI.01155-06

Wu, W., Liang, H., Kan, J., Liu, C., Yuan, M., Liang, C., Yang, K., and Pang, Y. (2008). Autographa californica multiple nucleopolyhedrovirus 38K is a novel nucleocapsid protein that interacts with VP1054, VP39, VP80, and itself. J. Virol. *82*, 12356–12364. https://doi.org/10.1128/JVI.00948-08

Wu, Y., and Carstens, E.B. (1998). A baculovirus single-stranded DNA binding protein, LEF-3, mediates the nuclear localization of the putative helicase P143. Virology *247*, 32–40. https://doi.org/10.1006/viro.1998.9235

Wu, Y., Liu, G., and Carstens, E.B. (1999). Replication, integration, and packaging of plasmid DNA following cotransfection with baculovirus viral DNA. J. Virol. *73*, 5473–5480.

Wu, Y.L., Wu, C.P., Huang, Y.H., Huang, S.P., Lo, H.R., Chang, H.S., Lin, P.H., Wu, M.C., Chang, C.J., and Chao, Y.C. (2014). Identification of a high-efficiency baculovirus DNA replication origin that functions in insect and mammalian cells. J. Virol. *88*, 13073–13085. https://doi.org/10.1128/JVI.01713-14

Xi, Q., Wang, J., Deng, R., and Wang, X. (2007). Characterization of AcMNPV with a deletion of me53 gene. Virus Genes *34*, 223–232. https://doi.org/10.1007/s11262-006-0045-0

Yamada, H., Kitaguchi, K., Hamajima, R., Kobayashi, M., and Ikeda, M. (2013). Novel apoptosis suppressor Apsup from the baculovirus Lymantria dispar multiple nucleopolyhedrovirus precludes apoptosis by preventing proteolytic processing of initiator caspase Dronc. J. Virol. *87*, 12925–12934. https://doi.org/10.1128/JVI.02065-13

Yang, S., and Miller, L.K. (1998). Control of baculovirus polyhedrin gene expression by very late factor 1. Virology *248*, 131–138. https://doi.org/10.1006/viro.1998.9272

Yin, F., Wang, M., Tan, Y., Deng, F., Vlak, J.M., Hu, Z., and Wang, H. (2014). Identification and functional analysis of inter-subunit disulfide bonds of the F protein of Helicoverpa armigera nucleopolyhedrovirus. J. Gen. Virol. *95*, 2820–2830. https://doi.org/10.1099/vir.0.068122-0

Yoo, S., and Guarino, L.A. (1994). The Autographa californica nuclear polyhedrosis virus ie2 gene encodes a transcriptional regulator. Virology *202*, 746–753. https://doi.org/10.1006/viro.1994.1396

Young, J.C., MacKinnon, E.A., and Faulkner, P. (1993). The architecture of the virogenic stroma in isolated nuclei of Spodoptera frugiperda cells in vitro infected by Autographa californica nuclear polyhedrosis virus. J. Struct. Biol *110*, 141–153.

Yu, I.L., Bray, D., Lin, Y.C., and Lung, O. (2009). Autographa californica multiple nucleopolyhedrovirus ORF 23 null mutant produces occlusion-derived virions with fewer nucleocapsids. J. Gen. Virol. *90*, 1499–1504. https://doi.org/10.1099/vir.0.009035-0

Yu, M., and Carstens, E.B. (2010). Identification of a domain of the baculovirus Autographa californica multiple nucleopolyhedrovirus single-strand DNA-binding protein LEF-3 essential for viral DNA replication. J. Virol. *84*, 6153–6162. https://doi.org/10.1128/JVI.00115-10

Yuan, C., Xing, L., Wang, M., Wang, X., Yin, M., Wang, Q., Hu, Z., and Zou, Z. (2017). Inhibition of melanization by serpin-5 and serpin-9 promotes baculovirus infection in cotton bollworm Helicoverpa armigera. PLOS Pathog. *13*, e1006645. https://doi.org/10.1371/journal.ppat.1006645

Yuan, M., Huang, Z., Wei, D., Hu, Z., Yang, K., and Pang, Y. (2011). Identification of Autographa californica nucleopolyhedrovirus ac93 as a core gene and its requirement for intranuclear microvesicle formation and nuclear egress of nucleocapsids. J. Virol. *85*, 11664–11674. https://doi.org/10.1128/JVI.05275-11

Yue, Q., Yu, Q., Yang, Q., Xu, Y., Guo, Y., Blissard, G.W., and Li, Z. (2018). Distinct roles of cellular ESCRT-I and ESCRT-III proteins in efficient entry and egress of budded virions of Autographa californica multiple nucleopolyhedrovirus. J. Virol. *92*, e01636-17. https://doi.org/10.1128/JVI.01636-17

Zhang, J., Dong, Z.Q., Zhang, C.D., He, Q., Chen, X.M., Cao, M.Y., Li, H.Q., Xiao, W.F., Lu, C., and Pan, M.H. (2014). Identification of a novel nuclear localization signal of baculovirus late expression factor 11. Virus Res. *184*, 111–119. https://doi.org/10.1016/j.virusres.2014.02.020

Zhang, S., An, S., Hoover, K., Li, Z., Li, X., Liu, X., Shen, Z., Fang, H., Ros, V.I.D., Zhang, Q., *et al.* (2018). Host miRNAs are involved in hormonal regulation of HaSNPV-triggered climbing behaviour in Helicoverpa armigera. Mol. Ecol. *27*, 459–475. https://doi.org/10.1111/mec.14457

Zhang, S.X., Han, Y., and Blissard, G.W. (2003). Palmitoylation of the Autographa californica multicapsid nucleopolyhedrovirus envelope glycoprotein GP64: mapping, functional studies, and lipid rafts. J. Virol. *77*, 6265–6273.

Zhang, Y., Hu, X., Mu, J., Hu, Y., Zhou, Y., Zhao, H., Wu, C., Pei, R., Chen, J., Chen, X., *et al.* (2018). Ac102 Participates in nuclear actin polymerization by modulating BV/ODV-C42 Ubiquitination during Autographa californica multiple nucleopolyhedrovirus infection. J. Virol. *92*, e00005-18. https://doi.org/10.1128/JVI.00005-18

Zheng, Q., Shen, Y., Kon, X., Zhang, J., Feng, M., and Wu, X. (2017). Protein-protein interactions of the baculovirus per os infectivity factors (PIFs) in the PIF complex. J. Gen. Virol. *98*, 853–861. https://doi.org/10.1099/jgv.0.000730

Zhou, J., and Blissard, G.W. (2008). Identification of a GP64 subdomain involved in receptor binding by budded virions of the baculovirus Autographica californica multicapsid nucleopolyhedrovirus. J. Virol. *82*, 4449–4460. https://doi.org/10.1128/JVI.02490-07

Zhu, Z., Wang, J., Wang, Q., Yin, F., Liu, X., Hou, D., Zhang, L., Liu, H., Li, J., Arif, B.M., *et al.* (2018). Genome Characteristics of the Cyclophragma Undans Nucleopolyhedrovirus: A Distinct Species in Group I of Alphabaculovirus. Virol. Sin. *33*, 359–368. https://doi.org/10.1007/s12250-018-0047-9

Recent Developments in the Use of Baculovirus Expression Vectors

10

Robert D. Possee[1,2], Adam C. Chambers[1], Leo P. Graves[1], Mine Aksular[1] and Linda A. King[2*]

[1]Oxford Expression Technologies Ltd, Bioinnovation Hub, Oxford, UK.
[2]Department of Biological and Medical Sciences, Oxford Brookes University, Oxford, UK.

*Correspondence: laking@brookes.ac.uk

https://doi.org/10.21775/9781912530083.10

Abstract

Over 35 years since it was established to make recombinant proteins, the baculovirus expression vector system continues to develop and improve. Early systems for recombinant virus selection were laborious, but better methods were rapidly devised that enabled non-virologists to use baculovirus vectors successfully in a wide range of applications. These applications include multiple gene expression for complex molecules, production of adeno-associated virus-like particles for gene therapy, the use of baculovirus budded virus for the same purpose, numerous potential human and animal vaccines, and for other therapeutic proteins. A number of products for human and veterinary use are now on the market, which attests to the utility of the systems. Despite these successes, baculovirus vectors essentially remain in a relatively primitive state of development. Many proteins, particularly membrane-bound or secreted products, continue to be difficult to produce. Various research groups are working to identify potential areas of improvement, which if combined into an ideal vector might offer considerable advances to the system. This chapter will review some of the most recent reports and highlight those that might have generic application for recombinant protein synthesis in insect cells. We also summarize parallel developments in host cells used for baculovirus expression and how culture conditions can influence protein production.

Introduction

It is more than 30 years since the first papers were published that described making recombinant proteins in insect cells using modified baculovirus genomes (Smith *et al.*, 1983; Pennock *et al.*, 1984). Since those early days, much progress has been made in making the baculovirus expression vector system (BEVS) more amenable for use in a wide variety of academic and industrial laboratories (van Oers *et al.*, 2015). This undoubtedly has contributed to the widespread use of the BEVS, which is now one of the most popular eukaryotic expression systems. However, the most commonly used commercially available vector platforms, Bac-to-Bac™ (Luckow *et al.*, 1993), in which foreign genes are inserted by targeted transposition from an intermediary transfer plasmid, has been unchanged since its launch as a commercial kit by Thermofisher. However, Mehalko and Esposito (2016) have produced Bac-2-the-Future, a modified version of the system. This version is claimed to improve the efficiency of recombinant virus production by reducing the number of false positives generated in the transposition process. For example, a site for Tn7 transposition was removed from the bacterial host cell used. It has yet to be developed into a commercial kit but clearly has potential. The second most popular BEVS, in which foreign genes are inserted into the virus genome by homologous recombination with a transfer plasmid, has evolved over time. It has spawned a number of commercially available kits such as *flash*BAC™ (Possee *et al.*, 2008; van Oers *et al.*, 2015), but the core technology of homologous recombination has not altered.

That does not mean, however, that there have been no improvements to the BEVS system in recent years. Rather it indicates that the considerable developments made have largely come from modifications to the intermediary transfer plasmids, to the cell lines used to propagate virus or produce recombinant proteins, or by the incorporation of new technology such as RNA interference (RNAi). Two exceptions are firstly, the development of the MultiBac system used to produce multiple proteins within a single recombinant virus (Berger *et al.*, 2004). Secondly, developments exemplified by the variants of the *flash*BAC system that have led to improvements in the yield and/or quality of particular types of proteins through deletion of non-essential baculovirus genes or have reduced the effort required to make recombinant viruses by removing the need to separate non-recombinant from recombinant viruses (Possee *et al.*, 2008; Hitchman *et al.*, 2010).

This chapter, therefore, focuses on recent improvements that have been made to the BEVS to either overcome some of the challenges posed by particularly difficult-to-express proteins or multi-protein complexes such as virus-like particles (VLPs). Novel uses of the expression vectors beyond the confines of insect cells, including enhancing gene delivery into mammalian cells, gene therapy and applications in vaccine production, are also considered. Table 10.1 summarizes what we consider to be the most significant improvements applicable across a wide range of applications of the baculovirus expression system. Those not included in the table have more specific uses.

Vector development

Multiple expression vectors and virus-like particle production

Baculovirus expression vectors have often been used to produce more than one recombinant protein, which can then interact to form highly complex

Table 10.1 Significant recent advances in baculovirus expression technology

Category	Name	Company	Description	Reference
Virus vectors	biGBac	N/A[1]	Gibson assembly and MultiBac to insert 25 genes into the baculovirus genome	Weissman *et al.* (2016); Weissman and Peters (2018)
	Top-Bac®	Alegenex	Overexpression of AcMNPV *ie1*, combined with tandem late/very late baculovirus gene promoters enhances recombinant protein production	Gómez-Sebastián *et al.* (2014); López-Vidal *et al.* (2015)
	Vankyrin	ParaTechs	Incorporating anti-apoptosis vankyrin genes into the baculovirus vector improves virus-infected cell viability 48–72 hours post infection	Steele *et al.*, (2017)
	Synthetic AcMNPV	N/A	Infectious virus genome assembled synthetically	Shang *et al.* (2017)
	Bac-2-the-Future	N/A	Updated version of Bac-to-Bac with improved efficiency of recombinant virus production	Mehalko, J.L. and Esposito, D. (2016)
Cell lines	Nodavirus-free	N/A	*Trichoplusia ni* cell line free of nodavirus	Chen *et al.* (2013)
	Rhabdovirus-free	N/A	*Spodoptera frugiperda* cell line free of rhabdovirus	Maghodia *et al.* (2016)
	Vankyrin	ParaTechs	Incorporating anti-apoptosis vankyrin genes into insect cell line improves protein yields	Fath-Goodin *et al.* (2006); Steele *et al.*, (2017)
	One Bac	N/A	Stably transformed Sf9 cells to produce AAV Rep and Cap proteins	Mietzsch *et al.* (2014, 2015, 2017)

[1]Not applicable.

structures such as virus-like particles (VLPs). One of the early examples of this application of the technology was the assembly of bluetongue VLPs (Roy, 1992). More recent studies have made further developments in multiple gene expression. Weissman *et al.* (2016) describe a method for assembling recombinant baculoviruses that could contain up to 25 individual genes of interest within 6 days. The method is based on the Gibson assembly reaction. The method is termed biGBac; biG referring to both the size of the achievable expression construct and the reverse of Gib from Gibson. Bac is a reference to the generation of the baculovirus vector, which is derived from the MultiBac system (Berger *et al.*, 2004). This method was used to generate and characterize recombinant forms of the anaphase-promoting complex/cyclosome, cohesin and kinetochore complexes. It should be noted that the genes of interest need to be cloned into an initial plasmid before being usable in the assembly reaction, which can be undertaken using the Gibson assembly reactions or conventional techniques. A detailed, recent review of this method is provided by Weissmann and Peters (2018).

By coincidence, Zhang *et al.* (2016) also targeted expression of the multi-subunit anaphase-promoting complex/cyclosome complex. They utilized a Uracil Specific Excision Reagent ligation-free cloning method to assemble recombinant MultiBac transfer vectors. Ultimately, 13 proteins were produced by coinfection of insect cells with two recombinant viruses with genes inserted at the polyhedrin gene (*polh*) or chitinase/cathepsin gene loci.

The assembly of the transfer vectors required for such complex expression projects is not a simple task. The approach builds on the MultiBac system (Berger *et al.*, 2004). Increasingly, the baculovirus expression system is being used to produce multi-subunit structures that require very careful design and production of gene constructs using synthetic biology.

Multiple expression vectors have also been developed with the goal of improving expression of individual recombinant proteins, which remain the most commonly used application of BEVS. Gómez-Sebastián *et al.* (2014) described a rearranged baculovirus expression vector cassette in which they over-produced the AcMNPV IE1/IE0 transcriptional transactivator using either the *polh* promoter or $pB2_9$, the latter comprising a lower strength promoter from *Trichoplusia ni* insects. In this vector, recombinant gene expression was driven by one of a number of promoters including some hybrid sequences formed from late and very late *Autographa californica multiple nucleopolyhedrovirus* (AcMNPV) genes. These modifications resulted in both enhanced recombinant protein production and an increase in virus-infected cell viability and integrity. While this initial study was performed using green fluorescent protein (GFP) as a reporter, a subsequent study showed that production of porcine circovirus type 2 and rabbit haemorrhagic disease calicivirus VLPs could also be enhanced using this system. A yield increase of 300% was quoted for each target (López-Vidal *et al.*, 2015). The system is known commercially as Top-Bac®.

Steele *et al.* (2017) described the incorporation of anti-apoptotic genes (vankyrin) into the baculovirus vector to extend the period of recombinant protein production in baculovirus-infected cells. Co-expression of vankyrin with a recombinant protein improved the viability of virus-infected cells at the critical 48–72 hours post-infection (hpi) period when most cultures are harvested. Importantly, the yield of recombinant protein was also higher when using these vectors. This approach is linked to the establishment of insect cell lines that have vankyrin genes inserted into their chromosomes for constitutive production (see below).

Vectors for recombinant adeno-associated virus production

Adeno-associated virus (AAV) is non-enveloped with a single-stranded DNA genome and is classified in the *Parvoviridae* (Hoggan *et al.*, 1966). AAV is unable to replicate autonomously and requires another virus helper, which is usually adenovirus or herpes simplex virus (HSV), both of which have large double-stranded DNA genomes. This dependence on another virus further classifies AAV as a member of genus *Dependovirus*. The AAV genome is 4.7 kb and contains two main open reading frames, *rep* and *cap*. The former encodes four regulatory proteins (Rep78, Rep68, Rep52, and Rep40) that play important roles in both replication and the encapsidation of viral DNA. The *cap* encodes three capsid proteins (VP1, VP2, and VP3) and assembly-activating protein (AAP) that promotes the formation of capsids. Each end of the virus genome is flanked by inverted terminal repeats (ITRs). These contain Rep recognition sequences important for AAV DNA replication and also for the packaging of the genome (Samulski and Muzycska, 2014). The latent state of AAV, where low numbers of genomes are integrated into the host chromosome, has led to the use of recombinant (r)AAV as a practical gene therapy vector and much effort has been made to optimize rAAV production (reviewed by Hastie and Samulski, 2015). In a recent review (Aponte-Ubillus *et al.*, 2018), it was concluded that use of the baculovirus expression platform to produce rAAV was one of the most promising, generating the highest yield of virus particles on a per cell basis.

In the first use of baculoviruses to produce rAAV vectors, *rep78* and *rep72* were placed under the control of the AcMNPV *ie1* and *polh* promoters, respectively. The *cap* was placed under the control of another copy of the *polh* promoter, with the VP1 translation initiation codon mutated to ACG to enable synthesis of VP2 and VP3. Finally, the transgene was linked with a gene promoter active in human cells, flanked by ITRs and inserted into a third baculovirus (Urabe *et al.*, 2002). Although production of rAAV vectors was successful with this approach, the use of three separate recombinant baculoviruses was undesirable. The requirement for coinfection of insect cells with each virus to achieve AAV formation was impractical for long term applications. Further developments of the baculovirus-AAV production system have seen the number of vectors required reduced to two. The Rep and Cap components have been incorporated into a single virus under the control of *polh* and *p10* promoters, with the transgene in the second virus flanked by ITRs (Chen, 2008; Smith *et al.*, 2009).

Wu *et al.* (2018) have reported the use of ribosome leaky-scanning to express AAV Rep and Cap proteins downstream of the *polh* and *p10* promoters, respectively, with the rAAV genome inserted between the two baculovirus promoters in a single recombinant virus. The yields of recombinant serotypes rAAV2, rAAV8, and rAAV9 derived from the single recombinant baculovirus-infected Sf9 cells exceeded 10^5 vector genomes (vg) per cell and the authors suggest that this system would be suitable for large-scale rAAV production. The single virus vector was shown to be stable and showed no apparent decrease of rAAV yield after at least four serial passages (p4). Although a decrease in yield was observed after the fourth passage, p4 is a workable limit for scale-up studies.

When using any rAAV production system, it is important that capsid stoichiometry is as close to wild-type as possible; imbalance results in reduced potency. Bosma *et al.* (2018) generated a library of recombinants with variations in the translation-initiation region of VP1, which resulted in the production of a potent rAAV serotype 5 with a balanced VP1/VP2/VP3 stoichiometry. They also demonstrated that excessive incorporation of VP3 had the highest decrease in potency. The authors suggest that modifications for other serotypes of VP1 could result in the development of high potency rAAV with capsid stoichiometry similar to wild-type virus.

An alternative rAAV production platform in insect cells, referred to as the OneBac system, uses a stably transformed Sf9 cell line to express *rep* and *cap* for specific serotypes. Infection with a single baculovirus containing the AAV vector backbone is, therefore, sufficient to induce rAAV production (Aslanidi et al., 2009; Mietzsch *et al.*, 2014, 2015). More recently, Mietzsch *et al.* (2017) have developed this system further by removing the Rep-binding element (RBE) from the *rep/cap* expression constructs, which results in a reduction in the packaging of unwanted foreign DNA into rAAV particles for four serotypes (AAV1, AAV2, AAV5 and AAV8). The RBE was previously believed to be essential for efficient rAAV packaging in Sf9 cells, but the authors' previous studies demonstrated this not to be the case (Mietzsch *et al.*, 2015).

Previous work has demonstrated that the use of truncated ITRs has had no impact on rAAV vector production in mammalian cells, hence the baculovirus production system has been based on using truncated ITRs (reviewed by Aponte-Ubillus *et al.*, 2018). Recent work by Savy *et al.* (2017) replaced the truncated ITRs with the wild-type version and additional wtAAV2 sequences, which improved the percentage of rAAV8 full capsids from 10% to 40% and demonstrated a 10-fold reduction in non-rAAV encapsidated baculovirus DNA. These factors were considered to mitigate the slight decrease in the overall rAAV titres observed with the constructions containing the wild-type ITRs. The authors also indicated that the position of the ITRs influences encapsulation, an outcome that was confirmed by the work of Penaud-Budloo *et al.* (2017).

Glybera was the first baculovirus-derived, rAAV-based treatment for a medical condition, a rare familial lipoprotein lipase deficiency (LPLD) (US National Library of Medicines, 2019). A 5-year authorization to market the drug in the European Union was granted in 2012. However, near the expiration of this term, an extension was not sought by the manufacturers owing to its failure to attract customers (Senior, 2017). Perhaps this was not surprising as the cost of the treatment per patient was estimated at around US$1M, although the product was never approved for use in the USA. This example highlights the perils for companies in producing a technically complex drug intended for a very limited market.

A further concern for developers of AAV-based therapies is the potential for contamination of the final product with protein, nucleic acids or other components of the respective host system utilized (Hastie and Samulski, 2015). rAAV particles produced in mammalian systems contained host cell and plasmid DNA (Hüser *et al.*, 2003; Nony *et al.*, 2003). This was a point raised in the European Medicines Agency (EMA) assessment report for Glybera prior to its approval for clinical use. There was concern about the characterization and elimination of baculovirus-derived impurities in the final product. While most studies examining the purity of AAV particles have focused on the use of quantitative polymerase chain reaction and Sanger sequencing, a recent report described the use of high throughput sequencing based on the Illumina platform (Penaud-Budloo *et al.*, 2017). These results showed that 2.1% of the sequence reads were baculovirus in origin, while Sf9 host cell DNA matched less than 0.03% of reads. The authors of this study suggest that their novel method, which they called single strand virus sequencing (SSV-seq), should enable effective DNA quality control for AAV vectors.

The Glybera example also suggests that costs of production need to be reduced considerably. While initial development costs have to be recovered, this has to be done at a rate that is compatible with a reasonable price for the drug. A very important factor in the reduction of costs of producing AAV is the yield of AAV particles. To take into account that some particles may be devoid of DNA, this is usually expressed as vector genomes (vg) per cell. In the original study (Urabe *et al.*, 2002) this was estimated at 5×10^4 vg/cell. In a more recent study, the yield had been increased to 5×10^5 vg/cell (Mietzsch *et al.*, 2014). For comparison, the host baculovirus vector produces about 100 infectious plaque forming units per cell; although total virion numbers may be higher given that some virus particles will be non-infectious.

BacMam vectors

Despite their successful use as expression vectors of foreign genes in insect cells, baculoviruses have also been developed as efficient vehicles for the delivery of transgenes into mammalian cells (Hofmann *et al.*, 1995; Boyce and Butcher, 1996). BacMam vectors utilize mammalian-specific transcription signals such as the human cytomegalovirus (CMV) immediate early gene promoter to drive recombinant protein production. Budded baculovirus particles are able to enter mammalian cells and then uncoat to release virus DNA in nuclei. However, the baculovirus cannot accomplish a complete cycle of replication in mammalian cells to produce progeny infectious virus. Consequently, the BacMam vector system is regarded as an inherently safe means of transducing mammalian, including human, cells. A recent overview of the system has been presented by Mansouri and Berger (2018).

One application of BacMam technology that is showing promise is their potential use as animal vaccines. Conventional inactivated or attenuated vaccines against *African swine fever virus* (AFSV) were unsuccessful or not practical (Neilan *et al.*, 2004). Argilaguet *et al.* (2013) constructed BacMam-sHAPQ, a vector expressing a fusion protein comprising ASFV P54, P30 and secretory haemagglutinin antigens from the E75 virus isolate using the CMV immediate early gene promoter. Promisingly, four out of six immunized pigs showed total protection against sub-lethal ASFV challenge and there was an induced, specific high T-cell response after *in vivo* immunization in surviving pigs.

Zhang *et al.* (2014) constructed a dual expressing BacMam vector (BV-Dual-s1). This produced S1 glycoprotein of avian infectious bronchitis virus (IBV) as a fusion with the AcMNPV GP64 major surface glycoprotein. The hybrid protein was placed under the control of both the CMV early gene promoter and the baculovirus *polh* promoter. This resulted in the display of the S1-GP64 on the budded virus surface when synthesized in insect cells. However, in avian cells, the S1-GP64 was also expressed owing to the presence of the CMV gene promoter. The BV-Dual-s1 virus also encoded the vesicular stomatitis virus (VSV) G glycoprotein under the control of the AcMNPV *p10* promoter. This resulted in the display of VSV-G on the surface of the budded virus, which was proposed to aid virus uptake into avian cells. Chickens immunized with BV-Dual-s1 produced a significantly higher antibody response when compared to the control groups. Immunized chickens challenged with virulent IBV showed a similar protection rate with BV-Dual-S1 BacMam (83%) when compared to chickens immunized with inactivated vaccine (89%). The BV-Dual-s1 also induced strong humoral and cell-mediated immune response.

Both of the above papers highlight stimulation of the cell mediated immune responses in vaccinated animals, which is often required for complete protection. This was also noted in a study by Heinimäki *et al.* (2017), where monomeric ovalbumin protein and oligomeric norovirus VLPs were inoculated into BALB/c mice with and without infectious budded baculovirus particles. Live baculovirus appeared to act as strong adjuvant for both humoral and cell-mediate responses. The response of the immune system appeared to be less effective if only the proteins or proteins plus inactivated baculovirus were injected. It is suggested that unmodified baculovirus can be an effective adjuvant for a subunit vaccine, which has profound implications for the further development of the system. Ono *et al.* (2018) concluded that the anti-baculoviral responses are induced by recognition of unmethylated CpG virus DNA in mammalian cells. This is mediated by the toll-like receptor (TLR) 9/MyD66/IRF7 in immune cells with the subsequent production of pro-inflammatory cytokines and type 1 interferon.

Vectors to optimise secreted and membrane protein production

The secretion of recombinant proteins by baculovirus expression vectors is often poor, relative to the total amount of protein synthesized in the cell. This is an issue that has long been recognized as a problem (Jarvis *et al.*, 1993). There was an encouraging report from Tessier *et al.* (1991) that secretion of plant propapain from baculovirus-infected insect cells could be improved by utilizing the signal peptide from the honey bee prepromellitin instead of its native sequence. However, this same study also showed that propapain linked with the *Drosophila* α-amylase signal peptide, and β-galactosidase fused with the honey bee prepromellitin signal peptide, failed to be secreted. Jarvis *et al.* (1993) assessed secretion of human tissue plasminogen activator (tPA) utilizing various signal peptides. These included those from the major baculovirus structural glycoprotein (GP64), the insect secretory protein (cecropin B), a luminal protein of the rough endoplasmic reticulum (GRP78/BiP) and the honey bee prepromellitin. None of these improved the secretion of tPA. The lack of secretion of β-galactosidase (Tessier *et al.*, 1991) emphasizes the difficulty of simply adding a signal peptide to another protein and expecting it to be secreted in a heterologous system. However, it is worth recalling that two of the first proteins to be produced using baculovirus vectors, interferon (Smith *et al.*, 1983) and influenza virus haemagglutinin (Possee, 1986) were secreted or plasma membrane-bound proteins that retained their native signal peptides.

Lou *et al.* (2018) approached the problem of secreting human thyroid peroxidase (hTPO) by testing a range of different signal peptides in a manner similar to Jarvis *et al.* (1993). Although hTPO can be produced using baculovirus expression vectors, it largely adopts an aggregated form in the cell lysate fraction despite having its own signal peptide (Fan *et al.*, 1996). Five alternative signal peptides were tested in conjunction with hTPO by Lou *et al.* (2018) using the *T. ni* High Five™ host cell for expression. These signal peptides comprised those from bee melittin, AcMNPV GP64, ecdysteroid UDP-glucosyl transferase (EGT), human peptidyl-glycine alpha-amidating monooxygenase (PAM) or human azurocidin. While EGT and azurocidin signal peptides decreased hTPO secretion, compared with the original native protein, the remaining three served to increase secretion of the protein. The best performer was the PAM signal peptide, which enhanced hTPO secretion about 2.5-fold. This modified form of hTPO in a baculovirus vector was then used in a wave-type bioreactor for scale-up production of active recombinant protein in insect cells, achieving a protein purity of

> 95% from the harvested culture medium. While the increase in hTPO secretion may appear modest, it can make all the difference when protein purification is required. What this study also emphasizes is that the selection of a suitable signal peptide sequence is still largely an empirical process. It requires investment in producing several different gene constructs to find the best combination of signal peptide and target protein to achieve optimal secretion.

Chakraborty *et al.* (2018) demonstrated that it is possible to adapt membrane-targeted proteins for secretion by adding signal peptides to the target protein extracellular domain. In this case, they produced secreted versions of *Arabidopsis thaliana* TDR and PRK3 plasma membrane receptors by adding either GP64 or hemolin signal peptide sequences to the extracellular domains. These proteins were subsequently purified and used in X-ray crystallography studies. The authors suggest that such modifications to recombinant proteins could be applied to other membrane proteins in the animal kingdom.

Some studies have attempted to improve production of secretory proteins by utilizing baculovirus promoters that are active earlier in the virus replication cycle, with the expectation that the host cell secretory apparatus is more effective than in the very late stages of infection. Early studies showed promise with this approach, for example, Lawrie *et al.* (1995) demonstrated an increased yield of urokinase using the AcMNPV *p6.9* promoter as did Bonning *et al.* (1994) for secreted juvenile hormone esterase. However, almost all recent studies use the standard *polh* promoter in combination with different approaches to optimize secretion. The *vp39* promoter has also been used in BmNPV expression systems to achieve reduced aggregation of several foreign proteins (Ishiyama and Ikeda, 2010).

Toth *et al.* (2011) conducted an extensive study with *Western equine encephalitis virus* (WEEV) glycoproteins comprising complete E1, E1 ectodomain, E26KE1 polyprotein precursor and an E2E1 chimera. These were placed under control of *ie1*, *p6.9* or *polh* promoters. No glycoprotein expression was observed using the *ie1* promoter and equal amounts of soluble, processed material were produced using the late and very late gene promoters. Thus, earlier expression did not result in large amounts of high quality, soluble recombinant glycoprotein. This study emphasizes the difficulty of making general recommendations about the optimal vector to use for gene expression.

Another approach to improve the secretion of recombinant proteins is to co-express protein chaperones. Ailor and Betenbaugh (1998) partially overcome the problem of insoluble immunoglobulin produced in baculovirus-infected cells by co-expressing cytosolic heat shock protein (hsp) 70 chaperones. The hsp70 was found to co-precipitate with the immunoglobulin, which indicated formation of a specific complex. Solubility of the immunoglobulin in the cell appeared to be increased by the presence of hsp70. Secretion of the immunoglobulin also appeared to be improved. Co-expression of immunoglobulin with BiP (murine chaperone immunoglobulin heavy chain binding protein) also increased the solubility of processed immunoglobulin chains.

More recently, the secretion of a fusion between enhanced GFP and secretory alkaline phosphatase (SEAP) was improved in insect cells by co-expression of the chaperone calreticulin (CALR) or translation initiation factor eIF4E (Teng *et al.*, 2013a). Another report showed that α-synuclein (α-syn) and β-synuclein (β-syn) also enhanced secreted protein production (Teng *et al.*, 2013b). When β-syn + CALR + eIF4E were co-expressed with SEAP, this resulted in a 1.8-fold increase in reporter protein production (Teng *et al.*, 2013c). The authors also noted a prolonged period of protein secretion, up to 10 days pi, using this combination of chaperones and reporter. It would be interesting to test this system with proteins that are poorly secreted when compared with SEAP.

A slightly different approach to studying the potential role of chaperones in the production of recombinant proteins in baculovirus-infected cells was taken by Imai *et al.* (2015). *Bombyx mori* cells were treated with double-stranded RNA specific for BiP, CNX, CRT, ERp57 and PDI to reduce expression of the respective protein chaperone via RNAi. Having demonstrated a reduction in mRNA for these genes in uninfected cells, other cultures were subsequently infected with BmNPV recombinant viruses encoding human erythropoietin (EPO), cat hepatocyte growth factor (HGF) or mouse wingless tumour virus (MMTV) integration site family of member 3A (Wnt3A). Protein secretion was assessed four days pi. Depleting BiP reduced the secretion of HGF and also decreased the insoluble fraction of EPO. Reducing CRT decreased soluble EPO and increased insoluble EPO. HGF soluble and insoluble forms were decreased by CRT RNAi. The silencing of PDI also reduced the insoluble levels of EPO and HGF.

In further experiments, Imai *et al.* (2015) produced cell lines overexpressing BiP, CNX and ERp57 chaperones. When these cell lines were used as hosts for the three recombinant viruses with their reporter genes the protein expression results when assessed three days pi were mixed. Secreted and soluble forms of EPO were higher in all three of the cell lines overproducing chaperone proteins. Soluble and insoluble forms of HGF were also higher. Soluble Wnt3A was increased if BiP but not CNX or ERp57 were overexpressed. None of the chaperones were able to increase secretion of Wnt3A. The authors concluded that although differences in expression levels were noted, these were not significant.

Fath-Goodin *et al.* (2006) described the benefits of co-expressing anti-apoptotic genes to extend the window of protein production and hence increase yields of proteins, particularly those secreted or membrane-bound. More recently, Steele *et al.* (2017) have described the production of several insect cell lines that incorporate one of these apoptotic genes, vankyrin. The cell lines included SfSWT-4 (Mabashi-Asazuma *et al.*, 2013), which can produce glycoproteins with complex-type N-glycans (see also 'Developments in cell lines to support recombinant protein production'). Vankyrin levels were further increased by also incorporating the *p-vank-1* gene into the baculovirus vector (see also 'Multiple expression vectors and virus-like particle production'). Cell lysis was delayed, and recombinant glycoprotein yield increased when SfSWT-4 cells were infected with a vankyrin-containing expression vector. A synergistic effect resulting in increased glycoprotein levels was observed when the vankyrin gene was expressed both constitutively in the cell line and from the incoming virus vector.

Many secreted and membrane-targeted proteins are glycosylated, and whilst insect cells are capable of producing biologically active and immunogenic proteins, the pattern of glycosylation differs from that in mammalian cells. Glycoproteins synthesized in insect cells produce more uniform and less complex glycans that normally terminate in mannose residues, in comparison to mammalian cell glycans that most frequently terminate with sialic acid (Harrison and Jarvis, 2006). Transformed insect cell lines, such as SfSWT-4 (Mabashi-Asazuma *et al.*, 2013) that have been modified to express mammalian N-glycan processing enzymes have met with some success, whilst an alternative approach has been to encode these enzymes alongside the target glycoprotein using MultiBac vectors (referred to as SweetBac; Palmberger *et al.*, 2012). In the latest developments, Moremen *et al.* (2018) describe an elegant modular approach in which they have created an expression vector library encoding all known human glycosyltransferases, glycoside hydrolases and other glycan-modifying enzymes. The enzymes were expressed as secreted catalytic domain fusion proteins in insect cells (and mammalian cells) and a subset were characterized. This library forms an important future resource for recombinant enzyme production.

Membrane proteins are still regarded as some of the most difficult to express in the baculovirus-insect cell system. Yields tend to be low and subsequent purification somewhat problematic. The efficient production of membrane proteins, particularly those destined for the plasma membrane, shares similar problems to those described above for secreted proteins. Both must enter the secretory pathway of the virus-infected cell and be directed to the plasma membrane. The passage of the plasma membrane bound protein is then halted by a hydrophobic domain normally at the carboxyl end of the sequence. One of the very first proteins made using the baculovirus expression system was the influenza virus haemagglutinin (Possee, 1986). This is moved efficiently to the plasma membrane and is also incorporated into budded virus particles as they leave the cells. Hence, baculoviruses can be used to display plasma membrane proteins (see Chapter 11).

The plasma membrane is not the only destination for membrane proteins. While the final maturation process for influenza and baculovirus virus particles is very similar, requiring the transit of ribonucleoprotein structures or capsids respectively to the plasma membrane and subsequent budding from the cell, others such as HSV are more complex. The molecular basis for the assembly of HSV particles is only beginning to be understood.

Virus capsids leave the nucleus via budding at the inner nuclear membrane to form primary enveloped virions in the perinuclear space. The outer leaflet of the nuclear membrane then fuses with the virion primary envelope to allow their release into the cytoplasm. More than 15 tegument and 10 envelope glycoproteins are added as the virus buds into Golgi-derived vesicles. The mature virions are released when the vesicle membrane fuses with the cell plasma membrane (Mettenleiter, 2006). One of the virus envelope components is the membrane fusion factor glycoprotein B (gB). This is a target for development as an HSV vaccine owing to its role in the fusion of the virus envelope with the host cell during virion uptake (Atanasiu *et al.*, 2010). Human cytomegalovirus (HCMV) is one of the herpesviruses currently targeted for the production of an effective vaccine.

HCMV gB is synthesized as a precursor polypeptide, which is cleaved by furin into the N-terminal surface (SU) and the C-terminal membrane-anchored (TM) domains. These are linked by a disulphide bridge (Heldwein *et al.*, 2006), and anchored to membranes by the hydrophobic helix of the TM chain. The protein is heavily glycosylated. It has 18 potential N-glycosides resulting in about 40% of its mass being comprised of sugars when expressed in human cells (Britt and Vugler, 1989). The structure of gB is a trimer of 360–420 kDa. This complex molecule has been difficult to produce in a recombinant form with particular regard to its use as a vaccine. An HCMV vaccine candidate based on a secreted, truncated form of gB produced in Chinese hamster ovary cells was reported to reduce by 50% the infection risk in a Phase II clinical trial (Pass *et al.*, 2009). This has stimulated further efforts to develop a vaccine.

Patrone *et al.* (2014) were able to produce a full length form of HCMV gB using the baculovirus-insect cell system. Yields of gB were 4–6-fold higher in High Five™ cells compared with Sf9 cultures. The authors noted that the latter cells were less able to tolerate high densities at the time of virus infection when compared with High Five™ cells. The addition of rapamycin was found to inhibit baculovirus-induced swelling that is commonly observed with this system. The drug also prolonged the intracellular accumulation of gB in the insect cells. Furthermore, the inhibition of cell swelling did not appear to interfere with budded virus assembly and infectivity. Supplementing the culture medium with cysteine further improved the gB yields by increasing the detergent-soluble protein fraction. This was compatible with the enhancements achieved with use of rapamycin. Overall, the use of rapamycin in combination with cysteine supplementation resulted in the production of gB at a milligram scale. The gB was purified in its trimeric post-fusion conformation and is likely a good candidate for use as a vaccine, more closely resembling the native protein. This study also illustrates how even very complex proteins may be expressed and purified using the baculovirus system.

Incorporating RNAi technologies to improve protein production

RNA interference (RNAi) has transformed the way we study gene function (Liu and Paroo, 2010). Although originally intended as a tool to study biological processes (Fire *et al.*, 1998), in recent years it has also been used to improve the yield and quality of recombinant proteins across different expression platforms including CHO (Herbert *et al.*, 2009) and insect cells (Chavez-Pena and Kamen, 2018). In insect virology, instead of making time-consuming deletions to study gene function, RNAi can be used to silence gene expression through targeted destruction of the relevant mRNA. For example, Terenius *et al.* (2011) used this approach to study the different processes involved in virus infection. Baculovirus expression vectors have also been used to deliver dsRNA into mammalian cells (Suzuki *et al.*, 2008).

The limiting factor in the application of RNAi technology in insect cells is the uptake of dsRNA from the culture medium into the cell cytoplasm (Yu *et al.*, 2013). Efforts are ongoing to make this step more efficient, for example, by making a transformed cell line expressing the *Caenorhabditis elegans sid-1* that facilities dsRNA uptake through transmembrane channel formation (Xu *et al.*, 2013a,b) without the need for transfection reagents. Others have examined insect cells for the presence of RNAi associated factors, which helped with understanding of some of the inconsistencies reported when silencing genes in insect cell lines. This also led to transformed cell lines with enhanced RNAi capabilities including the *B. mori* and Sf9 cell lines described by Xu *et al.* (2013a,b) and *B. mori* larvae and derived cell lines expressing *Bm-ago2* (Li *et al.*, 2015).

In terms of using RNAi technology to improve recombinant protein production, there are several advantages in deploying a system that is transient so that specific host or viral genes can be targeted at defined time points in the production process, especially where the gene to be silenced is essential for cell growth or virus vector replication. A good example of the use of RNAi technology to improve protein production is the silencing of apoptosis-related genes. Apoptosis, induced by nutrient limitation, shear or oxidative stress, or simply baculovirus infection, can lower the proportion of viable cells, leading to reduced yield and/or quality of target proteins. Although baculovirus-encoded anti-apoptotic genes like *p35* can inhibit the cellular response early in infection (Rohrmann, 2013), at later times it has been shown to be less effective (Lin *et al.*, 2007). Kim *et al.* (2007) showed that transfecting Sf9 cells with dsRNA against caspase-1 significantly decreased caspase-1 activity and since then others have produced transformed Sf9 (Lai *et al.*, 2012), *T. ni* High Five™ (Herbert *et al.*, 2009) and *B. mori* BmN (Wang *et al.*, 2016) cells that can stably down-regulate caspase-1 activity through RNAi. These studies reported that the cell lines were more resistant to apoptosis than the untransformed parental cell lines, enabling an extended window for protein production leading to higher yields – as much as 400% in some cases (Wang *et al.*, 2016).

When producing recombinant VLPs for early-stage vaccine studies, a common challenge is the unwanted co-purification of baculovirus particles that contaminate the final product. This is of concern to regulatory bodies that require VLPs to be free from contaminating proteins and/or nucleic acids. Lee *et al.* (2015) used RNAi to silence the baculovirus *gp64* that expresses the major virus glycoprotein GP64, which resulted in a significant, although not total, reduction in baculovirus particle formation. RNAi has also been used to silence viral cathepsin, whose presence may degrade recombinant proteins containing the appropriate target motif (Kim *et al.*, 2007); although use of an expression vector with *v-cath* (and viral *chitinase*) deleted from the virus genome (Hitchman *et al.*, 2010) is a more fool proof way of reducing proteolytic activity.

Engineering of insect cells to improve glycosylation of recombinant therapeutic proteins is an ongoing area of development for the baculovirus expression system. One approach has been to modify cell lines to express human glycosyltransferases, as described above. However, when these cells are inoculated with a virus expression vector, the shut-off of host cell genes means that as infection progresses, the supply of the required enzymes is reduced. An alternative approach, first described by Geisler *et al.* (2008), involves silencing of the gene (*fdl*) encoding N-acetylglucosaminidase, which very efficiently removes the terminal N-acetylglucosamine from the core oligosaccharide, thus preventing further elongation to form complex N-glycans. Whilst Geisler *et al.* (2008) used plasmids encoding an inverted repeat to down-regulate N-acetylglucosaminidase, more recent studies (Nagata *et al.*, 2013) have used RNAi to deplete the mRNA levels of the *Bmfdl* gene. In the latter study, almost all of the glycans synthesized were of the complex type.

Developments in cell lines to support recombinant protein production

Improvements and modifications to expression vectors involve just one half of the baculovirus-insect cell expression system. The other half comprises the host cell line, which in the past has received less attention than the virus vector. The most commonly used insect cell lines are derived from *T. ni*, the cabbage looper, or *S. frugiperda*, the fall armyworm. Most commercial products manufactured using the BEVS are made using the latter. The original cell line, IPLB-SF-21 (Sf21), was derived from pupal ovaries (Vaughan *et al.*, 1977) and the popular Sf9 cell line is a subclone of IPLB-SF-21. As discussed above, novel cell lines have been produced that have been engineered to improve recombinant protein quality and yield by either incorporating additional beneficial genes or deleting/silencing genes that may adversely affect protein production.

For recombinant products to be realized as commercial vaccines or therapeutic proteins, however, regulatory authorities require that the host cell is thoroughly characterized and, as far as is possible, shown to be free from adventitious agents that may contaminate the final product. Such studies led to the discovery of a novel rhabdovirus in *S. frugiperda* cell lines (Ma *et al.*, 2014). This has raised potential safety concerns with the commercial production of proteins in insect cells, which led Maghodia *et al.* (2016) to develop an Sf cell line that was demonstrated to be rhabdovirus-negative (Sf-RVN). Sf-RVN has similar properties to Sf9 cells in terms of doubling time, diameter and morphology. Further, three different recombinant proteins were produced using both Sf9 and Sf-RVN and yields were indistinguishable between the two cell lines. Interestingly, titres of recombinant viruses reached higher levels (5- to 10-fold) in the rhabdovirus-free cell line and thus this cell line may prove to be an attractive alternative to Sf9 cells for the commercial manufacture of proteins.

Similarly, *T. ni*-derived High Five™ cells have been shown to contain a latent infection of a newly described alphanodavirus, which was discovered when the cells were being used to produce hepatitis E virus-like particles, which were contaminated with nodavirus particles (Li *et al.*, 2007). The sequence of the genome of the nodavirus was identical to that of the natural latent infection. Subsequently, a nodavirus-free cell line (Tnms42) was derived from T.ni High Five™ cells (Chen *et al.*, 2013).

While mammalian stable cell lines used for protein production have generally been well characterized regarding their transcriptional regulatory or secretory bottlenecks, relatively little is known about the stress and metabolic loads in insect cells. Using the alphanodavirus-free cell line Tnms42, Koczka *et al.* (2018) undertook a transcriptome analysis after infection with viruses expressing the recombinant proteins mCherry or influenza HA. Their analysis suggests that the main cellular response is to the stress caused by viral infection and secreted protein production (HA), which results in the up-regulation of host factors responsible for correct folding, disulphide bridge formation and glycosylation of proteins. Due to the class of proteins that were upregulated during cell stress, the authors suggest that these proteins are promising targets to be used in molecular engineering of insect cell lines to improve yields and overcome the problems of secreted protein production (as described above).

Linked with the development of new insect cell lines for recombinant protein synthesis by baculoviruses is the role that modifications to growth can play in determining productivity. The use of rapamycin and cysteine to enhance production of HCMV gB (Patrone *et al.*, 2014) has already been described. Monteiro *et al.* (2016) investigated the addition of different supplements to the insect cell medium to improve infectious BV and influenza VLP yields. Overall, the addition of cholesterol provided the main increase to both BV (2.5-fold) and VLP (6-fold) cell-specific yields. Other supplements such as reduced glutathione, antioxidants and polyamines increased the cell-specific yields of baculovirus particles up to 3-fold. The addition of both polyamines and cholesterol improved infectious BV production by preventing the accumulation of non-infectious particles. What these studies emphasize is that significant improvements to recombinant protein production may be achieved using simple modifications to the production process.

Future developments

The two original publications on the use of baculoviruses as expression vectors were significant milestones in demonstrating that both secreted eukaryotic proteins as well as bacterial products could be synthesized in insect cells (Smith *et al.*, 1983; Pennock *et al.*, 1984). These were later followed by the use of the silkworm baculovirus (BmNPV) as a convenient vehicle for protein production in whole insect larvae (Maeda *et al.*, 1985). The complete genome sequences of both AcMNPV and BmNPV were subsequently published and have provided valuable information on virus gene function for further development of baculovirus expression vectors (Ayres *et al.*, 1994; Gomi *et al.*, 1999, respectively). However, to date such modifications made to baculovirus expression have been relatively minor in their scope. The deletion or silencing of a few virus genes have helped improve protein production in some cases. Conversely, the addition of various genes that encode 'helper' functions such as chaperones have also been beneficial.

Another very significant milestone was the production of a synthetic baculovirus genome based on AcMNPV (Shang *et al.*, 2017). This virus (AcMNPV-WIV-Syn1) was assembled using a combination of PCR and transformation-associated recombination in yeast. It was fully infectious for both cell cultures and insect larvae. This technically difficult feat opens the way to the design of minimal baculovirus replicons that might offer improved features as expression vectors. Baculovirus genes could be omitted or other genes added to optimize the synthesis of different recombinant proteins.

In this review, we have shown the advantages associated with various strategies for improvement of baculovirus expression vectors. What is apparent is how difficult it is to extrapolate results from one system to another. The challenge now is to draw the baculovirus expression vector modifications together in a rational way as well as paying attention to how these might be integrated with improvements in the host cells currently used for protein production.

References

Ailor, E., and Betenbaugh, M.J. (1998). Overexpression of a cytosolic chaperone to improve solubility and secretion of a recombinant IgG protein in insect cells. Biotechnol. Bioeng. *58*, 196–203. https://doi.org/10.1002/(SICI)1097-0290(19980420)58:2/3<196::AID-BIT12>3.0.CO;2-B

Aponte-Ubillus, J.J., Barajas, D., Peltier, J., Bardliving, C., Shamlou, P., and Gold, D. (2018). Molecular design for recombinant adeno-associated virus (rAAV) vector production. Appl. Microbiol. Biotechnol. *102*, 1045–1054. https://doi.org/10.1007/s00253-017-8670-1

Argilaguet, J.M., Pérez-Martín, E., López, S., Goethe, M., Escribano, J.M., Giesow, K., Keil, G.M., and Rodríguez, F. (2013). BacMam immunization partially protects pigs against sublethal challenge with African swine fever virus. Antiviral Res. *98*, 61–65. https://doi.org/10.1016/j.antiviral.2013.02.005

Aslanidi, G., Lamb, K., and Zolotukhin, S. (2009). An inducible system for highly efficient production of recombinant adeno-associated virus (rAAV) vectors in insect Sf9 cells. Proc. Natl. Acad. Sci. U.S.A. *106*, 5059–5064. https://doi.org/10.1073/pnas.0810614106

Atanasiu, D., Saw, W.T., Cohen, G.H., and Eisenberg, R.J. (2010). Cascade of events governing cell-cell fusion induced by herpes simplex virus glycoproteins gD, gH/gL, and gB. J. Virol. *84*, 12292–12299. https://doi.org/10.1128/JVI.01700-10

Ayres, M.D., Howard, S.C., Kuzio, J., Lopez-Ferber, M., and Possee, R.D. (1994). The complete DNA sequence of *Autographa californica* nuclear polyhedrosis virus. Virology *202*, 586–605. https://doi.org/10.1006/viro.1994.1380

Berger, I., Fitzgerald, D.J., and Richmond, T.J. (2004). Baculovirus expression system for heterologous multiprotein complexes. Nat. Biotechnol. *22*, 1583–1587. https://doi.org/10.1038/nbt1036

Bonning, B.C., Roelvinck, P.W., Vlak, J.M., Possee, R.D., and Hammock, B.D. (1994). Superior expression of juvenile hormone esterase and beta-galactosidase from the basic protein gene promoter of *Autographa californica* nuclear polyhedrosis virus compared to the p10 and polyhedrin gene promoters. J. Gen. Virol. *75*, 1551–1556. https://doi.org/10.1099/0022-1317-75-7-1551

Bosma, B., du Plessis, F., Ehlert, E., Nijmeijer, B., de Haan, M., Petry, H., and Lubelski, J. (2018). Optimization of viral protein ratios for production of rAAV serotype 5 in the baculovirus system. Gene Ther. *25*, 415–424. https://doi.org/10.1038/s41434-018-0034-7

Boyce, F.M., and Bucher, N.L. (1996). Baculovirus-mediated gene transfer into mammalian cells. Proc. Natl. Acad. Sci. U.S.A. *93*, 2348–2352. https://doi.org/10.1073/pnas.93.6.2348

Britt, W.J., and Vugler, L.G. (1989). Processing of the gp55-116 envelope glycoprotein complex (gB) of human cytomegalovirus. J. Virol. *63*, 403–410.

Chakraborty, S., Trihemasava, K., and Xu, G. (2018). Modifying baculovirus expression vectors to produce secreted plant proteins in insect cells. J. Vis. Exp. *138*. https://doi.org/10.3791/58283

Chavez-Pena, C., and Kamen, A.A. (2018). RNA interference technology to improve the baculovirus-insect cell expression system. Biotechnol. Adv. *36*, 443–451. https://doi.org/10.1016/j.biotechadv.2018.01.008

Chen, H. (2008). Intron splicing-mediated expression of AAV Rep and Cap genes and production of AAV vectors in insect cells. Mol. Ther. *16*, 924–930. https://doi.org/10.1038/mt.2008.35

Chen, Y.R., Zhong, S., Fei, Z., Hashimoto, Y., Xiang, J.Z., Zhang, S., and Blissard, G.W. (2013). The transcriptome of the baculovirus Autographa californica multiple nucleopolyhedrovirus in *Trichoplusia ni* cells. J. Virol. *87*, 6391–6405. https://doi.org/10.1128/JVI.00194-13

European Medicines Agency. Assessment report: Glybera. Available online: http://www.ema.europa.eu/en/documents/assessment-report/glybera-epar-public-assessment-report_en.pdf. Accessed 10 March 2019.

Fan, J.L., Patibandla, S.A., Kimura, S., Rao, T.N., Desai, R.K., Seetharamaiah, G.S., Kurosky, A., and Prabhakar, B.S. (1996). Purification and characterization of a recombinant human thyroid peroxidase expressed in insect cells. J. Autoimmun. *9*, 529–536. https://doi.org/10.1006/jaut.1996.0071

Fath-Goodin, A., Kroemer, J., Martin, S., Reeves, K., and Webb, B.A. (2006). Polydnavirus genes that enhance the baculovirus expression vector system. Adv. Virus Res. *68*, 75–90. https://doi.org/10.1016/S0065-3527(06)68002-0

Fire, A., Xu, S., Montgomery, M.K., Kostas, S.A., Driver, S.E., and Mello, C.C. (1998). Potent and specific genetic interference by double-stranded RNA in *Caenorhabditis elegans*. Nature *391*, 806–811. https://doi.org/10.1038/35888

Geisler, C., Aumiller, J.J., and Jarvis, D.L. (2008). A fused lobes gene encodes the processing beta-N-acetylglucosaminidase in Sf9 cells. J. Biol. Chem. *283*, 11330–11339. https://doi.org/10.1074/jbc.M710279200

Gómez-Sebastián, S., López-Vidal, J., and Escribano, J.M. (2014). Significant productivity improvement of the baculovirus expression vector system by engineering a novel expression cassette. PLOS ONE *9*, e96562. https://doi.org/10.1371/journal.pone.0096562

Gomi, S., Majima, K., and Maeda, S. (1999). Sequence analysis of the genome of *Bombyx mori* nucleopolyhedrovirus. J. Gen. Virol. *80*, 1323–1337. https://doi.org/10.1099/0022-1317-80-5-1323

Harrison, R.L., and Jarvis, D.L. (2006). Protein N-glycosylation in the baculovirus-insect cell expression system and engineering of insect cells to produce 'mammalianized' recombinant glycoproteins. Adv. Virus Res. *68*, 159–191. https://doi.org/10.1016/S0065-3527(06)68005-6

Hastie, E., and Samulski, R.J. (2015). Adeno-associated virus at 50: A golden anniversary of discovery, research, and gene therapy success—a personal perspective. Hum. Gene. Ther. *26*, 257–265. https://doi.org/10.1089/hum.2015.025

Heinimäki, S., Tamminen, K., Malm, M., Vesikari, T., and Blazevic, V. (2017). Live baculovirus acts as a strong B and T cell adjuvant for monomeric and oligomeric protein antigens. Virology *511*, 114–122. https://doi.org/10.1016/j.virol.2017.08.023

Heldwein, E.E., Lou, H., Bender, F.C., Cohen, G.H., Eisenberg, R.J., and Harrison, S.C. (2006). Crystal structure of glycoprotein B from herpes simplex virus 1. Science *313*, 217–220. https://doi.org/10.1126/science.1126548

Hebert, C.G., Valdes, J.J., and Bentley, W.E. (2009). In vitro and in vivo RNA interference mediated suppression of Tn-caspase-1 for improved recombinant protein production in High Five cell culture with the baculovirus expression vector system. Biotechnol. Bioeng. *104*, 390–399. https://doi.org/10.1002/bit.22411

Hitchman, R.B., Possee, R.D., Crombie, A.T., Chambers, A., Ho, K., Siaterli, E., Lissina, O., Sternard, H., Novy, R., Loomis, K., *et al.* (2010). Genetic modification of a baculovirus vector for increased expression in insect cells. Cell Biol. Toxicol. *26*, 57–68. https://doi.org/10.1007/s10565-009-9133-y

Hofmann, C., Sandig, V., Jennings, G., Rudolph, M., Schlag, P., and Strauss, M. (1995). Efficient gene transfer into human hepatocytes by baculovirus vectors. Proc. Natl. Acad. Sci. U.S.A. *92*, 10099–10103.

Hoggan, M.D., Blacklow, N.R., and Rowe, W.P. (1966). Studies of small DNA viruses found in various adenovirus preparations: physical, biological, and immunological characteristics. Proc. Natl. Acad. Sci. U.S.A. *55*, 1467–1474.

Hüser, D., Weger, S., and Heilbronn, R. (2003). Packaging of human chromosome 19-specific adeno-associated virus (AAV) integration sites in AAV virions during AAV wild-type and recombinant AAV vector production. J. Virol. *77*, 4881–4887. https://doi.org/10.1128/JVI.77.8.4881-4887.2003

Imai, S., Kusakabe, T., Xu, J., Li, Z., Shirai, S., Mon, H., Morokuma, D., and Lee, J.M. (2015). Roles of silkworm endoplasmic reticulum chaperones in the secretion of recombinant proteins expressed by baculovirus system. Mol. Cell. Biochem. *409*, 255–262. https://doi.org/10.1007/s11010-015-2529-5

Ishiyama, S., and Ikeda, M. (2010). High-level expression and improved folding of proteins by using the vp39 late promoter enhanced with homologous DNA regions. Biotechnol. Lett. *32*, 1637–1647. https://doi.org/10.1007/s10529-010-0340-7

Jarvis, D.L., Summers, M.D., Garcia, A., and Bohlmeyer, D.A. (1993). Influence of different signal peptides and prosequences on expression and secretion of human tissue plasminogen activator in the baculovirus system. J. Biol. Chem. *268*, 16754–16762.

Kim, E.J., Kramer, S.F., Herbert, C.G., Valdes, J.J., and Bentley, W.E. (2007). Metabolic engineering of the baculovirus-expression vector system via inverse shotgun genomic analysis and RNAinterference (dsRNA) increases product yield and cell longevity. Biotechnol. Bioeng. *98*, 645–654. https://doi.org/10.1002/bit.21353

Koczka, K., Peters, P., Ernst, W., Himmelbauer, H., Nika, L., and Grabherr, R. (2018). Comparative transcriptome analysis of a *Trichoplusia ni* cell line reveals distinct host responses to intracellular and secreted protein products expressed by recombinant baculoviruses. J. Biotechnol. *270*, 61–69. https://doi.org/10.1016/j.jbiotec.2018.02.001

Lai, Y.K., Hsu, J.T., Chu, C.C., Chang, T.Y., Pan, K.L., and Lin, C.C. (2012). Enhanced recombinant protein production and differential expression of molecular chaperones in sf-caspase-1-repressed stable cells after baculovirus infection. BMC Biotechnol. *12*, 83. https://doi.org/10.1186/1472-6750-12-83

Lawrie, A.M., King, L.A., and Ogden, J.E. (1995). High level synthesis and secretion of human urokinase using a late gene promoter of the Autographa californica nuclear polyhedrosis virus. J. Biotechnol. *39*, 1–8. https://doi.org/10.1016/0168-1656(94)00140-8.

Lee, H.S., Lee, H.Y., Kim, Y.J., Jung, H.D., Choi, K.J., Yang, J.M., Kim, S.S., and Kim, K. (2015). Small interfering (Si) RNA mediated baculovirus replication reduction without affecting target gene expression. Virus Res. *199*, 68–76. https://doi.org/10.1016/j.virusres.2015.01.015

Li, T.C., Scotti, P.D., Miyamura, T., and Takeda, N. (2007). Latent infection of a new alphanodavirus in an insect cell line. J. Virol. *81*, 10890–10896. https://doi.org/10.1128/JVI.00807-07

Li, Z., Zeng, B., Ling, L., Xu, J., You, L., Aslam, A.F., Tan, A., and Huang, Y. (2015). Enhancement of larval RNAi efficiency by over-expressing Argonaute2 in *Bombyx mori*. Int. J. Biol. Sci. *11*, 176–185. https://doi.org/10.7150/ijbs.10235

Lin, C.-C., Hus, J.T., Huang, K.L., Tang, H.-K., Shu, C.-W., and Lai, Y.-K. (2007). Sf-caspase-1-repressed stable cells: resistance to apoptosis and augmentation of recombinant of recombinant protein production. Biotechnol. Appl. Biochem. *48*, 11–19. https://doi.org/10.1042/BA20070044

Liu, Q., and Paroo, Z. (2010). Biochemical principles of small RNA pathways. Annu. Rev. Biochem. *79*, 295–319. https://doi.org/10.1146/annurev.biochem.052208.151733

López-Vidal, J., Gómez-Sebastián, S., Bárcena, J., Nuñez, M.d.e.l.C., Martínez-Alonso, D., Dudognon, B., Guijarro, E., and Escribano, J.M. (2015). Improved production efficiency of virus-like particles by the baculovirus expression vector system. PLOS ONE *10*, e0140039. https://doi.org/10.1371/journal.pone.0140039

Lou, Y., Ji, G., Liu, Q., Wang, P., Zhang, R., Zhang, Y., and Liu, X. (2018). Secretory expression and scale-up production of recombinant human thyroid peroxidase via baculovirus/insect cell system in a wave-type bioreactor. Protein Expr. Purif. *149*, 7–12. https://doi.org/10.1016/j.pep.2018.04.005

Luckow, V.A., Lee, S.C., Barry, G.F., and Olins, P.O. (1993). Efficient generation of infectious recombinant baculoviruses by site-specific transposon-mediated insertion of foreign genes into a baculovirus genome propagated in *Escherichia coli*. J. Virol. *67*, 4566–4579.

Ma, H., Galvin, T.A., Glasner, D.R., Shaheduzzaman, S., and Khan, A.S. (2014). Identification of a novel rhabdovirus in *Spodoptera frugiperda* cell lines. J. Virol. *88*, 6576–6585. https://doi.org/10.1128/JVI.00780-14

Mabashi-Asazuma, H., Shi, X., Geisler, C., Kuo, C.W., Khoo, K.H., and Jarvis, D.L. (2013). Impact of a human CMP-sialic acid transporter on recombinant glycoprotein sialylation in glycoengineered insect cells. Glycobiology *23*, 199–210. https://doi.org/10.1093/glycob/cws143

Maeda, S., Kawai, T., Obinata, M., Fujiwara, H., Horiuchi, T., Saeki, Y., Sato, Y., and Furusawa, M. (1985a). Production of human alpha-interferon in silkworm using a baculovirus vector. Nature *315*, 592–594. https://doi.org/10.1038/315592a0

Maghodia, A.B., Geisler, C., and Jarvis, D.L. (2016). Characterization of an Sf-rhabdovirus-negative Spodoptera *frugiperda* cell line as an alternative host for recombinant protein production in the baculovirus-insect cell system. Protein Expr. Purif. *122*, 45–55. https://doi.org/10.1016/j.pep.2016.02.014

Mansouri, M., and Berger, P. (2018). Baculovirus for gene delivery to mammalian cells: Past, present and future. Plasmid *98*, 1–7. https://doi.org/10.1016/j.plasmid.2018.05.002

Mehalko, J.L., and Esposito, D. (2016). Engineering the transposition-based baculovirus expression vector system for higher efficiency protein production from insect cells. J. Biotechnol. *238*, 1–8. https://doi.org/10.1016/j.jbiotec.2016.09.002

Mettenleiter, T.C. (2006). Intriguing interplay between viral proteins during herpesvirus assembly or: the herpesvirus assembly puzzle. Vet. Microbiol. *113*, 163–169. https://doi.org/10.1016/j.vetmic.2005.11.040

Mietzsch, M., Grasse, S., Zurawski, C., Weger, S., Bennett, A., Agbandje-McKenna, M., Muzyczka, N., Zolotukhin, S., and Heilbronn, R. (2014). OneBac: platform for scalable and high-titer production of adeno-associated virus serotype 1-12 vectors for gene therapy. Hum. Gene Ther. *25*, 212–222. https://doi.org/10.1089/hum.2013.184

Mietzsch, M., Casteleyn, V., Weger, S., Zolotukhin, S., and Heilbronn, R. (2015). OneBac 2.0: sf9 cell lines for production of AAV5 vectors with enhanced infectivity and minimal encapsidation of foreign DNA. Hum. Gene Ther. *26*, 688–697. https://doi.org/10.1089/hum.2015.050

Mietzsch, M., Hering, H., Hammer, E.M., Agbandje-McKenna, M., Zolotukhin, S., and Heilbronn, R. (2017). OneBac 2.0: Sf9 cell lines for production of AAV1, AAV2, and AAV8 vectors with minimal encapsidation of foreign DNA. Hum. Gene Ther. Methods *28*, 15–22. https://doi.org/10.1089/hgtb.2016.164

Monteiro, F., Bernal, V., Chaillet, M., Berger, I., and Alves, P.M. (2016). Targeted supplementation design for improved production and quality of enveloped viral particles in insect cell-baculovirus expression system. J. Biotechnol. *233*, 34–41. https://doi.org/10.1016/j.jbiotec.2016.06.029

Moremen, K.W., Ramiah, A., Stuart, M., Steel, J., Meng, L., Forouhar, F., Moniz, H.A., Gahlay, G., Gao, Z., Chapla, D., *et al.* (2018). Expression system for structural and functional studies of human glycosylation enzymes. Nat. Chem. Biol. *14*, 156–162. https://doi.org/10.1038/nchembio.2539

Nagata, Y., Lee, J.M., Mon, H., Imanishi, S., Hong, S.M., Komatsu, S., Oshima, Y., and Kusakabe, T. (2013). RNAi suppression of β-N-acetylglucosaminidase (BmFDL) for complex-type N-linked glycan synthesis in cultured silkworm cells. Biotechnol. Lett. *35*, 1009–1016. https://doi.org/10.1007/s10529-013-1183-9

Neilan, J.G., Zsak, L., Lu, Z., Burrage, T.G., Kutish, G.F., and Rock, D.L. (2004). Neutralizing antibodies to African swine fever virus proteins p30, p54, and p72 are not sufficient for antibody-mediated protection. Virology *319*, 337–342. https://doi.org/10.1016/j.virol.2003.11.011

Nony, P., Chadeuf, G., Tessier, J., Moullier, P., and Salvetti, A. (2003). Evidence for packaging of rep-cap sequences into adeno-associated virus (AAV) type 2 capsids in the absence of inverted terminal repeats: a model for generation of rep-positive AAV particles. J. Virol. *77*, 776–781. https://doi.org/10.1128/JVI.77.1.776-781.2003

Ono, C., Okamoto, T., Abe, T., and Matsuura, Y. (2018). Baculovirus as a Tool for Gene Delivery and Gene Therapy. Viruses *10*, E510. https://doi.org/10.3390/v10090510

Palmberger, D., Wilson, I.B., Berger, I., Grabherr, R., and Rendic, D. (2012). SweetBac: a new approach for the production of mammalianised glycoproteins in insect cells. PLOS ONE 7, e34226. https://doi.org/10.1371/journal.pone.0034226

Pass, R.F., Zhang, C., Evans, A., Simpson, T., Andrews, W., Huang, M.L., Corey, L., Hill, J., Davis, E., Flanigan, C., *et al.* (2009). Vaccine prevention of maternal cytomegalovirus infection. N. Engl. J. Med. *360*, 1191–1199. https://doi.org/10.1056/NEJMoa0804749

Patrone, M., Carinhas, N., Sousa, M.Q., Peixoto, C., Ciferri, C., Carfì, A., and Alves, P.M. (2014). Enhanced expression of full-length human cytomegalovirus fusion protein in non-swelling baculovirus-infected cells with a minimal fed-batch strategy. PLOS ONE *9*, e90753. https://doi.org/10.1371/journal.pone.0090753

Penaud-Budloo, M., Lecomte, E., Guy-Duché, A., Saleun, S., Roulet, A., Lopez-Roques, C., Tournaire, B., Cogné, B., Léger, A., Blouin, V., *et al.* (2017). Accurate Identification and Quantification of DNA Species by Next-Generation Sequencing in Adeno-Associated Viral Vectors Produced in Insect Cells. Hum. Gene Ther. Methods *28*, 148–162. https://doi.org/10.1089/hgtb.2016.185

Pennock, G.D., Shoemaker, C., and Miller, L.K. (1984). Strong and regulated expression of *Escherichia coli* beta-galactosidase in insect cells with a baculovirus vector. Mol. Cell. Biol. *4*, 399–406. https://doi.org/10.1128/MCB.4.3.399

Possee, R.D. (1986). Cell-surface expression of influenza virus haemagglutinin in insect cells using a baculovirus vector. Virus Res. *5*, 43–59.

Possee, R.D., Hitchman, R.B., Richards, K.S., Mann, S.G., Siaterli, E., Nixon, C.P., Irving, H., Assenberg, R., Alderton, D., Owens, R.J., *et al.* (2008). Generation of baculovirus vectors for the high-throughput production of proteins in insect cells. Biotechnol. Bioeng. *101*, 1115–1122. https://doi.org/10.1002/bit.22002

Rohrmann, G. (2013). Baculovirus Molecular Biology, 3rd Edition. (National Centre for Biotechnology, Bethesda, MD).

Roy, P. (1992). Bluetongue virus proteins. J. Gen. Virol. 73, 3051–3064. https://doi.org/10.1099/0022-1317-73-12-3051

Samulski, R.J., and Muzyczka, N. (2014). AAV-mediated gene therapy for research and therapeutic purposes. Annu. Rev. Virol. *1*, 427–451. https://doi.org/10.1146/annurev-virology-031413-085355

Savy, A., Dickx, Y., Nauwynck, L., Bonnin, D., Merten, O.W., and Galibert, L. (2017). Impact of Inverted Terminal Repeat Integrity on rAAV8 Production Using the Baculovirus/Sf9 Cells System. Hum. Gene Ther. Methods *28*, 277–289. https://doi.org/10.1089/hgtb.2016.133

Senior, M. (2017). After Glybera's withdrawal, what's next for gene therapy? Nat. Biotechnol. 35, 491–492. https://doi.org/10.1038/nbt0617-491

Shang, Y., Wang, M., Xiao, G., Wang, X., Hou, D., Pan, K., Liu, S., Li, J., Wang, J., Arif, B.M., *et al.* (2017). Construction and Rescue of a Functional Synthetic Baculovirus. ACS Synth. Biol. *6*, 1393–1402. https://doi.org/10.1021/acssynbio.7b00028

Smith, G.E., Summers, M.D., and Fraser, M.J. (1983). Production of human beta interferon in insect cells infected with a baculovirus expression vector. Mol. Cell. Biol. *3*, 2156–2165. https://doi.org/10.1128/MCB.3.12.2156

Smith, R.H., Levy, J.R., and Kotin, R.M. (2009). A simplified baculovirus-AAV expression vector system coupled with one-step affinity purification yields high-titer rAAV stocks from insect cells. Mol. Ther. *17*, 1888–1896. https://doi.org/10.1038/mt.2009.128

Steele, K.H., Stone, B.J., Franklin, K.M., Fath-Goodin, A., Zhang, X., Jiang, H., Webb, B.A., and Geisler, C. (2017). Improving the baculovirus expression vector system with vankyrin-enhanced technology. Biotechnol. Prog. 33, 1496–1507. https://doi.org/10.1002/btpr.2516

Suzuki, H., Tamai, N., Habu, Y., Chang, M.O., and Takaku, H. (2008). Suppression of hepatitis C virus replication by baculovirus vector-mediated short-hairpin RNA expression. FEBS Lett. *582*, 3085–3089. https://doi.org/10.1016/j.febslet.2008.07.056

Teng, C.Y., Chang, S.L., van Oers, M.M., and Wu, T.Y. (2013a). Enhanced protein secretion from insect cells by co-expression of the chaperone calreticulin and translation initiation factor eIF4E. Mol. Biotechnol. *54*, 68–78. https://doi.org/10.1007/s12033-012-9545-4

Teng, C.Y., Chang, S.L., Tsai, M.F., and Wu, T.Y. (2013b) α-Synuclein and β-synuclein enhance secretion protein production in baculovirus expression vector system. Appl. Microbiol. Biotechnol. *97*, 3875–3884. https://doi.org/10.1007/s00253-012-4679-7

Teng, C.Y., van Oers, M.M., and Wu, T.Y. (2013c). Additive effect of calreticulin and translation initiation factor eIF4E on secreted protein production in the baculovirus expression system Appl. Microbiol. Biotechnol. *97*, 8505. https://doi.org/10.1007/s00253-013-5098-0.

Terenius, O., Papanicolaou, A., Garbutt, J.S., Eleftherianos, I., Huvenne, H., Kanginakudru, S., Albrechtsen, M., An, C., Aymeric, J.L., Barthel, A., *et al.* (2011). RNA interference in Lepidoptera: an overview of successful and unsuccessful studies and implications for experimental design. J. Insect Physiol. *57*, 231–245. https://doi.org/10.1016/j.jinsphys.2010.11.006

Tessier, D.C., Thomas, D.Y., Khouri, H.E., Laliberté, F., and Vernet, T. (1991). Enhanced secretion from insect cells of a foreign protein fused to the honeybee melittin signal peptide. Gene *98*, 177–183. https://doi.org/10.1016/0378-1119(91)90171-7

Toth, A.M., Geisler, C., Aumiller, J.J., and Jarvis, D.L. (2011). Factors affecting recombinant Western equine encephalitis virus glycoprotein production in the baculovirus system. Protein Expr. Purif. *80*, 274–282. https://doi.org/10.1016/j.pep.2011.08.002

Urabe, M., Ding, C., and Kotin, R.M. (2002). Insect cells as a factory to produce adeno-associated virus type 2 vectors. Hum. Gene Ther. *13*, 1935–1943. https://doi.org/10.1089/10430340260355347

US National Library of Medicines. (2019). Familial lipoprotein lipase deficiency. Available online: https://ghr.nlm.nih.gov/condition/familial-lipoprotein-lipase-deficiency. Accessed 10 March 2019.

van Oers, M.M., Pijlman, G.P., and Vlak, J.M. (2015). Thirty-years of baculovirus-insect cell expression: from dark horse to mainstream technology. J. Gen. Virol. *96*, 6–23. https://doi.org/10.1099/vir.0.067108-0

Vaughn, J.L., Goodwin, R.H., Tompkins, G.J., and McCawley, P. (1977). The establishment of two cell lines from the insect *Spodoptera frugiperda* (Lepidoptera; *Noctuidae*). In Vitro *13*, 213–217.

Wang, Q., Zhou, Y., Chen, K., and Ju, X. (2016). Suppression of Bm-Caspase-1 expression in BmN Cells enhances recombinant protein production in a baculovirus expression vector system. Mol. Biotechnol. *58*, 319–327. https://doi.org/10.1007/s12033-016-9931-4

Weissmann, F., and Peters, J.M. (2018). Expressing Multi-subunit Complexes Using biGBac. Methods Mol. Biol. *1764*, 329–343. https://doi.org/10.1007/978-1-4939-7759-8_21

Weissmann, F., Petzold, G., VanderLinden, R., Huis In 't Veld, P.J., Brown, N.G., Lampert, F., Westermann, S., Stark, H., Schulman, B.A., and Peters, J.M. (2016). biGBac enables rapid gene assembly for the expression of large multisubunit protein complexes. Proc. Natl. Acad. Sci. U.S.A. *113*, E2564–9. https://doi.org/10.1073/pnas.1604935113

Wu, Y., Jiang, L., Geng, H., Yang, T., Han, Z., He, X., Lin, K., and Xu, F. (2018). A Recombinant Baculovirus efficiently generates recombinant adeno-associated virus vectors in cultured insect cells and larvae. Mol.

Ther. Methods Clin. Dev. *10*, 38–47. https://doi.org/10.1016/j.omtm.2018.05.005

Xu, J., Mon, H., Kusakabe, T., Li, Z., Zhu, L., Iiyama, K., Masuda, A., Mitsudome, T., and Lee, J.M. (2013a). Establishment of a soaking RNA interference and Bombyx mori nucleopolyhedrovirus (BmNPV)-hypersensitive cell line using Bme21 cell. Appl. Microbiol. Biotechnol. 97, 10435–10444. https://doi.org/10.1007/s00253-013-5279-x

Xu, J., Nagata, Y., Mon, H., Li, Z., Zhu, L., Iiyama, K., Kusakabe, T., and Lee, J.M. (2013b). Soaking RNAi-mediated modification of Sf9 cells for baculovirus expression system by ectopic expression of Caenorhabditis elegans SID-1. Appl. Microbiol. Biotechnol. *97*, 5921–5931. https://doi.org/10.1007/s00253-013-4785-1

Yu, N., Christiaens, O., Liu, J., Niu, J., Cappelle, K., Caccia, S., Huvenne, H., and Smagghe, G. (2013). Delivery of dsRNA for RNAi in insects: an overview and future directions. Insect Sci. *20*, 4–14. https://doi.org/10.1111/j.1744-7917.2012.01534.x

Zhang, J., Chen, X.W., Tong, T.Z., Ye, Y., Liao, M., and Fan, H.Y. (2014). BacMam virus-based surface display of the infectious bronchitis virus (IBV) S1 glycoprotein confers strong protection against virulent IBV challenge in chickens. Vaccine *32*, 664–670. https://doi.org/10.1016/j.vaccine.2013.12.006

Zhang, Z., Yang, J., and Barford, D. (2016). Recombinant expression and reconstitution of multiprotein complexes by the USER cloning method in the insect cell-baculovirus expression system. Methods *95*, 13–25. https://doi.org/10.1016/j.ymeth.2015.10.003

Baculovirus as Versatile Vectors for Protein Display and Biotechnological Applications

11

Chih-Hsuan Tsai[1,2,3], Sung-Chan Wei[2,3], Huei-Ru Lo[2] and Yu-Chan Chao[1,2,3,4,5*]

[1]Molecular and Cell Biology, Taiwan International Graduate Program, Academia Sinica and Graduate Institute of Life Science, National Defense Medical Center, Taipei, Taiwan, Republic of China.
[2]Institute of Molecular Biology, Academia Sinica, Taipei, Taiwan, Republic of China.
[3]Graduate Institute of Life Sciences, National Defense Medical Center, Taipei, Taiwan, Republic of China.
[4]Department of Plant Pathology and Microbiology, College of Bioresources and Agriculture, National Taiwan University, Taipei, Taiwan, Republic of China.
[5]Department of Life Sciences, College of Life Sciences, National Chung Hsing University, Taichung, Taiwan, Republic of China.

*Correspondence: mbycchao@gate.sinica.edu.tw

https://doi.org/10.21775/9781912530083.11

Abstract

The baculovirus–insect cell system has long been deployed for a variety of applications including for use as biopesticides, for recombinant protein production, transient transgene expression, tissue therapy, and for vaccine production. Apart from the advantage of large-scale heterologous protein production with appropriate eukaryotic post-translational modification, foreign proteins can also be displayed on the viral envelope. This surface-display technology preserves the native multimeric structure of the protein, thereby expanding the clinical and pharmaceutical utility of the baculovirus system. Recombinant baculoviruses displaying major antigens for human or animal viruses can serve as appropriate vaccines. They can also serve as effective diagnostic platforms and various cell-based assay systems. In this review, we discuss progress in applying baculovirus surface-display, including protein display on the envelope, capsid, and occlusion bodies of baculoviruses, as well as on cells. We will also describe strategies for improvement of this biotechnological approach.

Introduction

Surface display of foreign proteins

Display of foreign proteins or peptides on the surface of a cell or virus, which we refer to as 'surface-display', is one of the most valuable techniques for protein engineering. Surface display has been used for basic research for the characterization of protein function and identification of protein counterparts (Hartmann *et al.*, 2018; Nguyen *et al.*, 2018), in addition to applied research such as for the establishment of diagnostic tools for infectious disease and development of gene therapies (Rothe *et al.*, 2006; Sergeeva *et al.*, 2006; Brown, 2010; Aghebati-Maleki *et al.*, 2016; Goulart and Santos, 2016; Lee *et al.*, 2017). Thus far, the most widely applied surface-display technique is phage display (Smith *et al.*, 2015; Yang *et al.*, 2017), whereby a library of billions of bacterial phages displaying foreign peptides can be used for screening (Liu *et al.*, 2017a). However, being limited to bacterial hosts, phage surface-display systems fail to express eukaryotic proteins with post-translational modifications or proteins with complex folding or branched structures (Mäkelä and Oker-Blom, 2008; Grabherr and Ernst, 2010; Liu *et al.*, 2017a). Other surface-display techniques have been established using yeast (Kuroda and Ueda, 2011; Tanaka *et al.*, 2012), lentivirus (Taube *et al.*, 2008; Lei *et al.*, 2010), adenovirus (Meulenbroek *et al.*, 2004; Vujadinovic and Vellinga, 2018), and adeno-associated virus (AAV) (Adachi and Nakai, 2010; Varadi *et al.*, 2012; Münch *et al.*, 2013) to encompass post-translational modifications of expressed eukaryotic proteins. However, most of these systems can only express short peptide sequences (e.g. the yeast system and AAV) or are infectious to host cells and hazardous to researchers (Buchholz *et al.*, 2015; Levin *et al.*, 2016; Lee *et al.*, 2017). Surface-display using the baculovirus expression vector system (BEVS) does not have the limitations associated with these other systems, allowing for eukaryotic post-translational modifications and expression of large properly folded heterologous proteins, whilst not being infectious or a threat to human health. Hence BEVS has become an attractive platform for protein surface-display.

Characteristics of the baculovirus-insect cell system

The BEVS uses insect viruses from the family *Baculoviridae* and their hosts, allowing for large-scale recombinant protein expression (Palomares *et al.*, 2015). As discussed in Chapter 9, the *Baculoviridae* is a family of double-stranded DNA viruses that infect the larvae of Lepidoptera, Hymenoptera, and Diptera. These large, rod-shaped and enveloped viruses contain circular DNA genomes of approximately 80–180 kilobase pairs (kbp) (van Oers and Vlak, 2007). Based on the type of occlusion bodies (OBs) they produce, baculoviruses can be further divided into two genera, the nucleopolyhedroviruses (NPVs) and the granuloviruses (GVs). NPVs produce polyhedral-shaped OBs (the polyhedra) in the nucleus to generate multiple NPV virions, whereas GVs produce ovicylindrical OBs (granules) and typically generate a single virion that is usually found in infected cells with a disrupted nuclear membrane (Ikeda *et al.*, 2014). Both NPVs and GVs exhibit a biphasic life cycle, producing two forms of viral progeny, the intracellular occlusion-derived virus (ODV) and extracellular budded virus (BV). ODVs are the major forms for transmitting viruses between insects. ODV-containing OBs are dissolved in the insect host midgut, releasing the ODVs to initiate primary infection. BVs are produced in infected insect cells and are responsible for cell-to-cell secondary infection and ultimate systemic infection within the infected host (Clem and Passarelli, 2013). The BVs are typically used in insect cell cultures in the BEVS.

Autographa californica multiple nucleopolyhedrovirus (AcMNPV) is the type species of the *Baculoviridae*. AcMNPV expresses the major glycoprotein GP64 on the rod-shaped envelope of its budded virions (Fig. 11.1A). Capsid proteins inside the envelope enclose its circular DNA (134 kbp). A total of 156 open reading frames (ORFs) have been identified for AcMNPV, and each ORF is expressed at either early, late, or very late stages of infection (Ayres *et al.*, 1994). Early genes (e.g. *ie2*, *dnapol*, and *lef1–12*) are expressed 6–9 hours post-infection (hpi) to turn on viral replication and to stimulate the expression of late genes. Late genes mainly code for structural proteins such as *vp39*, *p24*, *p6.9* and *e25* and are expressed 6–12 hpi. Very late genes, including *p10* and *polyhedrin* (*polh*), are strongly expressed approximately 18–76 hpi, so their promoters are extensively used for recombinant protein expression in the BEVS. Although some genes are expressed at specific stages of infection, several genes are expressed from early to late stages, such as *ie1*, *pp31*, and *gp64* (Friesen, 2007).

The BEVS has long been applied as an efficient tool for protein expression and gene delivery in both insect and mammalian cell systems. The baculovirus genome allows for insertion of a large foreign DNA (at least 38 kbp) (Airenne *et al.*, 2013). Insertion of a foreign gene that is driven by the *p10* or *polh* promoter results in strong expression and the resulting recombinant protein production can be at the scale of milligrams per litre (Ikonomou *et al.*, 2001; Furuta *et al.*, 2010; de Pinheiro *et al.*, 2016). Moreover, the BEVS accommodates extensive eukaryotic post-translational modification and enables appropriate oligomerization of complex proteins, neither of which are possible through a bacterial expression system. Following the discovery that the AcMNPV glycoprotein, GP64, mediates entry of these insect viruses into mammalian cells, the BEVS has been further modified to transduce foreign genes into mammalian cells such as hepatocyte, Vero, CHO, and U-2OS (Hu, 2006;

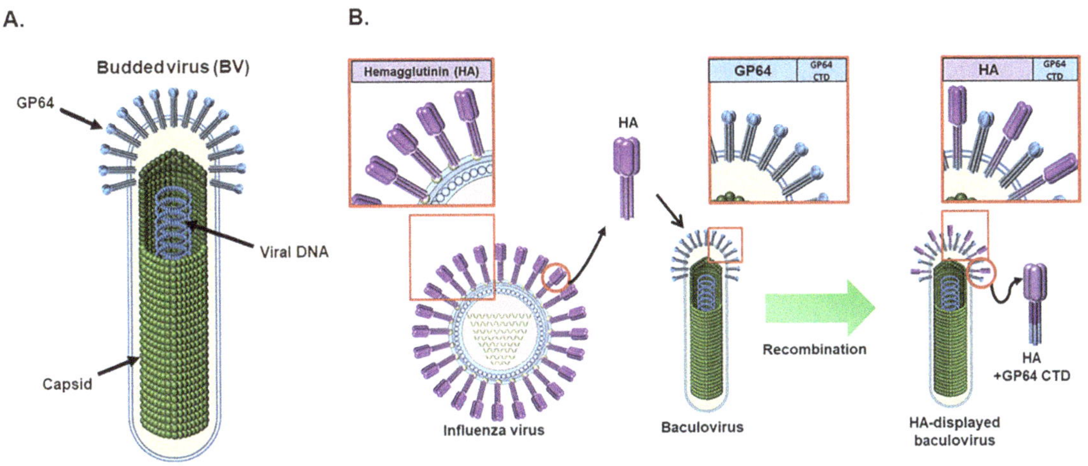

Figure 11.1 Baculovirus and display of foreign protein on the baculoviral surface envelope. (A) Schematic of the rod-shaped AcMNPV BV particle. GP64 is the major glycoprotein on the envelope responsible for cell entry and budding of the virus. Capsid proteins cover the circular viral DNA genome. (B) Display of HA from influenza virus on the baculovirus surface. The CTD of HA from influenza virus is truncated and fused with the CTD of GP64. The resulting recombinant virus displays the trimeric HA on the viral surface as well as the native GP64.

Liu *et al.*, 2010). Foreign genes in these transduced viruses are expressed with the additional insertion of mammal-appropriate promoters, e.g. *CMV*, *SV40*, and *RSV* promoters (Spenger *et al.*, 2004).

Principles and advantages of baculoviral surface-display techniques

Baculoviruses are typically used to display foreign proteins either on the viral surface or on the infected cell surface through fusion with viral glycoprotein GP64. In these cases, the foreign proteins [e.g. the haemagglutinin (HA) of *Influenza A virus*] are usually fused with the GP64 signal peptide (SP) and either its transmembrane (TMD), cytoplasmic tail (C-terminal domain, CTD), or both the TMD and CTD domains of GP64 (Fig. 11.1B) (Yang *et al.*, 2007). The resulting recombinant virus thus displays the foreign proteins on the envelope alongside wild-type (WT) GP64. WT GP64 allows the virus to propagate normally, so that recombinant viruses expressing foreign genes are readily amplified.

The BEVS surface-display technique has many advantages, such as allowing for post-translational modification, large insert capacity, preservation of the native multimeric structures of the fusion protein, the relatively easy purification process for recombinant virus compared to recombinant protein, and a safer operating system. Unlike other surface-display technologies using mammalian viruses (e.g. AAV and lentivirus), surface-display via BEVS is not hazardous to human health. In this chapter, we provide an overview of the surface-display technology using BEVS. We describe different platforms for surface-display, provide examples of current applications, suggest strategies for potential improvement, and present a final summary and future outlook for this versatile system.

Surface-display platforms

Several different strategies can be employed using BEVS to display a protein on a viral or cell surface (Fig. 11.2). Here, we discuss surface-display of foreign proteins on either baculovirus particles or infected/transduced cells.

Virus surface-display strategies

GP64 glycoprotein

The envelope glycoprotein GP64 is the major glycoprotein on the envelope of group I baculoviruses, including AcMNPV and *Bombyx mori nucleopolyhedrovirus* (BmNPV), and is the major fusion partner in most BEVS surface-display techniques. GP64 is a type I integral membrane protein consisting of 512 amino acids with a molecular mass of 64–67 kDa. GP64 is essential for AcMNPV infection as it

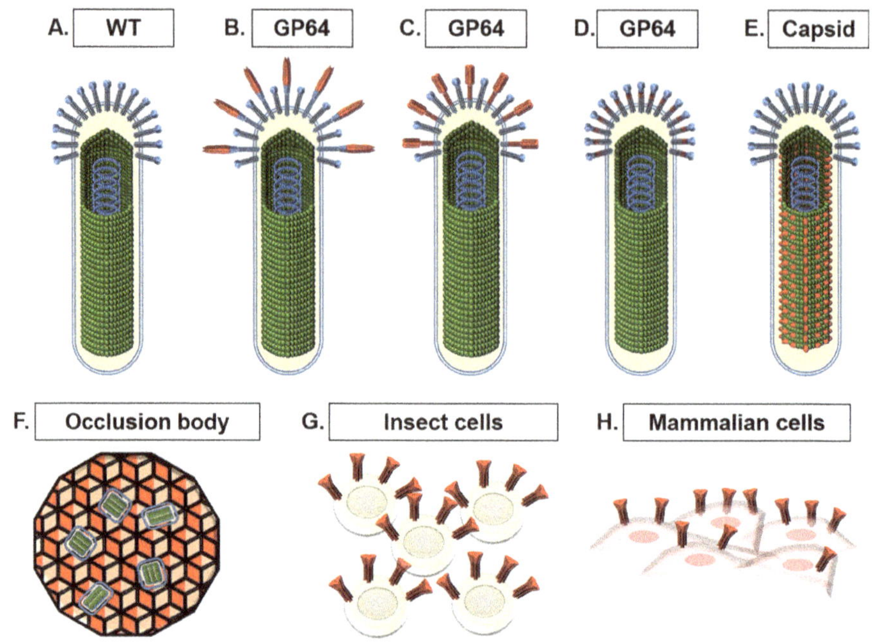

Figure 11.2 Different platforms for surface-display technologies. Schematic of different baculovirus surface-display technologies. The target foreign proteins for display are shown in red. (A) WT baculovirus with native GP64. (B) Baculoviral virion surface-display system with entire GP64 as a fusion partner. (C) Baculoviral virion surface-display system using the CTD or TMD+CTD of GP64. (D) Baculoviral virion surface-display system whereby the peptide for display is inserted into native GP64. (E) Baculoviral virion surface-display system using capsid fusion. (F) Occlusion body display. (G) Surface display on infected lepidopteran insect cells. (H) Surface display on baculovirus-transduced mammalian cells.

facilitates viral entry into host cells and membrane fusion upon endocytosis at the early stage of infection, as well as budding of BVs from infected cells at the final stage of infection. GP64 localizes to the envelope of both budded virions and infected insect cells, exhibiting a trimeric oligomeric structure thereby providing an outstanding fusion partner for heterologous proteins in the BEVS.

Several different forms of GP64 fusion can be deployed in baculovirus surface-display techniques. The first is to use the entire GP64 protein to fuse with the display targets (Fig. 11.2B). The target protein coding sequences are cloned into the baculovirus-expression vector, to produce N-terminal fusion to the GP64 SP and C-terminal fusion to GP64 (Grabherr and Ernst, 2010; Mäkelä *et al.*, 2010). Heterologous proteins as large as HIV-1 GP120 have been successfully displayed using this system and retained their protein bio-activities (Boublik *et al.*, 1995). Proteins displayed using this approach are highly accessible for immune-recognition using viral particles as immunogens for vaccine or monoclonal antibody production (Lindley *et al.*, 2000).

Another technique, now the most common approach, is to fuse the displayed target proteins with the SP, CTD or TMD+CTD of GP64 (Fig. 11.2C). This baculovirus surface-display strategy was first applied to HIV-1 envelope protein GP41 by fusing its ectodomain to the SP and the C-terminal 43 amino acids (covering the TMD and CTD) of GP64, resulting in successful display on the viral surface (Grabherr *et al.*, 1997). This strategy reduces the size of the transfer vector and simplifies the cloning procedure. However, whether a foreign protein can be displayed efficiently, either via fusion with the entire GP64 or partial GP64, is highly dependent on the properties of the target protein, size in particular (Oker-Blom *et al.*, 2003). It is generally easier to display shorter target proteins on the surface of virions or cells, and they are more likely to retain their biological activities. In some instances, the GP64 TMD may not incorporate the target protein on the surface as well as the native TMD (Tang *et al.*, 2010). In contrast, replacement of the GP64 TMD with the heterologous TMD and CTD of vesicular stomatitis virus glycoprotein (VSVG) has proven to be a promising alternative approach that increases both protein incorporation and production of virus progeny (Chapple and Jones, 2002; Zhou and Blissard, 2008). We provide a detailed discussion of peptide domain selection below.

In both of these above-described strategies, the fusion protein is expressed alongside the WT GP64 on the virion or cell envelope. As the virus randomly selects between the WT or engineered GP64 protein, the fusion target protein may not be efficiently incorporated onto virus particles (Mäkelä and Oker-Blom, 2008). An alternative baculovirus surface-display strategy is to insert the target peptide sequence directly into the native GP64 (Fig. 11.2D). The HIV-1 GP41 amino acid motif ELDKWA has been inserted into the naturally occurring *Not*I restriction site at residue 278 of GP64 (Ernst *et al.*, 2000), and another 13 insertion sites on GP64 have been described (Spenger *et al.*, 2002). However, as these recombinant baculoviruses lack WT GP64 to facilitate membrane fusion and virus budding, the insertions must be small to prevent dysfunction of the modified GP64. Thus far, the largest inserted peptide is the 23-amino-acid VP1 protein of *Foot-and-mouth disease virus* (FMDV), but the resulting recombinant virus exhibited decreased infectivity and low GP64 incorporation compared to virus with WT GP64 (Spenger *et al.*, 2002).

Capsid protein

The capsid protein of baculovirus has also been designed as a fusion partner in baculovirus surface-display systems and is called the baculovirus capsid display (BCD) system. VP39, the major and most abundant nucleocapsid protein of AcMNPV, is the main fusion partner used in BCD. VP39 is a 39-KDa protein monomer arranged in stacked rings around the baculoviral nucleoprotein core, so heterologous proteins can be displayed along the capsid surface by fusing them to VP39 (Fig. 11.2E). Kukkonen *et al.* (2003) first reported this strategy using VP39 to display enhanced green fluorescent protein (EGFP). EGFP display did not impact virus titre and virions displayed high quantities of EGFP, estimated at 590–860 EGFP molecules per virus particle.

The greatest advantages of BCD are the lack of interference with virus assembly and the high levels of protein displayed (Molinari *et al.*, 2011). Moreover, the protein displayed by BCD is uniformly distributed along the capsid surface (Fig. 11.2E), unlike the distribution of GP64-foreign protein constructs that are restricted to one end of the virus particle (Fig. 11.2B–D). Heterologous proteins as large as 41 kDa have been successfully displayed using BCD and display of even larger foreign proteins is possible as the capsid protein enables relatively unlimited length extension. BCD appears to accommodate both N- and C-terminal protein fusions, which may be due to the flexibility of the capsid protein (Kukkonen *et al.*, 2003).

VP39 has been linked to actin polymerization during viral entry into the cytoplasm and nucleus of infected insect cells (Charlton and Volkman, 1993; Ohkawa *et al.*, 2002), and is therefore a good candidate for anchoring target proteins onto the baculovirus surface for viral imaging in live cells. Liu *et al.* (2014) used an EGFP–VP39 fusion virus in co-localization and live-cell imaging experiments to study the transduction route of baculovirus in mammalian cells from the cell surface into the nucleus. Direct visualization of the BCD system in this way has helped elucidate the internalization and intracellular trafficking pathways of baculovirus in both insect and vertebrate cells.

Beyond imaging studies, Molinari *et al.* (2011) displayed ovalbumin (OVA) peptides on the capsid of baculovirus and found that they were delivered to MHC-I molecules in dendritic cells, which elicited a potent cytotoxic immune response in mice strong enough to reject implantation of tumour cells expressing OVA. Song *et al.* (2010) fused ZnO binding peptide to the N-terminus of VP39 and found that the virus not only retained its infectivity, but also gained the ability to bind nano-sized ZnO powders. A theoretical strategy for using capsid display to augment delivery of foreign proteins into the nuclei of mammalian cells has also been proposed (Kukkonen *et al.*, 2003). Unlike most viruses, baculoviruses transport their intact nucleocapsids through the nuclear pore into the cell nucleus. Accordingly, from a therapeutic point of view, the BCD system may enable nuclear delivery of therapeutic proteins to achieve the desired treatment effects as effectively as current protein transduction methods (Ford *et al.*, 2001).

Occlusion body

During the late phase of baculovirus infection, the ODV are produced. The ODVs are embedded with polyhedrin (29 kDa), which forms the crystalline protein matrix of polyhedral occlusion bodies (Harrap, 1972; Rohrmann, 1986). Polyhedrin accumulates in infected insects to very high levels, up to 30–50% of total protein (Seo *et al.*, 2003). The polyhedrin matrix, generated via extensive cross-linking of polyhedrin trimers to form cubic crystals (Coulibaly *et al.*, 2007), protects the viral particle and genome from damage and environmental threats, and assists in delivery of virus to the next target insect. Although abundant in infected cells, polyhedrin is not essential for the life cycle of budded baculovirus (Hu *et al.*, 1999).

Surface display of foreign proteins using baculoviral polyhedrin as a fusion partner (Fig. 11.2F), termed the occlusion body display (OBD) system, was first proposed by McLinden *et al.* (1992). In that study, recombinant polyhedrin proteins displaying a peptide with an influenza HA epitope were expressed on the surface of baculovirus, as revealed by immunoprecipitation with an anti-influenza monoclonal antibody (mAb) (McLinden *et al.*, 1992). Je *et al.* (2003) generated recombinant virus expressing a polyhedrin–GFP fusion construct and found that the recombinant polyhedrin only formed occlusion bodies upon co-expression with native polyhedrin. Lee *et al.* (2005) demonstrated that incorporation of foreign proteins does not require co-infection of WT virus or co-expression of native polyhedrin, as their polyhedrin-GFP and polyhedrin–VP1 constructs could both assemble occlusion bodies without co-infection with WT virus. Bae *et al.* (2013) determined that amino acids 19–110 and 32–110 of AcMNPV polyhedrin could localize the EGFP fusion in the nucleus with a 10-fold increase in protein yield, and Sampieri *et al.* (2015) identified polyhedrin residues 58–110 as being the minimal sequence necessary for efficient incorporation of foreign protein.

Polyhedrin from BmNPV has also been applied in the OBD system. Immobilization of EGFP and LacZ protein with fluorescence and β-galactosidase activity respectively into BmNPV polyhedral occlusion bodies has been demonstrated (Xiang *et al.*, 2012). The N-terminal 50 amino acids of polyhedrin represents the recognition signal necessary for foreign protein incorporation, and the N-terminal 100 amino acids of polyhedrin allowed for the most efficient protein incorporation into polyhedral occlusion bodies (Chen *et al.*, 2013). The protein encoded by ORF 134 of BmNPV has also been shown to be a suitable fusion partner for OBD (Guo *et al.*, 2017).

The major application of baculoviral OBD might be as a carrier of insecticidal toxins. Polyhedral occlusion bodies carrying such toxins could be delivered directly into the midgut of an insect host. This technology has been used to incorporate the *Bacillus thuringiensis* insecticidal toxins Cry1Ac (Chang *et al.*, 2003), Cry1-5 (Jung *et al.*, 2012; Shim *et al.*, 2013), and Cry1Ab (El-Menofy *et al.*, 2014) into host cells, resulting in enhanced speed of action and pathogenicity. An insect-specific cyto-insectotoxin (Cit1a) from the venom of the central Asian spider *Lachesana tarabaevi* has been fused to the polyhedrin of both AcMNPV and BmNPV, with improved insecticidal activity (Ali *et al.*, 2015). The OBD carrier can protect both the displayed toxin and the toxin-expressing baculovirus from ultraviolet destruction under field conditions (Mäkelä and Oker-Blom, 2008).

Cell surface-display strategies

Infected insect cells

Apart from displaying foreign protein on the surface of budded virions, baculoviruses can be used to display heterogeneous proteins on the surfaces of infected insect cells (Fig. 11.2G). Membrane-bound proteins with an appropriate SP and TMD can be expressed and translocated to the surfaces of infected host cells, such that foreign proteins are displayed on the surfaces of insect cells infected with the appropriate recombinant baculovirus. To do this, the coding sequence for the target protein for display can be cloned into an expression cassette with its full-length coding region (including the native SP and TMD), a truncated GP64-fused form (e.g. Fig. 11.2B–D), or a truncated form fused to foreign anchor domains (e.g. the C-terminus of VSVG).

Insect cell lines that are commonly used in BEVS include Sf21 (IPLB-Sf21AE) (Vaughn *et al.*, 1977) derived from *Spodoptera frugiperda* pupal ovarian tissue, and its subclone Sf9 (Smith *et al.*, 1983), High Five (BTI-TN-5B1-4) isolated from *Trichoplusia ni* embryos (Granados *et al.*, 1986, 1994), and Bm5 from *Bombyx mori* ovarian tissue. These insect cell lines are semi-adherent and can be adapted easily to serum-free media or suspension cultures for large-scale production. Sf9 cells are more resistant to physical shear stresses, and so are more suitable for suspension culture than Sf21 cells (Drugmand *et al.*, 2012). Both cell lines are commonly used to amplify recombinant AcMNPV and for the production of recombinant proteins. High Five cells are not used for virus propagation because these cells generate low titres of virus progeny, typically 100-fold lower than *S. frugiperda* cells (Kelly *et al.*, 2008; Wilde *et al.*, 2014), and rapidly accumulate mutations in the *fp25k* gene (Kelly *et al.*, 2006, 2008), which reduces the polyhedrin-based expression (Harrison *et al.*, 1996). However, High Five cells have a 50% larger cell volume than Sf9 cells, and recombinant protein expression by High Five cells can be 5- to 10-fold higher than either Sf21 or Sf9 cells (Saarenpää *et al.*, 2015). Hence High Five cells are excellent for large-scale protein production in some cases. While the *S. frugiperda* and *T. ni* cell lines are frequently used for AcMNPV protein production, they are not permissive to BmNPV, the virus used to infect the silkworm larvae or pupae for producing recombinant proteins. BmNPV only replicates in *B. mori* cell lines, such as Bm5 cell (Xu *et al.*, 2012b).

Apart from involving simpler culture conditions, a major advantage for use of insect cells for foreign protein expression is that these cells perform appropriate eukaryotic post-translational modifications such as glycosylation, phosphorylation, and fatty acid acylation. Moreover, complex mammal-derived proteins are folded and assembled in insect cells in a comparable way to mammalian cells. Although some differences between insect and mammalian expression systems remain, such as N-glycosylation (insect cells tend to produce simpler N-glycans with terminal mannose residues, whereas mammalian cells produce more complex N-glycans with terminal sialic acids), insect cells are still highly suited for mammalian protein expression (James *et al.*, 1995; Jarvis and Finn, 1995). Insect cell lines have been engineered with enzymes to produce sialic acid-terminating glycans for improved mammalian post-translational modification (Hollister *et al.*, 2002; Harrison and Jarvis, 2006; Aumiller *et al.*, 2012; see also Chapter 10).

As a baculovirus surface-display system, insect cells have proven ideal for over-expression of pharmaceutically important membrane proteins, particularly for functional studies and screening of interacting molecules. Membrane proteins, receptors, and transporters that are usually expressed in low amounts in mammalian cells can be over-expressed on insect cell surfaces through recombinant virus infection before being assayed for their substrates or inhibitors by cell-based platforms. The insect cell surface-display system circumvents tedious purification and immobilization processes and is easily optimized for high-throughput approaches. For study of protein function, display of foreign proteins by insect cells provides a system that prevents interference from homologous proteins in mammalian cells when mammalian proteins or mammalian virus proteins are being analysed. This approach has been used to study the pharmacological properties of G protein-coupled receptors (GPCR) for example (Schneider and Seifert, 2010; see below).

Mammalian and other non-lepidopteran cell surface display

Although baculoviruses do not replicate in mammalian cells, they can enter a variety of mammalian and several non-lepidopteran cells, and express heterologous target genes following insertion of mammal-specific or other appropriate promoters (such as *CMV* or *SV40*) in the expression cassette. These 'BacMam'-type viruses (Boyce and Bucher, 1996) are a powerful tool for transgene delivery into mammalian systems due to the ability of the baculovirus BV to enter diverse cell types. Baculovirus can transduce human, rabbit, non-human primate, rodent, porcine, bovine, and ovine mammalian cells (Kost and Condreay, 2002), as well as non-mammalian cells from fish (Kost *et al.*, 2005), fly (Lee *et al.*, 2000), and mosquito (Naik *et al.*, 2018), albeit with variable transduction efficiencies. Baculovirus transduction has been studied in many cell types, including hepatic cells (Hofmann *et al.*, 1995), A549 and HFL-1 (Chang *et al.*, 2004), chondrocytes (Ho *et al.*, 2004), kidney cells (Liang *et al.*, 2004), osteosarcoma cells (Song and Boyce, 2001), mesenchymal stem cells (Ho *et al.*, 2005), amongst many others. Uptake of baculovirus particles into mammalian cells greatly depends on the interaction of baculovirus GP64 with phospholipids on the mammalian cell surface (Tani *et al.*, 2001), but co-display of VSVG on the baculovirus surface has been shown to increase BacMam transgene expression efficiencies (Barsoum *et al.*, 1997).

The BacMam system provides additional attractive characteristics in that it has little cytopathic impact on the transduced cells and is safe for use by researchers. The transduction of baculovirus does not cause significant cellular toxicity or alter the gene expression profiles in mammalian cells (Kenoutis *et al.*, 2006). Although transduction of BacMam cells usually requires a high multiplicity of infection (MOI) (e.g. MOI of 100–1000), baculovirus can easily be amplified to high titres in insect cells. Several strategies have been developed to improve transgene efficiency (for details, see below).

The use of non-lepidopteran host cells for protein display (see Fig. 11.2H) is especially suited to pharmacological cell-based assays for cellular receptors, transporters, ion-channels, and viral targets (Condreay *et al.*, 2006). The combination of transient expression (e.g. transfection via a baculovirus expression cassette) and automatic assay platforms can further aid high-throughput screening of surface-display systems using non-lepidopteran host cells. Many pharmacological target proteins require co-expression of cellular components, and this requirement can be met by co-transducing multiple baculoviruses expressing the individual components (Condreay *et al.*, 1999). Co-transduction of different recombinant baculoviruses may limit reproducibility because of an uneven distribution of baculoviruses among cells. Consequently, the 'MultiBacMam' system was established to deliver multiexpression cassettes via a single baculovirus into mammalian cells. With the advantage of defined expression ratios (Trowitzsch *et al.*, 2011; Mansouri *et al.*, 2016), this technology has shown considerable potential for protein–protein interaction studies in mammalian and non-lepidopteran cells, such as in combination with bimolecular fluorescence complementation (BiFC) assays (Bellón-Echeverria *et al.*, 2018).

Applications for baculoviral surface-display systems

Viral-based assay

Baculovirus display of eukaryotic protein for functional study

The BEVS surface-display technology has long been favoured for functional studies of eukaryotic proteins including GPCR, the largest integral membrane protein family in the human genome (Cherezov *et al.*, 2010). GPCRs are important drug targets in the pharmaceutical industry, with 30–50% of currently marketed drugs acting on GPCRs (Gruber *et al.*, 2010). Display of GPCRs on Sf9 cell surfaces through the BEVS has proven an excellent system for GPCR studies (for details, see 'Insect cell surface display'). However, several GPCRs are expressed in relatively low amounts or are largely unglycosylated on Sf9 cell membranes (Masuda *et al.*, 2003). Unexpectedly, the surface of baculovirus virions was demonstrated to be a good alternative for GPCR display. The β_2-adrenergic receptor was expressed on the surface of recombinant baculovirus with homogenous receptor function and post-translational modification in contrast to display on the surface of the infected insect cell (Loisel *et al.*, 1997). This baculovirus virion display technique was subsequently applied to the study of target binding by the human leukotriene B_4 receptor (BLT1) (Masuda *et al.*, 2003), serotonin 1A receptors (5-HT_{1A}) (Tõntson *et al.*, 2014), melanocortin 4 receptors (MC4) (Veiksina *et al.*, 2014; Link *et al.*, 2017), and dopamine D_1 receptors (Sakihama *et al.*, 2008a; Allikalt and Rinken, 2017). The baculovirus particles were found in all of these studies to be a homogenous and far more stable system for membrane-bound GPCRs, allowing for their use in different experimental procedures and maintaining signal for long periods of time (Veiksina *et al.*, 2014; Allikalt and Rinken, 2017).

Apart from GPCRs, the γ-secretase complex implicated in Alzheimer's disease has also been displayed on the surface of baculovirus virions (Hayashi *et al.*, 2004). This complex requires at least four transmembrane proteins for protease function. Co-infection into Sf9 cells of four recombinant baculoviruses expressing the individual proteins achieved exclusive complex display on the budded virus particles while the Sf9 cell membranes retained the non-functional protein components. Consequently, γ-secretase activity was nearly 2.5-fold higher on budded baculovirus than on Sf9 cell membranes (Hayashi *et al.*, 2004). Similarly, display of human (pro)renin receptor (hPRR) on BmNPV virus particles resulted in easy amplification in silkworm larvae, and the virus particles could be subjected to an enzyme-linked immunosorbent assay (ELISA) system to detect protein–protein interactions and to screen for protein inhibitors (Kato *et al.*, 2009, 2011, 2012). Given the strong reconstitution of active forms of proteins and protein stability, surface-display systems using baculovirus virion particles are an attractive technology for studying protein–protein interactions and binding partners of complex eukaryotic transmembrane proteins (Sakihama *et al.*, 2008b; Kakutani *et al.*, 2011).

Baculovirus-mediated display of infectious viral proteins as pseudoviruses

A promising application for the baculovirus surface-display technique is the display of structural proteins from human-infectious viruses or pseudoviruses. Baculovirus vectors are safe and powerful tools in the study of such viruses. The first such pseudoviruses involved baculovirus virion display of HIV-1 glycoproteins GP120 (Boublik *et al.*, 1995) and GP41 (Grabherr *et al.*, 1997). The virion-displayed GP120 exhibited CD4-binding activity equivalent to WT HIV-1 (Boublik *et al.*, 1995). These baculovirus-based pseudoviruses are especially useful for functional studies of particular viral proteins as they eliminate cross-activities from other viral components. In a study of herpes simplex virus 1 (HSV-1) glycoprotein D (gD) (Zhou and Roizman, 2002), baculoviruses displaying different forms of gD were generated as pseudotyped HSV-1 and transduced into SK-N-SH cells to investigate how specific forms of gD block apoptosis and the reason for low efficiency of cell entry (Zhou and Roizman, 2002). Chang *et al.* (2004) used BEVS to display Spike (S) protein from SARS coronavirus (SARS-CoV) on the baculovirus and without interference from other SARS proteins, proved that the S protein triggers release of interleukin-8 (IL-8) in different lung cells to induce a cytokine storm (Fig. 11.3). Display of S protein fragments on virions further aided identification of specific functional regions, including one novel functional site.

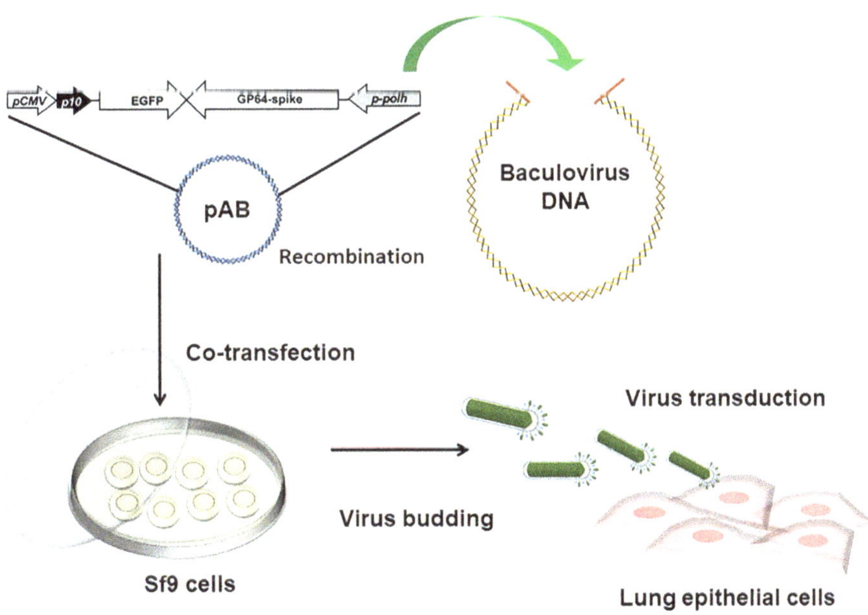

Figure 11.3 Baculovirus surface-display of spike (S) protein from SARS-CoV as a pseudovirus. SARS-S was fused with the baculovirus GP64 TMD and CTD and was then co-transfected with baculovirus DNA to generate recombinant virus that displayed SARS-S on its surface. These recombinant viruses can serve as pseudoviruses for the study of SARS-CoV.

These findings may support the development of vaccine immunogens and disease therapies (Chang *et al.*, 2004). Apart from their use in functional studies, baculovirus-based pseudoviruses could also serve as safe and efficient vaccine antigens for human pathogens (see below).

Baculovirus-mediated display of immunogens as vaccine antigens

Display of immunogenic proteins on baculovirus surfaces as vaccine antigens is one of the most popular ways of applying the BEVS technology (Premanand *et al.*, 2018). Immunogenic determinants (usually the membrane or capsid proteins) from infectious pathogens are displayed on the surface of the baculovirus envelope to serve as vaccine immunogens that can be administered to animal hosts. Many studies have also reported the immune responses induced by baculovirus-displayed antigens. Together, these studies (described further below) demonstrate the potential of generating next-generation vaccines.

Baculovirus as an immunizing reagent

Although baculovirus transduction does not have significant cytopathic impact on transduced mammalian cells, the internalization of live baculovirus still stimulates anti-viral responses. These responses probably result from the recognition of unmethylated CpG dinucleotides in viral DNA in either immune cells (e.g. macrophages or dendritic cells) or non-immune cells (e.g. fibroblasts) of mammals (Abe *et al.*, 2005; Ono *et al.*, 2018). The strong innate immune response triggered by inoculation with a WT baculovirus could sustain lethal challenge by influenza virus (Abe *et al.*, 2003). These responses on the other hand can be exploited as a valuable adjuvant activity in vaccination using baculovirus as immunogen. In addition to innate immunity, baculovirus can also induce B cell and T cell activation to improve adaptive immunity, with the removal or inactivation of baculovirus abolishing this adjuvant effect (Suzuki *et al.*, 2010; Heinimäki *et al.*, 2017).

Examples of using surface-display strategies in vaccination

HA protein, the major glycoprotein of influenza virus, is probably the best-studied immunogen that has been displayed on the surface of baculovirus. HA frequently mutates, resulting in new influenza outbreaks annually, and hence is an important target of influenza vaccines. Since HA is also a type I membrane protein (Veit and Thaa, 2011), it is particularly suited for display on the surface of baculovirus as a vaccine. Different HA subtypes have been displayed on baculovirus. These subtypes include H1N1 (Prabakaran *et al.*, 2011; Sim *et al.*, 2016), H5N1 (Yang *et al.*, 2007; Jin *et al.*, 2008; Prabakaran *et al.*, 2008; Wu *et al.*, 2009; Tang *et al.*, 2010; Hu *et al.*, 2012; Ge *et al.*, 2016), H6N8 (Musthaq *et al.*, 2014), H7N7 (Rajesh Kumar *et al.*, 2013), H7N9 (Prabakaran *et al.*, 2014), and H9N2 (Lin *et al.*, 2011). Other type I membrane proteins from human pathogens displayed on baculovirus as vaccine immunogens include envelope (E) glycoprotein from *Japanese encephalitis virus* (JEV) (Xu *et al.*, 2011) and Spike (S) protein from SARS-CoV (Feng *et al.*, 2006). A novel strategy has been applied to the display of VP1 capsid protein from enterovirus 71 (EV71) on baculovirus. As VP1 is not a membrane protein, it was fused to the SP, TMD, and CTD of baculovirus GP64 for display on baculovirus surface (Meng *et al.*, 2011; Premanand *et al.*, 2012; Kiener *et al.*, 2013; Premanand *et al.*, 2013). A series of human malaria vaccines have also been successfully developed using baculovirus surface-display technology (Yoshida *et al.*, 2003; Strauss *et al.*, 2007; Mlambo *et al.*, 2010; Iyori *et al.*, 2017).

Baculovirus surface-display techniques have also been applied to pathogens affecting non-human animals. Chang *et al.* (2018) displayed spike protein from *Porcine epidemic diarrhoea virus* (PEDV) on the surfaces of baculovirus to evaluate the immune responses and protective effects in mice and pig models (Fig. 11.4). In that study, recombinant viruses displaying spike protein elicited systemic anti-PEDV spike-specific IgG. Piglets immunized with spike-displayed baculovirus exhibited almost no or only mild diarrhoea after being challenged with PEDV, demonstrating the potential of spike-displayed baculovirus as an effective vaccine against PEDV infection. Similar approaches have been applied to generate vaccines against other swine fevers. These include the display of GP5 and glycoprotein ORFs of porcine reproductive

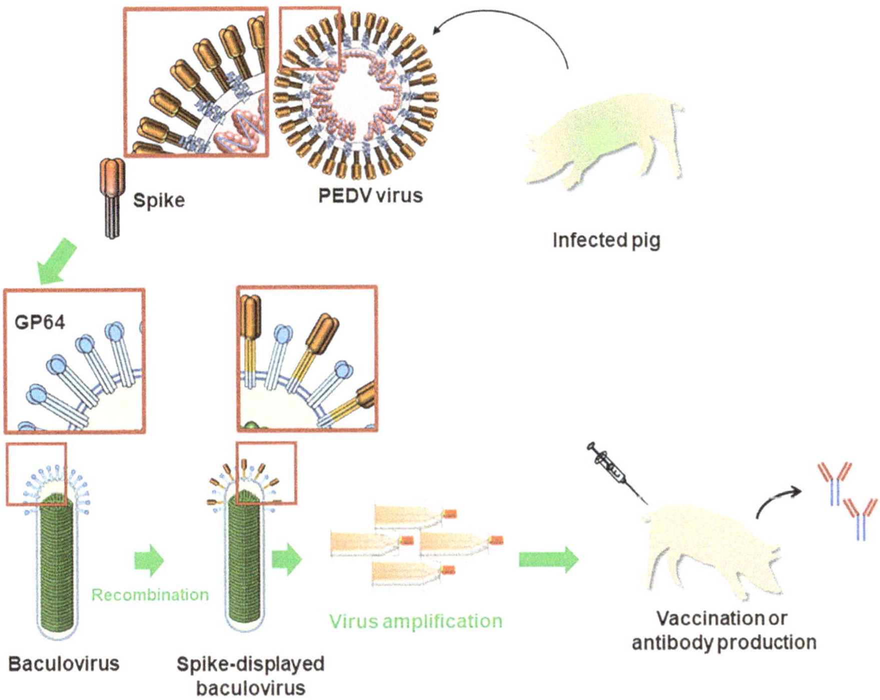

Figure 11.4 Baculovirus surface-display of spike protein from PEDV as a vaccine antigen. PEDV was isolated from infected pigs, and the DNA sequences encoding spike protein were amplified for fusion with baculovirus GP64 CTD. Recombinant viruses displaying spike were amplified and immunized into pigs as vaccination immunogens.

and respiratory syndrome virus (PRRSV) (Wang *et al.*, 2007; Xu *et al.*, 2012a; Karuppannan *et al.*, 2013), E2, E[rns] and NS3 proteins of classical swine fever virus (CSFV) (Xu and Liu, 2008; Xu *et al.*, 2008, 2009), and ORF2, the major capsid protein of *Porcine circovirus* 2 (PCV2) (Xu *et al.*, 2012a; Ye *et al.*, 2013). Although baculoviruses have not yet been approved for use as a vaccine antigen in humans, their use as animal vaccines is promising.

Improvements for baculovirus surface-display of immunogens

Chen *et al.* (2010) studied baculovirus as a vaccine vehicle by comparing the mouse immune responses induced by baculovirus that either displays the HA antigen on the surface or expresses the HA after entry into the host cell via a mammalian promoter. Baculovirus displaying HA on the viral surface triggered a stronger humoral and mucosal response upon intranasal and subcutaneous injection, whereas baculovirus expressing HA after cell entry triggered a more robust humoral response upon intramuscular injection. Accordingly, a baculovirus that both displays and expresses the HA after cell entry elicited more robust immune responses via either administration route (Chen *et al.*, 2010). If designed to enhance the mucosal immune response, co-administration of the mucosal adjuvant cholera toxin B subunit (CTB) with the antigen-displaying baculovirus (Prabakaran *et al.*, 2008), or co-display of the heat-labile enterotoxin B subunit (LTB) on the baculovirus surface (Zhang *et al.*, 2016), have proven effective.

Cell-based assays

Baculovirus can be used to display foreign proteins on the surface of both infected insect cells and transduced mammalian and non-lepidopteran cells. These cell surface displays provide ideal platforms for protein functional studies and protein-binding target identification. Different cell systems have their own advantages when applied to foreign protein display.

Insect cell surface display

Functional study of eukaryotic transmembrane proteins

Insect cell surface display mediated by baculovirus infection offers a heterologous display system for mammalian proteins especially suited to the establishment of screening platforms for pharmacologically interesting targets such as GPCRs, transporters, and ion channels. As mentioned above, GPCRs are important targets for drug development due to their involvement in many physiological responses on ligand binding. However, functional studies for GPCRs are difficult as expression of endogenous GPCRs is low in native tissue (Allikalt and Rinken, 2017). Heterogeneous overexpression of GPCRs in a mammalian cell system is confounded by the expression of endogenous GPCRs and the various G proteins produced in mammalian cells that may interfere with functional analyses. The display of GPCRs on insect cell surfaces hence provides a solution for these problems. Insect cells (e.g. Sf9 cells) have no active GPCRs and express only one form of G protein (Schneider and Seifert, 2010). Insect cells can also be used to enhance the signal-to-noise ratio in GPCR functional assays due to high level expression of heterologous proteins. The expression and activity of 16 different human GPCRs have been characterized in three insect cell lines. Albeit under different infection conditions, many of these GPCRs were expressed at high levels (greater than 20 pmol/mg determined by activity) on insect cell surfaces (Akermoun *et al.*, 2005). Many GPCRs have been studied by use of Sf9 cell surface display strategies, including neurotransmitter receptors (e.g. histamine receptors, serotonin receptors, and muscarinic acetylcholine receptors) (Schneider and Seifert, 2010), and human olfactory receptors (ORs) (Matarazzo *et al.*, 2005; Matarazzo and Ronin, 2013). In addition to GPCRs, several ion transporters have also been characterized by use of the insect cell surface display system. Sf9 cells were used to display a human cardiac Na^+/Ca^{2+} exchanger for functional study. As Sf9 cells lack native Na^+/Ca^{2+} exchanger activity, there was no background interference with the activity of the displayed target exchanger (Egger *et al.*, 1999). A Na/Ca-K exchanger has also been displayed on High Five cells for stoichiometric analysis (Szerencsei *et al.*, 2001).

Generation of protein library

The insect cell surface-display system has been exploited to generate an expression library in which the desired targets displayed on cells can be sorted by fluorescence-activated cell sorter (FACS) with specific antibodies. Granziero *et al.* (1997) reported a human placental cDNA library generated by baculovirus-infected Sf9 cells. Although display of cDNA products on the cells was not intended, the library allowed for identification of cDNAs encoding cell-surface molecules (Granziero *et al.*, 1997). Ernst *et al.* (1998) inserted a six amino-acid epitope (ELDKWA) from HIV-1 GP140 into the antigenic site of influenza virus HA and generated an insect cell surface display library expressing this peptide fusion adjacent to an additional three random amino acids. A single baculovirus clone was isolated by peptide-specific antibody from a pool harbouring approximately 8000 variants likely due to increased binding characteristics (Ernst *et al.*, 1998). Two display libraries were generated by Crawford *et al.* (2006) to display random peptides on insect cells by major histocompatibility complex (MHC) class I or class II fusion constructs and were screened to identify T-cell receptor mimotopes. Meller Harel *et al.* (2008) generated a maize cDNA library displayed on insect cells by fusing the cDNAs to SP of GP64 and TM/CTD of VSVG. Their library was estimated to contain 2.5×10^5 independent clones (Meller Harel *et al.*, 2008). However, no other insect cell surface-display libraries have been established. A possible obstacle elucidated by Xu *et al.* (2013), is that insect cells are easily co-infected by different baculoviruses. Hence each insect cell may harbour a mixture of proteins and a mixture of baculoviruses, making it difficult to recover the virus expressing the gene of interest. Another limitation for use of baculovirus to generate a cDNA library is the relatively low diversity (only 10^5–10^6 variants per library) compared to other display systems, e.g. 10^{10} variants per library for the phage display system (Bazan *et al.*, 2012).

Display of infectious viral protein

In addition to the baculovirus virion, the insect cell surface is a good platform to display infectious viral protein for functional study and diagnostic usage. Early in 1986, HA proteins of influenza virus were displayed on the insect cell surface by recombinant baculovirus infection (Kuroda *et al.*, 1986; Possee,

1986). In these studies, HA proteins were detected by antibody on the insect cell membrane, and exhibited HA activities such as hemadsorption and hemagglutination, indicating expression of the functional oligomeric conformation (Kuroda *et al.*, 1986; Possee, 1986). The insect cell surface-display technique of influenza HA has been modified slightly in expression constructs, with the replacement of the *polh* promoter with the *White spot syndrome virus* (WSSV) *ie1* promoter for example. The antigenicity of insect cell displayed HA was determined by hemagglutination and hemagglutination inhibition assays (Gadalla *et al.*, 2014). Apart from the influenza HAs, glycoprotein H (gH) of HSV-1 has also been displayed on the insect cell surface with authentic glycosylation. Intriguingly, gH expressed in mammalian cells requires other HSV-1 factors to be translocated to the cell surface, whereas in insect cells expression of gH alone was sufficient for transport (Ghiasi *et al.*, 1991). The insect cell surface-display of the haemagglutinin-neuraminidase (HN) glycoprotein of *Newcastle disease virus* (NDV) resulted in hemadsorption, hemagglutination, and neuraminidase activities (Niikura *et al.*, 1991; Murakami *et al.*, 1994; Ong *et al.*, 2000). Insect cells displaying HN have been deployed as immunogens to inoculate chickens and provided protection against NDV (Niikura *et al.*, 1991). As co-expression with NDV fusion (F) protein resulted in syncytium formation of the HN-displayed insect cells (Murakami *et al.*, 1994), these HN-displayed cells can be used for rapid detection of NDV infection. Similarly, Du *et al.* (2015) displayed the envelope (E) protein of JEV on the insect cell surface with retained fusion activity. The co-expression of JEV precursor membrane protein (prM) induced a considerable syncytial formation of infected insect cells applicable to the screening of neutralizing antibodies, or small antiviral molecules for JEV E protein (Du *et al.*, 2015).

Display on mammalian and other non-lepidopteran cells

Display of pharmacologically important transmembrane proteins

Baculovirus-mediated mammalian cell display technology is superior in cell-based assays for pharmacologically relevant proteins such as GPCRs, nuclear receptors, ion channels, and transporters. In these assays, target proteins are typically transiently expressed in case of cellular toxicity caused by over-expression (Su *et al.*, 2004; Mäkelä and Oker-Blom, 2008). The baculovirus transgene system (e.g. BacMam) meets the transient expression requirement as demonstrated by Pfohl *et al.* (2002). The expression of a multi-subunit ATP-sensitive K^+ (KATP) channel could be titrated by controlling the amount of BacMam virus added to mammalian cells. The co-display of two or more target proteins can be achieved by simply co-transducing with BacMam viruses expressing individual proteins (Hassan *et al.*, 2006). Assays conducted using the BacMam transduction system show high consistency relative to the standard DNA transfection approach, as reported by Clay *et al.* (2003) in their study to profile the agonists of an oestrogen receptor in human osteosarcoma cells. In a study of human ether-a-go-go-related gene (hERG) channel, Titus *et al.* (2009) developed a high-through-put screening assay using BacMam-transduced cells to profile the hERG inhibitors. The outcome IC_{50} values for 10 known hERG inhibitors correlated well with results determined by traditional manual patch clamp assays.

In terms of cell-based, high-through-put assays, the BacMam transduction system has several advantages over the stable cell line system; only one cell line needs to be maintained routinely when using BacMam transduction, in which transient expression can be developed quickly (Su *et al.*, 2004; Davenport *et al.*, 2009). Moreover, the BacMam transduced cells may adapt to different assays as reported by Shukla *et al.* (2012) in their study of ATP-binding cassette (ABC) drug transporters. They applied the BacMam transduced HeLa cells in both transport function and biochemical property assays, which previously had to be conducted using two different cell lines resulting in data inconsistency. As the BacMam transduced HeLa cells were able to adapt to both assays, it would be suitable for establishment of a robust assay for ABC transporters during the preclinical drug development process (Shukla *et al.*, 2012). The broad host tropism of baculovirus BV may further facilitate the application of BacMam in cell-based assays. The choice of cell line in cell-based assays for ABC transporters used to be limited as the cells have to polarize, but such cells usually show low transfection efficiency. Fung *et al.* (2016) found that the BacMam virus expressing the ABC transporters can efficiently transduce the polarized cell monolayers of LLC-PK1, MA104, and T84 cell lines, which is beneficial for establishment of ABC transporter drug screening methods (Fung *et al.*, 2016). As BacMam is such a versatile tool for pharmacologically relevant proteins for both characterization and drug discovery screening, the pharmaceutical company GSK has adapted BacMam to several of their GPCR cell-based assays (Davenport *et al.*, 2009).

Display of viral protein for high-through-put screening
Apart from the cellular transmembrane proteins, BacMam also facilitates study of infectious viruses by efficiently displaying the viral membrane proteins or antigens. The BacMam virus displaying the S1 glycoprotein of infectious bronchitis virus (IBV) gained the ability to enter chicken primary cells and increased IBV vaccine efficacy (Zhang *et al.*, 2014). As for the cell-based assays, Jenkinson *et al.* (2003) applied BacMam to develop a surrogate assay for measuring HIV viral/cell fusion. As M-tropic HIV mediates its cell entry by binding the viral envelope protein GP120 to the cellular CCR5/CD4 coreceptor, inhibition of this interaction can reduce viral fusion and entry into the cells. Jenkinson *et al.* (2003) generated a BacMam virus to mediate the surface display of HIV GP120 on HEK-293 cells. As the cell fusion event could be quantified by luciferase reporter between the GP120 displayed cells and CCR5/CD4 displayed cells, a safe high-through–put assay for screening of fusion inhibitors could thereby be established (Jenkinson *et al.*, 2003).

In addition to the mammalian systems, baculovirus are able to transduce the cells of several non-lepidopteran hosts. Naik *et al.* (2018) established a 'BacMos' system using baculovirus as a vector to deliver genes of interest into mosquito cells, larvae, and adults. They introduced a neuraminidase (NA) from H5N3 influenza virus into the baculovirus vector, and then transduced mosquito C6/36 cells. The NA protein was successfully displayed with activity on the surface of the mosquito cells by the native NA transmembrane domain (Fig. 11.5). Because many mosquito cell lines adhere to culture plates more tightly than lepidopteran cells and require a lower MOI for baculovirus transduction compared to mammalian cells, BacMos systems may prove to be a better approach for large-scale screens that involve intensive washing or physical treatments.

Strategies for improvements

Peptide domain selection

Signal peptide (SP)

Although the BEVS is suitable for the expression of complex secretory and membrane-bound glycoproteins, problems of insolubility or poor secretion may be encountered for some target proteins. Alteration of the baculovirus SP is one possible solution for overcoming this difficulty. SPs are required for protein entry into the cellular endoplasmic reticulum and subsequent secretion out of cells. Although many SPs possess similar amino acid compositions, heterologous SPs might be inefficiently recognized by the insect cell protein translocation machinery, thereby decreasing the yield of secretory protein (Jarvis *et al.*, 1993). The most frequently used native secretory SP from AcMNPV is GP64, but the SP from *Apis mellifera* prepromellitin has been shown to enhance secretion of plant propapain 5-fold in insect cells (Tessier *et al.*, 1991). The SPs derived from cecropin B (van Hofsten *et al.*, 1985) and GRP78/BiP (Gething and Sambrook, 1992) have also proved to be useful for BEVS. Finally, modifying the nucleotide sequence of SPs through codon optimization may also represent a feasible way of improving the expression and secretion of some proteins (Futatsumori-Sugai and Tsumoto, 2010).

Transmembrane (TMD) and cytoplasmic tail (CTD) domains

In a study by Yang *et al.* (2007) in which they displayed the influenza HA protein on the surface of baculovirus, they found that the CTD of GP64 in the expression construct increased the incorporation of HA protein on the viral surface. This also improved the transduction efficiency of the virus into mammalian cells and the immunogenicity of the HA-displayed baculovirus. It was later demonstrated that a construct comprising SP and CTD from GP64 with the TMD of HA resulted in the most efficient display of HA on baculoviral surfaces compared to other domain combinations

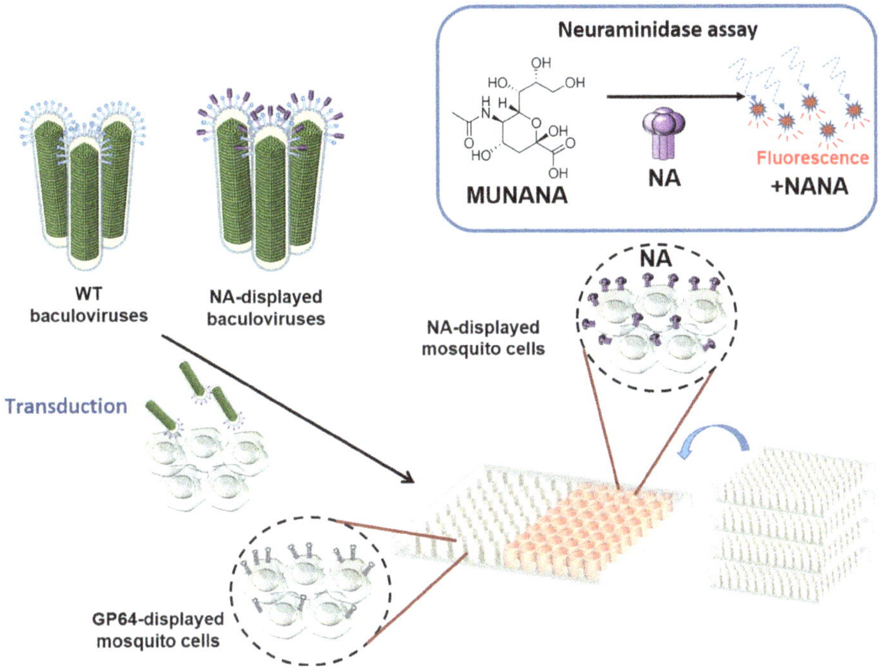

Figure 11.5 Application of baculovirus-transduced mosquito cells as a cell-based platform for influenza NA activity assay. Baculoviruses displaying NA were transduced into mosquito cells to induce surface-display of NA on cell surfaces, with the insertion driven by a mosquito-specific promoter. Wild-type (WT) baculoviruses were transduced in parallel as a negative control, resulting in mosquito cells displaying baculoviral GP64. The NA-displaying mosquito cells exhibited NA activity, as determined by a high-throughput fluorescence-based neuraminidase inhibition assay using 2′-(4-Methylumbelliferyl)-α-D-N-acetylneuraminic acid (MUNANA) as substrate.

(Tang *et al.*, 2010). As the target peptide domains largely influence the efficiency of protein display, many studies have also examined the potential for use of the TMDs and CTDs of other proteins. For example, the combination of the GP64 SP and VSVG TMD and CTD resulted in increased display efficiency (Chapple and Jones, 2002). Zhou and Blissard (2008) generated a G-stem construct containing nearly 90 amino acids from the C-terminus of VSVG (including a partial ectodomain, as well as the predicted TMD and CTD), and found that it increased HA incorporation on baculovirus. Apart from replacement of the peptide domains of type I membrane proteins, peptide domains from type II membrane proteins such as the NA of *Influenza A virus* have also been studied. Borg *et al.* (2004) determined the possibility of fusing EGFP with the SP and TMD of NA. In their construct, both SP and the TMD are located at the N-terminus of the foreign genes, thereby minimizing problems such as elimination of stop codons and frame-shift errors.

Promoters and enhancer elements used in the BEVS

Promoters

Promoter selection is an important issue in the BEVS, as it affects protein quantity and quality, as well as whether or not the target protein can be expressed by the cross-phylum baculovirus transduction system. Strong promoters, such as the *p10* or *polh* promoters, promote extremely high levels of foreign gene expression in insect cell systems, so they would be suitable for functional assays using either virus or insect cells that require high protein expression. Several baculovirus immediate-early promoters, e.g. the *ie1* and *gp64* promoters, have also been applied in baculovirus surface-display systems. One of the advantages of using an immediate-early promoter is that they utilize the cellular transcription system, so are able to express in different organisms. A *Heliothis zea nudivirus-1* (HzNV-1) early promoter (the *pag1* promoter) has been used in both Sf21 cells (Wu *et al.*, 2010) and mosquito C6/36 cells (Naik *et al.*, 2018). The *ie1* promoter of WSSV is also frequently used in baculovirus surface-display systems, especially for display of influenza HA protein. The WSSV-*ie1* promoter operates in both insect and mammalian systems, and hence is well suited for the display of proteins on baculovirus surfaces for use as gene therapy vectors and vaccine antigens (He *et al.*, 2008). Apart from these cross-phylum promoters, composite promoters that include two or more promoters with different features are also widely used. For expression in mammalian systems, foreign genes inserted into the viral genome can be turned on by composite mammalian promoters, such as *CMV* or *SV40*. Expression from three viral promoters (the *CMV*, *SV40* and *RSV* promoters), as well as one cellular promoter (*ubiquitin C*), each carried individually by baculoviruses was compared in three mammalian cell lines (CHO, COS-1 and HEK293). In this study, the *CMV* promoter resulted in the strongest expression levels (Spenger *et al.*, 2004).

Enhancer factors and elements

Several enhancer elements can be incorporated into vectors to assist in either expression of or function of a heterologous protein. When baculovirus vector is applied in mammalian systems, surface display of human or pig IgG Fc could specifically target cells of the immune system expressing Fc receptors (FcRs), thereby enhancing transduction and vaccination efficacy (Martyn *et al.*, 2009; Liu *et al.*, 2017b). Display of synthetic IgG-binding domains target the viral vector to desired cell types with IgG antibody (Ojala *et al.*, 2004). In addition, surface display of a 23 amino acid arginine–glycine–aspartic acid (RGD)-motif from the FMDV VP1 increased the uptake of baculovirus in mammalian cells (Ernst *et al.*, 2006). Co-expression of the baculovirus IE1 and IE2 proteins have been reported to trans-activate the *CMV* and *SV40* promoters in mammalian Vero E6 and U-2OS cells through both transient transfection and recombinant baculovirus transduction (Murges *et al.*, 1997; Liu *et al.*, 2007, 2009). This activation could be further augmented by the presence of a baculovirus enhancer element, i.e. the homologous region (*hr*) sequence (Lo *et al.*, 2002; Liu *et al.*, 2009). The *hr* sequence has been shown to activate *CMV* and *hsp70* promoters in several mammalian cell lines (Vero E6, COS-1 and HepG2); indeed, the *hr* sequence enhanced the *CMV* promoter activity more than 3-fold and over 122-fold upon co-expression of IE2 (Liu *et al.*, 2009). The Woodchuck hepatitis virus post-transcriptional regulatory element (WPRE) can enhance baculovirus-mediated transgene expression and has been incorporated as an element of BacMam systems (Mähönen *et al.*, 2007). Apart from co-display or co-expression of enhancer elements, histone deacetylase (HDAC) inhibitors such as sodium butyrate (NaBu) and trichostatin A (TSA) enhanced baculovirus transgene expression (Condreay *et al.*, 1999; Spenger *et al.*, 2004), and could be added to mammalian culture systems during baculovirus transduction to increase protein expression. Wang *et al.* (2017) performed a comprehensive screen of 176 known antiviral innate immune genes in mammalian cells and identified 43 host restriction factor genes that suppress baculovirus gene expression in a human A549 lung carcinoma cell line. They also found that by inhibiting some of these host restriction factor genes, baculovirus gene expression was greatly enhanced in mammalian cells.

Summary and future outlook

Baculovirus surface-display technology is one of the most creative and valuable applications of BEVS. This technology benefits from all of the advantage of BEVS, in terms of high biosafety, large insertion capacity, high-level recombinant protein production, and appropriate eukaryotic post-translational modifications. Heterologous proteins can be displayed on the surfaces of viral particles, occlusion bodies, infected cells, and even transduced cells. Different cloning strategies have been developed for appropriate anchoring of proteins on these various surfaces. Based on the literature, baculovirus-display technologies have been applied in a wide range of fields. The baculovirus virion-display strategy has been used to study the functions of complex eukaryotic transmembrane proteins, to serve as vaccine antigens, and to substitute human-infectious viruses in diagnostic or antiviral assays. The cell surface-display technologies, either in insect cells or in mammalian cells, have shown great potential for establishing high-throughput assay platforms. GP64 may still be the most frequently applied fusion partner for both virion and cell surface display strategies. By use of cryo-electron microscopy (cryo-EM), the baculovirus particle was shown to be an elongated ovoid shape with a lateral space between the envelope and nucleocapsid, rather than a rod-shaped particle (Wang *et al.*, 2016). More strikingly, the envelope proteins are located not only at one apical end, but are clustered at both polar ends (Fig. 11.6 and Wang *et al.*, 2016). These findings indicate that foreign proteins displayed by GP64 fusion may be more abundant and not restricted to one apical end on the virus particle. One of the major future applications of the baculovirus surface-display technology may be to generate vaccine antigens and diagnostic platforms for infectious pathogens, given the increasing incidence of emerging pathogens around the world. Another promising area for development is expansion of foreign protein display technologies in cells across different phyla in combination with the engineered baculovirus transgene delivery system. Exploitation of strategies to enhance transgene expression will thereby be substantial. Likewise, the engineering of baculovirus vectors to allow for transgene expression in other organisms is an area ripe for investigation. Finally, as for practical applications in high-throughput assays, use in conjunction with automated systems is worth considering.

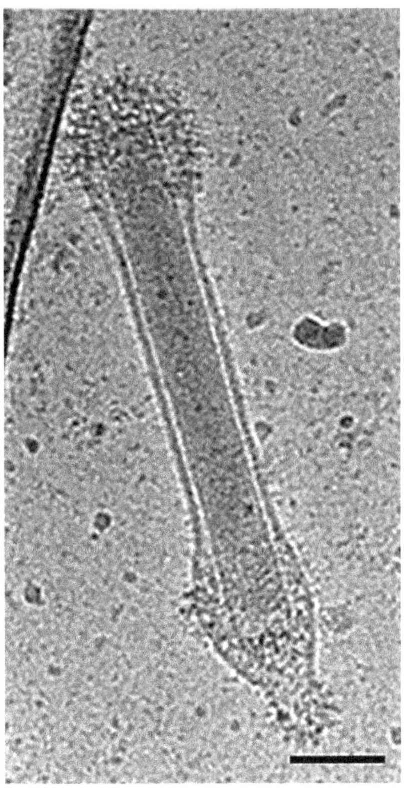

Figure 11.6 Cryo-electron microscopy of baculovirus particle. Baculovirus particle (budded virus) with envelope proteins located on both polar ends. Bar: 100 nm.

References

Abe, T., Takahashi, H., Hamazaki, H., Miyano-Kurosaki, N., Matsuura, Y., and Takaku, H. (2003). Baculovirus induces an innate immune response and confers protection from lethal influenza virus infection in mice. J. Immunol. *171*, 1133–1139.

Abe, T., Hemmi, H., Miyamoto, H., Moriishi, K., Tamura, S., Takaku, H., Akira, S., and Matsuura, Y. (2005). Involvement of the Toll-like receptor 9 signaling pathway in the induction of innate immunity by baculovirus. J. Virol. *79*, 2847–2858.

Adachi, K., and Nakai, H. (2010). A new recombinant adeno-associated virus (AAV)-based random peptide display library system: infection-defective AAV1.9-3 as a novel detargeted platform for vector evolution. Gene Ther. Regul. *5*, 31–55. https://doi.org/10.1142/S1568558610000197

Aghebati-Maleki, L., Bakhshinejad, B., Baradaran, B., Motallebnezhad, M., Aghebati-Maleki, A., Nickho, H., Yousefi, M., and Majidi, J. (2016). Phage display as a promising approach for vaccine development. J. Biomed. Sci. *23*, 66.

Airenne, K.J., Hu, Y.C., Kost, T.A., Smith, R.H., Kotin, R.M., Ono, C., Matsuura, Y., Wang, S., and Ylä-Herttuala, S. (2013). Baculovirus: an insect-derived vector for diverse gene transfer applications. Mol. Ther. *21*, 739–749. https://doi.org/10.1038/mt.2012.286

Akermoun, M., Koglin, M., Zvalova-Iooss, D., Folschweiller, N., Dowell, S.J., and Gearing, K.L. (2005). Characterization of 16 human G protein-coupled receptors expressed in baculovirus-infected insect cells. Protein Expr. Purif. *44*, 65–74.

Ali, M.P., Kato, T., and Park, E.Y. (2015). Improved insecticidal activity of a recombinant baculovirus expressing spider venom cyto-insectotoxin. Appl. Microbiol. Biotechnol. *99*, 10261–10269. https://doi.org/10.1007/s00253-015-6846-0

Allikalt, A., and Rinken, A. (2017). Budded baculovirus particles as a source of membrane proteins for radioligand binding assay: The case of dopamine D1 receptor. J. Pharmacol. Toxicol. Methods *86*, 81–86.

Aumiller, J.J., Mabashi-Asazuma, H., Hillar, A., Shi, X., and Jarvis, D.L. (2012). A new glycoengineered insect cell line with an inducibly mammalianized protein N-glycosylation pathway. Glycobiology *22*, 417–428. https://doi.org/10.1093/glycob/cwr160

Ayres, M.D., Howard, S.C., Kuzio, J., Lopez-Ferber, M., and Possee, R.D. (1994). The complete DNA sequence of *Autographa californica* nuclear polyhedrosis virus. Virology *202*, 586–605.

Bae, S.M., Kim, H.J., Lee, J.B., Choi, J.B., Shin, T.Y., Koo, H.N., Choi, J.Y., Lee, K.S., Je, Y.H., Jin, B.R., *et al.* (2013). Hyper-enhanced production of foreign recombinant protein by fusion with the partial polyhedrin of nucleopolyhedrovirus. PLOS ONE *8*, e60835. https://doi.org/10.1371/journal.pone.0060835

Barsoum, J., Brown, R., McKee, M., and Boyce, F.M. (1997). Efficient transduction of mammalian cells by a recombinant baculovirus having the vesicular stomatitis virus G glycoprotein. Hum. Gene Ther. *8*, 2011–2018. https://doi.org/10.1089/hum.1997.8.17-2011

Bazan, J., Całkosiński, I., and Gamian, A. (2012). Phage display – a powerful technique for immunotherapy: 1. Introduction and potential of therapeutic applications. Hum. Vaccin. Immunother. *8*, 1817–1828. https://doi.org/10.4161/hv.21703

Bellón-Echeverría, I., Carralot, J.P., Del Rosario, A.A., Kueng, S., Mauser, H., Schmid, G., Thoma, R., and Berger, I. (2018). MultiBacMam Bimolecular Fluorescence Complementation (BiFC) tool-kit identifies new small-molecule inhibitors of the CDK5-p25 protein-protein interaction (PPI). Sci. Rep. *8*, 5083. https://doi.org/10.1038/s41598-018-23516-x

Borg, J., Nevsten, P., Wallenberg, R., Stenstrom, M., Cardell, S., Falkenberg, C., and Holm, C. (2004). Amino-terminal anchored surface display in insect cells and budded baculovirus using the amino-terminal end of neuraminidase. J. Biotechnol. *114*, 21–30.

Boublik, Y., Di Bonito, P., and Jones, I.M. (1995). Eukaryotic virus display: engineering the major surface glycoprotein of the *Autographa californica* nuclear polyhedrosis virus (AcNPV) for the presentation of foreign proteins on the virus surface. Biotechnology *13*, 1079–1084.

Boyce, F.M., and Bucher, N.L. (1996). Baculovirus-mediated gene transfer into mammalian cells. Proc. Natl. Acad. Sci. U.S.A. *93*, 2348–2352.

Brown, K.C. (2010). Peptidic tumor targeting agents: the road from phage display peptide selections to clinical applications. Curr. Pharm. Des. *16*, 1040–1054.

Buchholz, C.J., Friedel, T., and Büning, H. (2015). Surface-engineered viral vectors for selective and cell type-specific gene delivery. Trends Biotechnol. *33*, 777–790.

Chang, C.Y., Hsu, W.T., Chao, Y.C., and Chang, H.W. (2018). Display of porcine epidemic diarrhea virus spike protein on baculovirus to improve immunogenicity and protective efficacy. Viruses *10*, E346.

Chang, J.H., Choi, J.Y., Jin, B.R., Roh, J.Y., Olszewski, J.A., Seo, S.J., O'Reilly, D.R., and Je, Y.H. (2003). An improved baculovirus insecticide producing occlusion bodies that contain *Bacillus thuringiensis* insect toxin. J. Invertebr. Pathol. *84*, 30–37.

Chang, Y.J., Liu, C.Y., Chiang, B.L., Chao, Y.C., and Chen, C.C. (2004). Induction of IL-8 release in lung cells via activator protein-1 by recombinant baculovirus displaying severe acute respiratory syndrome-coronavirus spike proteins: identification of two functional regions. J. Immunol. *173*, 7602–7614.

Chapple, S.D., and Jones, I.M. (2002). Non-polar distribution of green fluorescent protein on the surface of *Autographa californica* nucleopolyhedrovirus using a heterologous membrane anchor. J. Biotechnol. *95*, 269–275.

Charlton, C.A., and Volkman, L.E. (1993). Penetration of *Autographa californica* nuclear polyhedrosis virus nucleocapsids into IPLB Sf 21 cells induces actin cable formation. Virology *197*, 245–254.

Chen, C.Y., Liu, H.J., Tsai, C.P., Chung, C.Y., Shih, Y.S., Chang, P.C., Chiu, Y.T., and Hu, Y.C. (2010). Baculovirus as an avian influenza vaccine vector: differential immune responses elicited by different vector forms. Vaccine *28*, 7644–7651. https://doi.org/10.1016/j.vaccine.2010.09.048

Chen, L., Xiang, X., Yang, R., Hu, X., Cao, C., Malik, F.A., and Wu, X. (2013). Immobilization of foreign protein in BmNPV polyhedra by fusion expression with partial polyhedrin fragments. J. Virol. Methods *194*, 185–189. https://doi.org/10.1016/j.jviromet.2013.08.020

Cherezov, V., Abola, E., and Stevens, R.C. (2010). Toward drug design: recent progress in the structure determination of GPCRs, a membrane protein family with high potential as pharmaceutical targets. Methods Mol. Biol. *654*, 141–168.

Clay, W.C., Condreay, J.P., Moore, L.B., Weaver, S.L., Watson, M.A., Kost, T.A., and Lorenz, J.J. (2003). Recombinant baculoviruses used to study estrogen receptor function in human osteosarcoma cells. Assay Drug Dev. Technol. *1*, 801–810. https://doi.org/10.1089/154065803772613435

Clem, R.J., and Passarelli, A.L. (2013). Baculoviruses: sophisticated pathogens of insects. PLOS Pathog. *9*, e1003729. https://doi.org/10.1371/journal.ppat.1003729

Condreay, J.P., Witherspoon, S.M., Clay, W.C., and Kost, T.A. (1999). Transient and stable gene expression in mammalian cells transduced with a recombinant baculovirus vector. Proc. Natl. Acad. Sci. U.S.A. *96*, 127–132.

Condreay, J.P., Ames, R.S., Hassan, N.J., Kost, T.A., Merrihew, R.V., Mossakowska, D.E., Pountney, D.J., and Romanos, M.A. (2006). Baculoviruses and mammalian cell-based assays for drug screening. Adv. Virus Res. *68*, 255–286.

Coulibaly, F., Chiu, E., Ikeda, K., Gutmann, S., Haebel, P.W., Schulze-Briese, C., Mori, H., and Metcalf, P. (2007). The molecular organization of cypovirus polyhedra. Nature *446*, 97–101.

Crawford, F., Jordan, K.R., Stadinski, B., Wang, Y., Huseby, E., Marrack, P., Slansky, J.E., and Kappler, J.W. (2006). Use of baculovirus MHC/peptide display libraries to characterize T-cell receptor ligands. Immunol. Rev. *210*, 156–170.

Davenport, E.A., Nuthulaganti, P., and Ames, R.S. (2009). BacMam: versatile gene delivery technology for GPCR assays. Methods Mol. Biol. *552*, 199–211. https://doi.org/10.1007/978-1-60327-317-6_14

de Pinheiro, C.G., Pedrosa, M.d.e.O., Teixeira, N.C., Ano Bom, A.P., van Oers, M.M., and Oliveira, G.G. (2016). Optimization of canine interleukin-12 production using a baculovirus insect cell expression system. BMC Res. Notes *9*, 36. https://doi.org/10.1186/s13104-016-1843-7

Drugmand, J.C., Schneider, Y.J., and Agathos, S.N. (2012). Insect cells as factories for biomanufacturing. Biotechnol. Adv. *30*, 1140–1157. https://doi.org/10.1016/j.biotechadv.2011.09.014

Du, R., Yin, F., Wang, M., Hu, Z., Wang, H., and Deng, F. (2015). Glycoprotein E of the Japanese encephalitis virus forms virus-like particles and induces syncytia when expressed by a baculovirus. J. Gen. Virol. *96*, 1006–1014.

Egger, M., Ruknudin, A., Lipp, P., Kofuji, P., Lederer, W.J., Schulze, D.H., and Niggli, E. (1999). Functional expression of the human cardiac Na+/Ca2+ exchanger in Sf9 cells: rapid and specific Ni2+ transport. Cell Calcium *25*, 9–17.

El-Menofy, W., Osman, G., Assaeedi, A., and Salama, M. (2014). A novel recombinant baculovirus overexpressing a *Bacillus thuringiensis* Cry1Ab toxin enhances insecticidal activity. Biol. Proced. Online *16*, 7. https://doi.org/10.1186/1480-9222-16-7

Ernst, W., Grabherr, R., Wegner, D., Borth, N., Grassauer, A., and Katinger, H. (1998). Baculovirus surface display: construction and screening of a eukaryotic epitope library. Nucleic Acids Res. *26*, 1718–1723.

Ernst, W., Schinko, T., Spenger, A., Oker-Blom, C., and Grabherr, R. (2006). Improving baculovirus transduction of mammalian cells by surface display of a RGD-motif. J. Biotechnol. *126*, 237–240.

Ernst, W.J., Spenger, A., Toellner, L., Katinger, H., and Grabherr, R.M. (2000). Expanding baculovirus surface display. Modification of the native coat protein gp64 of *Autographa californica* NPV. Eur. J. Biochem. *267*, 4033–4039.

Feng, Q., Liu, Y., Qu, X., Deng, H., Ding, M., Lau, T.L., Yu, A.C., and Chen, J. (2006). Baculovirus surface display of SARS coronavirus (SARS-CoV) spike protein and immunogenicity of the displayed protein in mice models. DNA Cell Biol. *25*, 668–673. https://doi.org/10.1089/dna.2006.25.668

Ford, K.G., Souberbielle, B.E., Darling, D., and Farzaneh, F. (2001). Protein transduction: an alternative to genetic intervention? Gene Ther. *8*, 1–4. https://doi.org/10.1038/sj.gt.3301383

Friesen, P.D. (2007). Insect viruses. In Fields Virology, 5th edn, Knipe, D.M., Howley, P.M., Griffin, D.E., Lamb, R.A., Martin, M.A., Roizman, B., and Straus, S.E., eds. (Lippincott-Raven Publishers, Philadelphia, PA), pp. 707–736.

Fung, K.L., Kapoor, K., Pixley, J.N., Talbert, D.J., Kwit, A.D., Ambudkar, S.V., and Gottesman, M.M. (2016). Using the BacMam baculovirus system to study expression and function of recombinant efflux drug transporters in polarized epithelial cell monolayers. Drug Metab. Dispos. *44*, 180–188. https://doi.org/10.1124/dmd.115.066506

Furuta, T., Ogawa, T., Katsuda, T., Fujii, I., and Yamaji, H. (2010). Efficient production of an antibody Fab fragment using the baculovirus-insect cell system. J. Biosci. Bioeng. *110*, 577–581. https://doi.org/10.1016/j.jbiosc.2010.06.001

Futatsumori-Sugai, M., and Tsumoto, K. (2010). Signal peptide design for improving recombinant protein secretion in the baculovirus expression vector system. Biochem. Biophys. Res. Commun. *391*, 931–935. https://doi.org/10.1016/j.bbrc.2009.11.167

Gadalla, M.R., El-Deeb, A.H., Emara, M.M., and Hussein, H.A. (2014). Insect cell surface expression of hemagglutinin (HA) of Egyptian H5N1 avian influenza virus under transcriptional control of whispovirus immediate early-1 promoter. J. Microbiol. Biotechnol. *24*, 1719–1727.

Ge, J., An, Q., Gao, D., Liu, Y., and Ping, W. (2016). Construction of recombinant baculoviruses expressing hemagglutinin of H5N1 avian influenza and research on the immunogenicity. Sci. Rep. *6*, 24290. https://doi.org/10.1038/srep24290

Gething, M.J., and Sambrook, J. (1992). Protein folding in the cell. Nature *355*, 33–45.

Ghiasi, H., Nesburn, A.B., and Wechsler, S.L. (1991). Cell surface expression of herpes simplex virus type 1 glycoprotein H in recombinant baculovirus-infected cells. Virology *185*, 187–194.

Goulart, L.R., and Santos, P.d.e.S. (2016). Strategies for vaccine design using phage display-derived peptides. Methods Mol. Biol. *1404*, 423–435. https://doi.org/10.1007/978-1-4939-3389-1_28

Grabherr, R., and Ernst, W. (2010). Baculovirus for eukaryotic protein display. Curr. Gene Ther. *10*, 195–200.

Grabherr, R., Ernst, W., Doblhoff-Dier, O., Sara, M., and Katinger, H. (1997). Expression of foreign proteins on the surface of *Autographa californica* nuclear polyhedrosis virus. BioTechniques *22*, 730–735. https://doi.org/10.2144/97224rr02

Granados, R.R., Derksen, A.C., and Dwyer, K.G. (1986). Replication of the *Trichoplusia ni* granulosis and nuclear polyhedrosis viruses in cell cultures. Virology *152*, 472–476.

Granados, R.R., Guoxun, L., Derksen, A.C., and McKenna, K.A. (1994). A new insect cell line from *Trichoplusia ni* (BTI-Tn-5B1-4) susceptible to *Trichoplusia ni* single enveloped nuclear polyhedrosis virus. J. Invertebr. Pathol. *64*, 260–266.

Granziero, L., Nelboeck, P., Bedoucha, M., Lanzavecchia, A., and Reid, H.H. (1997). Baculovirus cDNA libraries for expression cloning of genes encoding cell-surface antigens. J. Immunol. Methods *203*, 131–139.

Gruber, C.W., Muttenthaler, M., and Freissmuth, M. (2010). Ligand-based peptide design and combinatorial peptide libraries to target G protein-coupled receptors. Curr. Pharm. Des. *16*, 3071–3088.

Guo, Z.J., Yu, M.H., Dong, X.Y., Wang, W.L., Tian, T., Yu, X.Y., and Tang, X.D. (2017). Protein composition analysis of polyhedra matrix of Bombyx mori nucleopolyhedrovirus (BmNPV) showed powerful capacity of polyhedra to encapsulate foreign proteins. Sci. Rep. *7*, 8768. https://doi.org/10.1038/s41598-017-08987-8

Harrap, K.A. (1972). The structure of nuclear polyhedrosis viruses. I. The inclusion body. Virology *50*, 114–123.

Harrison, R.L., and Jarvis, D.L. (2006). Protein N-glycosylation in the baculovirus-insect cell expression system and engineering of insect cells to produce 'mammalianized' recombinant glycoproteins. Adv. Virus Res. *68*, 159–191.

Harrison, R.L., Jarvis, D.L., and Summers, M.D. (1996). The role of the AcMNPV 25K gene, 'FP25,' in baculovirus *polh* and p10 expression. Virology *226*, 34–46.

Hartmann, J., Münch, R.C., Freiling, R.T., Schneider, I.C., Dreier, B., Samukange, W., Koch, J., Seeger, M.A., Plückthun, A., and Buchholz, C.J. (2018). A Library-Based Screening Strategy for the Identification of DARPins as Ligands for Receptor-Targeted AAV and Lentiviral Vectors. Mol. Ther. Methods Clin. Dev. *10*, 128–143. https://doi.org/10.1016/j.omtm.2018.07.001

Hassan, N.J., Pountney, D.J., Ellis, C., and Mossakowska, D.E. (2006). BacMam recombinant baculovirus in transporter expression: a study of BCRP and OATP1B1. Protein Expr. Purif. *47*, 591–598.

Hayashi, I., Urano, Y., Fukuda, R., Isoo, N., Kodama, T., Hamakubo, T., Tomita, T., and Iwatsubo, T. (2004). Selective reconstitution and recovery of functional gamma-secretase complex on budded baculovirus particles. J. Biol. Chem. *279*, 38040–38046. https://doi.org/10.1074/jbc.M405597200

He, F., Ho, Y., Yu, L., and Kwang, J. (2008). WSSV ie1 promoter is more efficient than CMV promoter to express H5 hemagglutinin from influenza virus in baculovirus as a chicken vaccine. BMC Microbiol. *8*, 238. https://doi.org/10.1186/1471-2180-8-238

Heinimäki, S., Tamminen, K., Malm, M., Vesikari, T., and Blazevic, V. (2017). Live baculovirus acts as a strong B and T cell adjuvant for monomeric and oligomeric protein antigens. Virology *511*, 114–122.

Ho, Y.C., Chen, H.C., Wang, K.C., and Hu, Y.C. (2004). Highly efficient baculovirus-mediated gene transfer into rat chondrocytes. Biotechnol. Bioeng. *88*, 643–651. https://doi.org/10.1002/bit.20239

Ho, Y.C., Chung, Y.C., Hwang, S.M., Wang, K.C., and Hu, Y.C. (2005). Transgene expression and differentiation of baculovirus-transduced human mesenchymal stem cells. J. Gene Med. *7*, 860–868. https://doi.org/10.1002/jgm.729

Hofmann, C., Sandig, V., Jennings, G., Rudolph, M., Schlag, P., and Strauss, M. (1995). Efficient gene transfer into human hepatocytes by baculovirus vectors. Proc. Natl. Acad. Sci. U.S.A. *92*, 10099–10103.

Hollister, J., Grabenhorst, E., Nimtz, M., Conradt, H., and Jarvis, D.L. (2002). Engineering the protein N-glycosylation pathway in insect cells for production of biantennary, complex N-glycans. Biochemistry *41*, 15093–15104.

Hu, Y.C. (2006). Baculovirus vectors for gene therapy. Adv. Virus Res. *68*, 287–320.

Hu, Z., Luijckx, T., van Dinten, L.C., van Oers, M.M., Hajós, J.P., Bianchi, F.J., van Lent, J.W., Zuidema, D., and Vlak, J.M. (1999). Specificity of polyhedrin in the generation of baculovirus occlusion bodies. J. Gen. Virol. *80*, 1045–1053. https://doi.org/10.1099/0022-1317-80-4-1045

Hu, Z.P., Yin, J., Zhang, Y.Y., Jia, S.Y., Chen, Z.J., and Zhong, J. (2012). Characterization of the immune responses elicited by baculovirus-based vector vaccines against influenza virus hemagglutinin. Acta Pharmacol. Sin. *33*, 783–790. https://doi.org/10.1038/aps.2012.23

Ikeda, M., Hamajima, R., and Kobayashi, M. (2014). Baculoviruses: diversity, evolution and manipulation of insects. Entomol. Sci. *18*, 1–20.

Ikonomou, L., Bastin, G., Schneider, Y.J., and Agathos, S.N. (2001). Design of an efficient medium for insect cell growth and recombinant protein production. In Vitro Cell. Dev. Biol. Anim. *37*, 549–559.

Iyori, M., Yamamoto, D.S., Sakaguchi, M., Mizutani, M., Ogata, S., Nishiura, H., Tamura, T., Matsuoka, H., and Yoshida, S. (2017). DAF-shielded baculovirus-vectored vaccine enhances protection against malaria sporozoite challenge in mice. Malar. J. *16*, 390. https://doi.org/10.1186/s12936-017-2039-x

James, D.C., Freedman, R.B., Hoare, M., Ogonah, O.W., Rooney, B.C., Larionov, O.A., Dobrovolsky, V.N., Lagutin, O.V., and Jenkins, N. (1995). N-glycosylation of recombinant human interferon-gamma produced in different animal expression systems. Biotechnology *13*, 592–596.

Jarvis, D.L., and Finn, E.E. (1995). Biochemical analysis of the N-glycosylation pathway in baculovirus-infected lepidopteran insect cells. Virology *212*, 500–511.

Jarvis, D.L., Summers, M.D., Garcia, A., and Bohlmeyer, D.A. (1993). Influence of different signal peptides and prosequences on expression and secretion of human tissue plasminogen activator in the baculovirus system. J. Biol. Chem. *268*, 16754–16762.

Je, Y.H., Jin, B.R., Park, H.W., Roh, J.Y., Chang, J.H., Seo, S.J., Olszewski, J.A., O'Reilly, D.R., and Kang, S.K. (2003). Baculovirus expression vectors that incorporate the foreign protein into viral occlusion bodies. BioTechniques *34*, 81–87. https://doi.org/10.2144/03341st04

Jenkinson, S., McCoy, D.C., Kerner, S.A., Ferris, R.G., Lawrence, W.K., Clay, W.C., Condreay, J.P., and Smith, C.D. (2003). Development of a novel high-throughput surrogate assay to measure HIV envelope/CCR5/CD4-mediated viral/cell fusion using BacMam baculovirus technology. J. Biomol. Screen. *8*, 463–470. https://doi.org/10.1177/1087057103255747

Jin, R., Lv, Z., Chen, Q., Quan, Y., Zhang, H., Li, S., Chen, G., Zheng, Q., Jin, L., Wu, X., *et al.* (2008). Safety and immunogenicity of H5N1 influenza vaccine based on baculovirus surface display system of *Bombyx mori*. PLOS ONE *3*, e3933. https://doi.org/10.1371/journal.pone.0003933

Jung, M.P., Choi, J.Y., Tao, X.Y., Jin, B.R., Je, Y.H., and Park, H.H. (2012). Insecticidal activity of recombinant baculovirus expressing both spider toxin isolated from *Araneus ventricosus* and *Bacillus thuringiensis* crystal protein fused to a viral polyhedrin. Entomol. Res. *42*, 339–346.

Kakutani, H., Takahashi, A., Kondoh, M., Saito, Y., Yamaura, T., Sakihama, T., Hamakubo, T., and Yagi, K. (2011). A novel screening system for claudin binder using baculoviral display. PLOS ONE *6*, e16611. https://doi.org/10.1371/journal.pone.0016611

Karuppannan, A.K., Qiang, J., Chang, C.C., and Kwang, J. (2013). A novel baculovirus vector shows efficient gene delivery of modified porcine reproductive and respiratory syndrome virus antigens and elicits specific immune response. Vaccine *31*, 5471–5478. https://doi.org/10.1016/j.vaccine.2013.08.101

Kato, T., Manoha, S.L., Tanaka, S., and Park, E.Y. (2009). High-titer preparation of *Bombyx mori* nucleopolyhedrovirus (BmNPV) displaying recombinant protein in silkworm larvae by size exclusion chromatography and its characterization. BMC Biotechnol. *9*, 55. https://doi.org/10.1186/1472-6750-9-55

Kato, T., Suzuki, F., and Park, E.Y. (2011). Purification of functional baculovirus particles from silkworm larval hemolymph and their use as nanoparticles for the detection of human prorenin receptor (PRR)

binding. BMC Biotechnol. *11*, 60. https://doi.org/10.1186/1472-6750-11-60

Kato, T., Suzuki, F., and Park, E.Y. (2012). Display of the human (pro)renin receptor on *Bombyx mori* nucleopolyhedrovirus (BmNPV) particles using Bm cells. J. Biosci. Bioeng. *114*, 564–569. https://doi.org/10.1016/j.jbiosc.2012.06.008

Kelly, B.J., King, L.A., Possee, R.D., and Chapple, S.D. (2006). Dual mutations in the *Autographa californica* nucleopolyhedrovirus FP-25 and p35 genes result in plasma-membrane blebbing in *Trichoplusia ni* cells. J. Gen. Virol. *87*, 531–536.

Kelly, B.J., Chapple, S.D., Allen, C., Pritchard, C., King, L.A., and Possee, R.D. (2008). Extended budded virus formation and induction of apoptosis by an AcMNPV FP-25/p35 double mutant in *Trichoplusia ni* cells. Virus Res. *133*, 157–166. https://doi.org/10.1016/j.virusres.2007.12.013

Kenoutis, C., Efrose, R.C., Swevers, L., Lavdas, A.A., Gaitanou, M., Matsas, R., and Iatrou, K. (2006). Baculovirus-mediated gene delivery into mammalian cells does not alter their transcriptional and differentiating potential but is accompanied by early viral gene expression. J. Virol. *80*, 4135–4146.

Kiener, T.K., Premanand, B., and Kwang, J. (2013). Immune responses to baculovirus-displayed enterovirus 71 VP1 antigen. Expert Rev. Vaccines *12*, 357–364. https://doi.org/10.1586/erv.13.18

Kost, T.A., and Condreay, J.P. (2002). Recombinant baculoviruses as mammalian cell gene-delivery vectors. Trends Biotechnol. *20*, 173–180.

Kost, T.A., Condreay, J.P., and Jarvis, D.L. (2005). Baculovirus as versatile vectors for protein expression in insect and mammalian cells. Nat. Biotechnol. *23*, 567–575.

Kukkonen, S.P., Airenne, K.J., Marjomäki, V., Laitinen, O.H., Lehtolainen, P., Kankaanpää, P., Mähönen, A.J., Räty, J.K., Nordlund, H.R., Oker-Blom, C., *et al.* (2003). Baculovirus capsid display: a novel tool for transduction imaging. Mol. Ther. *8*, 853–862.

Kuroda, K., and Ueda, M. (2011). Cell surface engineering of yeast for applications in white biotechnology. Biotechnol. Lett. *33*, 1–9. https://doi.org/10.1007/s10529-010-0403-9

Kuroda, K., Hauser, C., Rott, R., Klenk, H.D., and Doerfler, W. (1986). Expression of the influenza virus haemagglutinin in insect cells by a baculovirus vector. EMBO J. *5*, 1359–1365.

Lee, C.S., Bishop, E.S., Zhang, R., Yu, X., Farina, E.M., Yan, S., Zhao, C., Zheng, Z., Shu, Y., Wu, X., *et al.* (2017). Adenovirus-mediated gene delivery: potential applications for gene and cell-based therapies in the new era of personalized medicine. genes dis. *4*, 43–63. https://doi.org/10.1016/j.gendis.2017.04.001

Lee, D.F., Chen, C.C., Hsu, T.A., and Juang, J.L. (2000). A baculovirus superinfection system: efficient vehicle for gene transfer into *Drosophila* S2 cells. J. Virol. *74*, 11873–11880.

Lee, T.C., Yu, S.M., and Chao, Y.C. (2005). Fusion of foreign protein genes to the occlusion body gene of the baculovirus to assist protein isolation and antibody production. J. Genet. Mol. Biol. *16*, 138–150.

Lei, Y., Joo, K.I., Zarzar, J., Wong, C., and Wang, P. (2010). Targeting lentiviral vector to specific cell types through surface displayed single chain antibody and fusogenic molecule. Virol. J. 7, 35. https://doi.org/10.1186/1743-422X-7-35

Levin, E., Diekmann, H., and Fischer, D. (2016). Highly efficient transduction of primary adult CNS and PNS neurons. Sci. Rep. *6*, 38928. https://doi.org/10.1038/srep38928

Liang, C.Y., Wang, H.Z., Li, T.X., Hu, Z.H., and Chen, X.W. (2004). High efficiency gene transfer into mammalian kidney cells using baculovirus vectors. Arch. Virol. *149*, 51–60. https://doi.org/10.1007/s00705-003-0197-3

Lin, W., Fan, H., Cheng, X., Ye, Y., Chen, X., Ren, T., Qi, W., and Liao, M. (2011). A baculovirus dual expression system-based vaccine confers complete protection against lethal challenge with H9N2 avian influenza virus in mice. Virol. J. *8*, 273. https://doi.org/10.1186/1743-422X-8-273

Lindley, K.M., Su, J.L., Hodges, P.K., Wisely, G.B., Bledsoe, R.K., Condreay, J.P., Winegar, D.A., Hutchins, J.T., and Kost, T.A. (2000). Production of monoclonal antibodies using recombinant baculovirus displaying gp64-fusion proteins. J. Immunol. Methods *234*, 123–135.

Link, R., Veiksina, S., Rinken, A., and Kopanchuk, S. (2017). Characterization of ligand binding to melanocortin 4 receptors using fluorescent peptides with improved kinetic properties. Eur. J. Pharmacol. *799*, 58–66.

Liu, C.Y., Wang, C.H., Wang, J.C., and Chao, Y.C. (2007). Stimulation of baculovirus transcriptome expression in mammalian cells by baculoviral transcriptional activators. J. Gen. Virol. *88*, 2176–2184.

Liu, C.Y., Wang, C.H., Hsiao, W.K., Lo, H.R., Wu, C.P., and Chao, Y.C. (2009). RING and coiled-coil domains of baculovirus IE2 are critical in strong activation of the cytomegalovirus major immediate-early promoter in mammalian cells. J. Virol. *83*, 3604–3616. https://doi.org/10.1128/JVI.01778-08

Liu, C.Y., Chen, H.Z., and Chao, Y.C. (2010). Maximizing baculovirus-mediated foreign proteins expression in mammalian cells. Curr. Gene Ther. *10*, 232–241.

Liu, R., Li, X., Xiao, W., and Lam, K.S. (2017). Tumor-targeting peptides from combinatorial libraries. Adv. Drug Deliv. Rev. *110-111*, 13–37.

Liu, Y., Joo, K.I., Lei, Y., and Wang, P. (2014). Visualization of intracellular pathways of engineered baculovirus in mammalian cells. Virus Res. *181*, 81–91. https://doi.org/10.1016/j.virusres.2014.01.006

Liu, Z., Liu, Y., Zhang, Y., Yang, Y., Ren, J., Zhang, X., and Du, E. (2017). Surface displaying of swine IgG1 Fc enhances baculovirus-vectored vaccine efficacy by facilitating viral complement escape and mammalian cell transduction. Vet. Res. *48*, 29. https://doi.org/10.1186/s13567-017-0434-5

Lo, H.R., Chou, C.C., Wu, T.Y., Yuen, J.P., and Chao, Y.C. (2002). Novel baculovirus DNA elements strongly stimulate activities of exogenous and endogenous promoters. J. Biol. Chem. *277*, 5256–5264. https://doi.org/10.1074/jbc.M108895200

Loisel, T.P., Ansanay, H., St-Onge, S., Gay, B., Boulanger, P., Strosberg, A.D., Marullo, S., and Bouvier, M. (1997). Recovery of homogeneous and functional beta 2-adrenergic receptors from extracellular baculovirus particles. Nat. Biotechnol. *15*, 1300–1304. https://doi.org/10.1038/nbt1197-1300

Mähönen, A.J., Airenne, K.J., Purola, S., Peltomaa, E., Kaikkonen, M.U., Riekkinen, M.S., Heikura, T., Kinnunen, K., Roschier, M.M., Wirth, T., *et al.* (2007). Post-transcriptional regulatory element boosts baculovirus-mediated gene expression in vertebrate cells. J. Biotechnol. *131*, 1–8.

Mäkelä, A.R., and Oker-Blom, C. (2008). The baculovirus display technology – an evolving instrument for molecular screening and drug delivery. Comb. Chem. High Throughput Screen. *11*, 86–98.

Mäkelä, A.R., Ernst, W., Grabherr, R., and Oker-Blom, C. (2010). Baculovirus-based display and gene delivery systems. Cold Spring Harb. Protoc. *2010*, pdb.top72. https://doi.org/10.1101/pdb.top72

Mansouri, M., Bellon-Echeverria, I., Rizk, A., Ehsaei, Z., Cianciolo Cosentino, C., Silva, C.S., Xie, Y., Boyce, F.M., Davis, M.W., Neuhauss, S.C., *et al.* (2016). Highly efficient baculovirus-mediated multigene delivery in primary cells. Nat. Commun. *7*, 11529. https://doi.org/10.1038/ncomms11529

Martyn, J.C., Cardin, A.J., Wines, B.D., Cendron, A., Li, S., Mackenzie, J., Powell, M., and Gowans, E.J. (2009). Surface display of IgG Fc on baculovirus vectors enhances binding to antigen-presenting cells and cell lines expressing Fc receptors. Arch. Virol. *154*, 1129–1138. https://doi.org/10.1007/s00705-009-0423-8

Masuda, K., Itoh, H., Sakihama, T., Akiyama, C., Takahashi, K., Fukuda, R., Yokomizo, T., Shimizu, T., Kodama, T., and Hamakubo, T. (2003). A combinatorial G protein-coupled receptor reconstitution system on budded baculovirus. Evidence for Galpha and Galphao coupling to a human leukotriene B4 receptor. J. Biol. Chem. *278*, 24552–24562. https://doi.org/10.1074/jbc.M302801200

Matarazzo, V., and Ronin, C. (2013). Human olfactory receptors: recombinant expression in the baculovirus/Sf9 insect cell system, functional characterization, and odorant identification. Methods Mol. Biol. *1003*, 109–122. https://doi.org/10.1007/978-1-62703-377-0_8

Matarazzo, V., Clot-Faybesse, O., Marcet, B., Guiraudie-Capraz, G., Atanasova, B., Devauchelle, G., Cerutti, M., Etiévant, P., and Ronin, C. (2005). Functional characterization of two human olfactory receptors expressed in the baculovirus Sf9 insect cell system. Chem. Senses *30*, 195–207.

McLinden, J.H., Ploplis, V.A., Fraser, M.J., and Rosen, E.D. (1992). Expression of foreign epitopes on recombinant occlusion bodies of baculoviruses. Vaccine *10*, 231–237.

Meller Harel, H.Y., Fontaine, V., Chen, H., Jones, I.M., and Millner, P.A. (2008). Display of a maize cDNA library on baculovirus infected insect cells. BMC Biotechnol. *8*, 64. https://doi.org/10.1186/1472-6750-8-64

Meng, T., Kolpe, A.B., Kiener, T.K., Chow, V.T., and Kwang, J. (2011). Display of VP1 on the surface of baculovirus and its immunogenicity against heterologous human enterovirus 71 strains in mice. PLOS ONE *6*, e21757. https://doi.org/10.1371/journal.pone.0021757

Meulenbroek, R.A., Sargent, K.L., Lunde, J., Jasmin, B.J., and Parks, R.J. (2004). Use of adenovirus protein IX (pIX) to display large polypeptides on the virion – generation of fluorescent virus through the incorporation of pIX-GFP. Mol. Ther. *9*, 617–624. https://doi.org/10.1016/j.ymthe.2004.01.012

Mlambo, G., Kumar, N., and Yoshida, S. (2010). Functional immunogenicity of baculovirus expressing Pfs25, a human malaria transmission-blocking vaccine candidate antigen. Vaccine *28*, 7025–7029. https://doi.org/10.1016/j.vaccine.2010.08.022

Molinari, P., Crespo, M.I., Gravisaco, M.J., Taboga, O., and Morón, G. (2011). Baculovirus capsid display potentiates OVA cytotoxic and innate immune responses. PLOS ONE *6*, e24108. https://doi.org/10.1371/journal.pone.0024108

Münch, R.C., Janicki, H., Völker, I., Rasbach, A., Hallek, M., Büning, H., and Buchholz, C.J. (2013). Displaying high-affinity ligands on adeno-associated viral vectors enables tumor cell-specific and safe gene transfer. Mol. Ther. *21*, 109–118. https://doi.org/10.1038/mt.2012.186

Murakami, Y., Kagino, T., Niikura, M., Mikami, T., Ishii, K., and Matsuura, Y. (1994). Characterization of Newcastle disease virus envelope glycoproteins expressed in insect cells. Virus Res. *33*, 123–137.

Murges, D., Kremer, A., and Knebel-Mörsdorf, D. (1997). Baculovirus transactivator IE1 is functional in mammalian cells. J. Gen. Virol. *78*, 1507–1510. https://doi.org/10.1099/0022-1317-78-6-1507

Musthaq, S.K., Kumar, S.R., Szyporta, M., and Kwang, J. (2014). Immunization with baculovirus displayed H6 hemagglutinin vaccine protects mice against lethal H6 influenza virus challenge. Antiviral Res. *109*, 42–53. https://doi.org/10.1016/j.antiviral.2014.06.002

Nguyen, A.W., Le, K.C., and Maynard, J.A. (2018). Identification of high affinity HER2 binding antibodies using CHO Fab surface display. Protein Eng. Des. Sel. *31*, 91–101. https://doi.org/10.1093/protein/gzy004

Niikura, M., Matsuura, Y., Hattori, M., Onuma, M., and Mikami, T. (1991). Characterization of haemagglutinin-neuraminidase glycoprotein of Newcastle disease virus expressed by a recombinant baculovirus. Virus Res. *20*, 31–43.

Naik, N.G., Lo, Y.W., Wu, T.Y., Lin, C.C., Kuo, S.C., and Chao, Y.C. (2018). Baculovirus as an efficient vector for gene delivery into mosquitoes. Sci. Rep. *8*, 17778. https://doi.org/10.1038/s41598-018-35463-8

Ohkawa, T., Rowe, A.R., and Volkman, L.E. (2002). Identification of six *Autographa californica* multicapsid nucleopolyhedrovirus early genes that mediate nuclear localization of G-actin. J. Virol. *76*, 12281–12289.

Ojala, K., Koski, J., Ernst, W., Grabherr, R., Jones, I., and Oker-Blom, C. (2004). Improved display of synthetic IgG-binding domains on the baculovirus surface. Technol. Cancer Res. Treat. 3, 77–84.

Oker-Blom, C., Airenne, K.J., and Grabherr, R. (2003). Baculovirus display strategies: Emerging tools for eukaryotic libraries and gene delivery. Brief. Funct. Genomic. Proteomic. 2, 244–253.

Ong, H.K., Ali, A.M., Omar, A.R., and Yusoff, K. (2000). Cloning and expression of the HN gene from the velogenic viscerotropic Newcastle disease virus strain AF2240 in Sf9 insect cells. Cytotechnology *32*, 243–251. https://doi.org/10.1023/A:1008136326756

Ono, C., Okamoto, T., Abe, T., and Matsuura, Y. (2018). Baculovirus as a tool for gene delivery and gene therapy. Viruses *10*, E510.

Palomares, L.A., Realpe, M., and Ramírez, O.T. (2015). An overview of cell culture engineering for the insect cell-baculovirus expression vector system (BEVS). In Animal Cell Culture, Al-Rubeai, M., ed. (Springer International Publishing, Cham), pp. 501–519.

Pfohl, J.L., Worley, J.F., Condreay, J.P., An, G., Apolito, C.J., Kost, T.A., and Truax, J.F. (2002). Titration of KATP channel expression in mammalian cells utilizing recombinant baculovirus transduction. Recept. Channels *8*, 99–111.

Possee, R.D. (1986). Cell-surface expression of influenza virus haemagglutinin in insect cells using a baculovirus vector. Virus Res. *5*, 43–59.

Prabakaran, M., Velumani, S., He, F., Karuppannan, A.K., Geng, G.Y., Yin, L.K., and Kwang, J. (2008). Protective immunity against influenza H5N1 virus challenge in mice by intranasal co-administration of baculovirus surface-displayed HA and recombinant CTB as an adjuvant. Virology *380*, 412–420. https://doi.org/10.1016/j.virol.2008.08.002

Prabakaran, M., Meng, T., He, F., Yunrui, T., Qiang, J., Lin, R.T., and Kwang, J. (2011). Subcutaneous immunization with baculovirus surface-displayed hemagglutinin of pandemic H1N1 Influenza A virus induces protective immunity in mice. Clin. Vaccine Immunol. *18*, 1582–1585. https://doi.org/10.1128/CVI.05114-11

Prabakaran, M., Kumar, S.R., Raj, K.V., Wu, X., He, F., Zhou, J., and Kwang, J. (2014). Cross-protective efficacy of baculovirus displayed hemagglutinin against highly pathogenic influenza H7 subtypes. Antiviral Res. *109*, 149–159. https://doi.org/10.1016/j.antiviral.2014.06.017

Premanand, B., Kiener, T.K., Meng, T., Tan, Y.R., Jia, Q., Chow, V.T., and Kwang, J. (2012). Induction of protective immune responses against EV71 in mice by baculovirus encoding a novel expression cassette for capsid protein VP1. Antiviral Res. *95*, 311–315. https://doi.org/10.1016/j.antiviral.2012.05.017

Premanand, B., Prabakaran, M., Kiener, T.K., and Kwang, J. (2013). Recombinant baculovirus associated with bilosomes as an oral vaccine candidate against HEV71 infection in mice. PLOS ONE *8*, e55536. https://doi.org/10.1371/journal.pone.0055536

Premanand, B., Zhong Wee, P., and Prabakaran, M. (2018). Baculovirus surface display of immunogenic proteins for vaccine development. Viruses *10*, E298.

Rajesh Kumar, S., Syed Khader, S.M., Kiener, T.K., Szyporta, M., and Kwang, J. (2013). Intranasal immunization of baculovirus displayed hemagglutinin confers complete protection against mouse adapted highly pathogenic H7N7 reassortant influenza virus. PLOS ONE *8*, e63856. https://doi.org/10.1371/journal.pone.0063856

Rohrmann, G.F. (1986). Polyhedrin structure. J. Gen. Virol. *67*, 1499–1513. https://doi.org/10.1099/0022-1317-67-8-1499

Rothe, A., Hosse, R.J., and Power, B.E. (2006). Ribosome display for improved biotherapeutic molecules. Expert Opin. Biol. Ther. *6*, 177–187. https://doi.org/10.1517/14712598.6.2.177

Saarenpää, T., Jaakola, V.P., and Goldman, A. (2015). Baculovirus-mediated expression of GPCRs in insect cells. Meth. Enzymol. *556*, 185–218. https://doi.org/10.1016/bs.mie.2014.12.033

Sakihama, T., Masuda, K., Sato, T., Doi, T., Kodama, T., and Hamakubo, T. (2008a). Functional reconstitution of G-protein-coupled receptor-mediated adenylyl cyclase activation by a baculoviral co-display system. J. Biotechnol. *135*, 28–33. https://doi.org/10.1016/j.jbiotec.2008.02.022

Sakihama, T., Sato, T., Iwanari, H., Kitamura, T., Sakaguchi, S., Kodama, T., and Hamakubo, T. (2008b). A simple detection method for low-affinity membrane protein interactions by baculoviral display. PLOS ONE *3*, e4024. https://doi.org/10.1371/journal.pone.0004024

Sampieri, A., Luz-Madrigal, A., Zepeda, J., and Vaca, L. (2015). Identification of fragments from *Autographa californica* polyhedrin protein essential for self-aggregation and exogenous protein incorporation. BMC Biochem. *16*, 5. https://doi.org/10.1186/s12858-015-0034-9

Schneider, E.H., and Seifert, R. (2010). Sf9 cells: a versatile model system to investigate the pharmacological properties of G protein-coupled receptors. Pharmacol. Ther. *128*, 387–418. https://doi.org/10.1016/j.pharmthera.2010.07.005

Seo, J.H., Li, L., Yeo, J.S., and Cha, H.J. (2003). Baculoviral polyhedrin as a novel fusion partner for formation of inclusion body in *Escherichia coli*. Biotechnol. Bioeng. *84*, 467–473. https://doi.org/10.1002/bit.10798

Sergeeva, A., Kolonin, M.G., Molldrem, J.J., Pasqualini, R., and Arap, W. (2006). Display technologies: application for the discovery of drug and gene delivery agents. Adv. Drug Deliv. Rev. *58*, 1622–1654.

Shim, H.J., Choi, J.Y., Wang, Y., Tao, X.Y., Liu, Q., Roh, J.Y., Kim, J.S., Kim, W.J., Woo, S.D., Jin, B.R., *et al.* (2013). NeuroBactrus, a novel, highly effective, and environmentally friendly recombinant baculovirus insecticide. Appl. Environ. Microbiol. *79*, 141–149. https://doi.org/10.1128/AEM.02781-12

Shukla, S., Schwartz, C., Kapoor, K., Kouanda, A., and Ambudkar, S.V. (2012). Use of baculovirus BacMam vectors for expression of ABC drug transporters in mammalian cells. Drug Metab. Dispos. *40*, 304–312. https://doi.org/10.1124/dmd.111.042721

Sim, S.H., Kim, J.Y., Seong, B.L., Nguyen, H.H., and Chang, J. (2016). Baculovirus displaying hemagglutinin elicits broad cross-protection against influenza in mice. PLOS ONE *11*, e0152485. https://doi.org/10.1371/journal.pone.0152485

Smith, G.E., Summers, M.D., and Fraser, M.J. (1983). Production of human beta interferon in insect cells infected with a baculovirus expression vector. Mol. Cell. Biol. *3*, 2156–2165.

Smith, M.R., Khera, E., and Wen, F. (2015). Engineering novel and improved biocatalysts by cell surface display. Ind. Eng. Chem. Res. *54*, 4021–4032. https://doi.org/10.1021/ie504071f

Song, S.U., and Boyce, F.M. (2001). Combination treatment for osteosarcoma with baculoviral vector mediated gene therapy (p53) and chemotherapy (adriamycin). Exp. Mol. Med. 33, 46–53. https://doi.org/10.1038/emm.2001.9

Song, W., Zhang, J., Guo, J., Zhang, J., Ding, F., Li, L., and Sun, Z. (2010). Role of the dissolved zinc ion and reactive oxygen species in cytotoxicity of ZnO nanoparticles. Toxicol. Lett. *199*, 389–397. https://doi.org/10.1016/j.toxlet.2010.10.003

Spenger, A., Grabherr, R., Töllner, L., Katinger, H., and Ernst, W. (2002). Altering the surface properties of baculovirus *Autographa californica* NPV by insertional mutagenesis of the envelope protein gp64. Eur. J. Biochem. *269*, 4458–4467.

Spenger, A., Ernst, W., Condreay, J.P., Kost, T.A., and Grabherr, R. (2004). Influence of promoter choice and trichostatin A treatment on expression of baculovirus delivered genes in mammalian cells. Protein Expr. Purif. *38*, 17–23.

Strauss, R., Hüser, A., Ni, S., Tuve, S., Kiviat, N., Sow, P.S., Hofmann, C., and Lieber, A. (2007). Baculovirus-based vaccination vectors allow for efficient induction of immune responses against *plasmodium falciparum* circumsporozoite protein. Mol. Ther. *15*, 193–202.

Su, J.L., Fornwald, J., Rivers, P., Goldsworthy, S., Looney, N.A., Hanvey, J., Plumpton, C., Parham, J., Romanos, M., Kost, T.A., *et al.* (2004). A cell-based time-resolved fluorescence assay for selection of antibody reagents for G protein-coupled receptor immunohistochemistry. J. Immunol. Methods *291*, 123–135. https://doi.org/10.1016/j.jim.2004.05.006

Suzuki, T., Chang, M.O., Kitajima, M., and Takaku, H. (2010). Baculovirus activates murine dendritic cells and induces non-specific NK cell and T cell immune responses. Cell. Immunol. *262*, 35–43. https://doi.org/10.1016/j.cellimm.2009.12.005

Szerencsei, R.T., Prinsen, C.F., and Schnetkamp, P.P. (2001). Stoichiometry of the retinal cone Na/Ca-K exchanger heterologously expressed in insect cells: comparison with the bovine heart Na/Ca exchanger. Biochemistry *40*, 6009–6015.

Tanaka, T., Yamada, R., Ogino, C., and Kondo, A. (2012). Recent developments in yeast cell surface display toward extended applications in biotechnology. Appl. Microbiol. Biotechnol. *95*, 577–591. https://doi.org/10.1007/s00253-012-4175-0

Tang, X.C., Lu, H.R., and Ross, T.M. (2010). Hemagglutinin displayed baculovirus protects against highly pathogenic influenza. Vaccine *28*, 6821–6831. https://doi.org/10.1016/j.vaccine.2010.08.040

Tani, H., Nishijima, M., Ushijima, H., Miyamura, T., and Matsuura, Y. (2001). Characterization of cell-surface determinants important for baculovirus infection. Virology *279*, 343–353. https://doi.org/10.1006/viro.2000.0699

Taube, R., Zhu, Q., Xu, C., Diaz-Griffero, F., Sui, J., Kamau, E., Dwyer, M., Aird, D., and Marasco, W.A. (2008). Lentivirus display: stable expression of human antibodies on the surface of human cells and virus particles. PLOS ONE *3*, e3181. https://doi.org/10.1371/journal.pone.0003181

Tessier, D.C., Thomas, D.Y., Khouri, H.E., Laliberté, F., and Vernet, T. (1991). Enhanced secretion from insect cells of a foreign protein fused to the honeybee melittin signal peptide. Gene *98*, 177–183.

Titus, S.A., Beacham, D., Shahane, S.A., Southall, N., Xia, M., Huang, R., Hooten, E., Zhao, Y., Shou, L., Austin, C.P., *et al.* (2009). A new homogeneous high-throughput screening assay for profiling compound activity on the human ether-a-go-go-related gene channel. Anal. Biochem. *394*, 30–38. https://doi.org/10.1016/j.ab.2009.07.003

Tõntson, L., Kopanchuk, S., and Rinken, A. (2014). Characterization of 5-HT1A receptors and their complexes with G-proteins in budded baculovirus particles using fluorescence anisotropy of Bodipy-FL-NAN-190. Neurochem. Int. *67*, 32–38.

Trowitzsch, S., Klumpp, M., Thoma, R., Carralot, J.P., and Berger, I. (2011). Light it up: highly efficient multigene delivery in mammalian cells. Bioessays *33*, 946–955. https://doi.org/10.1002/bies.201100109

van Hofsten, P., Faye, I., Kockum, K., Lee, J.Y., Xanthopoulos, K.G., Boman, I.A., Boman, H.G., Engström, A., Andreu, D., and Merrifield, R.B. (1985). Molecular cloning, cDNA sequencing, and chemical synthesis of cecropin B from *Hyalophora cecropia*. Proc. Natl. Acad. Sci. U.S.A. *82*, 2240–2243.

van Oers, M., M., and Vlak, J., M. (2007). Baculovirus genomics. Curr. Drug Targets *8*, 1051–1068.

Varadi, K., Michelfelder, S., Korff, T., Hecker, M., Trepel, M., Katus, H.A., Kleinschmidt, J.A., and Müller, O.J. (2012). Novel random peptide libraries displayed on AAV serotype 9 for selection of endothelial cell-directed gene transfer vectors. Gene Ther. *19*, 800–809. https://doi.org/10.1038/gt.2011.143

Vaughn, J.L., Goodwin, R.H., Tompkins, G.J., and McCawley, P. (1977). The establishment of two cell lines from the insect *Spodoptera frugiperda* (Lepidoptera; Noctuidae). In Vitro *13*, 213–217.

Veiksina, S., Kopanchuk, S., and Rinken, A. (2014). Budded baculoviruses as a tool for a homogeneous fluorescence anisotropy-based assay of ligand binding to G protein-coupled receptors: The case of melanocortin 4 receptors. Biochim. Biophys. Acta *1838*, 372–381.

Veit, M., and Thaa, B. (2011). Association of influenza virus proteins with membrane rafts. Adv. Virol. *2011*, 370606. https://doi.org/10.1155/2011/370606

Vujadinovic, M., and Vellinga, J. (2018). Progress in adenoviral capsid-display vaccines. Biomedicines *6*, E81.

Wang, C.H., Naik, N.G., Liao, L.L., Wei, S.C., and Chao, Y.C. (2017). Global screening of antiviral genes that suppress baculovirus transgene expression in mammalian cells. Mol. Ther. Methods Clin. Dev. *6*, 194–206. https://doi.org/10.1016/j.omtm.2017.07.002

Wang, Q., Bosch, B.J., Vlak, J.M., van Oers, M.M., Rottier, P.J., and van Lent, J.W.M. (2016). Budded baculovirus particle structure revisited. J. Invertebr. Pathol. *134*, 15–22.

Wang, S., Fang, L., Fan, H., Jiang, Y., Pan, Y., Luo, R., Zhao, Q., Chen, H., and Xiao, S. (2007). Construction and immunogenicity of pseudotype baculovirus expressing GP5 and M protein of porcine reproductive and respiratory syndrome virus. Vaccine *25*, 8220–8227.

Wilde, M., Klausberger, M., Palmberger, D., Ernst, W., and Grabherr, R. (2014). Tnao38, high five and Sf9 – evaluation of host-virus interactions in three different insect cell lines: baculovirus production and recombinant protein expression. Biotechnol. Lett. *36*, 743–749. https://doi.org/10.1007/s10529-013-1429-6

Wu, Q., Fang, L., Wu, X., Li, B., Luo, R., Yu, Z., Jin, M., Chen, H., and Xiao, S. (2009). A pseudotype baculovirus-mediated vaccine confers protective immunity against lethal challenge with H5N1 avian influenza virus in mice and chickens. Mol. Immunol. *46*, 2210–2217. https://doi.org/10.1016/j.molimm.2009.04.017

Wu, Y.L., Wu, C.P., Lee, S.T., Tang, H., Chang, C.H., Chen, H.H., and Chao, Y.C. (2010). The early gene hhi1 reactivates *Heliothis zea* nudivirus 1 in latently infected cells. J. Virol. *84*, 1057–1065. https://doi.org/10.1128/JVI.01548-09

Xiang, X.W., Yang, R., Chen, L., Hu, X.L., Yu, S.F., Cao, C.P., and Wu, X.F. (2012). Immobilization of foreign protein into polyhedra of *Bombyx mori* nucleopolyhedrovirus (BmNPV). J. Zhejiang Univ. Sci. B *13*, 111–117. https://doi.org/10.1631/jzus.B1100131

Xu, X., Chen, Y., Zhao, Y., Liu, X., Dong, B., Jones, I.M., and Chen, H. (2013). Baculovirus superinfection: a probable restriction factor on the surface display of proteins for library screening. PLOS ONE *8*, e54631. https://doi.org/10.1371/journal.pone.0054631

Xu, X.G., and Liu, H.J. (2008). Baculovirus surface display of E2 envelope glycoprotein of classical swine fever virus and immunogenicity of the displayed proteins in a mouse model. Vaccine *26*, 5455–5460. https://doi.org/10.1016/j.vaccine.2008.07.090

Xu, X.-G., Chiou, M.-T., Zhang, Y.-M., Tong, D.-W., Hu, J.-H., Zhang, M.-T., and Liu, H. J. (2008). Baculovirus surface display of Erns envelope glycoprotein of classical swine fever virus. J. Virol. Methods *153*, 149–155.

Xu, X.G., Tong, D.W., Chiou, M.T., Hsieh, Y.C., Shih, W.L., Chang, C.D., Liao, M.H., Zhang, Y.M., and Liu, H.J. (2009). Baculovirus surface display of NS3 nonstructural protein of classical swine fever virus. J. Virol. Methods *159*, 259–264. https://doi.org/10.1016/j.jviromet.2009.04.013

Xu, X.G., Wang, Z.S., Zhang, Q., Li, Z.C., Zhao, H.N., Li, W., Tong, D.W., and Liu, H.J. (2011). Baculovirus surface display of E envelope glycoprotein of Japanese encephalitis virus and its immunogenicity of the displayed proteins in mouse and swine models. Vaccine *29*, 636–643. https://doi.org/10.1016/j.vaccine.2010.11.045

Xu, X.G., Wang, Z.S., Zhang, Q., Li, Z.C., Ding, L., Li, W., Wu, H.Y., Chang, C.D., Lee, L.H., Tong, D.W., *et al.* (2012). Baculovirus as a PRRSV and PCV2 bivalent vaccine vector: baculovirus virions displaying simultaneously GP5 glycoprotein of PRRSV and capsid protein of PCV2. J. Virol. Methods *179*, 359–366. https://doi.org/10.1016/j.jviromet.2011.11.023

Xu, Y.P., Gu, L.Z., Lou, Y.H., Cheng, R.L., Xu, H.J., Wang, W.B., and Zhang, C.X. (2012). A baculovirus isolated from wild silkworm encompasses the host ranges of *Bombyx mori* nucleopolyhedrosis virus and *Autographa californica* multiple nucleopolyhedrovirus in cultured cells. J. Gen. Virol. *93*, 2480–2489.

Yang, D.G., Chung, Y.C., Lai, Y.K., Lai, C.W., Liu, H.J., and Hu, Y.C. (2007). Avian influenza virus hemagglutinin display on baculovirus envelope: cytoplasmic domain affects virus properties and vaccine potential. Mol. Ther. *15*, 989–996.

Yang, M., Sunderland, K., and Mao, C. (2017). Virus-derived peptides for clinical applications. Chem. Rev. *117*, 10377–10402. https://doi.org/10.1021/acs.chemrev.7b00100

Ye, Y., Cheng, X., Zhang, J., Tong, T., Lin, W., Liao, M., and Fan, H. (2013). Induction of robust immunity response in mice by dual-expression-system-based recombinant baculovirus expressing the capsid protein of porcine circovirus type 2. Virol. J. *10*, 316. https://doi.org/10.1186/1743-422X-10-316

Yoshida, S., Kondoh, D., Arai, E., Matsuoka, H., Seki, C., Tanaka, T., Okada, M., and Ishii, A. (2003). Baculovirus virions displaying *Plasmodium berghei* circumsporozoite protein protect mice against malaria sporozoite infection. Virology *316*, 161–170.

Zhang, J., Chen, X.W., Tong, T.Z., Ye, Y., Liao, M., and Fan, H.Y. (2014). BacMam virus-based surface display of the infectious bronchitis virus (IBV) S1 glycoprotein confers strong protection against virulent IBV challenge in chickens. Vaccine *32*, 664–670. https://doi.org/10.1016/j.vaccine.2013.12.006

Zhang, J., Fan, H.Y., Zhang, Z., Zhang, J., Zhang, J., Huang, J.N., Ye, Y., and Liao, M. (2016). Recombinant baculovirus vaccine containing multiple M2e and adjuvant LTB induces T cell dependent, cross-clade protection against H5N1 influenza virus in mice. Vaccine *34*, 622–629.

Zhou, G., and Roizman, B. (2002). Truncated forms of glycoprotein D of herpes simplex virus 1 capable of blocking apoptosis and of low-efficiency entry into cells form a heterodimer dependent on the presence of a cysteine located in the shared transmembrane domains. J. Virol. *76*, 11469–11475.

Zhou, J., and Blissard, G.W. (2008). Display of heterologous proteins on *gp64* null baculovirus virions and enhanced budding mediated by a vesicular stomatitis virus G-stem construct. J. Virol. *82*, 1368.

Index

Q

R

S

W

X

Y

Z

www.ingramcontent.com/pod-product-compliance
Ingram Content Group UK Ltd.
Pitfield, Milton Keynes, MK11 3LW, UK
UKHW062000041026
15591UKWH00016B/1095

9 781912 530083